Lecture Notes in Networks and Systems

Volume 1181

The series "Lecture Notes in Networks and Systems" publishes the latest developments in Networks and Systems—quickly, informally and with high quality. Original research reported in proceedings and post-proceedings represents the core of LNNS.

Volumes published in LNNS embrace all aspects and subfields of, as well as new challenges in, Networks and Systems.

The series contains proceedings and edited volumes in systems and networks, spanning the areas of Cyber-Physical Systems, Autonomous Systems, Sensor Networks, Control Systems, Energy Systems, Automotive Systems, Biological Systems, Vehicular Networking and Connected Vehicles, Aerospace Systems, Automation, Manufacturing, Smart Grids, Nonlinear Systems, Power Systems, Robotics, Social Systems, Economic Systems and other. Of particular value to both the contributors and the readership are the short publication timeframe and the world-wide distribution and exposure which enable both a wide and rapid dissemination of research output.

The series covers the theory, applications, and perspectives on the state of the art and future developments relevant to systems and networks, decision making, control, complex processes and related areas, as embedded in the fields of interdisciplinary and applied sciences, engineering, computer science, physics, economics, social, and life sciences, as well as the paradigms and methodologies behind them.

Indexed by SCOPUS, EI Compendex, INSPEC, WTI Frankfurt eG, zbMATH, SCImago.

All books published in the series are submitted for consideration in Web of Science.

For proposals from Asia please contact Aninda Bose (aninda.bose@springer.com).

Vinit Kumar Gunjan · Jacek M. Zurada
Editors

Proceedings of 5th International Conference on Recent Trends in Machine Learning, IoT, Smart Cities and Applications

ICMISC 2024, Volume 1

Springer

Editors
Vinit Kumar Gunjan
Department of Computer Science and Engineering
CMR Institute of Technology
Hyderabad, Telangana, India

Jacek M. Zurada
Department of Electrical and Computer Engineering
University of Louisville
Louisville, KY, USA

ISSN 2367-3370 ISSN 2367-3389 (electronic)
Lecture Notes in Networks and Systems
ISBN 978-981-97-8860-6 ISBN 978-981-97-8861-3 (eBook)
https://doi.org/10.1007/978-981-97-8861-3

This Springer imprint is published by the registered company Springer Nature Singapore Pte Ltd.
The registered company address is: 152 Beach Road, #21-01/04 Gateway East, Singapore 189721, Singapore

Preface

We are pleased to present the 5th International Conference on Recent Trends in Machine Learning, IoT, Smart Cities, and Applications, ICMISC 2024 proceedings. It has emerged as one of the most prestigious annual conferences at the CMR Institute of Technology, Hyderabad, India, where many researchers, academicians, and industry professionals gathered together to present and discuss recent advances in the respective cutting-edge domains.

ICMISC 2024, in continuation, showcases points of intersection between Machine Learning and the Internet of Things issues that are now driving the face of many modern industries and societies regarding Smart City technologies. The solutions sought become increasingly crucial as more cities become connected and intelligent. Machine learning and IoT are integral to such smart infrastructures, opening new avenues to improve urban living, sustainability, security, and resource management.

This year's conference has attracted contributions from a wide variety of participants worldwide, representing developments in theory, applications, and technology. The technical program encompasses machine learning algorithms, IoT architectures, smart city infrastructures, data-driven decision-making processes, cybersecurity, and artificial intelligence in urban systems. These contributions reflect the fundamental elements of up-and-coming field trends and provide an overview of future developments and applications.

We would also like to express our profound appreciation to the keynote speakers, session chairs, and program committee members for all their effort, time, and expertise in contributing to the high quality of this year's conference program. Special thanks also goes out to the authors for the presentation of their latest research contributions and to those attending the conference for their active participation and discussions, which are so essential for the success of ICMISC 2024.

We trust that the experiences from this conference will inspire further creativity and cooperation for years to come. May ideas ignited here at our meeting enrich the development of machine learning, IoT, and innovative city technologies, making the world even more connected, clever, and sustainable.

On behalf of the Organising Committee, please accept our hearty thanks for participating and supporting the same. We look forward to meeting you at future editions of ICMISC.

Conference Chairs

5th International Conference on Recent Trends in Machine Learning, IoT, Smart Cities, and Applications, ICMISC 2024.

Hyderabad, India
Louisville, USA

Vinit Kumar Gunjan
Jacek M. Zurada

Editors and Contributors

About the Editors

Vinit Kumar Gunjan is the Associate Professor in the Department of Computer Science and Engineering at CMR Institute of Technology Hyderabad (Affiliated to Jawaharlal Nehru Technological University, Hyderabad). Vinit Kumar Gunjan is an Researcher, published research papers in IEEE, Elsevier & Springer Conferences, authored several books, and edited volumes of Springer series, most of which are indexed in SCOPUS database. Vinit Kumar Gunjan was awarded with the prestigious Early Career Research Award in the year 2016 by the Science Engineering Research Board, Department of Science & Technology Government of India.

Dr. Jacek M. Zurada (SM'85, F'96, LF'14) is the Professor of Electrical and Computer Engineering and Director of the Computational Intelligence Laboratory at the University of Louisville, Louisville, Kentucky, USA, where he served as Department Chair and Distinguished University Scholar. He was Professor at Princeton University, Northeastern University, Auburn University, the National University of Singapore, Nanyang Technological University in Singapore, Chinese University of Hong Kong, the University of Chile, Santiago, Toyohashi University of Technology, Japan, the University of Stellenbosch, South Africa, the University of Marie-Curie, Paris, France, and was Postdoctoral Fellow at the Swiss Federal Institute of Technology, Zurich, Switzerland.

Contributors

Abhinav Yogananda School of AI, Computers and Data Science, Shoolini University, Solan, Himachal Pradesh, India

C. M. Abhishek Jain College of Engineering, VTU, Belagavi, Karnataka, India

Prachi Acharya School of Computer Science and Engineering, Vellore Institute of Technology, Chennai, India

G. K. Aishwarya Electronics and Communication Department, KLE Technological University, Hubballi, Karnataka, India

Meghana Alagawadi KLE Technological University, Hubli, Karnataka, India

Mohammad Shabbir Alam Department of Computer Science, College of Computer Science and Information Technology, Jazan University, Jazan, Saudi Arabia

Sanower Alam Department of Computer Science, Jamia Millia Islamia, New Delhi, India

C. M. Amar Jain College of Engineering, VTU, Belagavi, Karnataka, India

S. Aravindaraj Rajalakshmi Engineering College, Chennai, India

A. Aswini Rajalakshmi Engineering College, Chennai, India

Aliabbas Attarwala Computer Engineering Department, CSPIT, CHARUSAT, Changa, Gujarat, India

Deepti Ayyanagoudar KLE Technological University, Hubli, Karnataka, India

Megha M. Bagodi Jain College of Engineering, VTU, Belagavi, Karnataka, India

Rishabh Balse Computer Engineering Department, CSPIT, CHARUSAT, Changa, Gujarat, India

Nalini Kanta Barpanda Department of Electronics, Sambalpur University, Burla, Odisha, India

Santi Kumari Behera Department of Computer Science and Engineering, VSSUT, Burla, India

Deepika Jaya Sri Bezawada Velagapudi Ramakrishna Siddhartha Engineering College, Vijayawada, Andhra Pradesh, India

G. J. Bharat Kumar Department of Computer Science and Engineering, Sreyas Institute of Engineering and Technology, Hyderabad, Telangana, India

Mylu Rohit Bhargav KLE Technological University, Hubli, Karnataka, India

N. P. G. Bhavani ECE Department, Saveetha Institute of Medical Sciences, Chennai, India

Kiran Bhowmick Dwarkadas J. Sanghvi College of Engineering, Mumbai, Maharashtra, India

S. Bhuvaneswari EEE Department, S.A. Engineering College, Chennai, India

Sarthak Bindal Information Technology Department, SCTR'S Pune Institute of Computer Technology, Pune, Maharashtra, India

Tarannum Alimahmad Bloch Marwadi University, Rajkot, Gujarat, India

Suneeta V. Budihal School of Electronics and Communication Engineering, KLE Technological University Hubballi, Hubballi, Karnataka, India

Satish Chikkamath Electronics and Communication Department, School of Electronics and Communication Engineering, KLE Technological University Hubballi, Hubballi, Karnataka, India

Dayanand V. Daddi Jain College of Engineering, VTU, Belagavi, Karnataka, India

Reva Dalal Electronics and Telecommunications Department, SCTR'S Pune Institute of Computer Technology, Pune, Maharashtra, India

Siddhi Darekar Department of Electronics and Telecommunication Engineering, Fr. Conceicao Rodrigues Institute of Technology, Navi Mumbai, India

Vidit Dave Computer Engineering Department, CSPIT, CHARUSAT, Changa, Gujarat, India

Nikunj Delavadiya Computer Engineering Department, CSPIT, CHARUSAT, Changa, Gujarat, India

Anirudh Deshpande School of Electronics and Communication Engineering, KLE Technological University, Hubballi, India

Sahana Devali Jain College of Engineering, VTU, Belagavi, Karnataka, India

Aditya A. Dhamanekar Jain College of Engineering, VTU, Belagavi, Karnataka, India

Soham Dhumane Information Technology Department, SCTR'S Pune Institute of Computer Technology, Pune, Maharashtra, India

Aniket Digole Army Institute of Technology, Pune, Maharashtra, India

Anushka Dudhane Department of Electronics and Telecommunication Engineering, Fr. Conceicao Rodrigues Institute of Technology, Navi Mumbai, India

G. Durga Devi ECE Department, New Prince Shri Bhavani College of Engineering and Technology, Chennai, India

T. Nikhil Eshwar Tamil Nadu, Chennai, India

Raj Gandhi Dwarkadas J. Sanghvi College of Engineering, Mumbai, Maharashtra, India

M. A. Gangarde SCTR's Pune Institute of Computer Technology Pune, Pune, India

Anjali Gangone Department of Computer Science and Engineering, CMR Institute of Technology, Hyderabad, India

Pallavi P. Gaonkar Jain College of Engineering, VTU, Belagavi, Karnataka, India

Roshani Sunil Gavande Savitribai Phule Pune University, Pune, India

Pratik R. Gopale School of Mechatronics Engineering, Symbiosis Skills and Professional University, Pune, Maharashtra, India

Reshma Gopinathan Department of Computer Engineering, Pillai College of Engineering, Panvel, Maharashtra, India

Karanbir Singh Grover Department of CSE, Chandigarh University, Patiala, Punjab, India

R. Gunavathi Christ (Deemed to be) University, Bengaluru, India

Abhishek Hadimani Jain College of Engineering, VTU, Belagavi, Karnataka, India

Mustafizul Haque Dr. D. Y. Patil Vidyapeeth's Centre for Online Learning, Dr. D. Y. Patil Vidyapeeth, Pune, India

Kaveri S. Haramagatti School of Electronics and Communication Engineering, KLE Technological University Hubballi, Hubballi, Karnataka, India

Archana Haridas Department of Computer Engineering, Pillai College of Engineering, Panvel, Maharashtra, India

Roshith Hegde School of Electronics and Communication Engineering, KLE Technological University, Hubballi, India

Soumya H. Hooli School of Electronics and Communication Engineering, KLE Technological University Hubballi, Hubballi, Karnataka, India

Shivashankar A. Huddar KLE Technological University, Hubli, Karnataka, India

Pramod Hullur School of Electronics and Communication Engineering, KLE Technological University Hubballi, Hubballi, Karnataka, India

Nandan Illur School of Electronics and Communication Engineering, KLE Technological University, Hubballi, India

Ibrahim Inusa Universiti Tenaga Nasional, Kajang, Selangor, Malaysia

Nalini C. Iyer KLE Technological University, Hubli, Karnataka, India

Vrushali Jadhav Information Technology Department, SCTR'S Pune Institute of Computer Technology, Pune, Maharashtra, India

Jahiruddin Department of Computer Science, Jamia Millia Islamia, New Delhi, India

Vinayak Jainapur School of Electronics and Communication Engineering, KLE Technological University Hubballi, Hubballi, Karnataka, India

N. Janaki Vels Institute of Science, Technology and Advanced Studies, Chennai, India

Stefani Jeyaseelan School of Computer Science and Engineering, Vellore Institute of Technology, Chennai, India

Amrita Jha Department of Mathematics, Chandigarh University, Gharuan, India

Amala Johnson Christ (Deemed to be) University, Bengaluru, India

Sarang L. Joshi School of Mechatronics Engineering, Symbiosis Skills and Professional University, Pune, Maharashtra, India

Sharvari Joshi Dwarkadas J. Sanghvi College of Engineering, Mumbai, Maharashtra, India

Bhakti D. Kadam Department of Electronics and Telecommunication Engineering, SCTR's Pune Institute of Computer Technology, Pune, India

Swapnil Kadam Electronics and Telecommunications Department, SCTR'S Pune Institute of Computer Technology, Pune, Maharashtra, India

Pooja Kanabur Jain College of Engineering, VTU, Belagavi, Karnataka, India

Chandan P. Kangralkar Jain College of Engineering, VTU, Belagavi, Karnataka, India

Riddhi Kanse Department of Electronics and Telecommunication Engineering, Fr. Conceicao Rodrigues Institute of Technology, Navi Mumbai, India

Nitika Kapoor Department of CSE, Chandigarh University, Patiala, Punjab, India

Venkata Mamatha Kasireddy Velagapudi Ramakrishna Siddhartha Engineering College, Vijayawada, Andhra Pradesh, India

Harpreet Kaur Department of Mathematics, Chandigarh University, Gharuan, India

Kimaya Kerkar Vishwakarma Institute of Technology, Pune, India

Sai Keshav I Department of Information Technology-Data Science, Rajalakshmi Engineering College, Chennai, India

Manish Khodaskar Information Technology Department, SCTR'S Pune Institute of Computer Technology, Pune, Maharashtra, India

P. Surya Kiran Department of computational Intelligence, SRM Institute of Science and Technology, Chennai, India

A. Kiruthika Department of Information Technology, Rajalakshmi Engineering College, Chennai, India

Nandita Kottargi Electronics and Communication Department, KLE Technological University, Hubballi, Karnataka, India

K. Krishnendhu Christ (Deemed to be) University, Bengaluru, India

Dhanya Kulkarni School of Electronics and Communication Engineering, KLE Technological University Hubballi, Hubballi, Karnataka, India

Hrushikesh B. Kulkarni School of Mechatronics Engineering, Symbiosis Skills and Professional University, Pune, Maharashtra, India

Shailaja Satishrao Kulkarni Savitribai Phule Pune University, Pune, India

K. Nithish Kumar Tamil Nadu, Chennai, India

P. Kumar Department of Computer Science and Engineering, Rajalakshmi Engineering College, Chennai, India

Megha Kumari Army Institute of Technology, Pune, Maharashtra, India

B. Latha Physics Department, Dr. M.G.R. Educational and Research Institute, Chennai, India

A. Jackulin Mahariba Department of computational Intelligence, SRM Institute of Science and Technology, Chennai, India

Kaushik Mallibhat Electronics and Communication Department, School of Electronics and Communication Engineering, KLE Technological University Hubballi, Hubballi, Karnataka, India

Sanjana Mamde SCTR's Pune Institute of Computer Technology Pune, Pune, India

A. C. Akshitha Mary School of Computer Science and Engineering, Vellore Institute of Technology, Chennai, India

Neha Elizabeth Mathew Christ (Deemed to be) University, Bengaluru, India

Amruta Mhatre Department of Electronics and Telecommunication Engineering, Fr. Conceicao Rodrigues Institute of Technology, Navi Mumbai, India

Yashaswini S. Murgod School of Electronics and Communication Engineering, KLE Technological University Hubballi, Hubballi, Karnataka, India

Thanvitha Nandakumar School of Computer Science and Engineering, Vellore Institute of Technology, Chennai, India

S. R. Nirmala Electronics and Communication Department, School of Electronics and Communication Engineering, KLE Technological University Hubballi, Hubballi, Karnataka, India

Ritika Pandey Army Institute of Technology, Pune, Maharashtra, India

Aayush Parikh Dwarkadas J. Sanghvi College of Engineering, Mumbai, Maharashtra, India

Ronakkumar Patel Computer Engineering Department, CSPIT, CHARUSAT, Changa, Gujarat, India

Trusha Patel Computer Engineering Department, CSPIT, CHARUSAT, Changa, Gujarat, India

Keshav Patil Jain College of Engineering, VTU, Belagavi, Karnataka, India

Rakshita M. Patil Jain College of Engineering, VTU, Belagavi, Karnataka, India

Shahak Patil Jain College of Engineering, VTU, Belagavi, Karnataka, India

Vaishnavi Rajan Pol Department of Computer Engineering, Pillai College of Engineering, Panvel, Maharashtra, India

R. S. Ponmagal CSE Department, SRM Institute of Science and Technology, Chennai, India

K. Poornimathi Department of Computer Science and Business Systems, Rajalakshmi Engineering College, Chennai, India

P. Prakash School of Computer Science and Engineering, Vellore Institute of Technology, Chennai, India

J. Pratiksha KLE Technological University, Hubli, Karnataka, India

Malavika Preeth Department of Electronics and Telecommunication Engineering, Fr. Conceicao Rodrigues Institute of Technology, Navi Mumbai, India

L. Priya Department of Information Technology, Rajalakshmi Engineering College, Chennai, India

Vaishnavi Khandu Purkar Savitribai Phule Pune University, Pune, India

H. P. Rajani Jain College of Engineering, VTU, Belagavi, Karnataka, India

Singamaneni Rajesh Velagapudi Ramakrishna Siddhartha Engineering College, Vijayawada, Andhra Pradesh, India

R. Rakshinee School of Computer Science and Engineering, Vellore Institute of Technology, Chennai, India

N. Gokul Ram Tamil Nadu, Chennai, India

Darshan Ramesh Department of computational Intelligence, SRM Institute of Science and Technology, Chennai, India

Poojitha Ramesh Department of Information Technology-Data Science, Rajalakshmi Engineering College, Chennai, India

Krupa Rasane Jain College of Engineering, VTU, Belagavi, Karnataka, India

Ashoka Kumar Ratha Department of Electronics, Sambalpur University, Burla, Odisha, India

Tejas K. Rayangoudar Jain College of Engineering, VTU, Belagavi, Karnataka, India

G. Rohini EEE Department, S.A. Engineering College, Chennai, India

Dinesh Prasad Sahu School of Computer Science Engineering and Technology, Bennett University, Greater Noida, India

V. Sakthivel School of Computer Science and Engineering, Vellore Institute of Technology, Chennai, India

Nikita M. Sambrekar Jain College of Engineering, VTU, Belagavi, Karnataka, India

M. Sandhiya Department of Information Technology, Rajalakshmi Engineering College, Chennai, India

V. C. Sai Santhosh Tamil Nadu, Chennai, India

Shravani S. Sawant Jain College of Engineering, VTU, Belagavi, Karnataka, India

A. Selvi Rajalakshmi Engineering College, Chennai, India

Prabira Kumar Sethy Department of Electronics, Sambalpur University, Burla, Odisha, India

Mohak Shah Department of Electronics and Telecommunication Engineering, SCTR's Pune Institute of Computer Technology, Pune, India

Salma S. Shahapur Jain College of Engineering, VTU, Belagavi, Karnataka, India

Nazeer Shaik Department of Computer Science and Engineering, Srinivasa Ramanujan Institute of Technology—Autonomous, Anantapur, India

S. N. Shashidhar KLE Technological University, Hubli, Karnataka, India

Apurva Shinde SCTR's Pune Institute of Computer Technology Pune, Pune, India

Deepak M. Shinde Electronics and Telecommunications Department, SCTR'S Pune Institute of Computer Technology, Pune, Maharashtra, India

Tejal M. Shirodkar Electronics and Communication Department, KLE Technological University, Hubballi, Karnataka, India

V. Shruthy Tamil Nadu, Chennai, India

Abhijit Singh Army Institute of Technology, Pune, Maharashtra, India

Jagendra Singh School of Computer Science Engineering and Technology, Bennett University, Greater Noida, India

P. Sivakumar Rajalakshmi Engineering College, Chennai, India

Kartik Solanki Jain College of Engineering, VTU, Belagavi, Karnataka, India

Shilpa Sondkar Vishwakarma Institute of Technology, Pune, India

V. Srividhya EEE Department, Meenakshi College of Engineering, Chennai, India

Riya Elizabet Stanly Department of Computer Engineering, Pillai College of Engineering, Panvel, Maharashtra, India

B. Sudheer Department of computational Intelligence, SRM Institute of Science and Technology, Chennai, India

K. Sujatha EEE Department, Dr. M.G.R. Educational and Research Institute, Chennai, India

Swapna Mandu Department of Computer Science and Engineering, Sreyas Institute of Engineering and Technology, Hyderabad, Telangana, India

Arya Pulakeshi Talukar Jain College of Engineering, VTU, Belagavi, Karnataka, India

H. S. Thakar Savitribai Phule Pune University, Pune, India

Bharti Thakur Yogananda School of AI, Computers and Data Science, Shoolini University, Solan, Himachal Pradesh, India

Payel Thakur Department of Computer Engineering, Pillai College of Engineering, Panvel, Maharashtra, India

Yashraj Thakur Information Technology Department, SCTR'S Pune Institute of Computer Technology, Pune, Maharashtra, India

Mohit Tiwari Department of Computer Science and Engineering, Bharati Vidyapeeth's College of Engineering, Delhi, India

K. Uday Kiran Jain College of Engineering, VTU, Belagavi, Karnataka, India

Suyash Udchan Department of Electronics and Telecommunication Engineering, SCTR's Pune Institute of Computer Technology, Pune, India

Gouri Umate SCTR's Pune Institute of Computer Technology Pune, Pune, India

Varad Unhale Department of Electronics and Telecommunication Engineering, SCTR's Pune Institute of Computer Technology, Pune, India

Kamal Upreti Department of Computer Science, CHRIST (Deemed to Be University), Ghaziabad, India

Priya Vijay Department of Information Technology, Rajalakshmi Engineering College, Chennai, India

Komal Wajantri Jain College of Engineering, VTU, Belagavi, Karnataka, India

Balayya Walaginamath Jain College of Engineering, VTU, Belagavi, Karnataka, India

Sita Yadav Army Institute of Technology, Pune, Maharashtra, India

GPT-3-Based AI Cover Letter Generator: A Feasibility Study & Implementation

Tarannum Alimahmad Bloch and Ibrahim Inusa

Abstract Natural Language Processing (NLP) is the emerging field research studies of the interaction between human and computing systems. With advancement of NLP techniques, machines are becoming increasingly proficient in understanding, generating, and manipulating natural language. This study conducts in depth analysis on using the advanced natural language processing systems using Generative Pre-trained Transformer 3 (GPT-3) in AI cover letter generator, additionally; the paper discusses the ethical implications of using GPT-3-based AI cover letter generators in the job application process, as well as the potential benefits and drawbacks of using GPT-3-based AI cover letter generators for job applicants. The objective of this research paper is to assess the potential effects of incorporating GPT-3 in AI cover letter generation as well as any ethical issues that may arise. Therefore we used research methodology that combines both methods: qualitative and quantitative research methods. The study was carried out in two phases. In the early phase, a qualitative content analysis was performed to examine the ethical implications of using GPT-3-based AI cover letter generators in the job application process and also highlight the potential benefits and drawbacks of using GPT-3-based AI cover letter generators for job applicants. In the second phase, a survey was conducted to generate various cover letters using GPT-3-based cover letter generators with a small sample size of data. To provide data to support that explore incorrect information that may occur when using a GPT-3-based cover letter generator, particularly if the company's culture is not provided in the prompt. The information in this paper will help researchers better understand how artificial intelligence is used in the hiring process and potential effects it might have. The paper provides an introduction to cover letters and their importance in job applications, as well as an overview of natural language processing, generative pre-trained transformer models and development of

T. A. Bloch (✉)
Marwadi University, Rajkot 360003, Gujarat, India
e-mail: tarannum.bloch@marwadieducation.edu.in

I. Inusa
Universiti Tenaga Nasional, Kajang 43000, Selangor, Malaysia
e-mail: talk2ajmilala@yahoo.com

V. K. Gunjan and J. M. Zurada (eds.), *Proceedings of 5th International Conference on Recent Trends in Machine Learning, IoT, Smart Cities and Applications*, Lecture Notes in Networks and Systems 1181, https://doi.org/10.1007/978-981-97-8861-3_1

GPT-3. The paper will be of interest to AI researchers, and anyone interested in the intersection of AI and the job application process.

Keywords GPT-3 · AI · Cover letter generator · OpenAI · Ethical implication

1 Introduction

A cover letter is a document that accompanies a job seeker resume during job or admission application and it addressed the reason for application, how the applicant fits the role, it also highlight the applicant's relevant skills. However, cover letters are not always required but are still necessary, and in a competitive market they can give you a serious edge if they are written and presented effectively [1]. The cover letter can introduce the candidate, serves as a preliminary screening tool, and present a convincing case for their candidature all at once if it is written correctly, conversely, a dull and unimpressive cover letter lacking originality, riddled with errors, can significantly raise the likelihood of rejection for the job application, potentially permanently preventing the applicant from securing their desired position [2].

In another aspect, a cover letter is also an important aspect of the manuscript compliance process, particularly at the time of submitting to a journal. It serves not only as a way to introduce your work to the editor and demonstrate its relevance to the journal's readers, but also as an opportunity to "sell" your manuscript and increase the chances of its acceptance for publication. Furthermore, the cover letter is an important tool for clearly outlining the significance and originality of your research and focusing on any potential issues or concerns that the editor may have. Therefore, when submitting the journals, it is vital to take the time and effort to create an excellently written, convincing, and informative cover letter [3].

Table 1 summarizes the results of two studies on employer preferences for cover letters. According to the Job Seeker Nation Study conducted by Jobvite in 2017, 26% of employers read cover letters and judge the applications critically in their decision to hire new employee [4]. Meanwhile, the study by Schullery, Ickes, and Schullery in 2009 found that 56% of recruiters want applicants to attach a cover letter to their resume [5]. These indicates the significance of a well-written cover letter in the job application process and the need for job seekers to attach a cover letter to their resumes to increase their chances of getting hired. However during the current era of smart automated content generation, this area remains unexplored by researchers, making it difficult to find applied research papers on the mentioned research topic.

With the rapid advancement of emerging technologies, particularly the artificial intelligence, the landscape of trends is constantly changing as proven by the inception of new applications of AI such as such ChatGPT, Elicit, and LEX a etc. The possibility that AI may displace some of the academic workforce and individual professional activities is currently one of the main worries. Writing a cover letter and other documents that are required for job applications and also are crucial in determining the qualifications and suitability of a candidate for a job sometimes.

Table 1 Employer preferences for cover letters

Percentage (%)	Source	Type of study
26	Jobvite (2017)	Job seeker nation study
56	Schullery, Ickes, L. N. M. and S. E. (2009)	Employer preferences for resume and cover letters

1.1 Natural Language Processing

Natural Language Processing is to group computational methods that aim to analyze and represent texts that occur naturally. These techniques are driven by theoretical principles and are used to achieve language processing that resembles human capabilities. NLP is employed across various tasks or applications, involving various linguistic levels of analysis [6]. "Natural language processing" is a discipline that comprises a number of theories and methods designed to help computers comprehend and interact with human language. Resolution of co-references among the NLP tasks that have been studied include autonomous summary, machine translation, morphological categorization, part-of-speech labeling, recognized entities, recognition of optical characters, analysis of discourse, and others [7].

1.2 Generative Pre-Trained Transformer (GPT)

Advanced natural language processing system called the Generative Pre-trained Transformer (GPT) is frequently trained in two steps. They are initially trained to anticipate the subsequent word using a sizable dataset of manuscript from the Internet. In order to provide outputs that are favored by human labelers, the models are then refined with fresh data utilizing the technique referred to as RLHF (reinforcement learning from human feedback). The utilization of extensive text datasets to train language models has led to the development of several advancements, including few-shot learning and the capacity to perform various natural language tasks across diverse domains. These tasks encompass question answering, arithmetic calculations, and classification [8].

The use of GPT-3, one of the most cutting-edge methods for natural language processing available today and the biggest neural network ever trained on 175 billion parameters [9] has prompted additional discussion over the moral ramifications of employing GPT-3-based AI cover letter generators for job applications. The motivation behind this study is rooted in its inception; moreover the study is organized into different sections (methodology, background, findings, and conclusion) to ensure cohesiveness in the subject matter.

1.3 Research Objectives

Major two research objects are to evaluate the ethical implications of using GPT-3-based AI cover letter generators in the job application process and to assess the potential advantages and disadvantages of employing GPT-3-based AI cover letter generators for job applicants.

1.4 Scope and Limitations

This research will focus solely on the use of GPT-3-based AI cover letter generators in the job application process. The research will only consider the perspective of job applicants and will not explore the viewpoint of employers or other stakeholders in the job application process. The research will be limited to the analysis of the effectiveness of AI based cover letters and will not include an examination of other factors that may impact the success of a job application.

2 Research Methodology

2.1 Research Approach

Utilizing both qualitative and quantitative research approaches, this study uses a mixed method research methodology. There will be two phases to the investigation. The implementation of GPT-3-based AI cover letter generators in the job application process will be ethically scrutinized in the initial stage utilizing qualitative content analysis. In the second phase, various cover letters, generated by GPT-3-based cover letter generators will be gather by looking on to the potential benefits and drawbacks of using GPT-3-based AI cover letter generators for job applicants with different levels of education and professional experience.

2.2 Data Collection & Implemetation

Data will be gathered for the qualitative content analysis from online sources including academic publications, news articles, and online discussions about the moral ramifications of employing GPT-3-based AI cover letter generators for job applications. For the survey, we will use a GPT-3-based cover letter generator which is proposed in this paper to generate random cover letters using same and various prompt by observing the spelling errors and incorrect information in the letters. These observations are not to draw conclusion on the performance of the GPT-3 based cover

Table 2 Datasets used in training the GPT-3 model

Dataset	Quantity (tokens)	Weight in training mix (%)
Common crawl (filtered)	410 billion	60
WebText2	19 billion	22
Books1	12 billion	8
Books2	55 billion	8
Wikipedia	3 billion	3

letter generator but to provide a data which will support the claim of GPT-3 based cover letter generator can carry potential incorrect information in terms of skills, company culture, and working experience, while generating a cover letter.

3 Background Work

3.1 Overview of GPT-3

The GPT-3 produces and comprehends text in a natural manner. By giving it text suggestions like phrases or words, developers can "teach" GPT-3. GPT-3 can generate writing that resembles that of a human being and perform a variety of language-based activities, including language translation, question-answering, and text completion.

The GPT-3 API is designed to be user-friendly for everyone, while it is also robust enough to enable machine learning teams to operate more efficiently. Nine months after its debut, more than 300 applications are using GPT-3, and tens of thousands of developers around the world use the platform [10]. Approximately 4.5 billion words are being produced daily by OpenAI, and this number is rising [11].

The datasets used to train the GPT-3 are listed in Table 2, along with the relevant weights in the training mix, or the percentage of instances used in training that come from each dataset. Common Crawl, WebText2, Books1, Books2, and Wikipedia are the datasets that were used. Some datasets are seen up to 3.4 times during training, while others are only seen a single time since the weights in the training mix purposefully do not represent the size of the dataset. [12]

3.2 The Capabilities of GPT-3

The approach has numerous potential uses in both education and business, and it may be used to doing a range of things involving NLP, including a couple of instances like:

- Text generation: GPT-3 can generate new text based on a given topic or theme

- Language translation: GPT-3 can translate text between languages
- Text completion: GPT-3 can complete a given sentence or paragraph with relevant text
- Text summarization: GPT-3 can summarize a given text by extracting key points
- Question answering: GPT-3 can answer questions based on a given context [13]

3.3 *GPT-3 Model Series*

With increasing degrees of processing power, the OpenAI offers four main models: text-davinci-003, text-curie-001, text-babbage-001, and text-ada-001 that are appropriate for various applications. The most capable model is Davinci, while the fastest is Ada. However, there are a number of alternative GPT-3 series models that are mentioned in OpenAI publications and platforms, which you can fine-tune for your unique use case [14].

Table 3 above presents information about various models in the GPT-3 series, which are offered by OpenAI. It provides the model name as mentioned in the research paper and API, the number of parameters in billions, and the publication date of the paper.

4 The Findingss

4.1 *Experimental Setup*

Machine Configurations:

Processor (CPU):	Intel Core i5 (sixth generation)
Memory:	8 GB RAM
Storage:	500 GB internal storage drive
Network adapter:	802.11ac 2.4/5 GHz wireless adapter

System Configurations:

GPT-3 Series: Third Generation Model, Model (API) Name: Text-Davinci -003, Maximum Tokens: 1024, Temperature: 0.7, Presence Penalty: 0, Frequency penalty: 0

Maximum Tokens: The maximum amount of tokens (words or subwords) the model can generate for its response is determined by this parameter. The maximum number of tokens in this situation is 1024, which is the highest number that can be used with GPT-3.

Table 3 Empirical review data for the support of the research objective

Index	Model name in paper	Model name in API	Parameters in (billions)	Paper	Published
1	GPT-3 175B	Davinci	175B	[2005.14165] Language models are few-shot learners	22-Jul-20
2	GPT-3 6.7B	Curie	6.7B		
3	GPT-3 1B	babbage	1B		
4	Codex 12B	Code-cushman-0013	12B	[2107.03374] Evaluating large language models trained on code	14-Jul-21
5	GPT-3 unsupervised cpt-text 175B	Text-similarity-Davinci-001	175B	[2201.10005] Text and code embeddings by contrastive pre-training	14-Jan-22
6	GPT-3 unsupervised cpt-text 6B		6B		
7	GPT-3 unsupervised cpt-text 1.2B	No close matching model on API	1.2B		
8	GPT-3 6.7B pretrain		6.7B	[2009.01325] Learning to summarize from human feedback	15-Feb-22
9	GPT-3 2.7B pretrain		2.7B		15-Feb-22
10	GPT-3 1.3B pretrain		1.3B		15-Feb-22
11	InstructGPT-3 175B SFT	Davinci-instruct-beta	175B	[2203.02155] Training language models to follow instructions with human feedback	04-Mar-22
12	InstructGPT-3 175B	No close matching model on API	175B		04-Mar-22
13	InstructGPT-3 6B		6B		04-Mar-22
14	InstructGPT-3 1.3B		1.3B		04-Mar-22
15	GPT-3 unsupervised cpt-text 1.2B		1.2B	[2201.10005] Text and code embeddings by contrastive pre-training	2022

Temperature: The hyper parameter temperature regulates the randomness and originality of the model's responses. More varied and unexpected responses result from higher temperatures, whereas more conservative and predictable responses result from lower temperatures. In this case, the temperature value is set to 0.7, which is a relatively low value that will result in more predictable responses.

Presence Penalty: Presence penalty is a factor that deters the model from repeating words or phrases that were present in the input prompt. A higher presence penalty value will result in fewer repetitions in the model's responses, while a lower value will allow for more repetitions. In this case, the presence penalty is set to 0, meaning that the model is not penalized for repeating words or phrases from the input prompt.

Frequency Penalty: Frequency penalty is a factor that deters the model from generating high-frequency words or phrases in its responses. A higher frequency penalty value will result in fewer high-frequency words in the model's responses, while a lower value will allow for more high-frequency words. In this case, the frequency penalty is set to 0, meaning that the model is not penalized for generating high-frequency words or phrases.

4.2 Experimental Analysis

The above observations are based on a small sample size of data from cover letters generated by GPT-3 using the same and different prompts up to twenty-five (25) times (Fig. 1).

The purpose of this observation is not to draw a conclusion on the performance of the GPT-3 model on cover letter generation, but rather to provide empirical data to support the statement that a GPT-3-based cover letter generator can potentially include incorrect information when generating a letter, especially when the company's culture is not provided in the prompt. The "Incorrect information" graph

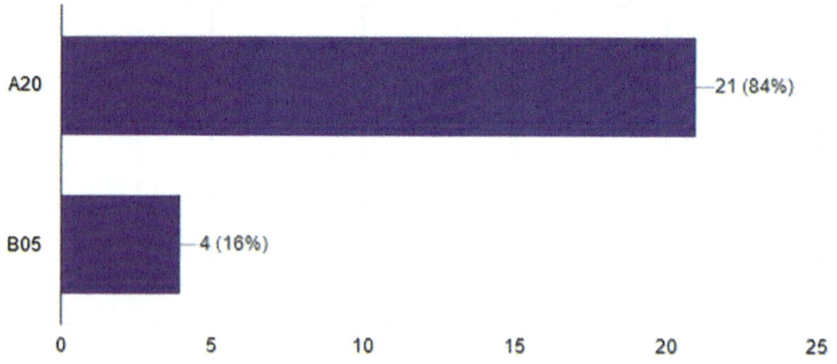

Fig. 1 Prompt used

Fig. 2 Spelling errors

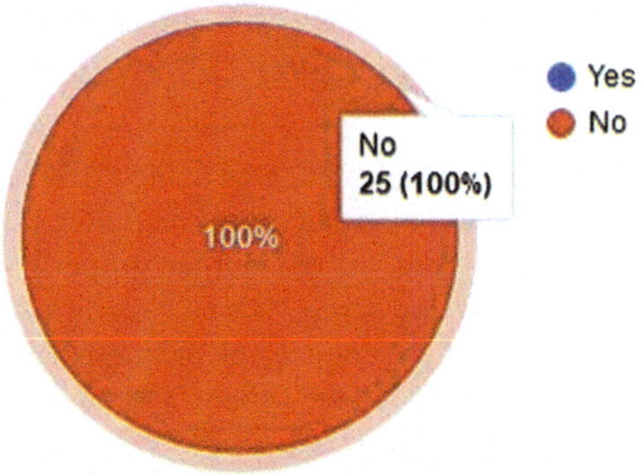

evaluates twenty-five (25) GPT-3-based cover letters and considers any false information in the output as the "Yes" variable and "No" when there is no false information (Fig. 2).

The false information in the cover letter uses the assumption of GPT-3 to enhance the quality of the letter, but the user may not possess those qualities, making it false information or a lie on their cover letter. The prompts used are two: A020, which is used 20 times to generate different cover letters, and B05, which is also used 5 times to generate different cover letters. A prompt in the GPT-3 API refers to the initial text or instructions that are given to the model to generate a continuation or output. The prompt can be a sentence, paragraph, or even a longer text document, and it acts as the model's starting point to start producing fresh content (Fig. 3).

The spelling errors are determined by examining each word in each generated cover letter from these prompts. The "Incorrect information" variables indicate that there may be some inaccuracies in the cover letters generated by GPT-3. This is not surprising, as GPT-3 is a language model and is not specifically trained on cover letters, so it may make errors or inaccuracies just it was shown above in Fig. 3 contains 88% of inaccuracies.

Fig. 3 Incorrect information

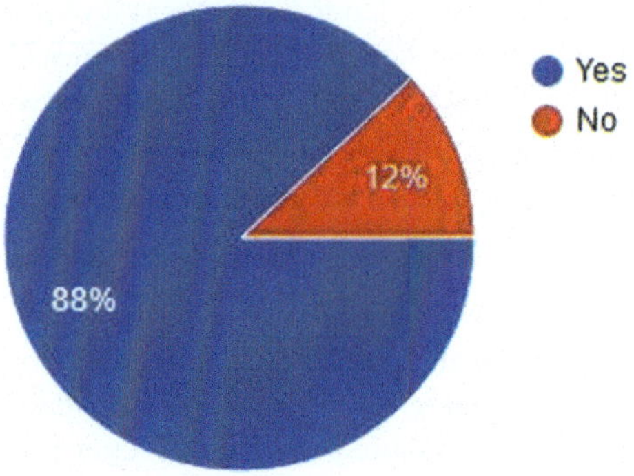

4.3 Potential Benefits and Drawbacks of Using GPT-3-Based AI Cover Letter Generators

A potential disadvantage of using GPT-3 for cover letter generation is that it could lead to a lack of creativity in the letter. As the generator relies on prompts defined to it, if the prompt is not defined properly by indicating keywords that will make the letter sound like it was written by a human, it may not be able to capture the unique qualities and experiences of the candidate. This could result in a generic letter that blends in with all the others, rather than standing out and making a memorable impression [15].

Another disadvantage of using GPT-3 for cover letter generation is that it may struggle to produce a high-quality cover letter in languages other than English. The model is not yet sophisticated enough to accurately translate to other languages; it may result in poor quality letters that may be riddled with mistranslations, spelling errors, inappropriate wording, and addressing issues [16]. Since a static prompt will be defined in the GPT-3 base cover letter generator which will allowed the users to provide an input through user interface without changing the prompt defined at the backend then GPT-3 will generated a letter which may carry several potential dangers. Some of these include:

Lack of understanding of the job requirements: since the prompt is static the GPT-3 will not able to reflect the candidate's understanding of the job requirements, and will generate a cover letter which will seem like the candidate is not a good fit for the job [17].

Lack of familiarity with the organization's culture and workplace: The GPT-3 will not accurately reflect the candidate's familiarity with the organization's culture and workplace, and will generate a cover letter which can make it seem like the candidate hasn't done their research.

Incorrect information: Generated cover letters may contain inaccurate information or formatting errors, which can lead to confusion or mistrust from the employer.

GPT-3 cover letter generator is a powerful and cutting-edge tool that can greatly benefit job seekers by providing them with a fast, efficient, and highly personalized way to create a cover letter. There are several key advantages of using GPT-3 for cover letter generation, including:

Time-saving: GPT-3 can generate a cover letter template tailored to particular job with relevant skills to the job, in a matter of minutes, saving job seekers time and effort in crafting a letter from scratch.

Personalization: GPT-3 can be trained to understand specific keywords and phrases that will tailor the letter to the candidate's unique experiences and qualifications, resulting in a more personalized letter [10].

Increased efficiency: GPT-3 can be used as sample generator, as reviewer, fixing grammatical errors, paraphrasing and shorten the length of the letter.

4.4 *The Ethical Implications of Using GPT-3-Based AI Cover Letter Generators in the Job Application Process*

Modern neural language model GPT-3 is not immune to the problem of bias in machine learning. Its training data, which primarily comprises of Internet text, has led to the model replicating various biases present in human-generated online content. This has been highlighted in a research study conducted by Kris McGuffie and Alex Newhouse at the Middlebury Institute of International Studies, a centre dedicated to preventing terrorism, extremism, and related issues and found that GPT-3 has a tendency to generate text that aligns with extremist ideologies, such as those promoting conspiracy theories and white supremacist discourses [4]. Considering the ethical implications of relying on ChatGPT to craft a document that is typically seen as a personally created message, can be used for sample generator, as reviewer, fixing grammatical errors, paraphrasing and shorten the length of the letter [18].

Among the ethical implications to consider, is the potential for the model to replicate biases present in its training data and generate a cover letter that contains lack of understanding of the job requirements, unawareness of the company's culture and working environment and incorrect information [19]. Another ethical implication may involve the negative implications for the hiring industry. As jobs become easier to apply to with tailored resumes in seconds, it may become too overwhelming for hiring channels to look through every application. This can lead to the increased in use of Applicant tracking system (ATS) which is being used to sort through applications and can result in candidates being rejected before a human even views their application [20]. It can also could lead to a workload that is too costly for companies to hire through job boards and may drive hiring to focus more on referrals, which could impact accessibility of obtaining jobs for those who are not well-connected or breaking into a new industry [21].

5 Conclusion

GPT-3 cover letter generator can be a powerful and cutting-edge tool that can greatly benefit job seekers by providing them with a fast, efficient, and highly personalized way to create a cover letter. Benefits include time-saving, personalization, and increased efficiency. However, there are also ethical implications to consider, such as the potential for the model to replicate biases present in its training data and generate a cover letter that contain lack of understanding of the job requirements, unawareness of the company's culture and working environment and Incorrect information. Additionally, the ease of using AI-based cover letter generators may lead to an overwhelming number of applications for companies, resulting in reliance on Applicant tracking system (ATS) and potentially leading to a more exclusive hiring process. Therefore it is important for job seekers to carefully review and customize their cover letters before using them and for companies to be aware of the potential ethical implications when using AI-based cover letter generators in the hiring process. Our findings lead us to the following recommendations for further study: Improving the language and translation capabilities of GPT-3-based AI cover letter generators to enhance their versatility for multilingual job seekers. Investigating ways to minimize the potential biases present in GPT-3's training data to ensure that the cover letters generated are fair and will not produce incorrect information.

References

1. Brandt PM, Herzberg PY (2020) Is a cover letter still needed? Using LIWC to predict application success. In: IEEE international joint conference on artificial intelligence (IJCAI), pp 1–6. https://doi.org/10.1111/ijsa.12299
2. Ten simple rules for writing a cover letter to accompany a job application for an academic position by Lubomir Tomaska, Jozef Nosek (2018). https://doi.org/10.1371/journal.pcbi.1006132
3. McDowell EE (1987) Perceptions of the ideal cover letter and ideal resume. J Tech Writ Commun 17(2):179–191. https://doi.org/10.2190/BK23-74U3-333Q-0T86
4. Jobvite (2017) Job seeker nation study: finding the fault lines in the American workforce. https://web.jobvite.com/rs/328-BQS-080/images/2023-01-2017JobSeekerNationStudy.pdf
5. Schullery NM, Ickes L, Schullery SE (2009) Employer preferences for résumés and cover letters. Bus Commun Q 72(2):163–176. https://doi.org/10.1177/1080569909334015
6. Liddy ED (2001) Natural language processing. In: Encyclopedia of library and information science, 2nd Ed. NY. Marcel Decker, Inc
7. Khurana D, Koli A, Khatter K et al (2023) Natural language processing: state of the art, current trends and challenges. Multimed Tools Appl 82:3713–3744. https://doi.org/10.1007/s11042-022-13428-4
8. OpenAI (2023) GPT-4 technical report. arXiv [cs.CL]
9. Zhang M, Li J (2021) A commentary of GPT-3 in MIT technology review 2021. Fundam Res 1(6):831–833. https://doi.org/10.1016/j.fmre.2021.11.011
10. OpenAi (2021) GPT-3 Powers the next generation of apps
11. Zhang N, Li L, Chen X, Deng S, Bi Z, Tan C, Huang F, Chen H (2022) Differentiable prompt makes pre-trained language models better few-shot learners. In: Proceeding international conference on learning representations (ICLR), pp 1–16. arXiv:2108.13161v7 [cs.CL]
12. Brown TB, Mann B, Ryder N, Subbiah M, Kaplan J, Dhariwal P, Nee-lakantan A, Shyam P, Sastry G, Askell A, Agarwal S, Herbert-Voss A, Krueger G, Henighan T, Child R, Ramesh A, Ziegler DM, Wu J, Winter C, Hesse C, Chen M, Sigler E, Litwin M, Gray S, Chess B, Clark J, Berner C, McCandlish S, Radford A, Sutskever I, Amodei D (2020) Language models are few-shot learners
13. Lecler A, Duron L, Soyer P (2023) Revolutionizing radiology with GPT-based models: current applications, future possibilities and limitations of ChatGPT. Diagn Interv Imaging 104:307–314. https://doi.org/10.1016/j.diii.2023.02.003
14. OpenAI Index for researchers. https://beta.openai.com/docs/model-index-for-researchers/model-index-for-researchers
15. Short E, Short JC (2023) The artificially intelligent entrepreneur: ChatGPT, prompt engineering, and entrepreneurial rhetoric creation. J Bus Ventur Insights 19(e00388):1–7. https://doi.org/10.1016/j.jbvi.2023.e00388
16. Doshi S (2020) Understanding the GPT-3 research paper. Towards data science
17. Cole E, Short JC (2023) The artificially intelligent entrepreneur: ChatGPT, prompt engineering, and entrepreneurial rhetoric creation. J Bus Ventur Insights 19:(e00388). https://doi.org/10.1016/j.jbvi.2023.e00388. ISSN: 2352-6734

18. Eke DO (2023) ChatGPT and the rise of generative AI: threat to academic integrity? J Responsible Technol13:100060. https://doi.org/10.1016/j.jrt.2023.100060
19. Lim WM, Gunasekara A, Pallant JL, Pallant JI, Pechenkina E (2023) Generative AI and the future of education: Ragnarök or reformation? A paradoxical perspective from management educators. Int J Manag Educ 21(2):100790
20. Askari M (2023) Here's how to get an AI to write a cover letter for every position you apply to, UX collective
21. Leippold M (2023) Sentiment spin: attacking financial sentiment with GPT-3. Financ Res Lett 43:103957. https://doi.org/10.1016/j.frl.2023.103957.ISSN: 1544-6123

Meteorology Driven Air Quality Prediction

Sharvari Joshi, Aayush Parikh, Raj Gandhi, and Kiran Bhowmick

Abstract In today's world, there is an increasing concern over air pollution. It is obligatory to monitor the quality of the air, based on the Air Quality Index (AQI), in times like these since the air quality is continuously deteriorating. Enhancing the ability to monitor and forecast air quality encourages individuals to make informed decisions concerning outdoor activities, thereby mitigating the potential exposure to harmful pollutants. This paper was undertaken to examine the utilization of machine learning models to the prediction of AQI values. The proposed model aims to examine 11 different meteorological parameters that influence the air quality in a given geographical setting. The degree to which these factors directly impact human health and welfare determines their significance. After the required preprocessing, three classification models are applied and evaluated to train and test the dataset to predict the AQI for the city of Pune. After conducting extensive analysis, it was concluded that for the given dataset, the Random Forest Classifier gives the least Mean Square Error (MSE).

Keywords Air quality · AQI · Prediction · Meteorology · Classification models · Random forest · SVC · LightGBM

1 Introduction

Air pollution plagues the economy and health. Worldwide, air pollution causes 7 million early deaths, according to WHO data. Also note that air pollution can cause respiratory infections, cardiovascular illness, cerebrovascular accidents, and malignant neoplasms.

S. Joshi (✉) · A. Parikh · R. Gandhi · K. Bhowmick
Dwarkadas J. Sanghvi College of Engineering, Mumbai, Maharashtra, India
e-mail: sharvarijoshi288@gmail.com

K. Bhowmick
e-mail: kiran.bhowmick@djsce.ac.in

V. K. Gunjan and J. M. Zurada (eds.), *Proceedings of 5th International Conference on Recent Trends in Machine Learning, IoT, Smart Cities and Applications*, Lecture Notes in Networks and Systems 1181, https://doi.org/10.1007/978-981-97-8861-3_2

Modern air pollution has had global effects, necessitating a detailed study of air quality prediction. Air pollution is the main cause of cardiovascular and respiratory diseases, which kill millions annually. By improving the ability to predict air quality, individuals and communities may make informed choices regarding outdoor activities, reducing their chance of exposure to harmful pollutants. Air quality forecasting is linked to environmental sustainability goals. Poor air quality reduces biodiversity and degrades ecosystems, endangering human health.

Air pollution can be estimated using the air quality index (AQI). Several atmospheric pollutant concentrations were examined for the computation. In this study, Machine Learning techniques are investigated and tested to predict the AQI accurately. In recent years, machine learning and other prediction tasks have yielded success. The purpose of this research is to investigate AQI prediction using machine learning models. This enhances AQI forecast accuracy and reliability. This model builds AQI prediction models, evaluates the current research and does practical evaluations to reveal the potential benefits of machine learning in this field. Machine Learning can identify complex relationships between the AQI and weather, transportation, and industrial discharges. Compared to statistical models, machine learning models make more accurate predictions. Machine learning systems can be taught using textual data factors linked to air quality. After training, the system can assess fresh parameter input environmental quality.

2 Literature Review

The Air Quality Index (AQI) is crucial for measuring and reporting air quality. Predictive models for AQI have garnered attention in recent years to aid rapid reaction and intervention. The methods employed include data-driven methods that use neural networks and Random Forests and hybrid models that combine statistical and machine learning techniques. Most of these models' data comes from meteorological, chemical, and satellite observations.

Research by Mahanta et al. [1] examines how well different prediction models forecast Air Quality Index (AQI) values based on pollution and meteorological data in New Delhi, India. A regression analysis is performed to identify which meteorological conditions have a stronger influence on AQI values and how predictive models can help with air quality forecasting. This is done to determine which meteorological conditions have the most impact. According to the findings, the majority of models have an accuracy of approximately 85%, with the Extra Trees regression model achieving the highest; this demonstrates the utility of the models as predictors. In the future, upgrades will include merging historical and real-time traffic data with weather data to enhance the accuracy of the AQI prediction.

Ensemble learning techniques are used by Liu et al. [2]. Data relating to air quality and weather can be found on a website dedicated to air quality and are included in the training datasets. After that, Leveraging Bagging is used in the dataset to create a model that has capacity of accurately predicting the air quality grade. The first a forecasting model is employed for the purpose of generating predictions based on the prediction dataset. According to the findings, the prediction model can produce more accurate grades for estimating the air quality in cities and stations.

The method that has been proposed by Kulkarni et al. [3] keeps track of the air quality over time and makes predictions about the AQI for the next 15 h. In this research, a Support Vector Machine (SVM) model is presented to forecast the air quality using the AQI for the next 15 h. When compared to linear prediction models in terms of RMSE, the suggested model demonstrates superior performance. It is able to predict the amount of pollutants such as PM2.5, PM10, NO, NO_2, and NH_3 and shows higher performance overall.

Historical sensor data gathered across the Port of LA from 4 major sites is utilized by Nagrecha et al. [4]. The data consists of hourly samples of measurements from six different sensors, capturing levels of O_3, CO, SO_2, NO_2, and PM2.5. A Convolutional Neural Network and a Long Short-Term Memory (LSTM) are the two interconnected parts that make up the prediction model. Their CNN model consisted of a straightforward single-layer system with 12 filters and a kernel of size 5X1. The output of this system was then flattened and passed to an LSTM for time series-based prediction. Data is updated each hour. It does not use satellite imaging but rather sensors based on the ground.

A study by Bekkar et al. [5] predicts PM2.5 hourly scale concentrations using meteorological data and PM2.5 concentrations from surrounding stations using Decision Tree (DT), Random Forest (RF), and Gradient Boosting Regressor (GBR) machine learning algorithms. Study data was obtained in Beijing, China. The experimental findings suggest that the gradient boosting regressor model outperforms the other proposed models in terms of predictive accuracy for estimating hourly PM2.5 concentrations, with R2 values ranging from 0.9 to 0.97. Gradient boosting regression performs best with Re as the endpoint, followed by Random Forest and Decision Tree Regression.

In order to offer multi-step-ahead air quality forecasts for 24 and 48 h, Sharma et al. [6] combine data on contaminants, meteorology, and transportation with statistical temporal-spatial feature engineering. To model and predict the levels of PM2.5, PM10, and NO_2 pollution at three air quality monitoring stations located in India, a multivariate time series technique is investigated. Thus, it is thought that the data-driven strategy makes a wonderful addition to the knowledge-driven model.

In a study, Muthukumar et al. [7] propose a two-stage model that has the ability to learn spatiotemporal trends using remote-sensing satellite imagery of air pollution, as well as data from ground-based sensors that monitor air contaminants and meteorological parameters. The approach's objectives include using the GCN architecture to learn spatial correlations in meteorological data, the ConvLSTM model to use both spatial and temporal correlations within satellite and remote-sensing data, and sequentially integrating the GCN and ConvLSTM models.

The survey of this paper also entails analysis of various machine learning models used for classification and prediction. Research by Su et al. [8] suggests, the Light Gradient Boosting Machine (Light GBM) and the extreme Gradient Boosting (XGB) methods to extract and forecast the features of the air monitoring data in Beijing. At last, the conclusion that the accuracy and operational efficiency of Light GBM was significantly greater than that of XGB was achieved.

Training and Testing phases were included in the research paper that was carried out by Madan et al. [9], which made use of Kaggle datasets and air quality monitoring stations. Linear Regression, Decision Tree, Random Forest, Artificial Neural Network, and Support Vector Machine are all examples of algorithms that are utilized in Machine Learning.

Dragomir [10] uses a database with values recorded for sulfur dioxide, nitrogen oxides, carbon monoxide, and ozone to apply the k-nearest neighbor (kNN) technique to estimate the value of the air quality index. The k-nearest neighbor technique may be utilized for classification research since the domain of air quality analysis occasionally includes little or the user does not possess any prior knowledge regarding the distribution of the data. Out of the total 29 occurrences, 19 instances were found to have a prediction error of zero. This accounts for approximately 65.51% of accurate predictions made by the model. If a novel data measurement becomes accessible, there exists a 65.51% likelihood that this model will accurately forecast the overall air quality index.

Comprehensive experimental comparisons are carried out by Zheng et al. [11] to investigate the performance of ensemble methods. A range of well-known models, including the Long Short-Term Memory (LSTM) neural network, the Autoregressive Integrated Moving Average (ARIMA) model, and nine additional base-level models, such as Support Vector Machine (SVM), Random Forest (RF), Logistic Regression (LR), and multiple boosting models, make up the contrast models. All of the models are trained and validated using data sets obtained from the cities of Hong Kong and Beijing. Hong Kong is a coastline city, whereas Beijing is an inland city. The authors have found that ensemble approaches outperform most individual models in experiments, with the best results coming from stacking with probability distributions combined with engineering unique features.

The exponential smoothing method was used by Ma et al. [12] in the ambient air quality data mining system to forecast the medium- and long-term ambient air quality. This was accomplished by mining the data. The research provides a data mining system for ambient air quality that is based on computers and networks. The system can process and transform massive volumes of information regarding air quality, air pollutants, and emissions in urban areas. The findings of the trials provided empirical evidence that the prediction model had a high degree of predictive accuracy.

Some research gaps and drawbacks were observed in the literature survey. Weather, emissions, chemical reactions, and geographic features are just a few of the many variables that affect air quality. Creating thorough models that take into account these complexities is a continuous challenge. Researching the ways in which climate change can impact air quality and developing forecasting models that take into account these complex interrelationships is becoming an increasingly pressing concern.

3 Proposed Methodology

Sequential measures were taken to complete the intended research article. First, the Indian Meteorological Department data was collected and examined. This information was in text format. After examining 11 parameters that affect air quality, the findings are calculated. On adding the 'Air Quality' column, the dataset is cleaned and pre-processed before use. The model's prediction accuracy will be calculated next, as detailed further in this section. An outline of the proposed flow of the paper is shown in Fig. 1.

3.1 Calculating Air Quality of Acquired Dataset

A meteorological parameter-based air quality function is created. The function calculates air quality based on these parameters and returns a label from 'Good,' 'Poor,' 'Moderate,' or 'Unhealthy,' indicating air quality at a given location and time. Inhaled PM2.5, microscopic particles that can harm health, is a key indication of air quality. Air sulfate and nitrate calculate PM2.5. Criteria are used to calculate air quality label after parameters and threshold levels. Following are guidelines: If precipitation exceeds 10 mm, air quality is "Good." Thus, rain cleaned the air. When pH is 0–6 or 7–8, air quality is "Poor." High or low pH may indicate environmental issues. At 500 conductivity, air quality is "Poor." Overconductivity frequently indicates pollution. Sulphate, nitrate, or pm25 (sulphate plus nitrate) over requirements make air quality "Unhealthy." This shows high contamination. When ammonium exceeds 5, air is "unhealthy." High ammonium levels may indicate poor air quality. Not meeting these requirements means air quality is "Moderate." This category applies if "Good," "Poor," or "Unhealthy" are not met.

Fig. 1 Proposed methology

3.2 Data Pre-processing

LabelEncoder preprocesses 'Air Quality' labels before encoding them with numbers. The information in the 'Air Quality' column is changed and saved in a new 'air_quality_numeric' column.

3.3 Model Classification

After the dataset had been preprocessed, it was then used as an input for 3 different machine learning models known as the Random Forest, LightGBM and SVC classifiers.

Random Forest Statisticians and machine learners employ Random Forest ensemble learning. Random forest training requires three hyperparameters. Classification tasks predict classes using a majority vote with the most common category variable. The prediction is completed via out-of-bag cross-validation.

LightGBM Open-source, distributed, and high-performance gradient-boosting framework. It solves several machine-learning problems well. LightGBM is a gradient boosting technique extension that boosts samples with greater gradients and applies automatic feature selection. It has become the standard approach for machine learning contests using tabular data for regression and classification predictive modeling.

Support Vector Classifier Pattern identification and classification using SVM is common. After converting data points to high-dimensional space, SVC identifies the optimum hyperplane to divide them into two classes. Data is classified into four air quality categories using this model. The dataset is 20% test, 80% training. Combine decision tree results to improve ensemble learning prediction. GridSearchCV adjusts hyperparameters. CV hyperparameter grid-search. Gradient-boosting speeds Train LightGBM classifier. This study classifies binary and multi-class using SVC SVMs.

3.4 Model Comparison

After training classification models with 80% of the processed dataset, empirical findings were collected, reviewed, and compared to select the best model for each classification technique. These models were compared using accuracy, precision, recall, and F1 score. Precision-recall (PR) curves and histograms are offered later in this study to help analyze and compare these models.

4 Experimentation

4.1 Dataset Description

Pune's urban environmental and water quality data is being studied [13]. The dataset is arranged by rows and columns for observations and variables or parameters. The 11 parameters considered are: rainfall, pH, conductivity, sulphate, nitrate, chloride, ammonium, calcium, magnesium, sodium, and potassium. The dataset is useful for environmental monitoring, water quality assessment, and air quality correlation studies. The data must be examined more thoroughly to draw conclusions and better understand the relationship between environmental factors and air quality in the study area. A correlation between two variables 'Rainfall' and 'Air Quality' is visualized by creating a scatter plot with a regression line in Fig. 2.

Similarly, another scatter plot between an important attribute 'ph' and 'Air Quality' is visualized in Fig. 3.

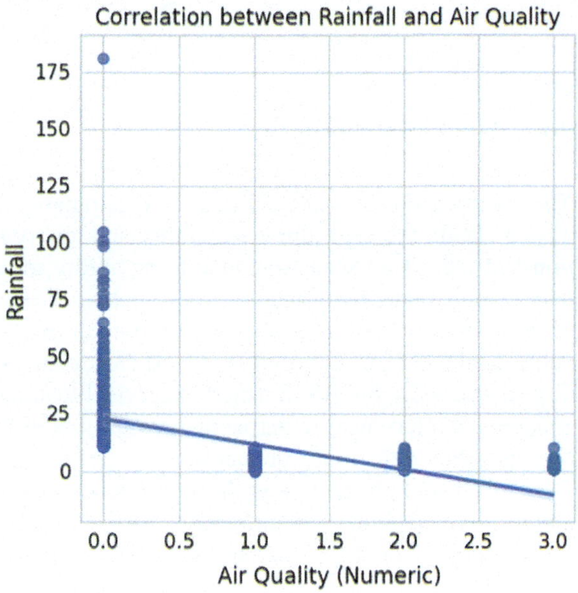

Fig. 2 Scatter plot correlation between rainfall and air quality

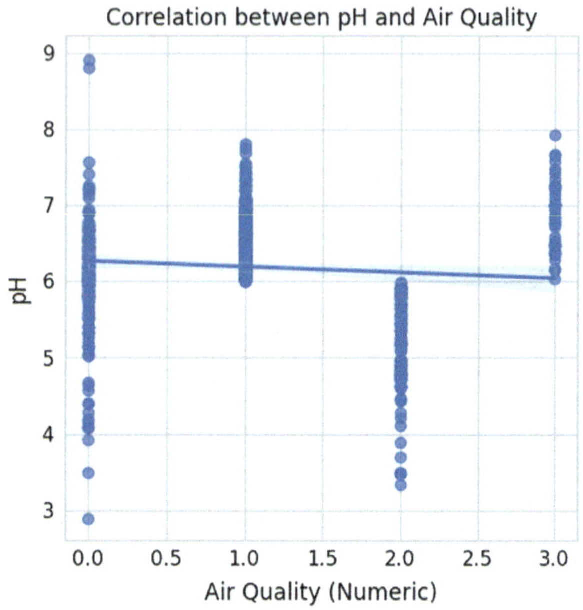

Fig. 3 Scatter plot of correlation between pH and air quality

4.2 Model Advantages

The Random Forest method has gained popularity in machine learning recently. This is due to the algorithm's versatility and ability to be used in many classification and regression scenarios. Due to its ability to manage big datasets and evaluate feature value, the Random Forest technique was used to solve the overfitting problem. Reason for doing this was to exploit the method's capacity to analyze traits' value. LightGBM was chosen for its efficiency and ability to prevent overfitting in large datasets. LightGBM handles categorical features directly, eliminating one-hot encoding. It offers many hyperparameters and model fine-tuning options to customize the algorithm to the user's demands.

In this model, the SVC was chosen for its domain flexibility, outlier robust- ness, and low hyperparameter requirements. SVC may also balance datasets by changing class weights. When one class has much more samples than the other, this helps.

4.3 Parameter Tuning

This experiment employed the following RF model parameters: Number of estimators, 'n_estimators,' was 50, 100, or 200. Additionally,'max_depth' was set to [None, 10], 'min_samples_split' to [2, 5], and 'n_jobs' to − 1. GridSearchCV's 'cv = 3' argument indicates threefold cross-validation. LGBM model parameters were 0.1

for 'learning rate', 2000 for 'n_estimators', and -1 for 'n_jobs'. SVC random state was 42.

5 Results

As previously stated, RF, LGBM, and SVC models were trained and evaluated for Air Quality Label classification. LGBM had 0.0945 MSE and 97.63% accuracy. Additionally, the SVC classifier has an MSE of 0.37 and 95.27% accuracy. A bar graph was plotted to compare the accuracies of different models in Fig. 4. Based on the 11 chemical characteristics in this article, RF was the best model for air quality classification into 4 groups.

After the individual analysis of the results for Air Quality classification, the empirical class-wise classification accuracies have been compared in Table 1 as shown.

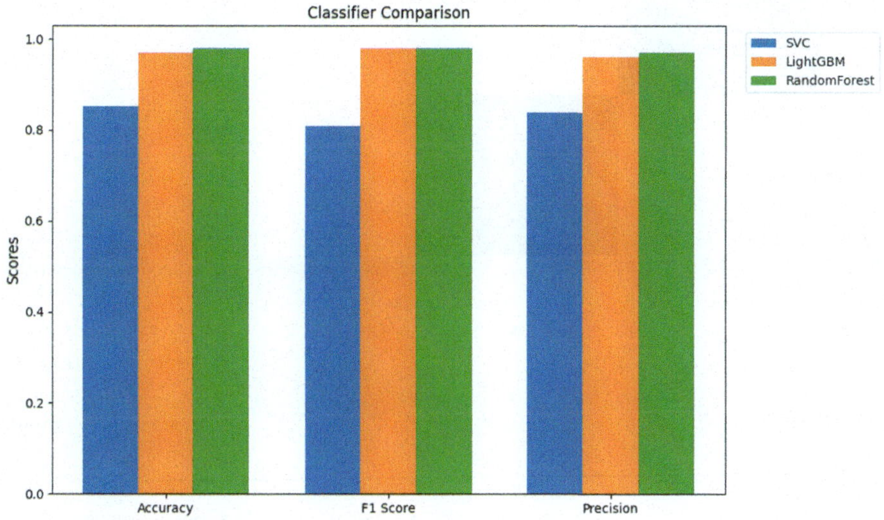

Fig. 4 Perfomance comparisons for proposed models

Table 1 Comparison of accuracies of SVC, LGBM, RF for each air quality class

Air quality class	SVC	LGBM	RF
Unhealthy	0.0002	1.0000	1.0000
Poor	0.9811	0.9623	0.9811
Moderate	0.1852	1.0000	1.0000
Good	0.2500	0.9167	0.9167

This article finds similar findings for RF and LGBM because sensor limitations, measurement errors, or quick environmental changes can make air quality data noisy. Noise-resilient Random Forest and LightGBM may explain their similar performance. LGBM uses gradient boosting with decision trees, while RF uses ensemble learning with multiple trees. Both models can capture many data patterns and relationships due to their ensemble nature, which may lead to similar air quality estimates.

With an MSE of 0.0629 and an accuracy of 98.42%, RF was the most accurate model among the three. Figure 5 shows a confusion matrix for the 4 Air Quality labels.

Because a dataset with fewer dimensions was utilized for this paper, there is a possibility that the predicted data were overfitted during the training of the RF classifier.

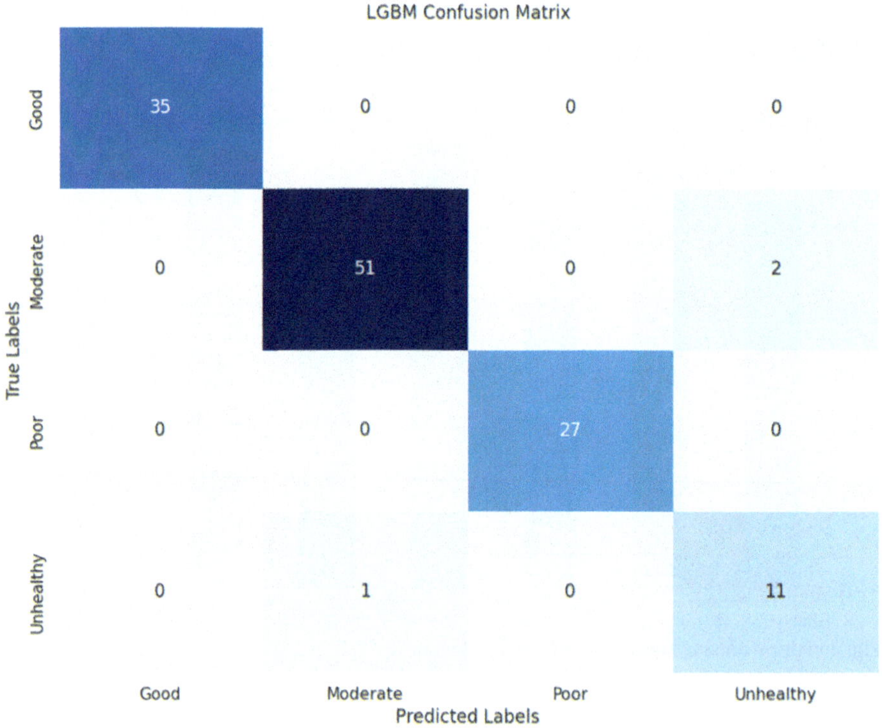

Fig. 5 Confusion matrix for RF

6 Conclusion

Nowadays, air quality is important since toxins and pollutants are more than ever. Recent interest has focused on meteorological data-based AQI predictions. Its importance for public health and environmental management is the main reason. Three steps are given to assign classes that will affect atmospheric conditions. IMD meteorological data can be cleaned and label encoded to classify air quality into four categories. Three machine learning models—LGBM, SVC, and RF—have been used to anticipate air quality in different regions. RF has the greatest accuracy of these three classifiers at 98.42.

This paper's long-term goal is to employ a more diverse dataset with more geographical attributes. The data analyzed in this research fit the Machine Learning models. If more data were used, geostatistical models like Kriging or Spatial Autoregressive could estimate air quality at unobserved sites using meteorological and satellite data. This would improve the understanding and classification of the findings.

References

1. Mahanta S, Ramakrishnudu T, Jha RR, Tailor N (2019) Urban air quality prediction using regression analysis. In: TENCON 2019–2019 IEEE region 10 conference (TENCON). IEEE, pp 1118–1123
2. Liu W, Zhang H, Liu Q (2019) An air quality grade forecasting approach based on ensemble learning. In: 2019 International conference on artificial intelligence and advanced manufacturing (AIAM). IEEE, pp 87–91
3. Kulkarni M, Raut A, Chavan S, Rajule N, Pawar S (2022) Air quality monitoring and prediction using SVM. In: 2022 6th International conference on computing, communication, control and automation (ICCUBEA). IEEE, pp 1–4
4. Nagrecha K, Muthukumar P, Cocom E, Holm J, Comer D, Burga I, Pourhomayoun M (2020) Sensor-based air pollution prediction using deep CNN-LSTM. In: 2020 International conference on computational science and computational intelligence (CSCI). IEEE, pp 694–696
5. Bekkar A, Hssina B, Douzi S, Douzi K (2022) Air quality forecasting using decision trees algorithms. In: 2022 2nd International conference on innovative research in applied science, engineering and technology (IRASET). IEEE, pp 1–4
6. Sharma S, Tanwar P, Yadav A, Sairam BK, Jaswal S (2022) Critical review of air quality prediction using machine learning techniques. In: 2022 Sixth international conference on I-SMAC (IoT in social, mobile, analytics and cloud) (I-SMAC). IEEE, pp 1–7
7. Muthukumar P, Cocom E, Nagrecha K, Comer D, Burga I, Taub J, Pourhomayoun M (2021) Predicting PM2. 5 atmospheric air pollution using deep learning with meteorological data and ground-based observations and remote- sensing satellite big data. Air Qual Atmos Health, 1–14
8. Su Y (2020) Prediction of air quality based on gradient boosting machine method. In: 2020 International conference on big data and informatization education (ICBDIE). IEEE, pp 395–397
9. Madan T, Sagar S, Virmani D (2020) Air quality prediction using machine learning algorithms–a review. In: 2020 2nd International conference on advances in computing, communication control and networking (ICACCCN). IEEE, pp 140–145
10. Dragomir EG (2010) Air quality index prediction using K-nearest neighbor technique. Bull PG Univ Ploiesti Ser Math Inform Phys LXII 1(2010):103–108

11. Zheng H, Cheng Y, Li H (2020) Investigation of model ensemble for fine- grained air quality prediction. China Commun 17(7):207–223
12. Ma L, Gao Y, Zhao C (2020) Research on machine learning pre- diction of air quality index based on spss. In: 2020 International conference on computer network, electronic and automation (ICCNEA). IEEE, pp 1–5
13. Indian meteorological data set (Pune) (2023) Precipitation chemistry, IMD

Video Enhancement by Progressive Use of Denoising and Super-Resolution

Rishabh Balse, Aliabbas Attarwala, Vidit Dave, Nikunj Delavadiya, Trusha Patel, and Ronakkumar Patel

Abstract Watching a low-resolution video is the troublesome. Now a days this problem is faced due to unavailability of internet. Moreover, there are thousands of old films and videos which are available only in low-quality and low-resolution due to the absence of advanced technology and high-resolution cameras during those time. This is where the concept of video enhancement arrives. Video Enhancement is widely explored yet a challenging domain due to its inability to recreate the content same as original one. The results of using two video improvement techniques, namely denoising and video super-resolution, in sequence is demonstrated in this paper. Two distinct models from denoising and super-resolution are utilized for experimentation. The impacts of four permutations of the said models is tested on the UAV123 dataset, which is one of the largest datasets for high-resolution aerial drone videos.

Keywords Video enhancement · Video denoising · Video super-resolution

1 Introduction

Video enhancement is the process of applying different image processing techniques like denoising, deblocking, upscaling, frame interpolation, brightness adjustments, saturation adjustments, improving stability, etc. on each frame of video to boost its visual quality. Researchers have worked on all those techniques individually.

This paper shows the implementation results of two such techniques viz. denoising and super resolution, in a consecutive order respectively. These techniques does enhancement on pixel-by-pixel bases though CNN (Convolution Neural Network) [1].Two different models for each of the above techniques namely VNLNet [2] and FastDVDNet [3] for denoising and ESPCN [4] and VSR_DUF [5] for super-resolution are used in experiment. Experimentation results are collected on the

R. Balse · A. Attarwala · V. Dave · N. Delavadiya · T. Patel (✉) · R. Patel
Computer Engineering Department, CSPIT, CHARUSAT, Changa, Gujarat, India
e-mail: trushapatel.ce@gmail.com

V. K. Gunjan and J. M. Zurada (eds.), *Proceedings of 5th International Conference on Recent Trends in Machine Learning, IoT, Smart Cities and Applications*, Lecture Notes in Networks and Systems 1181, https://doi.org/10.1007/978-981-97-8861-3_3

UAV123 [6] dataset. The task sequence is determined as denoising and super-resolution, respectively, and then four distinct combinations of the proposed four models are constructed. The UAV123 (Unmanned Aerial Vehicle) dataset, which comprises of high resolution (1280 × 720) low altitude aerial films, was chosen for testing and implementation of the said models. After generating the findings for all four combinations, a decision is made based on many image testing parameters such as PSNR (Peak Signal to Noise Ratio), SSIM (Structural Similarity Index Measure), and VMAF (Video Multi-Method Assessment Fusion) to determine which the combination performs best on the given dataset.

2 Video Enhancement Techniques

2.1 *Video Denoising*

The process of eliminating noise from videos is known as video denoising. Video denoising has additional complexities as compared to Image denoising, hence a video is much more than just a set of frames [7].Video denoising algorithms need to evolve themselves from standard Image denoising algorithms and be able to take into consideration temporal coherence as well. Below are descriptions of two such algorithms: (1) VNLNet and (2) FastDVDNet.

VNLNet: VNLNet [2] presents a video denoising CNN approach that incorporates non-local information in a simple way. Traditional CNN architectures are incompatible with searching for self-similarities. VNLNet would incorporate non-locality into the network by finding similar patches in the search region, assigning feature vector to each frame pixel and passing the information to a CNN to predict a clean image. VNLNet is supposed to be the first of the successful algorithms to use CNN in video denoising. Fig. 1 show the architecture of VNLNet.

FastDVDNet: The FastDVDNet [3] was specially designed towards embedded systems. A system with restricted computing power would certainly appreciate a denoising algorithm with faster run time. FastDVDNet promises to achieve temporal

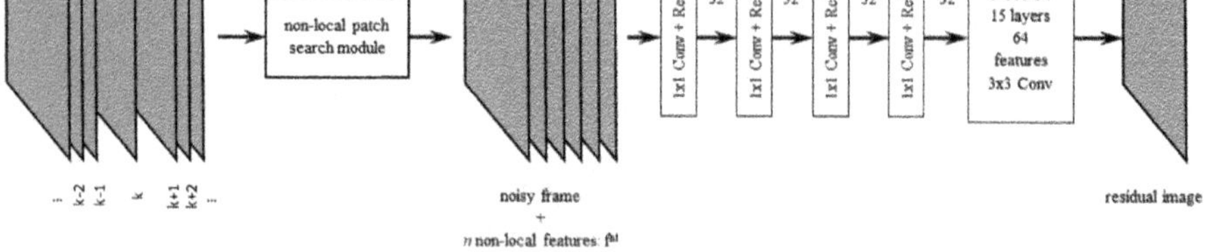

Fig. 1 Architecture of VNLNet [2]

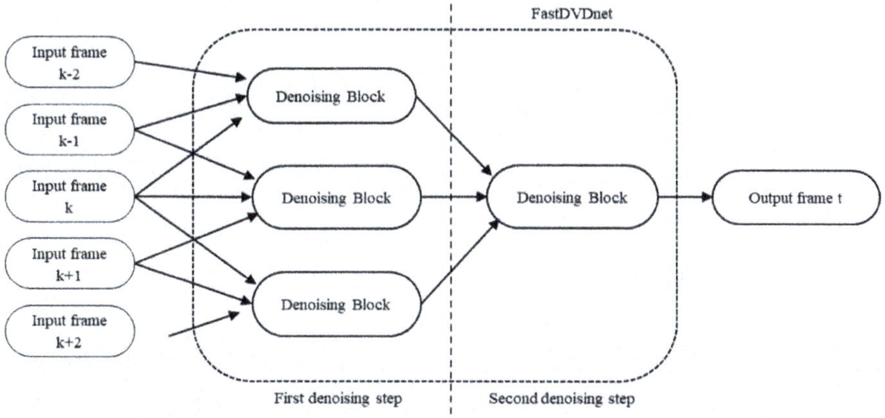

Fig. 2 Architecture of FastDVDNet [3]

stability and get rid of flickering effects between frames by making use of multiple input frames instead of just one and deploying multiscale denoising blocks. Unlike other advanced approaches, FastDVDNet tries to cut on runtime costs by not estimating the optical flow. The architecture of FastDVTNet is given in Fig. 2.

2.2 *Super Resolution*

Super resolution is used for generating high resolution video from a low resolution one. This algorithm assumes that the low-resolution video is passed through a low pass filter (i.e., blurring) and then down-sampled. However, the problem becomes more complicated as there is a loss of high frequency information. Majority of the redundant high frequency data is reconstructed from low resolution components.

ESPCN: The ESPCN [4] aims to train LR (low resolution) images in order to recover its corresponding HR (high resolution) images. The ESPCN model can be taken one step further and trained on the frames of a video. HR data is created from the network's LR feature maps. As a result, the upscaling method is handled by the network's last layer. The network receives the LR frame directly, and feature extraction takes place via non-linear convolutions within LR space. ESPCN architecture is specified in Fig. 3.

VSR_DUF: The use of Dynamic Upscaling Filters in video super resolution has become important to produce ultra high definition displays [5]. VSR_DUF proposes a new deep neural network with end-to-end connectivity. The HR frame is directly constructed by doing local filtering to the input center frame with the help of created up-sampling filters. Figure 4 illustrates the architecture of VSR_DUF.

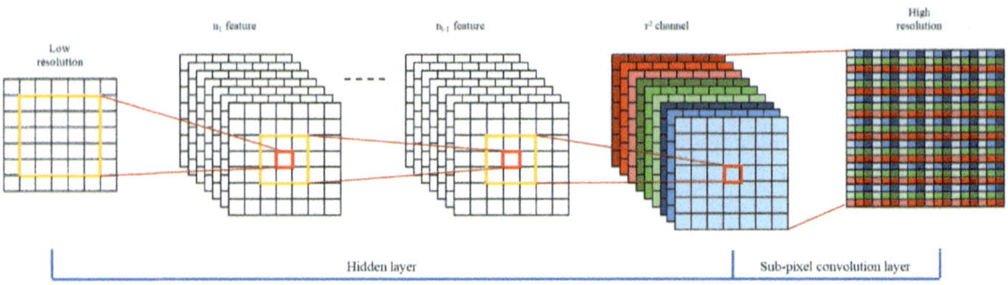

Fig. 3 Architecture of ESPCN [4]

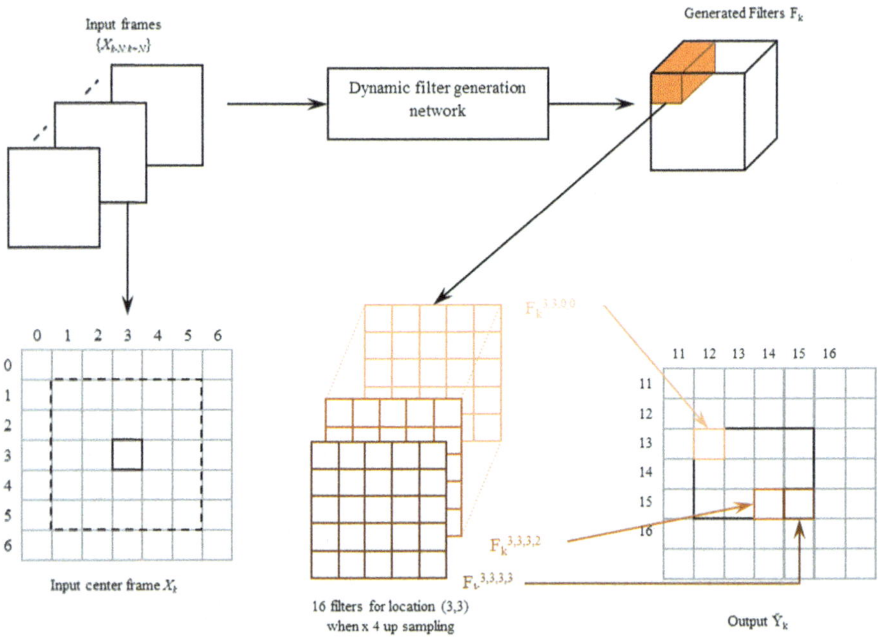

Fig. 4 Architecture of VSR_DUF [5]

3 Evaluation Metrics

According to [8], the two types of metrics for evaluating video quality are (1) objective video quality Metrics (2) subjective video quality metrics. Furthermore, objective video quality metrics can be segregated into three types (1) NR (No-Reference) (2) RR (Reduced Reference) and (3) FR (Full Reference). To evaluate the resulting video by comparing it to the original, full reference video quality metrics need the original video or images. SSIM and PSNR are two such widely used FR objective video quality metrics which are taken into consideration for evaluation of our model combinations. These metrics evaluate the obtained videos by comparing them with the original ones. The higher PSNR value suggests better video quality. Similarly, having the value of SSIM closer to 1 ($-1 \leq \text{SSIM} \leq 1$), indicates its close resemblance with the original video. However, the PSNR is an equation-based approach which does not use the concept of human perception. VMAF, which is a subjective video quality metric.

3.1 PSNR

MSE (Mean Squared Error) [9] is widely used full reference quality metric to compare the quality of two images by finding the value of cumulative squared error between the degraded image and original image. It makes use of a decoder for the evaluation of the degree of image reconstruction, whose values range from 0 (no difference) to 65025 (maximum difference at 8-bit color depth). Formula of MSE is given below.

$$MSE = \frac{1}{mn}\sum_{i=0}^{m-1}\sum_{j=0}^{n-1}(O(i,j) - D(i,j))^2 \tag{1}$$

where O and D refers to original and degraded images respectively. m and n are row and column count in the input images. i and j are the row and column indices, respectively.

PSNR [8, 10] is the most commonly used evaluation metric whose values do not exceed 50dB in practical scenarios. The bigger the PSNR, the better. Computed by using MSE, the equation for PSNR can be given below:

$$PSNR = 10log_{10}\left(\frac{(L-1)^2}{MSE}\right) = 20log_{10}\left(\frac{L-1}{MSE}\right) \tag{2}$$

where L represents total number possible intensity levels in an image ranges from 0 to L-1, and RMSE represents Root Mean Squared Error.

3.2 SSIM

The SSIM [11] shows the similarity between the obtained image and original image by comparing image elements perceived by the human rather than pixel-by-pixel comparison. As mentioned in [11], structural distortions are used for the evaluation of perceptual distortion based on the luminance values' difference. It includes three types of distortion: luminance, contrast and texture. SSIM values range from -1 (huge difference) to 1 (no difference). Therefore, the closer the SSIM to 1, the better.

$$SSIM(x, y) = [l(x, y)]^{\alpha} \bullet [c(x, y)]^{\beta} \bullet [s(x, y)]^{\gamma} \tag{3}$$

$$l(x, y) = \frac{2\mu_x\mu_y + C_1}{\mu_x{}^2 + \mu_y{}^2 + C_1\prime} \tag{4}$$

$$c(x, y) = \frac{2\sigma_x\sigma_y + C_2}{\sigma_x{}^2 + \sigma_y{}^2 + C_2\prime} \tag{5}$$

$$s(x, y) = \frac{\sigma_{xy} + C_3}{\sigma_x\sigma_y + C_3\prime} \tag{6}$$

$$C_3 = \frac{C_2}{2} \tag{7}$$

when $\alpha = \beta = \gamma = 1$ then

$$SSIM\,(x, y) = \frac{(2\mu_x\mu_y + C_1)(2\sigma_x + C_2)}{(\mu_x{}^2 + \mu_y{}^2 + C_1)(\sigma_x{}^2 + \sigma_y{}^2 + C_2)} \tag{8}$$

$$SSIM_{ij} = W_Y \bullet SSIM_{ij}^{Y} + W_{Cb} \bullet SSIM_{ij}^{Cb} + W_{Cr} \bullet SSIM_{ij}^{Cr} \tag{9}$$

where:

α, β, γ: importance coefficients C_1, C_2: stabilizing constants.

μ_i: mean of sample i W_Y: weight of Y.

σ_i: deviation of sample i W_{Cb}: weight of Cb.

σ_{ij}: covariance of sample i and j W_{Cr}: weight of Cr.

3.3 VMAF

VMAF [12] was built on the need of a better perceptual metric. Compared to PSNR and SSIM, it is more accurate on measuring the human perception of video quality. VMAF uses HVS (Human Visual System) [13] modelling to collect the perceptual evidence of how good the quality of the video really is. This evidence is then fed to a Machine Leaning Algorithm (SVM Regression [14]) to align the features with subjective scores trained on ACR (Absolute Category Rating) scale collected from lab tests.

4 Datasets

All four models used in experimentation are pretrained models. The details of training dataset of each of them is given in Table 1.

Table 1 Training dataset of used pretrained models

Type	Model name	Training dataset	Resolution	Video category	Number of videos
Denoising	FastDVDNet [3]	Davis [15]	854 × 480	All categories	50
	VNLNet [2]	HD videos from YouTube	1280 × 720	Antelope, cars, factory, etc	1068
Super-resolution	ESPCN [4]	Xiph	1920 × 1080	Random	10
	VSR_DUF [5]	Internet videos with landscape, wildlife and other contents	144 × 144	Landscape, wildlife etc	351

In this paper, UAV123 [6] dataset is used for the testing. The UAV123 dataset is a subset of a much larger UAV20L dataset meant for long-term aerial tracking. It includes 123 high-resolution video sequences with over 110,000 frames shot from an aerial view. Each frame has resolution of 1280 × 720. This makes UAV123, the second-largest object tracking dataset after ALOV300++ [16].

5 Combined Implementation

5.1 Preprocessing

From the UAV123 dataset, 10 random videos were selected for the. From the chosen videos, 50 consecutive frames were randomly extracted. Next, all the extracted frames are down sampled from resolution 1280 × 720, to 640 × 360. Furthermore, Gaussian noise with 0.2 mean and 0.002 variance was added to make the frames appropriate for testing purposes the same can be depicted in Fig. 5.

5.2 Testing and Results

Considering 2 models each for denoising and super-resolution, 4 permutations can be generated viz. (1) FastDVDNet + ESPCN (2) FastDVDNet + VSR_DUF (3) VNLNet + ESPCN (4) VNLNet + VSR_DUF. The 50 noisy frames of each of the 10 videos were passed into the denoising models (FastDVDNet and VNLNet) as an input and the generated denoised frames, obtained as an output, were given as an input into the super-resolution models (ESPCN and VSR_DUF) following which,

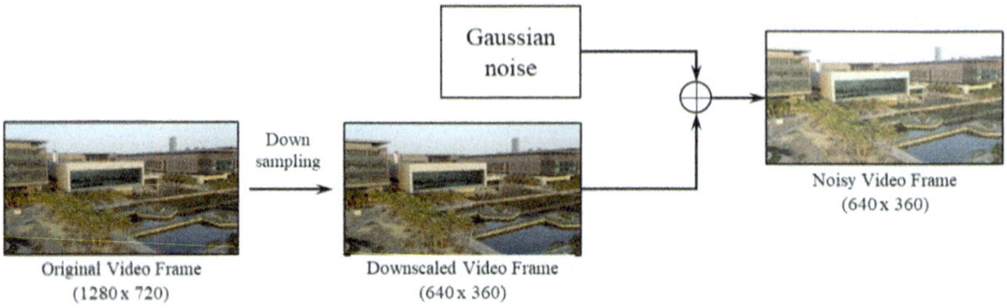

Fig. 5 Video preprocessing stages

the PSNR, SSIM and VMAF of the obtained output were calculated. Figure 6 shows the result of all thesaid permutations of the chosen denoising and super-resolution models for the sample noisy image.

To compare the performance of the said four permutations of the chosen denoising and super-resolution models, SSIM, PSNR and VMAF metrics are used. Details of all of them are discussed in Sect. 3. Table 2 contains PSNR, SSIM and VMAF values for each clip along with average values. Figures 7, 8 and 9 are plots of average values

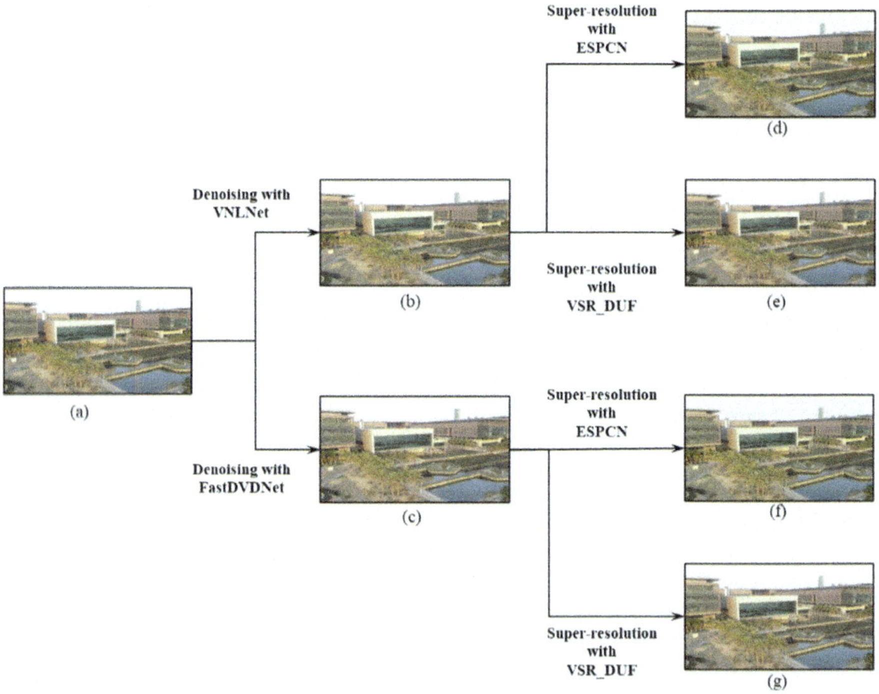

Fig. 6 Results of said four combination of denoising and supre-resolution **a** noisy and low resolution image **b** result of denoising of 'a' using VNLNet (**c**) result on denosing of 'a' using

of each metric. For determining the best combination, all the values of PSNR, SSIM and VMAF must be taken into consideration. Hence, they are scaled and merged into one single graph, given in Fig. 10. It is clearly apparent that the model permutation VNLNet + ESPCN is the superior among all the tested permutations as the values of SSIM and VMAF are the highest among the others.

6 Conclusion

Combining various video enhancement models is certainly a great idea, however, the choosing models also play a major role in determining the quality of our output. VNLNet + ESPCN might work the best for UAV123 dataset compared to other tested combinations. Factors like 'Model Architecture' and 'Model Training' should be taken into consideration before choosing the best combination. A simple analysis like the one in this paper, could help de-mystify a bigger question—"Which model combination is the best for me?"

Table 2 Results obtained by testing all different combinations (largest values are highlighted in red color)

Model Combina-tion	Metric	Clip name										Average
		Bike1	Boat3	Boat5	Building1	Person23	Truck1	Wake board1	Person 18	Car 3_s	Peron 3_s	
VNLNet + VSR_DUF	PSNR	16.92	17.07	16.90	17.49	16.97	16.97	17.06	17.01	16.54	16.56	16.74
	SSIM	0.89	0.86	0.83	0.87	0.86	0.89	0.90	0.88	0.77	0.75	0.82
	VMAF	73.51	59.46	60.99	66.10	63.72	72.51	68.21	65.45	61.80	61.83	67.67
VNLNet + ESPCN	PSNR	17.41	17.44	17.33	17.99	17.37	17.43	17.50	17.48	17.08	17.02	17.22
	SSIM	0.89	0.86	0.83	0.87	0.86	0.89	0.91	0.89	0.78	0.76	0.82
	VMAF	78.57	62.37	73.52	76.64	69.47	82.64	76.24	77.01	81.48	75.83	77.20
FastDVDNet + VSR_ DUF	PSNR	17.19	17.35	17.14	17.76	17.21	17.23	17.35	17.27	16.70	16.75	16.97
	SSIM	0.86	0.84	0.79	0.84	0.82	0.87	0.88	0.86	0.72	0.71	0.78
	VMAF	61.44	52.70	53.01	57.96	53.37	63.30	61.15	57.62	52.98	50.98	56.21
FastDVDNet + ESPCN	PSNR	17.67	17.73	17.58	18.25	17.61	17.72	17.81	17.74	17.23	17.20	17.44
	SSIM	0.87	0.84	0.80	0.85	0.83	0.88	0.89	0.87	0.73	0.72	0.79
	VMAF	67.32	57.40	63.52	67.64	59.23	72.53	70.41	69.28	68.30	61.28	64.30

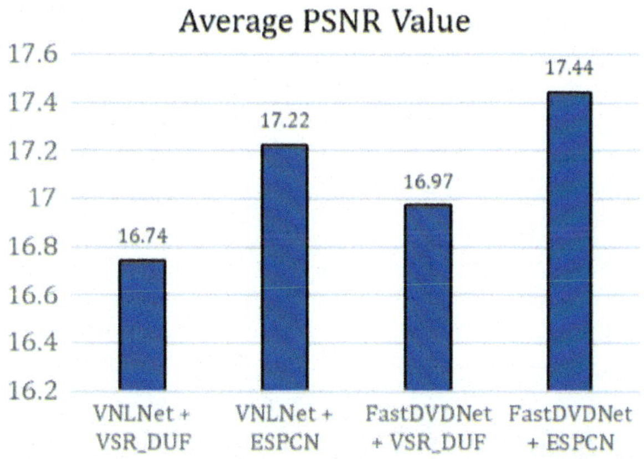

Fig. 7 Qualitative comparison of PSNR

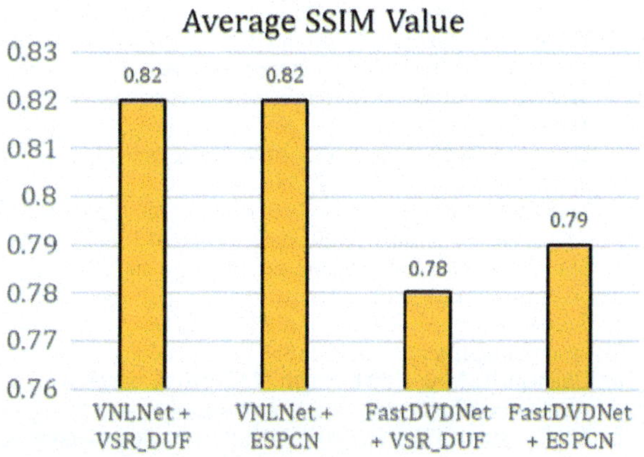

Fig. 8 Qualitative comparison of SSIM

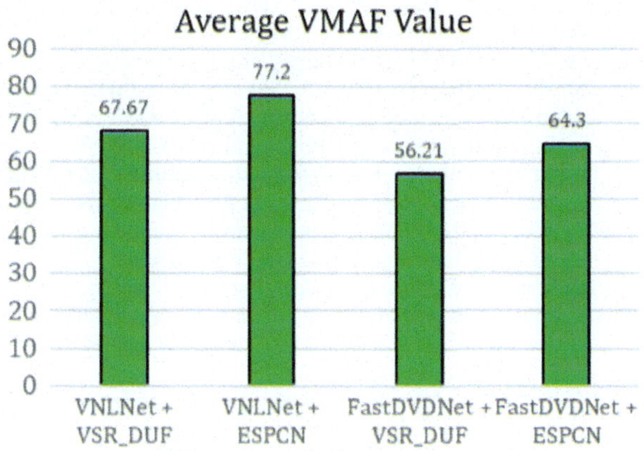

Fig. 9 Qualitative comparison of VMAF

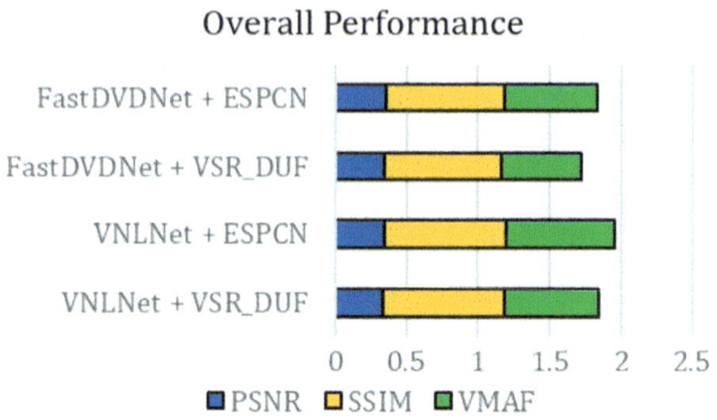

Fig. 10 Qualitative comparison of all metrics

References

1. Bayat O, Aljawarneh S, Carlak HF (2017) International association of researchers, institute of electrical and electronics engineers, and Akdeniz Üniversitesi. In: Proceedings of 2017 international conference on engineering & technology (ICET'2017): Akdeniz University, Antalya, Turkey, pp 21–23
2. Davy A, Ehret T, Morel J-M, Arias P, Facciolo G (2018) Non-local video denoising by CNN. http://arxiv.org/abs/1811.12758
3. Tassano M, Delon J, Veit T (2019) FastDVDnet: towards real-time deep video denoising without flow estimation. http://arxiv.org/abs/1907.01361
4. Shi W et al (2016) Real-time single image and video super-resolution using an efficient subpixel convolutional neural network, [Online]. Available: http://arxiv.org/abs/1609.05158
5. Seoung YJ, Oh W, Kang J, Kim SJ (2018) Deep video super-resolution network using dynamic upsampling filters without explicit motion compensation
6. Ghanem B, Mueller M, Smith N (2016) A benchmark and simulator for UAV tracking. In: Computer vision—ECCV 2016, pp 445–461
7. Liu C, Freeman WT (2010) A high-quality video denoising algorithm based on reliable motion estimation
8. Jung A (2017) Comparison of video quality assessment methods
9. Root mean square error (RMSE) or mean absolute error (MAE)? (2014) Geosci Model Dev Discuss 7(1):1525–1534. https://doi.org/10.5194/gmdd-7-1525-2014
10. Antsiferova A, Yakovenko A, Safonov N, Kulikov D, Gushin A, Vatolin D (2021) Objective video quality metrics application to video codecs comparisons: choosing the best for subjective quality estimation. http://arxiv.org/abs/2107.10220
11. Wang Z, Bovik AC, Sheikh HR, Simoncelli EP (2004) Image quality assessment: from error visibility to structural similarity. http://www.cns.nyu.edu/~lcv/ssim/
12. Rassool R (2017) VMAF reproducibility: validating a perceptual practical video quality metric. http://www.elemental.com/resources/4k-test
13. Thorpe SJ (1991) Image processing by the human visual system
14. Evgeniou T, Pontil M (2001) Support vector machines: theory and applications. Lecture Notes in Computer Science (including subseries lecture notes in artificial intelligence and lecture notes in bioinformatics) LNAI, 2049:249–257. https://doi.org/10.1007/3-540-44673-7_12
15. Khoreva A, Rohrbach A, Schiele B (2018) Video object segmentation with language referring expressions. http://arxiv.org/abs/1803.08006
16. Smeulders AWM, Chu DM, Cucchiara R, Calderara S, Dehghan A, Shah M (2014) Visual tracking: an experimental survey. IEEE Trans Pattern Anal Mach Intell 36(7):1442–1468. https://doi.org/10.1109/TPAMI.2013.230

Integrating GPS Technology and Smart Payment Systems for Enhanced Public Transportation Efficiency in India

S. N. Shashidhar, Shivashankar A. Huddar, Deepti Ayyanagoudar, J. Pratiksha, Meghana Alagawadi, Mylu Rohit Bhargav, Nalini C. Iyer, and Suneeta V. Budihal

Abstract Buses are the integral means of transport sources in India. In metropolitan cities, over millions of people travel through public transport every day. The major problem faced by the passengers is unnecessary waiting time at stations. Hence to overcome this, tracking of a vehicle is much necessary. Because of the accuracy and affordability, GPS technology has become one of the most popular solutions for such applications. On the other side, smart payment is trending everywhere. Smart payment can offer an easy and convenient automated bill payment system. Implementing this in transportation, would help the passengers to travel even when they don't have enough cash with them. This project mainly deals with developing a website which will help in smart transport and smart payment system. Tracking and fetching of the location of a bus is done using the GPS system. This helps in knowing the location of vehicle, which in return helps in proper planning and managing time. The tracking of the bus is done using the GPS system, by which the passenger can come to know the location of that particular bus in which he desires to travel by entering the required information in the website. Smart payment is also done through the website developed in this project, where a passenger needs to sign in and enter the required information for the payment. This will even help in displaying the balance amount in his account.

Keywords HTML · Public transportation · GPS technology · Smart payment system · Vehicle tracking

S. N. Shashidhar (✉) · S. A. Huddar · D. Ayyanagoudar · J. Pratiksha · M. Alagawadi · M. R. Bhargav · N. C. Iyer · S. V. Budihal
KLE Technological University, Hubli, Karnataka, India
e-mail: shashisn.2009@gmail.com

V. K. Gunjan and J. M. Zurada (eds.), *Proceedings of 5th International Conference on Recent Trends in Machine Learning, IoT, Smart Cities and Applications*, Lecture Notes in Networks and Systems 1181, https://doi.org/10.1007/978-981-97-8861-3_4

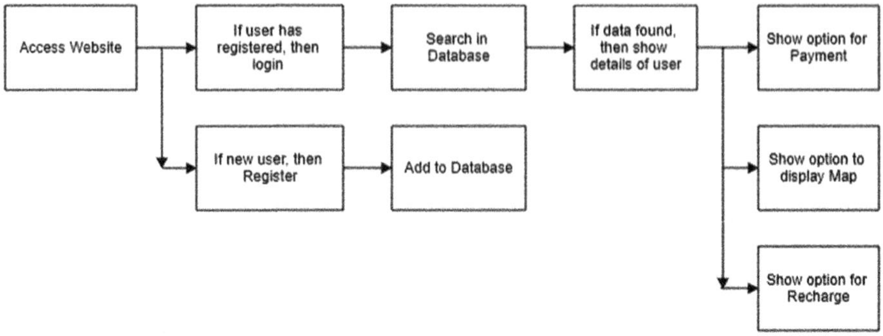

Fig. 1 Functional block diagram

1 Introduction

With the development of technology in various fields and with inventions like television, mobile phones, computers, the internet, there has been a drastic change in human life. A person's life is completely influenced by technology for transportation, entertainment, social synergy, and medical aid. A person's morning begins with these technology and also ends with them. Smart transportation helps in improving the quality of life, in monitoring and tracking of the public transportation network through which action could be taken quickly to incidents and emergencies and also in improving travel time.

The functional block diagram of the system goes as follows:

The user who wishes to travel by bus has to access the website which is been developed. If the user has registered himself already, then he can login. On authentication, the user gets access to his details like his phone number, username, address, RFID number and balance amount. If authentication fails, the user can retry by logging in again. If the user is a new user, then the user has to register himself by adding his details like his username, password, mobile number, address. Then the user can login and do things of his choice. On logging in, the user has options to recharge his account, for making payment and also to know the location of the bus. In order to recharge, the user has to enter the registered email-id, his bank details and the amount he wants to store charge. For the payment, the user has to select the destination stop according to which the prescribed amount will be deducted from this account. To know the location of the bus, the user can use the maps option. Knowing the location of the bus would help the user to plan his journey in a more effective manner (Fig. 1)

2 Objectives

The main objectives are:

- To enable the user to do online payment.
- To enable the user to recharge through online.
- To enable user to know location of the bus.
- To enable user to know his balance.
- To update the locations dynamically.
- To host the website.

3 Literature Survey

Smart solutions for urban transportation may encompass:

(1) Implementation of Smart Ticketing and Automated Fare Collection
(2) Integration of Bus Rapid Transit systems
(3) Deployment of Smart GPS-based buses and bus-stops.

Smart ticketing technology stands out as a solution that not only reduces operational costs but also simplifies the experience for public transport passengers. It involves electronically storing a travel ticket on a microchip, typically embedded in a smart card. This innovative approach proves advantageous for public transportation sectors and government entities as it enables easy tracking of routes and travel times taken by passengers. This data serves as valuable input for the future development and planning of public transportation systems and routes. Moreover, the security features of smart tickets contribute to fraud prevention, allowing for quick deactivation in case of theft, with the option to transfer the remaining credit to a new card, akin to the functionality of an ATM card [1].

The adoption of smart ticketing not only eliminates the need for printed ticket machines but also lays the groundwork for the business case for mobile ticketing. Meanwhile, the integration of high-end GPS technology in buses provides real-time location information. An Intelligent Passenger Information System, operating on cloud-based GPS tracking, incorporates prediction algorithms to offer up-to-the-minute bus service details. This system is capable of pinpointing the exact location of a bus at any given time and predicting its arrival time at specific bus stops.

In the paper [2], the GSM and GPS module have been used in order to know the arrival time, delay time and the location of the bus. The GPS modules end the latitude and the longitude values of the bus. The processor processes the data and sends the information to the terminal which is displayed. This technology enables the travelers with real time passenger information like the availability of seats of particular bus service, the location of the bus, via websites, smart phone apps, and bus stop displays.

Cemil SUNGUR along with Ismail BABAOGLU and Aysegul SUNGUR describe in the paper about the smart bus station-passenger information system. The system described here, makes use of Friendly ARM Micro2440, Windows CE 6.0 operating system made use on Friendly ARM Micro2440, Microsoft Visual Studio 2008 platform, Oracle database management system, GPS module with SiRFstar GPS chip and LCD indicator. The LCD screens installed at the bus stops present the people the information of the route for the public transport vehicle they intend to get on and their arrival time at the bus stop. The safe and easy data exchange on web services made major contributions to the progress of the system [3].

4 Implementation Details

A. *Front-end and Back-end Development of website*

The development of the website starts with creating of basic HTML pages. Hypertext Markup Language [HTML] is a quite simple language made up of elements represented as tags, which can be applied to pieces of text to give them different meaning in a document. These HTML elements tell the browser how to display the content. HTML describes the structure of a Web page. HTML helps in creating hyperlinks and in linking documents together. The style of the web page is defined by CSS [Cascading Style Sheet]. JavaScript adds logic and inter activity to the webpage. All the web pages are redirected using routes. The website is been hosted using Amazon EC2 [Amazon Elastic Compute Cloud] cloud server which provides scalable computing capacity in the Amazon Web Services [AWS] cloud. The ports accessed for hosting are 3000 and 8080 for node JS applications. Through these ports, traffic will be controlled by AWS. If needed, we can have a secured http i.e. https protocol easily by obtaining AWS certificates and installing them. Node.js is used for back-end API services. Node.js is a platform built on Chrome's JavaScript runtime for easily building fast and scalable network applications (Fig. 2).

B. *Creating login functionality*

In order to login, the user needs to have an account. In order to register, the user has to enter the required details. He must also enter the password of his choice which he will be using to login in the future. On registration, the details of the user will be stored in the mongodb database. In order to provide password security, the password entered by the user will be verified and then stored in the data base. After registering, the user can login. For logging in, the user name and password are to be entered which are used to find user's details in the database. If found, the user has successfully logged in, else the entered details might be wrong. The user can retry using the correct credentials.

C. *Display user details*

On logging in, the user details is been first searched in the database. If found, the details of the user such as user name, RFID number, password, address and balance will be displayed. This page has option for map display using which the user will be redirected to the maps page which shows the real-time location of the bus and also has option to recharge his account.

D. *Creating payment and recharge functionality*

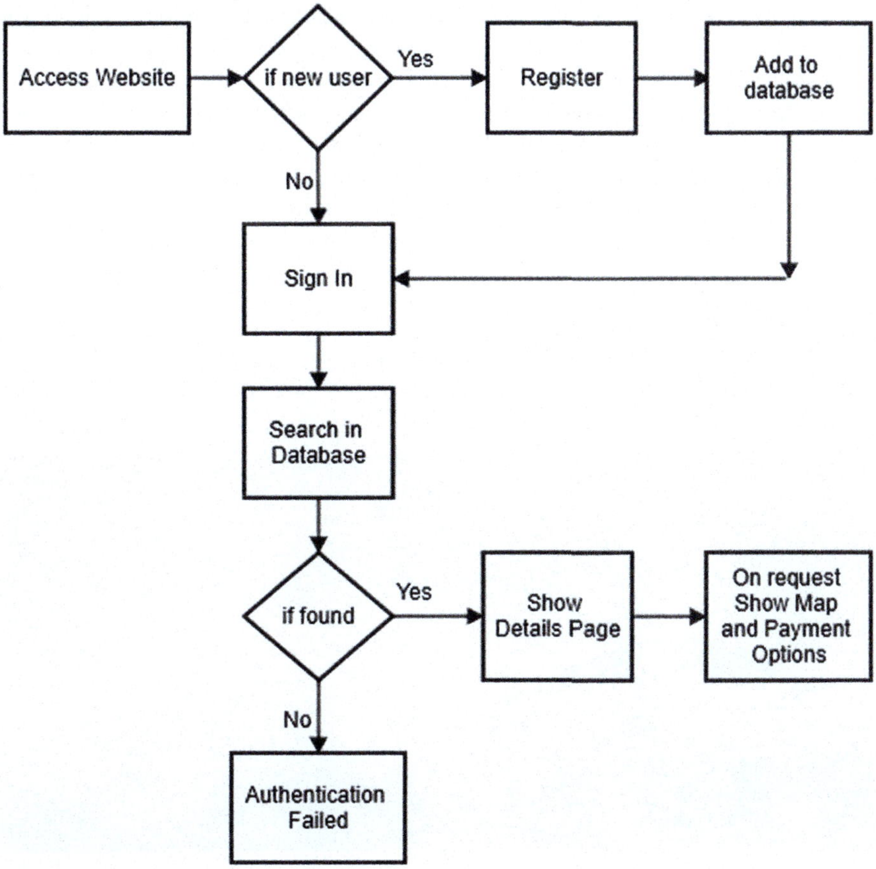

Fig. 2 Flow chart

In order to make payment, the user has to have balance in his account.If there is no sufficient balance, the user can recharge by availing he recharge option which is been displayed in the details page which will be shown on authentication. For recharge, the user has to provide his credit/debit card details, the registered email-id and the amount he wants to recharge. On doing so, the prescribed account will be deducted from his bank account and will be added to the user's account. On having sufficient balance, [4] the user can make payment by first entering the RFID number through which the user is been identified and then selecting the destination stop. Depending on the destination stop, the correct amount will be deducted from the user's account. This facility helps in avoiding ticket counters.

E. *Display dynamic location of bus*

In order to know the location of the bus, the user has to first login. On authentication, the user will be redirected to the details page which has the option to display map. It is a link which further redirects to the maps page. Mapbox API is used for Maps. GPS fetches the location and sends it in terms of latitudes and longitudes. These values are sent to Mapbox API, where necessary changes have been made and the user is able to see the exact location of the bus. Knowing the location helps in providing personalized travel information.

5 Snapshots

See Figs. 3, 4, 5, 6, 7 and 8.

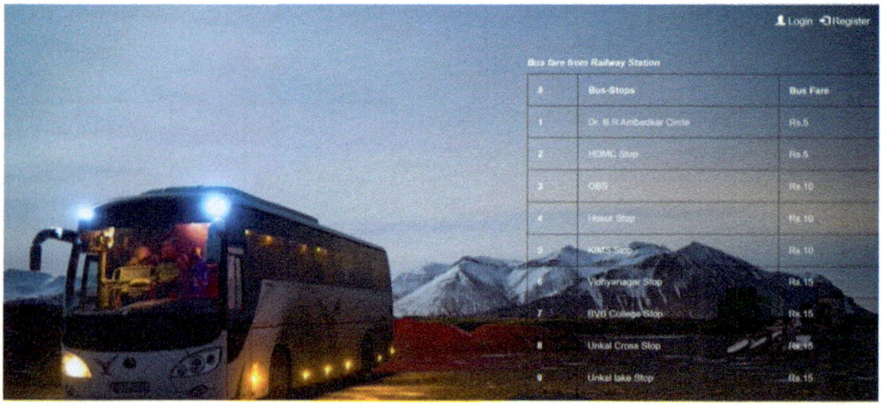

Fig. 3 Home page

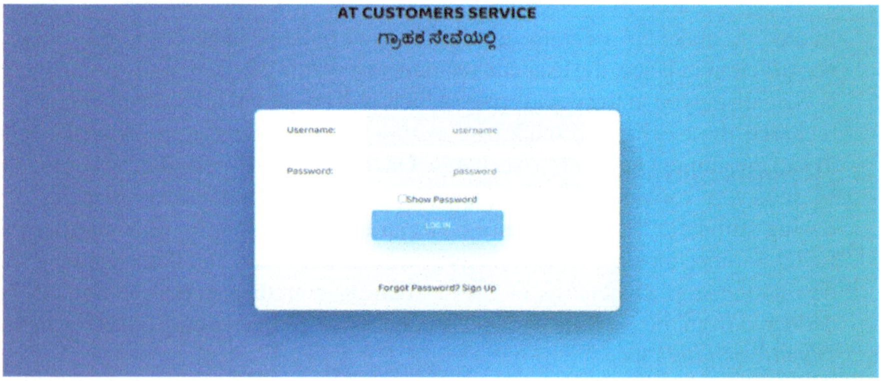

Fig. 4 Login page

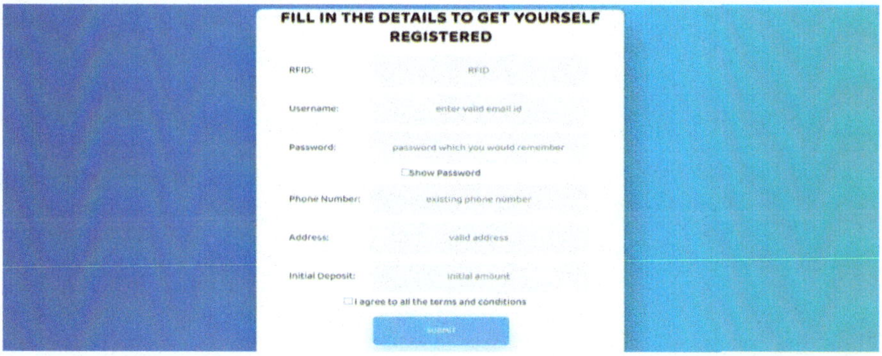

Fig. 5 Register page

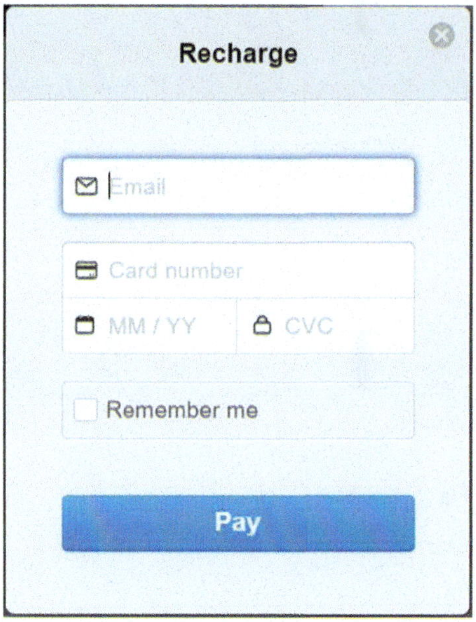

Fig. 6 Recharge page

6 Results and Conclusion

The password security is been taken care of as the password is been hashed and then stored in the database. Whenever the user has no sufficient balance, he can recharge his account through online anytime and from anywhere. The user can get to know the exact location of the bus using the Maps option which can be accessed only on authentication.

Transport in India is a major activity therefore boosting the Indian economy. Main activities of the nation cannot take place without transport. Therefore, the means of

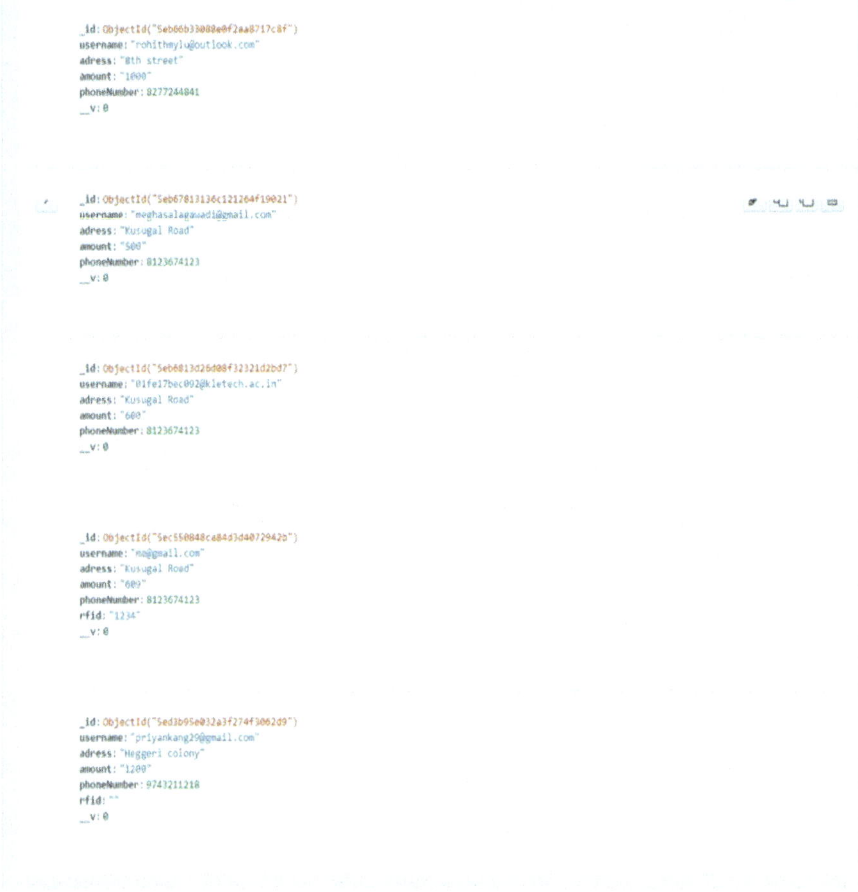

Fig. 7 Database

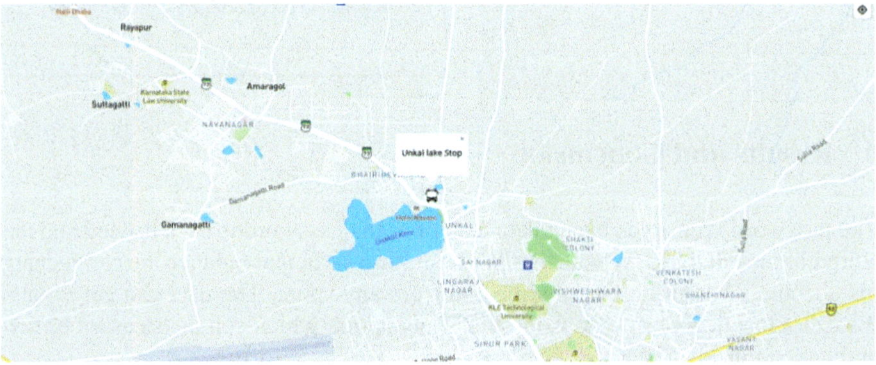

Fig. 8 Map page

transport should be improved in order to make numerous milestones. The design proposed in this report is a user-friendly vehicle-tracking and digital payment system which can be implemented easily.

7 Future Scope

The system can be implemented in large scale by using GPS modules which give accurate values of latitudes and longitudes. The website can be added with extra applications like displaying the nearest bus to the station so that the passenger can plan accordingly and also giving the information about the number of seats available in the bus. Also motion sensors can be installed in the bus so that even the speed of the bus can be calculated along with the location of the bus.

References

1. Vakula D, Raviteja B (2017) Smart public transport for smart cities. In: Proceedings of the international conference on intelligent sustainable systems (ICISS 2017)
2. Megalingam RK, Raj N, Soman AL, Prakash L, Satheesh N, Vijay D (2014) Smart, public buses information system. In: International conference on communication and signal processing, pp 3–5
3. Sungur C, Babaoglu I, Sungur A (2015) Smart bus station-passenger information system. In: 2015 2nd International conference on information science and control engineering
4. Git Hub repository. https://github.com/myluRohith/minor

A Machine Learning-Driven Approach for Early Leaf Disease Detection and Classification: A Survey

Sita Yadav, Ritika Pandey, Abhijit Singh, Aniket Digole, and Megha Kumari

Abstract Mangifera Indica, popularly known as mangoes, is India's highly important fruit crop. However, the widespread cultivation of mangoes in India exposes them and makes them vulnerable to different types of diseases, resulting in lower yield and economic losses. Anthracnose, Powdery Mildew, and Sooty Mould are the most common diseases that harm mango leaves. Many researchers have been working for control of disease on plants including mango trees to reduce loss of farmers' production and indirectly reducing the stress levels for average mango farmers. The popular methodologies adopted by researchers are based on CNN. In the proposed sur vey work major challenges are covered to prevent diseases on mango trees. This study provides an in-depth examination of the strategies used to diagnose plant leaf diseases. It entails an investigation of several categorization systems useful for detecting plant leaf disease. Several scholars have made contributions to the topic, providing various ideas and implementations for recognising and managing plant leaf diseases.

Keywords Leaf disease · Disease detection · Image processing · Convolutional neural network and artificial neural network

S. Yadav · R. Pandey (✉) · A. Singh · A. Digole · M. Kumari
Army Institute of Technology, Pune, Maharashtra, India
e-mail: ritikapandey20061@aitpune.edu.in

S. Yadav
e-mail: syadav@aitpune.edu.in

A. Singh
e-mail: abhijitsingh20230@aitpune.edu.in

A. Digole
e-mail: aniketdigole20241@aitpune.edu.in

M. Kumari
e-mail: meghakumari20472@aitpune.edu.in

V. K. Gunjan and J. M. Zurada (eds.), *Proceedings of 5th International Conference on Recent Trends in Machine Learning, IoT, Smart Cities and Applications*, Lecture Notes in Networks and Systems 1181, https://doi.org/10.1007/978-981-97-8861-3_5

1 Introduction

Plant diseases have far-reaching consequences on agriculture, impacting plant growth and crop yield. These effects extend to the realms of society, ecology, and economics. Recent research has shed light on the detrimental effects of leaf diseases on plants, underscoring their significance [1]. Leaf diseases also inflict substantial economic losses on farmers. Therefore, there is a pressing need to emphasize early disease detection [2]. The existing body of literature primarily focuses on the biological aspects of plant diseases, often making predictions based on the observable characteristics of plants and leaves. Detecting diseases promptly upon their onset is crucial for effective disease management. Traditionally, this detection task challenged human experts who visually identify diseases [3]. In India, we have many different types of mango trees like Hapus, Totapuri, Himsagar, Langda, Chausa, Dashari, Sindura, Raspuri, Kesar and many more. Additionally, there are various diseases that can affect these mango trees like Anthracnose, Bacterial Canker, Cutting Weevil, Die Back, Gall Midge, Powdery Mildew, and Sooty Mould [4]. After conducting some research, we found that one particular disease called "Anthracnose" is a significant concern as it can harm almost all varieties. So, our research is not focused on a specific mango tree type; it applies to all of them [5]. To diagnose these diseases, manual observation and pathogen detection are employed, which are time-consuming, expensive, and less accurate [6]. Fungal infections not only have an economic impact on mango plants and their products, but they also damage their natural balance. The aim of this research survey paper is to produce a resource efficient method for detecting these diseases and associated symptoms, allowing for an pre-affecting stage and financially meaningful treatment.

In recent years, computer vision and deep learning have become increasingly popular methods for categorizing different types of fungal infections [6]. These methods have been shown to be more accurate and efficient than traditional methods [7]. In this study, we present the ResNet-50, a system that detects diseased mango leaves using a pre-trained Convolutional Neural Network architecture. The collection includes photos of mango leaves that have been classified as healthy or unhealthy. The images are pre-processed and the discussion of the model that extracts the features by the use of the ResNet-50 model followed by in-place data augmen tation upon the dataset [8].

2 Literature Survey

Smart farming revolutionizes agricultural management by offering timely and dependable data on field conditions, particularly diseases. This eliminates the inefficiencies of manual monitoring, ensuring precision and saving time for farmers, researchers, and specialists [9].

The first step in identifying diseases from images is to pinpoint the unique features of the affected areas. These can vary depending on the plant's leaf disease and may include things like color, shape, edges, texture and size [10].

Even though technology plays a critical role in many fields today, we still rely on traditional farming practices. However, incorrect disease detection can cost money, time, and productivity [11].

That's where image processing techniques come in. Notably, India stands as the global leader in mango production, responsible for approximately 50% of the world's mangoes and contributing to 52% of the global total production, producing an impressive 12 million tonnes compared to the global output of 23 million tonnes [12].

In this research paper, we delve into the intricate process of diagnosing and predicting diseases in mango leaves through the submission of phone-captured leaf images to our analysis engine system. In Table 2, we have compared Algorithms for leaf disease detection and faced some limitations, to tackle this, we introduce an innovative approach founded on Convolutional Neural Networks (CNNs) designed to discern between healthy and diseased leaf images. Our utilization of CNNs serves the dual purpose of enhancing classification accuracy for diseases and yielding promis ing outcomes [7].

3 Literature Work Related to Mango Leaf

In Table 2, we dig deeper into the specific context of mango leaf disease detection, highlighting various methods employed and datasets utilized to enhance the precision and reliability of the detection process.

Traditional CNNs require extensive datasets and significant computational resources, with concerns of overfitting, as suggested by [13]. Transfer learning, while promising, can lead to negative transfer and increased overfitting [14]. Lightweight custom CNNs often sacrifice accuracy, according to [15], while the complex tuning process and suboptimal performance of SVMs with image data have been highlighted by [16]. These findings underscore the need for innovative strategies to address these challenges effectively in leaf disease detection research (Fig. 1).

4 Research Techniques and Proceduress

The dataset is manually collected, containing over 4000 labelled images of mango leaves by mobile phone camera (50 MP Main Camera), encompassing eight categories: (1) healthy, (2) Anthracnose, (3) Bacterial Canker, (4) Cutting Weevil, (5) Die Back, (6) Gall Midge, (7) Powdery Mildew, and (8) Sooty Mould [17]. Our study focused primarily on Anthracnose, utilizing images from category (2), with

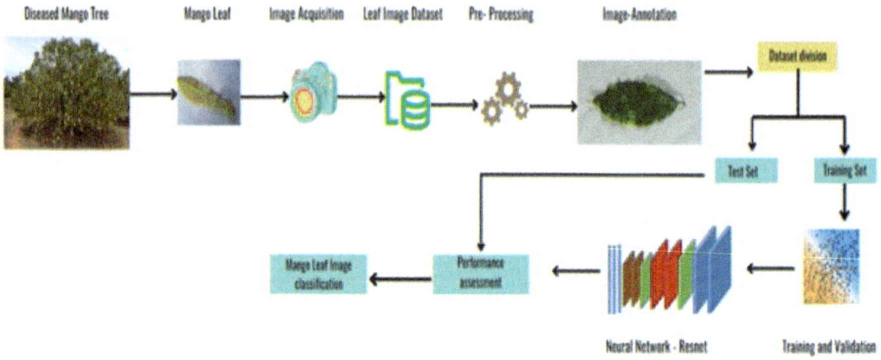

Fig. 1 Methodology

Fig. 2 Various types of diseases and their visual representation

each category comprising 500 images for analysis. Figure 2 provide visual representations of infected leaf quantities at different stages and select sample images from each category (Table 1).

Powdery Mildew, white superficial powdery fungal growth on leaves, stalks, flowers, and young fruits, accompanied by the premature dropping of affected flowers and fruits [17]. Anthracnose causes damage to young shoots, flowers, and fruits

Leading to leaf spot, blossom blight. It survives in conditions of high humidity, frequent rains, and temperatures of 24–32 °C [23]. Dieback is characterized by the drying back of twigs, withering leaves, and internal browning of wood tissue. Pathogens responsible for this disease survive in plant debris, and occur in high humidity and moist conditions [26]. Bacterial Canker, manifests as water-soaked lesions on leaves, stems, twigs, and fruits that develop into cankerous spots [16]. Bacteria survive in lesions and diseased plant tissues, with the spring season being conducive to disease development. Sooty Mold is recognized by a black velvety coating on leaves and twigs, resulting from honeydew secretion by insects. It thrives in high humidity and moist environments [14]. The path for our project is shown in Fig. 1, which shows the in-depth procedure of the flow of project.

Table 1 Comparison of algorithms for leaf disease detection

No	Authors (year)	Methods used	Observations	Limitations
1	Sujatha et al. [9]	SVM, Inception net	Comparison was made between ML models like SVM, Random Forest, and Stochastic Gradient Descent, and Deep learning models including VGG-16, InceptionNet, and ResNet for the purpose of citrus disease detection. The results revealed that DL outperformed ML in terms of accuracy, with the highest accuracy of 89.5%	Analysis done for citrus disease dataset. Limited number of Architecture
2	Too et al. [10]	CNN	Comprehensive review of deep learning techniques for plant disease detection. Examination of datasets, strategies for data augmentation, and methods for visualizing the results are discussed	There's no critical method analysis, a lack of technique comparisons
3	Zhang et al. [11]	RCNN, Resnet	Faster RCNN with ResNet used for the detection of tomato diseases. They also employed k-means clustering to optimize anchor boxes, leading to a 2.71% increase in accuracy compared to the original Faster RCNN method	Model tested on a dataset related to tomato diseases. Analysis of model on various ResNet architectures not done
4	Pandian et al. [18]	Resnet—197	Research deals with weevil on different mango varieties. They found that the Ashwina variety was the most preferred by these weevils. Showed that different mango varieties can impact the behavior and success of these evil attacks	The experiment was conducted solely under controlled lab conditions, involved only four mango varieties, and lacked an analysis of nutritional components
5	Maldonado-Celis et al. [19]	Regression, Multi-label classification	Deep learning excels over traditional methods like SVM and KNN for plant disease detection, consistently delivering superior performance	They used only one dataset of apple leaf images, which might not cover the full spectrum of possibilities

(continued)

Table 1 (continued)

No	Authors (year)	Methods used	Observations	Limitations
6	Jackulin and Murugavalli [20]	SVM, KNN	CNN used for plant disease detection. Deep learning models have the ability to bypass the need for manual feature extraction, resulting more accuracy	Implemented on small dataset
7	Panchal et al. [21]	CNN	Highlighting advantages in automating feature extraction, models consistently exhibit superior accuracy in classifying plant diseases	There are a few issues with existing models, such as limited generalization capabilities and a lack of explainability
8	Shoaib et al. [15]	CNN, GAN	Challenges like data bias and model interpretability were discussed. The paper propose potential solutions, the use of Generative Adversarial Networks (GAN's) and the implementation of attention models to address these challenges	Limited datasets, the risks of models overfitting and the complexity of deep learning models
9	Applalanaidu and Kumaravelan [22]	KNN, SVM	The paper surveys various machine learning approaches used in plant disease detection, emphasizing image preprocessing and the use of classifiers like SVM and KNN for leaf image analysis	Manual feature engineering, less effective than deep learning techniques
10	Saleem et al. [13]	DL, CNN	Convolutional Neural Networks (CNN) outperformed traditional models in accuracy. It emphasized the efficacy of data augmentation methods in improving the performance of these models	Limited datasets, model overfitting risk, and increased computa tional demands for deep learning
11	Manoharan et al. [17]	SVM, NN, NB	A study compared SVM, NN, and NB for plant disease classification, favoring deep learning for its automated feature extraction. However, the reliance on curated datasets like PlantVillage for model training and validation was noted	Classifying mango diseases is complex due to similar visual symptoms, limited data for deep learning models. Urgent requirement for real-time image datasets

Table 2 Deep research on methods and datasets for mango leaf disease detection

No	Year	Objectives	Algorithms	Data used
1	2020	Mango leaf disease detection and identification [16]	Neural network ensemble (SVM) + K-means clustering	Onsite hand clicked images
2	2021	Mango leaf disease detection and identification [17]	CCA + cubic SVM	Hand clicked RGB images using various gadgets & areas
3	2020	Mangoes quality grading[14]	The deep learning models selected are AlexNet, VGGs, and ResNets + transfer learning	6,400 single mango images, labeled as quality grade A, B, or C, based on color uniformity & defect severity
4	2020	Mango sorting and grading according to their quality [23]	SqueezeNet deep CNN	Approx. 10,000 images of mango fruits were clicked by researchers
5	2021	Detection of mangoes infected with anthracnose [5]	It uses a CNN model with AlexNet as the base architecture	It uses a CNN model with AlexNet as the base architecture
6	2021	Identify the diseases (Anthracnose and apical nacrosis) on the mango leaf [24]	CNN based Fully-convolutional network (FrCNnet) model	Onsite hand clicked 2286 images taken using various image capturing gadgets
7	2021	Mango leaf disease detection and identification [25]	Neural network ensemble (SVM) + K-means clustering	Hand clicked pictures
8	2021	Mango leaf disease detection and identification [26]	CCA and cubic SVM	29 hand clicked RGB images using various gadgets and areas. 135 images were retrieved after pre-processing
9	2020	Mangoes quality grading[14]	The deep learning models selected for our classification task are AlexNet, VGGs, and ResNets + transfer learning	6,400 single mango images, graded A, B, or C based on color evenness and defect severity
10	2020	Mango sorting and grading according to their quality [23]	SqueezeNet deep CNN	About 10.000 mango fruits hand clicked photos using phone, camera and Thermal camera
11	2021	Segment and identify the diseases (Anthracnose and apical necrosis) on the mango leaves [24]	CNN based Fully-convolutional network (FrCNnet) model	Onsite hand clicked 2286 images taken using various image capturing gadgets

4.1 Image and Data Sets

We have created a dataset of mango leaves encompassing eight categories: (1) healthy, (2) Anthracnose, (3) Bacterial Canker, (4) Cutting Weevil, (5) Die Back, (6) Gall Midge, (7) Powdery Mildew, and (8) Sooty Mould [41]. Procedure of dataset creation step by step:

1. **Defined our Data Needs**: Create a dataset for training a machine learning model that can classify mango leaves into eight categories based on their health condition or disease.
2. **Data Collection**: Visit mango orchards, agricultural research institutions, or local farms to collect mango leaves. Collaborate with experts in the field who can provide authenticated samples.
3. **Data Pre-processing**: Clean the collected data by removing any leaves with physical damage or contamination. Standardize the size and format of the mango leaf images by resizing them to a consistent resolution (e.g., 200 × 200 pixels).
4. **Data Annotation**: Manually label each mango leaf image with the corresponding category (e.g., "healthy," "Anthracnose," etc.) [42].
5. **Data Augmentation**: Rotate, flip, or introduce slight variations in color and brightness to the images to increase dataset diversity.
6. **Dataset Split**: Split the dataset into three subsets: training, validation, and testing, using a common split ratio like 80% for training, 20% for testing.

4.2 Image Pre-processing

This process is all about enhancing the qualities of an image. We do this to make the image easier to work with and analyze. The various process and methods that take place in the image pre-processing:

1. **Applying Filters**: One of the first things we do is apply filters to the image. These filters can be like different tools to improve the picture. For example highpass, low-pass, or median filters [27].
2. **Cropping**: Cutting image down to focus on the important part and remove any distracting or unnecessary background [28].
3. **Resizing**: Resize images to make them all the same size, helpful when comparing or working with multiple images.
4. **Denoising**: Getting rid of unwanted speckles or imperfections that can make a picture look grainy or unclear [29]. In the context of denoising, we have some specific goals:

   a. We want the flat areas in the image to look smooth
   b. We don't want to blur the edges or important details.
   c. We should preserve the textures in the picture.
   d. We don't want to introduce any new artifacts or strange things into the image.

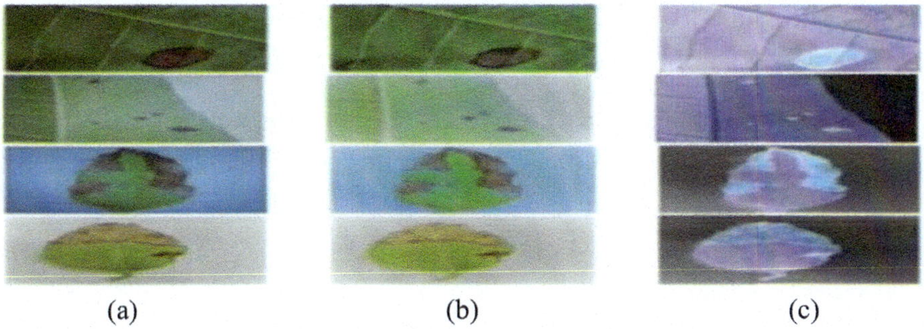

Fig. 3 Image pre-processing and segmentation procedures **a** input image **b** denoised image **c** segmented image [32]

The provided Fig. 3 offer invaluable visual representation in identifying and understanding various Pre-processing and segmentation procedures.

4.3 Feature Extraction

Feature extraction is the process of identifying the most important characteristics of an image, such as its color, shape, and texture, followed by a few dense layers for classification, specifically in the context of analyzing plant leaves [30]. Texture is particularly emphasized as a key factor in understanding the structure of plant leaves. Additionally, the text mentions that the region of interest (ROI) in an image refers to the part that is of particular interest or importance, in this case, the area of the leaf that shows signs of damage or disease [31].

ResNet-50 Architecture: Key points of Resnet -50 is discussed in following points:

1. Base Model: Our approach leverages the proven feature extraction ability of ResNet-50, setting the stage for precise disease identification.
2. Modifications: Three extra dense layers empowered ResNet-50 to learn intricate mango leaf disease patterns.
3. Activation Function: Final softmax layer classifies diseases by predicting probabilities, enhancing identification accuracy
4. Frozen Layers: Freezing specific layers harnessed pre-trained knowledge, curbed overfitting, and protected key features during training.
5. Average Pooling Layer: An average pooling layer reduced feature map size, easing computation while maintaining performance

These are some advantages of Resnet:

1. Enhanced Accuracy: The integration of additional dense layers and the strategic utilization of the ResNet-50 architecture significantly improves the accuracy.

2. Robust Feature Extraction: The utilization of the ResNet-50 model facilitates robust feature extraction, allowing the system to capture intricate patterns and subtle variations in mango leaf diseases.
3. Reduced Overfitting Risks: By freezing specific layers and incorporating an average pooling layer.

5 Applications

Mango leaf disease detection plays a pivotal role in various aspects of mango cultivation, offering a multitude of benefits. Firstly, it enables the early identification of diseases like Anthracnose, Powdery Mildew, Bacterial Canker, and more [5]. This early detection is instrumental in implementing timely interventions and treatments, effectively curtailing the spread of diseases and minimizing crop losses. Beyond this, it empowers farmers and orchard managers with valuable insights for informed decision-making regarding disease management strategies. These strategies encompass precise pesticide application, targeted pruning of infected branches, and adjustments to irrigation and fertilization practices [12]. Importantly, it promotes sustainability by reducing the overuse of pesticides and fungicides, leading to cost savings and more environmentally friendly farming practices.

6 Conclusion

Mango, a popular fruit crop in India, is important for agriculture, demanding strong protection and early disease detection techniques. This research paper gives an outline of deep learning core ideas as well as a detailed assessment of current advances that are going on respectively for the field of plant leaf disease identification recognised by help of deep learning models. Deep learning approaches demonstrate outstanding accuracy in diagnosing plant leaf diseases when a huge dataset is available for training purpose, taking into account morphological parameters such as color, intensity, and leaf dimensions. Furthermore, used for clustering the mango leaf diseases, this process offers an ensemble stacking model that integrates deep learning with traditional machine learning, with a special focus on fungal and bacterial diseases such as Anthracnose, Powdery Mildew, and sooty mould. Our country is largely rural, with agriculture employing more than half of the workforce. The planned work will be improved in the future using updated technologies such as IoT and sensor sensing technology that give the highest accuracy output.

References

1. Veling PS, Kalelkar RS, Ajgaonkar LV, Mestry NV (2019) Mango disease detection by using image processing. Int J Res Appl Sci Eng Technol 7(4):3717–3726
2. Gautam V, Ranjan RK, Dahiya P et al (2023) ESDNN: a novel ensembled stack deep neural network for mango leaf disease classification and detection. Multimed Tools Appl.https://doi.org/10.1007/s11042-023-16012-6
3. Ganatra N, Patel A (2020) A multiclass plant leaf disease detection using image processing and machine learning techniques. Int J Emerg Technol 11(2):1082–1086
4. S. (2022). A swarm intelligence-based model for disease detection in mango crops. In: 2022 1st International conference on computational science and technology (ICCST). IEEE, pp 845–850
5. Singh UP, Chouhan SS, Jain S, Jain S (2019) Multilayer convolution neural network for the classification of mango leaves infected by anthracnose disease. IEEE Access 7:43721–43729. https://doi.org/10.1109/ACCESS.2019.2907383
6. Sandhya S, Balasundaram A, Arun S (2022) Deep learning and computer vision based model for detection of diseased mango leaves. Int J Recent Innov Trends Comput Commun 10:70–79. https://doi.org/10.17762/ijritcc.v10i6.5555
7. Barburiceanu S, Meza S, Orza B, Malutan R, Terebes R (2021) Convolutional neural net- works for texture feature extraction. Applications to leaf disease classification in precision agriculture. IEEE Access 9:160085–160103. https://doi.org/10.1109/ACCESS.2021.3131002
8. Ahmed SI, Ibrahim M, Nadim M, Rahman M, Shejunti M, Jabid T, Ali MS (2023) MangoLeafBD: a comprehensive image dataset to classify diseased and healthy mango leaves. Data Brief 47:108941. https://doi.org/10.1016/j.dib.2023.108941
9. Sujatha R, Chatterjee JM, Jhanjhi NZ, Brohi SN (2021) Performance of deep learning versus machine learning in plant leaf disease detection. Microprocess MicrosystS 80
10. EC Too, Yujian L, Njuki S, Yingchun L (2019) A comparative study of fine-tuning deep learning models for plant disease identification. Comput Electron Agric 161:272–279. ISSN 0168-1699
11. Zhang Y, Song C, Zhang D (2020) Deep learning-based object detection improvement for tomato disease. IEEE Access 8:56607–56614. https://doi.org/10.1109/AC-CESS.2020.2982456
12. Bhole V, Kumar A (2020) Mango quality grading using deep learning technique: perspectives from agriculture and food industry. In: Proceedings of the 21st annual conference on information technology education (SIGITE '20). Association for computing machinery, New York, NY, USA, pp 180–186
13. Saleem R, Shah JH, Sharif M, Ansari GJ (2021) Mango leaf disease identification using fully resolution convolutional network. Comput Mater Contin 69:3581–3601
14. Mia MR, Roy S, Das SK et al (2020) Mango leaf disease recognition using neural network and support vector machine. Iran J Comput Sci 3:185–193. https://doi.org/10.1007/s42044-020-00057-z
15. Shoaib M, Shah B, El-Sappagh S, Ali A, Ullah A, Alenezi F, Gechev T, Hussain T, Ali F (2023) An advanced deep learning models- based plant disease detection: a review of recent research. Front Plant Sci 14
16. Saleem R, Shah JH, Sharif M, Yasmin M, Yong H-S, Cha J (2021) Mango leaf disease recognition and classification us ing novel segmentation and vein pattern technique. Appl Sci 11(24):11901. https://doi.org/10.3390/app112411901
17. Manoharan N, Thomas VJ, Anto Sahaya Dhas D (2021) Identification of mango leaf disease using deep learning. In: 2021 Asian conference on innovation in technology (ASIAN- CON), Pune, India, pp 1–8. https://doi.org/10.1109/ASIANCON51346.2021.9544689
18. Pandian JA, K K, Rajalakshmi NR, Arulkumaran G (2022) An improved deep residual convolutional neural network for plant leaf disease detection. Comput Intell Neurosci 2022(2). https://doi.org/10.1155/2022/5102290. PMID: 36156945, PMCID: PMC9492343
19. Maldonado-Celis ME, Yahia EM, Bedoya R, Landa´zuri P, Loango N, Aguillo´n J, Restrepo B, Guerrero Ospina JC (2019) Chemical composition of mango (Mangifera indica L.) fruit:

nutritional and phytochemical compounds. Front Plant Sci 17(10):1073. https://doi.org/10.3389/fpls.2019.01073. PMID: 31681339, PMCID: PMC6807195
20. Jackulin C, Murugavalli S (2022) A comprehensive review on detection of plant disease using machine learning and deep learning approaches. Meas: SensS 24:100441. ISSN 2665-9174
21. Panchal AV, Patel SC, Bagyalakshmi K, Kumar P, Khan IR, Soni M (2023) Image-based plant diseases detection using deep learning. Mater Today: Proc 80(3):3500–3506. ISSN 2214-7853
22. Applalanaidu MV, Kumaravelan G (2021) A review of machine learning approaches in plant leaf disease detection and classification. In: 2021 Third international conference on intelligent communication technologies and virtual mobile networks (ICICV), Tirunelveli, India, pp 716–724. https://doi.org/10.1109/ICICV50876.2021.9388488
23. Gupta C, Tewari VK, Machavaram R, Shrivastava P (2022) An image processing approach for measurement of chili plant height and width under field conditions. J Saudi Soc Agric Sci 21(3):171–179
24. Thenmozhi S, Lakshmi RJ, Kumudavalli M, Ibrahim I, Mohan R (2021) A novel plant leaf ailment recognition method using image processing algorithms. J Sci Ind Res (India) 80(11):979–984
25. Barbedo JGA, Koenigkan LV, Santos TT (2016) Identifying multiple plant diseases using digital image processing. Biosyst Eng 147:104–116
26. Hessane A, El Youssefi A, Farhaoui Y, Aghoutane B, Amounas F (2023) A machine learning based framework for a stage-wise classification of date palm white scale disease. Big Data Min Anal 6(3):263–272. https://doi.org/10.26599/BDMA.2022.9020022
27. Sandino JD, Ramos-Sandoval OL, Amaya-Hurtado D (2016) Method for estimating leaf coverage in strawberry plants using digital image processing [Me´todo para estimar a cobertura foliar das plantas de morango usando processamento de imagem digital]. Rev Bras Eng Agric e Ambient 20(8):716–721
28. Vamsidhar E, Rani PJ, Babu KR (2019) Plant disease identification and classification using image processing. Int J Eng Adv Technol 8(3):442–446
29. Roy K et al (2023) Detection of tomato leaf diseases for agro-based industries using novel PCA DeepNet. IEEE Access 11:14983–15001. https://doi.org/10.1109/AC-CESS.2023.3244499
30. Vijayakumar J, Arumugam S (2014) Odium piperis fungus identification for piper betel plants using digital image processing. Int J Soft Comput 9(4):207–212
31. Pham TN, Van Tran L, Dao SVT (2020) Early disease classification of mango leaves using feed-forward neural network and hybrid metaheuristic feature selection. IEEE Access 8:189960–189973
32. Garg R, Sandhu AK (2022) Comparison of various proposed techniques for processing of different images. In: 2022 2nd International conference on advance computing and innovative technologies in engineering (ICACITE). IEEE, pp 1028–1032

Software Defect Code Analyzer Using Cosine Similarity

P. Surya Kiran, A. Jackulin Mahariba, Darshan Ramesh, and B. Sudheer

Abstract The software testing process is an essential phase in the software life cycle. However, the vast size of code and the ambiguity present in defect reports often pose challenges in identifying defects within the source code. This paper aims to propose a method that automatically identifies defect-causing modules in faulty source code. Software developers can utilize this tool in product testing to enhance performance and save time. In our proposed approach, we focus on analyzing the similarity metrics among keywords sourced from Defect Descriptions, commonly found in feedback reports, and the specific Module or Function names embedded within the software's source code. Employing the Cosine similarity technique, we generate precise similarity constants between these linguistic elements. This innovative tool emerges as a potential time-saving asset for software developers, offering the capability to significantly reduce debugging timelines. With its implementation, we anticipate the potential for saving weeks of laborious debugging efforts, empowering developers to tackle defects efficiently and effectively within the source code.

Keywords Cosine similarity · Defect analyzer · Software testing

P. S. Kiran (✉) · A. J. Mahariba · D. Ramesh · B. Sudheer
Department of computational Intelligence, SRM Institute of Science and Technology, Chennai, India
e-mail: pk4395@srmist.edu.in

A. J. Mahariba
e-mail: jackulia@srmist.edu.in

D. Ramesh
e-mail: dr7625@srmist.edu.in

B. Sudheer
e-mail: bs9012@srmist.edu.in

V. K. Gunjan and J. M. Zurada (eds.), *Proceedings of 5th International Conference on Recent Trends in Machine Learning, IoT, Smart Cities and Applications*, Lecture Notes in Networks and Systems 1181, https://doi.org/10.1007/978-981-97-8861-3_6

1 Introduction

Traceability of root cause defects is too difficult due to clarity defined in defect report. Too complex code to understand to find root cause of the defect. No Debugger or Tool available to trace back to Source code to find possible causes of the defect raised by Test. Defect Analysis takes more time from Bug raised to close due to Knowledge if the engineer insufficient logs or inappropriate logs or traceability code is too complex. Defect Description/ Summary is irrelevant to module/model used for test. Defects can be in any type of testing say Unit Test, Integration Test or System Test or other format of test which are identified as defect. Software defect code analyzer is very helpful in different testing methods like unit testing, integration testing etc.

1.1 Cosine Similarity

The cosine similarity between two numerical sequences is a metric used in data analysis. It is defined by viewing the numerical sequences as vectors and it is defined as the product of the vector's dot products divided by their products of lengths. It is also known as the cosine of the angle between them. Because of this, the cosine similarity is only dependent on the vectors' direction and angle, not their magnitudes. The cosine similarity lies in the interval $[-1, 1]$. For example, the similarity between two orthogonal vectors is 0, the similarity between two opposing vectors is 1, and the similarity between two proportional vectors is 1. Positive space is the region where the semantic similarity is most frequently used when the result is neatly restricted in $[0, 1]$.

A subject text is represented as a vector representing the frequency of each word occurring in the document, as is the case in information retrieval and text mining, where each word is assigned a distinct coordinate. When two texts are expected to be similar in terms of the themes they cover, no matter how long of the documents, cosine similarity gives a good signal.

The Euclidean dot product formula can be used to get the cosine of angle between the two given non-zero vectors:

$$\mathrm{A.B} = \|\mathrm{A}\|\,\|\mathrm{B}\|\cos\theta \tag{1}$$

The cosine similarity $\cos(\theta)$ of vectors A and B is represented as follows:

$$\cos\theta = \frac{A.B}{||A||||B||} \tag{2}$$

The similarity ranges from 1 (which indicates the same) to 0 (which indicates decorrelation or orthogonality), with values in between indicating either intermediate similarity or dissimilarity. One way to express the resulting similarity is as a ratio between 1 and 1, where 1 denotes exact equality. The vectors A and B for text matching are the term frequency vectors of the documents. One way to normalize document length when comparing documents is to use cosine similarity. The cosine similarity of two texts will be between 0 and 1 because word frequencies cannot be negative. Even with the use of tf-idf weights, this remains true. No angle greater than 90° may exist between two term frequency vectors.

1.2 Motivation

It is very complex to identify the defects in source code from a defect log and is very common in software testing methods. The source code analysis also takes more time for testing in Product Lifetime Management (PLM) and Turn Around Time (TAT).

Tracing defects in source code is relatively easy in small code but needs a considerable workforce and time in vast lines of code with multiple modules and functionalities. The tool we aim to build can save software developers from this problem by automizing the process and saving time. This tool can be used for an efficient and swift software cycle in the software development domain.

1.3 Scope

The project aims to detect the defects in the source code with confidence factors. The data gathering is fulfilled by making some source codes of our own and having a defect log as feedback from the source code we generated. The plan is to use open-source libraries for fast and efficient working of our tool.

2 Proposed Methodology

The proposed methodology is to build a tool that automatically identifies defects in faulty CPP, Python, and Java code. Our approach is to analyze cosine similarity constants between the keywords extracted from the log description report and the module names or function names in the source code. Keywords are extracted from the given log file using Natural Language Processing technique POS tagging. The essential keywords extracted are used to trace back to the source code for defects in the source code.

A data structure called "Function Table" is built to store the information of filename, module name, and the span of lines in the file. The data of module names from the function is processed to find the similarity between the keywords extracted from the log description. The Cosine similarity technique is used to find the similarity constant. The module containing the defect is traced using the cosine similarity metric computed using the keywords extracted and module names and finally reports the module name containing the defect. The tool can also trace the possible lines from the source code instead of only the module name by using the data of the span of lines in the function table.

3 Previous Work Reference

This study suggests a technique called CGCN that combines the benefits of CNN and GCN which extract structural and semantical data for much precise SDP [1]. In particular, ASTs are initially created from the source codes. The semantic data in the ASTs is then captured using CNN, while the structural data in the software network is captured using GCN. Incorporating CNN and GCN, the innovative defect prediction technique CGCN is an improved supervised ML approach.

They describe a unique method for predicting source code defects inside projects in this study. They treat bug analysis as anomaly detection because datasets for defect prediction are frequently unbalanced and include few examples of defects. The Reconstruction Error Probability Distribution (REPD) model is capable of handling both individual and group abnormalities treating the source code bug prediction binary classification problem as an anomaly detection issue [2].

This approach locates Abstract Syntax Tree n-grams of source code that are linked to flawed methods [3]. They gathered information on which methods were flawed. To determine whether AST n-grams are strongly connected with 6 defective and non-defective ways of code, they employed statistical approaches. The AST n-grams that are strongly linked to inaccurate code to prediction models of defects are ultimately included.

The most important statistic in the dataset is identified using MANOVA in this study, which uses the NASA software quality dataset. Using the smaller dataset, a fuzzy logic model based on data mining is created [4]. Gini decision tree is the data mining algorithm employed. The hybrid model successfully detects software quality defects which is combination of fuzzy logic and data mining algorithm, as from the findings of ROC analysis. Static code metrics and quality metrics are the foundations of quality control studies, and each study employs a unique set of metrics throughout the process.

In order to solve the ongoing issue, they proposed a hybrid machine learning strategy in this study that combines PCA and SVM. The principle components are discovered for feature optimization using PCA, which lowers the time complexity. SVM was then used for classification since it has several advantages over the traditional and conventional approaches. Additionally, they used the GridSearchCV technique to fine-tune the hyperparameters [5].

The software bug prediction issue is defined as a classification challenge in this research, and the performance of several ensemble approaches is then examined. In order to address the unbalanced distribution problem, the top ensemble classifier will also be chosen and retrained on oversampled datasets using the SMOTE method

[6]. The suggested hybrid technique is assessed using four datasets for software problems. Experimental findings show that the suggested strategy may significantly improve fault prediction accuracy.

In this research, they present a unique method for predicting the quantity of software flaws using deep learning techniques. They first do preprocessing on a dataset that is openly accessible, which includes data normalization and log transformation. To prepare the input data for the deep neural network model, they do data modelling second. Third, they give the modelled data as input to a deep convolutional neural network model that was specifically created to estimate the number of faults [7]. They use two well-known datasets to assess the suggested technique as well. The evaluation's findings show that the suggested strategy can outperform cutting-edge methods while still being accurate.

They suggested a method in this research where characteristics are chosen using a genetic algorithm (GA) [8]. The model is then trained using various Neural Network (NN) methods, including: FNN, RNN, DNN, before the cluster of the chosen features is created using Particle 8 Swarm Optimization (PSO). Finally, accuracy score, specificity, recall, precision, negative prediction value, sensitivity, Matthew's correlation coefficient and F1-score are calculated. For their research, they employ five distinct datasets from the NASA Promise soft- ware engineering repository. According to our research, deep neural networks produce the greatest accuracy results.

In this study, the authors used an LSTM network to use source code to automatically extract the semantics and contextual aspects. They specifically analyze what and how much information they can keep for various node kinds after extracting the program's ASTs, which are composed of AST nodes. Each file's AST is traversed, and the results are put into an LSTM network [9], which uses automated semantic and contextual factors to assess if a file is faulty. Experimental findings on a number of opensource projects demonstrated the excellence of the suggested LSTM technique among other state-of-the art approaches.

This study uses the (LTSA-SVM) method that is Local Tangent Space Alignment - Support Vector Machine to provide a unique model that resolves the software fault prediction problem [10]. As the primary classifier for the defect prediction model, the model uses the SVM algorithm. Then, by combining the grid search approach with 10-fold cross validation, the model parameters are optimized. The accuracy of SVM is decreased in classic dimensionality reduction techniques due to data loss brought on by the low characteristics of data non-linearity. This work tackles this issue by performing efficient dimension 9 reduction and extracting the fundamental structure of low-dimensional data using an LTSA method. The reduced dimensional data is used to train the SVM algorithm. The prediction model's viability is then confirmed.

In this study, they examine 4 log parsers and put them together into a toolkit so that they may be used again. Additionally, they analyze the log parsers' performance on five datasets totaling more than 10 million unprocessed log messages in order to derive six enlightening conclusions, and their performance on a challenging log mining job has also been carefully studied [11].

On 10 NASA datasets, this research examines and contrasts both supervised learning and ensemble classifiers. Classifiers such as decision trees (DS), random forests (RF), support vector machines (SVM) and bagging are renowned for their effectiveness in predicting flaws [12]. The experimental findings demonstrated that, when compared to all the other classifiers, RF performed the best in most instances.

The widely studied supervised defect prediction approaches have one drawback: they depend on labelled software modules, which are occasionally unavailable. Clustering based Unsupervised Defect Prediction (CUDP) is an alternative approach that uses unsupervised 10 models based on clustering to analyze unlabeled defect data. Few research has examined the effects of clustering algorithms on the accuracy of software bug prediction, though [13]. To close this gap, they conducted sizable empirical research on unsupervised models in this article. They selected an open-source dataset with 27 project versions and three different feature kinds. The experimental findings demonstrate that various clustering-based models function significantly differently.

Logs are often the first and only source of information used by software developers to understand and troubleshoot their systems. Software developers can use automated log-analysis algorithms to learn from massive log data sets. These strategies go through several phases, but the most important one is log abstraction because it transforms raw log data into sophisticated knowledge. Software and research engineers can perform additional analytics with the aid of log abstraction. There are large differences in the designs and performances of the various log-abstraction techniques currently in use [14].

The usefulness of the two-similarity metrics utilized by the Automated Essay Type Grading System the cosine similarity metric and Euclidean distance is examined in this study (AETGS). With the use of N-gram terms, a Document Term Matrix (DTM) was created. The sparseness of the DTM was reduced using a modified PCA approach to produce a vector representation [15]. The cosine similarity metric and the 11 Euclidean distance measurement were used to grade the reduced word vectors of the student's responses in accordance with the mark allocated to every question in the marking scheme. The Pearson Correlation coefficient was used to determine how the similarity metrics affected the system that had been constructed.

4 Results and Discussions

- The tool can successfully trace probable modules for the root cause of the defect.
- The outcome of the defect is not 100% accurate, but it gives us the most probable modules for causing the defect.
- From Fig. 1, the output in Pane 4 contains filename, function name and confidence factor of probable error locations.
- The tool developed works for faulty python, java and CPP code and can be enhanced for several other programming languages.

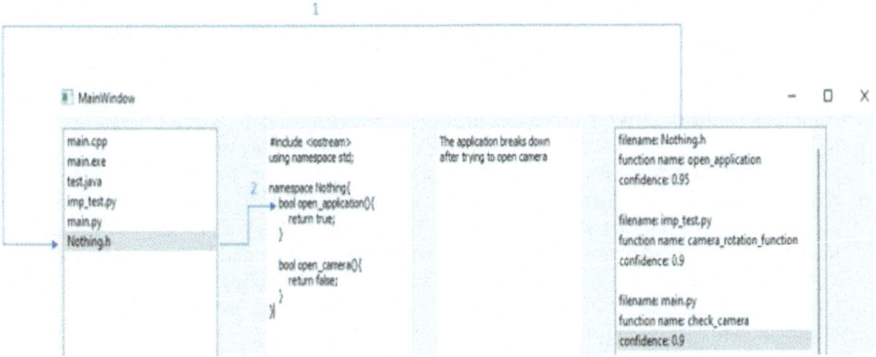

Fig. 1 Result of the tool

- Tool developed works fine for hybrid code environments, that is projects with more than one programming language. Refer to Fig. 1 Pane 1 for the source code files of test project.
- From Fig. 1, the output from Pane 2 is the text editor for the selected source file. The output from Pane 3 is the defect description that is given as input.

5 Conclusion

The significant objective of our work is to build a tool that automatically identifies defects in faulty CPP and python code and use this method in product testing to improve performance and save time. Our approach is to analyze the cosine similarity constants between the keywords in the log description report and the module names in the source code. The keywords are extracted from the log file, which plays a crucial role in our model to trace the source code defect. The model processes the source code stores the information of module names, and computes cosine similarity between the keywords extracted to module names in the source code. Computed cosine similarity constants help to trace back to defects in source code. This tool helps to trace the possible locations of the root cause from source code from the Defect Log for the test release software.

References

1. Zhou C, He P, Zeng C, Ma J (2022) Software defect prediction with semantic and structural information of codes based on graph neural networks. Inf Softw Technol 152:107057. ISSN 0950-5849
2. Afric P, Sikic L, Kurdija AS, Silic M (2020) REPD: source code defect prediction as anomaly detection. J Syst Softw 168:110641. ISSN 0164-1212

3. Shippey T, Bowes D, Hall T (2019) Automatically identifying code features for software defect prediction: using AST N-grams. Inf Softw Technol 106:142–160. ISSN 0950-5849
4. Adak MF (2018) Software defect detection by using data mining based fuzzy logic. In: 2018 Sixth international conference on digital information, networking, and wireless communications (DINWC), pp 65–69. https://doi.org/10.1109/DINWC.2018.8356997
5. Mustaqeem M, Saqib M (2021) Principal component based support vector machine (PC-SVM): a hybrid technique for software defect detection. Cluster Comput 24:2581–2595
6. Alsawalqah H, Faris H, Aljarah I, Alnemer L, Alhindawi N (2017) Hybrid SMOTE-ensemble approach for software defect predicion. In: Silhavy R, Silhavy P, Prokopova Z, Senkerik R, Kominkova Oplatkova Z (eds) Software engineering trends and techniques in intelligent systems CSOC 2017. Advances in intelligent systems and computing. Springer, Cham, vol 575
7. Qiao L, Li X, Umer Q, Guo P (2020) Deep learning based software defect prediction. Neurocomputing 385:100–110. ISSN 0925-2312
8. Ayon SI (2019) Neural network based software defect prediction using genetic algorithm and particle swarm optimization. In: 2019 1st International conference on advances in science, engineering and robotics technology (ICASERT), pp 1–4. https://doi.org/10.1109/ICASERT.2019.8934642
9. Deng J, Lu L, Qiu S (2020) Software defect prediction via LSTM. Institute of engineering and technology
10. Wei H, Hu C, Chen S, Xue Y, Zhang Q (2019) Establishing a software defect prediction model via effective dimension reduction. Inf Sci 477:399–409. ISSN 0020-0255
11. He P, Zhu J, He S, Li J, Lyu MR (2016) An evaluation study on log parsing and its use in log mining. In: 2016 46th annual IEEE/IFIP international conference on dependable systems and networks (DSN), pp 654–661. https://doi.org/10.1109/DSN.2016.66
12. Alsaeedi A, Khan M (2019) Software defect prediction using supervised machine learning and ensemble techniques: a comparative study. J Softw Eng Appl 12
13. Xu Z, Li L, Yan M, Liu J, Luo X, Grundy J, Zhang Y, Zhang X (2021) A comprehensive comparative study of clustering-based unsupervised defect prediction models. J Syst Softw 172:110862. ISSN 0164-1212
14. El-Masri D, Petrillo F, Guéhéneuc Y-G, Hamou-Lhadj A, Bouziane A (2020) A systematic literature review on automated log abstraction techniques. Inf Softw Technol
15. Odunayo O, Ibrahim A, Falohun AS, Obe O (2018) A comparative analysis of euclidean distance and cosine similarity measure for automated essay-type grading. J Eng Appl Sci

Unleashing Agricultural Precision: A Deep Learning Paradigm for Papaya (Carica Papaya L.) Variety Discrimination and Yield Optimization

Ashoka Kumar Ratha, Nalini Kanta Barpanda, Prabira Kumar Sethy, and Santi Kumari Behera

Abstract The categorization of papaya fruit varieties is important within the agricultural industry. There is a need for a system that can accurately evaluate the different varieties of papaya fruits to save labor expenses associated with discarding specific types of papaya fruits throughout the packaging and distribution process. The present work introduces a novel approach for the categorization of papaya fruit varieties with the aim of enhancing efficiency and accuracy. The proposed method comprises a series of interconnected steps. Once the data on papaya fruits are collected, they undergo preprocessing techniques, such as color uniformization, image scaling, augmentation, and image labelling. Subsequently, the AlexNet architecture was employed, incorporating 08 layers, comprising 05 conv layers and 03 fully connected layers. Simultaneously, the process of transfer learning and fine-tuning of the convolutional neural network was executed. During the concluding phase, a SoftMax classifier was employed to perform the ultimate classification. Detailed simulations were conducted using self-generated datasets. The proposed model demonstrated the highest classification accuracy of 98.0%. Furthermore, our developed model exhibits computational efficiency with an average processing time of 8 ms to generate the ultimate classification outcome. The obtained results enhance the comprehensibility and reliability of convolutional neural network (CNN) models, thereby offering valuable insights for the prospective utilization of deep learning techniques in agricultural fields.

Keywords CNN · Papaya varieties · Transfer learning · Classification · Interpretable model

A. K. Ratha · N. K. Barpanda · P. K. Sethy (✉)
Department of Electronics, Sambalpur University, Burla, Odisha 768019, India
e-mail: prabirsethy.05@gmail.com

N. K. Barpanda
e-mail: nkbarpanda@suniv.ac.in

S. K. Behera
Department of Computer Science and Engineering, VSSUT, Burla 768018, India

V. K. Gunjan and J. M. Zurada (eds.), *Proceedings of 5th International Conference on Recent Trends in Machine Learning, IoT, Smart Cities and Applications*, Lecture Notes in Networks and Systems 1181, https://doi.org/10.1007/978-981-97-8861-3_7

1 Introduction

Papaya (Carica papaya L.), a tropical fruit, has tremendous commercial significance owing to its considerable nutritional and therapeutic value. Papaya cultivation originated in the regions of Costa Rica and Mexico. The estimated global annual production of fruits is 6 million tons. India is the global leader in papaya production, with an annual yield of over 3 million tons. Brazil, China, Mexico, Indonesia, Nigeria, Peru, the Philippines, and Thailand are among the most prominent producers in this industry. Papaya is considered a lucrative agricultural commodity in the Indian market. The primary factors contributing to its significant achievements are its ease of cultivation and versatile applications. Papayas have the potential to be cultivated in large-scale agricultural enterprises, as well as in domestic settings such as personal gardens. Additionally, it has the potential to serve as a source of both fruit and latex. It is abundant in vitamin C and possesses beneficial properties for the digestive system. Hence, engaging in papaya trading proves profitable. India holds the top position in papaya production among fruits, with a share of 39.30%, as reported by the Food and Agriculture Organization (FAO) in 2021 [1]. A notable characteristic of these plants is their rapid growth rate. Owing to its tropical nature, this plant requires a moderate level of humidity to survive. According to data obtained from the National Horticultural Board (NHB), the projected aggregate yield of papayas for 2021–22 was recorded at 5744.26 million metric tons, indicating a 3.55% increase compared to the previous year's production of 5539.80 million metric tons [2]. This positive attribute leads to a comparatively elevated market valuation of papaya fruit [3]. Between 2009 and 2020, India consistently engaged in the exportation of papaya to various destinations across the globe, with an estimated value ranging from 3 to 5 million dollars. In 2020, despite prolonged restrictions on exports due to the COVID-19 pandemic, India managed to achieve a successful shipment of papaya valued at $4.36 million. Papaya quality sorting is a postharvest issue that has not yet been solved, as it is still performed manually. Undoubtedly, there are constraints and room for development for this approach. Manual sorting is inefficient not only because it takes a lot of people, but also because people's perceptions of quality differ, and human consistency is sometimes unreliable due to fatigue [4]. Many previous studies have used ANN models [5], which are continuously implemented in digital image processing techniques for sorting the quality of papaya, whereas other studies have used techniques such as Fuzzy Mamdani [6], Image Processing [7], and K-nearest neighbor [8].

The application of deep learning has proven to be effective in the nondestructive identification, classification, and detection of fruits and vegetables. This approach offers several benefits, including rapid processing, ease of use, affordability, and high accuracy [9]. Convolutional Neural Networks, which are deep learning-based frameworks known for their ability to automatically learn features from images, have demonstrated remarkable performance in a range of food and agricultural applications [10, 11]. An automated papaya variety classification system can promote smart farming in orchards, agricultural sorting systems, and supermarket checkout systems

[12]. In recent years, significant progress has been made in the development of several Convolutional Neural Network designs that have exhibited exceptional efficacy in the domain of picture categorization. Transfer learning is a prominent methodology that leverages pre-existing structures to address diverse problem domains, thereby enhancing the precision and efficacy of the outcomes. The utilization of DL and ML methodologies in conjunction with TL provides an effective approach for image classification. This study examines the potential of TL in the context of convolutional neural network models for the classification of papaya fruits as an alternative to utilizing CNN architectures from scratch. In practical contexts, the classification of fruits is typically conducted by individuals whose efficacy is perceived as inadequate by fruit producers and retailers. Hence, there is a need to create a precise classification methodology that may effectively minimize human labor and expenses. The implementation of a robust system for classifying fruits can lead to a reduction in production time within the agricultural industry, as it accurately identifies the variety of fruits. Hence, this study introduces a novel approach for automatically classifying fruit varieties by employing fine-tuning and transfer learning techniques of AlexNet. The efficacy of the suggested methodology was verified using internally created papaya fruit datasets. The primary contributions of this study are as follows.

- **Enhanced Efficiency in Papaya Fruit Categorization**: The proposed approach significantly improves the efficiency of papaya fruit categorization in the agricultural industry. The combination of sophisticated preprocessing methods and transfer learning with AlexNet results in a 98% classification accuracy with much lower computational complexity.
- **Optimized Preprocessing Techniques**: The introduction of comprehensive preprocessing techniques, including color uniformization, image scaling, augmentation, and image labeling, contribute to the robustness of the model. These preprocessing steps ensure that the input data are well-prepared for the subsequent stages of the classification process, leading to improved model performance and accuracy.
- **Effective Transfer Learning and Fine-Tuning**: The incorporation of transfer learning and fine-tuning in a convolutional neural network (CNN) AlexNet allows the proposed method to adapt and specialize in the unique characteristics of papaya fruit varieties. This approach accelerated the training process and enhanced the overall performance of the method.
- **Superior Classification Accuracy**: The achieved classification accuracy of 98.0% is a significant contribution to the field of papaya fruit categorization. This level of accuracy is crucial for reducing labor expenses associated with discarding specific types of papaya fruits during the packaging and distribution process. The model's ability to distinguish between different varieties with high precision adds value to the agricultural industry by minimizing errors and optimizing resource utilization.
- **Computational Efficiency**: The proposed approach not only excels in accuracy but also demonstrates notable computational efficiency. With an average processing time of 8 ms to generate the final classification outcome, the model

proved practical for real-world applications. This efficiency is essential for seamless integration of the proposed system into agricultural processes, ensuring timely decision-making and operational effectiveness.

The remainder of this study is structured as follows. In the Sect. 2, we provide a detailed description of the proposed method. Section 3 presents the simulation results and comparisons, and Sect. 4 concludes our findings and provides suggestions for further research.

2 Materials and Methods

This section deals with the details of the dataset and adapted methodology.

2.1 *Sample Collection and Preprocessing of Image*

We investigated using some of the currently available data by utilizing existing datasets, such as Fruits 360. However, after evaluating the available free datasets, we realized that some new data would need to be collected from scratch to make this effective in the Indian context and for the quality prediction use case. To address this, we bought various papaya subfamilies from the market, took images of them, and developed the project's own necessary dataset as shown in Fig. 1. Using image augmentation techniques such as horizontal right-flip, horizontal left-flip, rotate right 90°, and rotate left 90° operations, we constructed more than 3,000 images of the papaya dataset from these images. A total of 3500 samples were used to perform the experiments, with the number of images varying depending on the class of papayas, and the training and testing datasets were chosen in an 80:20 ratio, as shown in Table 1.

2.2 *Methodology*

Figure 2 illustrates the Generic flow of the proposed method for papaya variety classification, which consists of several interconnected components. In the next section, we provide an in-depth discussion of the modules employed in the algorithmic framework.

The model under consideration is a weighted model consisting of eight layers, with the initial five levels being convolutional layers and the last three layers being fully connected layers [13]. In the AlexNet architecture, each of the five layers performs batch normalization in conjunction with a 2×2 maximum pooling operation. The careful choice of an appropriate activation function enhances the accuracy of our

Fig. 1 Images of Papaya datasets

Table 1 Statistics of papaya fruit datasets

Types of papaya	Number of images	Enhanced images	Training datasets	Testing datasets
Red lady	80	400	320	80
Pusa majesty	100	500	400	100
Pusa dwarf	100	500	400	100
Pusa delicious	100	500	400	100
Maradol	80	400	320	80
Ice berry	80	400	320	80
Arka surya	80	400	320	80
Arka Pabhath	80	400	320	80

methodology. Therefore, when selecting an activation function, it is imperative to ensure that the gradient function exhibits rapid convergence, and that the activation function does not approach zero at infinity. Equations (1) and (2) depict the Re-LU activation function and its corresponding derivative, respectively.

$$f(x) = \max(0, x), \tag{1}$$

$$f\prime(x) = \begin{cases} x, \ if \ x \geq 0 \\ 0, \ if \ x < 0 \end{cases} \tag{2}$$

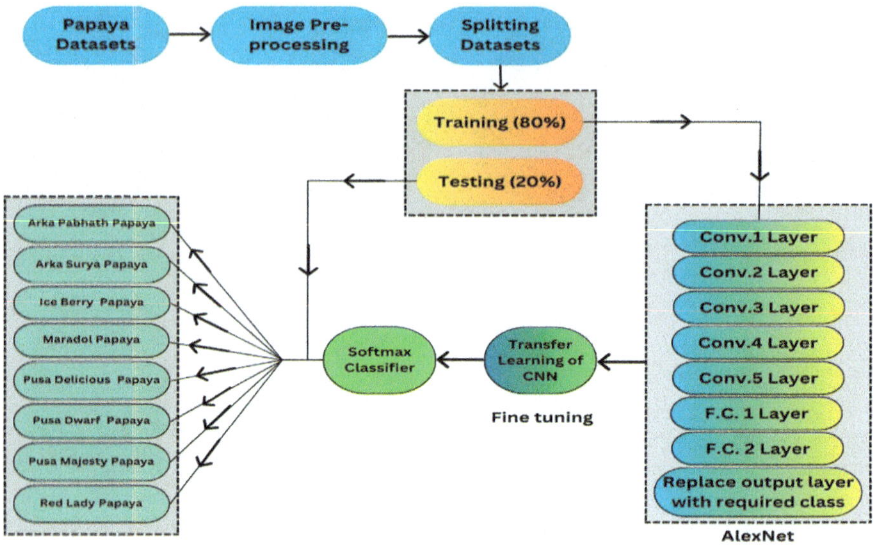

Fig. 2 Generic flow of the proposed method for papaya variety classification

According to Eq. (2), the derivative of the Rectified Linear Unit (Re-LU) function remains constant at x for all positive values of x, whereas it remains constant at 0 for all negative values of x. In the first five convolutional layers, a rectified linear unit (Re-LU) function was used. In addition, the output from the first five layers was routed to the next three levels, which were all fully connected. These three levels are made up of two fully connected layers (4096 units each) and a third fully connected layer (1000 units). In the final output layer, the SoftMax activation technique is used. Following AlexNet's data processing, the next phase involves the application of transfer learning, which is further discussed.

2.3 Transfer Learning (TL)

Typically, Convolutional Neural Networks (CNNs) are trained using extensive datasets comprising over one million images, such as ImageNet. These networks exhibit optimal performance when equipped with deeper layers that are densely coupled. Convolutional neural networks (CNNs) used for fruit and vegetable classification have been trained on a limited set of categories and small datasets. This limited scope can result in overfitting of deep networks, compromising the theoretical accuracy and effectiveness of the outcomes [14]. In addition, the training of these large Convolutional Neural Networks (CNNs) might result in notable disadvantages, including heightened computing expenses and slower execution times [15]. Transfer learning has been employed to address these issues. Transfer learning is a deep learning technique that applies data obtained from a pre-existing model to initialize the training process for a distinct problem [16]. Using AlexNet as an example, the transfer learning process involves splitting the overall architecture of AlexNet into two distinct components: the pre-trained segment and the replacement segment. The substituted portion constitutes a minor component of the overall network architecture; hence, it requires a relatively modest training dataset for effective transfer learning. In the context of deep networks, transfer learning has been identified as a potential approach for reducing dependence on computer hardware and training time [17]. Further elaboration on this topic can be found in our prior research endeavors [18, 19]. Transfer learning, also known as knowledge transfer, involves the utilization of a pre-trained neural network as an initial framework for addressing specific categorization tasks. During the transfer learning phase, we adjusted the upper tiers of an existing model while leaving the lower layers unchanged. In addition, new layers were added to the model architecture. The number of epochs, model size, and learning rate were adjusted during fine-tuning to improve performance. Additionally, the last layer of the output was replaced with relevant categories.

2.4 *Object Classifier -SoftMax*

Multinomial logistic regression, or SoftMax, possesses a distinct benefit in handling vectors of N dimensions. The SoftMax function is extensively employed across a range of disciplines, including deep learning, to facilitate the classification of distinct objects [20]. The SoftMax classifier is utilized to estimate the probability of classification of the extracted vectors. Equation (3) demonstrates that, for a given dataset, the sum of the probabilities across all vectors is equal to 1.

$$f_j(z) = \frac{e^{zj}}{\sum_k e^{zk}} \tag{3}$$

The input vector utilized by the fine-tuned network in the preceding stage was denoted as z. The results were converted from the exponential to probability domains. The fruit classification uses a SoftMax loss function. Our investigation shows that the SoftMax function performs well and converges quickly. Equation (4) mathematically represents the SoftMax function.

$$Loss_{softmax} = \frac{1}{\mathrm{N}}\sum_i - log\frac{e^{zyi}}{\sum_j e^{zj}} \tag{4}$$

where N is the output of the final fully connected layer for the correct class (yi). Additionally, "zj" denotes the output value from the last fully connected layer for the jth class. Equations (3) and (4) demonstrate that the $Loss_{Softmax}$ function leverages the information derived from accurate labels to optimize the probability of the data.

Furthermore, it fails to acknowledge the data provided by frequent false classifications. The pseudocode of the developed approach is presented in the following algorithm. According to the algorithm, the fruit images sourced from the self-generated dataset underwent pre-processing, as detailed in the previous section. The algorithm involves the processing of data obtained from rows to 2–7, which are subsequently utilized as input for the deep learning models. As shown in rows 8–13, the AlexNet model continues to process the data that was first addressed. All eight levels were utilized in the process of running AlexNet. The Re-LU activation function was used for the initial five convolutional layers. The transfer learning process for the convolutional neural network (CNN) was executed on rows 14–17. In this study, the CNN model was optimized by adjusting the batch size, learning rate, and epochs, as indicated in row 16 of the algorithm.

Algorithm: Pseudocode for classifying different kinds of papaya fruits

1. **Input:** Obtained images of papaya fruits
2. **do:**
3. **Evaluate** the data collected in step (1).
4. Execute the preprocessing steps.
5. Perform image resizing.
6. Perform data splitting randomly (Training-80% & Testing-20%)
7. End
8. Begin
9. **Load** the AlexNet model.
10. Enable 8 layers, including both the 5 convolution and 3 fully connected.
11. For the first 5 conv levels, use Re-LU.
12. The process and use of the eight layers are described in Section 2.2.
13. End
14. Begin
15. CNN Transfer learning
16. Use the technique of fine tuning and establish the settings as: Epochs = 10 and learn rate = 0.0001 and Batch size = 32
17. end
18. Use SoftMax as an object classifier.
19. Classify using Equations (1) and (2).
20. **Output:** The concluding outcome of the classification: papaya fruit varieties

In the last stages, the utilization of the SoftMax classifier is employed for the purpose of classifying the different varieties of papaya fruit. The subsequent section will show that the system we created exhibits robustness and achieves fast and accurate classification of fruit varieties. Furthermore, the processes and directions presented in the algorithm are straightforward and easily accessible. In the subsequent section, we present an in-depth overview of our research supported by visually informative images. Furthermore, we analyzed our findings during various stages of algorithm development.

Table 2 Configuration setup during experiment

Parameter	Simulation environment
Resolution of training/test image	227 × 227
Optimizer	ADAM
Learning rate	10^{-4}
Epochs	10
Validation frequency	30
Shuffle	Every-epoch
Momentum	0.9

3 Results and Discussions

In this section, we describe how we set up our experiments, present the results of fruit classification, and provide a detailed discussion along with observations.

3.1 Experimental Setup and System Specifications

Our algorithm was executed on a computer featuring an Intel Core i5 processor running at 2.50 GHz, complemented by an NVIDIA GTX3050 graphics card. With a robust 16 GB RAM, this setup sufficiently accommodated the exploration of papaya fruit variety classification results generated by our algorithm. Table 2 shows the experimental setup and settings that were consistently applied throughout CNN's transfer learning phase. The ADAM optimizer was used in all simulations, with a learning rate of 0.0001. In addition, the validation frequency was set to 30, and the number of batch sizes and epochs were set to 32 and 10, respectively. The dataset was shuffled at the completion of each epoch. To accomplish the classification output within a reasonable timeframe, the momentum rate was set to 0.9. The training and test images had a resolution of 227 × 227 pixels.

3.2 Evaluating the Performance Parameters

In our approach, we use the renowned parameters of accuracy and confusion matrix to assess the effectiveness of our papaya fruit variety classification. The confusion matrix allows us to analyze predicted versus actual classes, enabling the computation of various classification performance metrics [21]. Accuracy, a widely used parameter, is employed to measure the correctly classified pixels in our developed model, and its mathematical formulation is expressed by the Eq. (5)

$$\text{Accuracy} = \frac{[\text{True Positive } + \text{True Negetive}]}{[\text{True Positive} + \text{ False Positive} + \text{True Negetive} + \text{False Negetive}]} \tag{5}$$

3.3 Analysis of Papaya Fruit Classification

To determine the effectiveness of the proposed approach, validation and evaluation were performed. After 40 iterations, the minimum classification error is 0.02. The confusion matrices and ROC curves for all eight papaya classes are shown in the following Fig. 3.

It has been observed that the overall classification accuracy for our developed method is 98.0% with AUC of 1. Our tests show that it takes about 8 ms to process the test image sets and get the final classification result. In the confusion matrix, the diagonal represents the true positives, which are the correct classifications. So, the diagonal gives us a measure of how accurate our method is in classifying images. As shown in the diagonal of the Fig. 3 that for each of Arka Pabhath Papaya, Ice Berry Papaya, Maradol Papaya, Pusa Delicious Papaya, Pusa Dwarf Papaya, Red Ledy Papaya, our developed method obtains 100% accuracy. Similarly, for Arka Surya Papaya and Pusa Majesty Papaya our proposed algorithm obtained 87.5% and 96% of classification accuracy respectively.

We compared our method with others using the same datasets. The VGGNet method had the lowest validation accuracy at 23.57% and a test accuracy of 75.10%. The ResNet-50 method also had low validation accuracy at 13.57%, but a higher test accuracy of 87.90%. On the other hand, the GoogleNet and DenseNet methods performed well with individual validation accuracies of 87.86% and 79.29%, and test

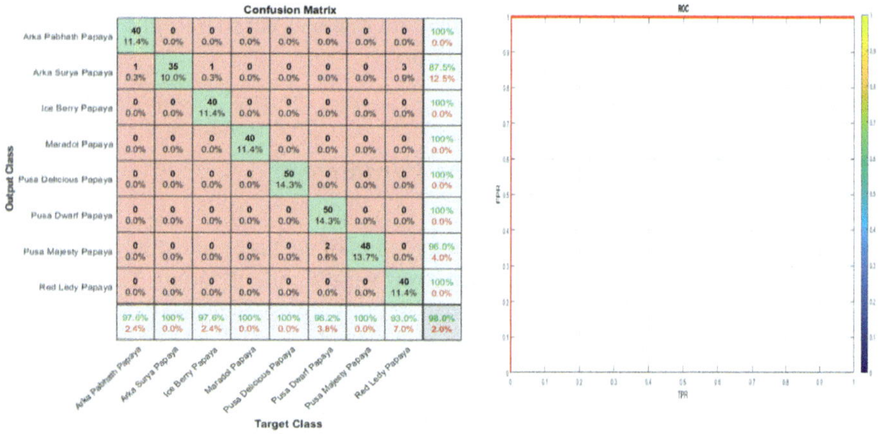

Fig. 3 Confusion matrix with ROC for classification of sub-family of papaya fruit

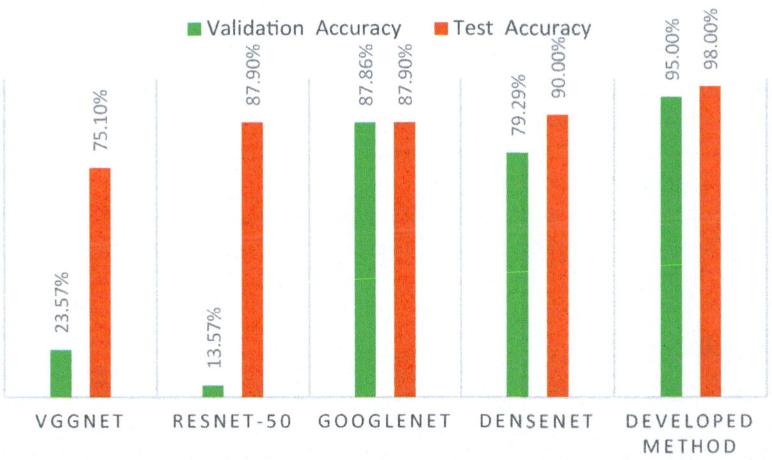

Fig. 4 Comparison of Accuracy of Developed method with that of other methods

accuracies of 87.90% and 90%, respectively. Our developed method outperformed all these methods, achieving a high 98% classification accuracy. This is illustrated in Fig. 4, showing that our method is superior in comparison.

3.4 Discussion on the Developed Method

These points provide valuable insights into the classification performance of our method for categorizing papaya fruit varieties. However, the subsequent discussion provides further insights into the efficacy of the proposed algorithm.

- Our study found that different networks perform differently on various algorithms. VGGNet had the lowest validation accuracy (23.57%) and a test accuracy of 75.10%. ResNet-50 also showed low validation accuracy (13.57%) but a higher test accuracy (87.90%). Meanwhile, GoogleNet and DenseNet had good validation accuracies (87.86% and 79.29%, respectively) and high-test accuracies (87.90% and 90%). This indicates that each network has its strengths and weaknesses in classification accuracy.
- Our method performs exceptionally well with all the papaya datasets, boasting an impressive 98% classification accuracy. This achievement can be attributed to the clever design in the intelligence section of the AlexNet Architecture.
- Our method performs exceptionally well because we employed data augmentation during preprocessing, which helps prevent overfitting on smaller datasets. Additionally, the improved classification results can be attributed to using a pretrained Convolutional Neural Network (CNN) framework and modifying the final layer to match the specific characteristics of the class we are focusing on.

- Our study findings indicate that AlexNet takes less time to train compared to other architectures with more layers. Furthermore, it is important to acknowledge that the AlexNet architecture exhibits computational efficiency and does not require the use of high-performance workstations [22]. AlexNet utilizes the Rectified Linear Unit (Re-LU) activation function to effectively mitigate the problem of vanishing gradients.
- AlexNet layers have more filters and a pooling layer after each convolutional layer. These qualities have led us to use AlexNet for fruit classification.

4 Conclusion

Our research focused on using advanced computer networks to automatically classify different types of papaya fruits. Usually, people check the quality of fruits by looking at them, but we wanted to make this process more efficient. In our study, we used a powerful computer model called AlexNet, and through a process called transfer learning, we fine-tuned it to become good at identifying different papaya varieties. We made some adjustments to improve its performance, like changing certain settings. With these changes, our model reached an impressive average accuracy of 98% when classifying papaya fruits in various datasets. In the future, our focus is to diversify papaya varieties, aiding customers in discerning fruit types for better quality purchases. Additionally, we plan to launch a user-friendly mobile app displaying classification results for a broader range of fruits and vegetables. Research efforts will explore the impact of parameters like activation functions, pooling optimizations, and loss functions. To facilitate the execution of intricate machine learning and deep learning methods, we aim to deploy our approach in a cloud-based framework.

Acknowledgements We would like to thank Prof. N. K. Barpanda for his expert advice and encouragement throughout the preparation of the research as well as Dr. S. K. Behera for his brilliance in adding a final touch to the manuscript.

References

1. https://apeda.gov.in/apedawebsite/six_head_product/FFV.htm. 21 Sep 2023
2. https://agriexchange.apeda.gov.in/India%20Production/India_Productions.aspx?cat=fruit&hscode=1051. 21 Sep 2023
3. Salinas I, Hueso JJ, Cuevas J (2021) Active control of greenhouse climate enhances papaya growth and yield at an affordable cost. Agronomy 11(2):378. https://doi.org/10.3390/agronomy11020378
4. Patel KK (2020) Studies on destructive and non-destructive quality evaluations of mango fruit. Notion Press
5. Naranjo-Torres J, Mora M, Hernández-García R, Barrientos RJ, Fredes C, Valenzuela A (2020) A review of convolutional neural network applied to fruit image processing. Appl Sci 10(10):3443. https://doi.org/10.3390/app10103443

6. Al Rivan ME, Suherman J (2020) Determining the quality of California papaya Fruit (Carica Papaya L.) using fuzzy Mamdani. ELKHA: J Electr Eng 12(2):76–83
7. Cahyono WD, Wirawan D (2015) KAJIAN SIFAT FISIK BUAH PEPAYA (Carica papaya L.) MENGGUNAKAN PENGOLAHAN CITRA
8. Wibowo F, Hakim DK, Sugiyanto S (2018) Estimation of papaya fruit quality class based on GLCM texture characteristics using the K-nearest neighbors algorithm. Natl J Inform Eng Educ JANAPATI 7(1):100–106
9. Abasi S, Minaei S, Jamshidi B, Fathi D (2018) Dedicated non-destructive devices for food quality measurement: a review. Trends Food Sci Technol 78:197–205
10. Gikunda PK, Jouandeau N (2019) State-of-the-art convolutional neural networks for smart farms: A review. In: Proceedings of the intelligent computing-computing conference, London, UK, 16–17 July 2019; pp 763–775
11. Hamid Y, Wani S, Soomro AB, Alwan AA, Gulzar, Y (2022) Smart seed classification system based on mobileNetV2 architecture. In: Proceedings of the 2nd international conference on computing and information technology (ICCIT), Tabuk, Saudi Arabia, pp 217–222
12. Hossain MS, Al-Hammadi M, Muhammad G (2018) Automatic fruit classification using deep learning for industrial applications. IEEE Trans Ind Inform 15:1027–1034
13. Sultana N, Jahan M, Uddin MS (2022) An extensive dataset for successful recognition of fresh and rotten fruits. Data Brief 44:108552
14. Hameed K, Chai D, Rassau A (2018) A comprehensive review of fruit and vegetable classification techniques. Image Vis Comput 80:24–44
15. Canziani A, Paszke A, Culurciello E (2016) An analysis of deep neural network models for practical applications. arXiv 2016. arXiv:1605.07678
16. Ratha AK, Barpanda NK, Sethy PK, Behera SK (2023) Papaya fruit maturity estimation using wavelet and ConvNET. Ingenierie des Systemes d'Information 28(1):175
17. Lu S, Lu Z, Zhang Y-D (2019) Pathological brain detection based on AlexNet and transfer learning. J Comput Sci 30:41–47
18. Yu F, Lu T, Han B, Xue C (2021) A quantitative study of aggregation behaviour and integrity of spray-dried microcapsules using three deep convolutional neural networks with transfer learning. J Food Eng 300:110515
19. Lu T, Yu F, Xue C, Han B (2021) Identification, classification, and quantification of three physical mechanisms in oil-in-water emulsions using AlexNet with transfer learning. J Food Eng 288:110220
20. Enciso-Aragón CJ, Pachón-Suescún CG, Jimenez-Moreno R (2018) Quality control system by means of CNN and fuzzy systems. Int J Appl Eng Res 13:12846–12853
21. Haghighi S, Jasemi M, Hessabi S, Zolanvari A (2018) PyCM: multiclass confusion matrix library in Python. J Open Source Softw 3:729
22. Hosny KM, Kassem MA, Foaud MM (2019) Classification of skin lesions using transfer learning and augmentation with Alex-net. PLoS ONE 14:e0217293

Fish Farm Monitoring and Controlled System Using LoRaWAN Network

Reva Dalal, Swapnil Kadam, and Deepak M. Shinde

Abstract Fish farm automated monitoring systems are becoming increasingly popular because of their numerous advantages, including better fish health and survival rates, lower labor costs, increased production, and enhanced sustainability. These systems track variables related to water quality, including temperature, pH, dissolved oxygen, and water level, using a variety of sensors. The central processing unit can be used to monitor the environment of the fish farm and make appropriate adjustments after receiving the data collected by the sensors. Automated monitoring systems for fish farms can be developed using various technologies, including wireless sensor networks, the Internet of Things (IoT), and artificial intelligence (AI). Although the use of automated monitoring systems for fish farms is still in its infancy, the potential advantages are substantial. Automated monitoring systems will probably be used more and more in the aquaculture sector as technology advances and costs come down.

Keywords Temperature sensor · TDS sensor · pH sensor · ORP sensor · LoRa module · ESP32 · Microcontroller · Cloud

1 Introduction

Demand for dietary fish protein has been increasing due to the increase in global population as well as the awareness generated amongst people about the health benefits it can provide. Roughly around 1 trillion to 3 trillion fish are killed every year for human consumption; causing overfishing and can hamper wild fisheries as well as disturb the natural food chain. Fish welfare is primarily affected by the water

R. Dalal (✉) · S. Kadam · D. M. Shinde
Electronics and Telecommunications Department, SCTR'S Pune Institute of Computer Technology, Pune, Maharashtra 411043, India
e-mail: reva.dalal.197167@gmail.com

D. M. Shinde
e-mail: dmshinde@pict.edu

V. K. Gunjan and J. M. Zurada (eds.), *Proceedings of 5th International Conference on Recent Trends in Machine Learning, IoT, Smart Cities and Applications*, Lecture Notes in Networks and Systems 1181, https://doi.org/10.1007/978-981-97-8861-3_8

quality wherein the non-ideal condition of the water quality, such as the dissolved oxygen (DO), pH, temperature, ammonia, nitrites, and nitrates, will cause health problems, and fish kills. Uneaten feeds and feces are the primary solid wastes that can cause to accelerate the deterioration of water quality. The fishes can also have underlying health issues that may go unnoticed by the consumer. Thus to tackle such problems, the introduction of sensing technologies, IoT, automation, and intelligent systems are growing in popularity to effectively manage needs towards precision fish farming. Fish farming or aquaculture involves the commercial breeding of fish, usually for food in such fish tanks or artificial enclosures. It involves controlled cultivation and harvesting of aquatic animals. Fish farming allows the establishment of artificial fish colonies that are provided with sufficient feeding, protection from natural predators and competitive threats and easier harvesting when needed and does not impact the sustainable yields of wild fish populations. And to do so timely monitoring is required and using our system we can ease and automate this process.

2 Literature Survey

To monitor and manage the environment, this study offers an automated monitoring system for fish farms that makes use of wireless sensor networks and a mobile device. The system uses MCU processing to capture the physical sensor signal and combines temperature, dissolved oxygen, pH value, and water level sensing modules. The data is brought via the ZigBee wireless sensor network to a central processing core, which has a WIFI interface to provide the data to the user terminal device. Through the terminal device, which uses Android software to construct the user interface, the user may manage every aspect of the fish farm environment. A low-power MSP430 series MCU provides the system's power, which can come from batteries, a regular power source, or solar batteries. The system has been created [1].

This study describes a wireless transmission-based remote fish farming monitoring system. The system gathers and transmits information on environmental parameters like temperature, dissolved oxygen, and pH via wireless sensor networks. The information is subsequently transmitted to a computer network for fish farm monitoring and management in real time. The technology is inexpensive, energy-efficient, and simple to use. As a backup power source, it also uses solar panels. Visual Basic programing is used to design the interface's features and manage the sensor module. The technology is a dependable and effective method for remotely monitoring fish aquaculture [2]. The use of wireless sensor networks for managing and monitoring fish farms is covered in this study. The suggested system makes use of battery-operated wireless sensor nodes and the ZigBee protocol for communication over short distances. It can also be simply expanded to monitor other farms. The system can be expanded with extra sensor made possible by data logging and IoT characteristics. The benefits of an IoT-enabled DC Nano grid for offshore fish farms are highlighted in the article [3].

Using the Internet of Things (IoT) idea, this study suggests a smart fish pond system for contemporary catfish farming. Using sensors, the system keeps an eye on and regulates variables affecting water quality, including temperature, pH, turbidity, and water level. An application for smartphones can access and track the data after it has been uploaded to a cloud database. Additionally, the system has a function that enables cooperation among catfish growers to meet consumer demand. With the technology in place, pond conditions can be easily monitored, fish quality is improved, and farmer revenue is raised [4]. The paper examines various automatic recognition systems and provides an overview of computer vision techniques utilized in this sector. The suggested method applies convolutional neural networks (CNN) for disease classification and makes use of several color spaces. The authors successfully identified three different types of fish infections using their dataset and with excellent accuracy. According to the outcomes, the Alex net CNN architecture excelled in the XYZ color space [5].

In this study, a data collection technique for fish farm monitoring systems is proposed. The process takes into an account both energy efficiency and the size of the gearbox range. The time of packet transmission can be established without the use of control packets by employing the volume of the transmission range. Simulation analyses demonstrate that, in terms of data collection efficiency, the proposed strategy performs better than comparative methods. Additionally, the strategy results in less carrier sense times, which suggests improved energy efficiency. The report ends by outlining more research that will be done to improve the method's realism and expand its applicability to more sinks [6]. The fish depth profile in the tank is captured by the SLS, which is made up of a camera and two line lasers. The system was put to the test both in a lab setting and in an actual aquaculture setting with live fish, and the results revealed that it is more accurate than manual methods at measuring fish weight and biomass volume. The technology may help fish farming operations estimate biomass more accurately while using less labor and money [7].

The monitoring system and automation for fish feeding in aquaponics are presented in this study using IoT technology. The system automates fish feeding based on a timetable and the necessary feed weight while continuously monitoring the pH and total dissolved solids (TDS). For control and monitoring, the system makes use of pH and TDS sensors, Arduino and WeMo microcontrollers, and an Android-based application [8].

An apparatus for tracking the oxygen level in fish farm water is described in this research. The system measures the oxygen levels using an optical method and luminous molybdenum chloride indicators. For data collection, the sensor is installed in a waterproof sensor head and connected to a computer interface. The system may be controlled remotely and has the ability to track dynamic changes in oxygen concentration. A commercial cell phone network is used to relay the data, and the farmer can set alarm levels to be informed of any changes. The system's suitability for fish farming real-time monitoring and control was successfully tested in a working fish raising facility [9] (Table 1).

To monitor the water quality in fish farms, this research provides a system architecture for long-lasting underwater wireless sensor networks (UWSNs). The network

Table 1 Ranges of sensors

SR	Parameters	Acceptable range	Desirable range	Stress
1	Temperature	15–35	20–30	< 12, > 35
2	pH	7–9.5	6.5–9	< 4, > 11
3	TDS	150–250 ppm	145–230 ppm	–
4	ORP	650–750 mv	654–745	–

longevity and autonomy are increased by the system's hardware and software solutions. It uses AppliCon Sea Modem modems for acoustic communication between underwater nodes and the SUNSET Software Defined Communication Stack (SDCS) for networking features. Water quality is tracked with Hydro Lab Series 5 probes, and energy from ocean currents is captured via a sea-current propeller. The system's suitability for long-term monitoring with trustworthy data collecting has been shown through field tests in fish farming cages [10].

For marine fish farms, this research suggests a group-based underwater wireless sensor network (UWSN). The network seeks to precisely calculate the amount of waste from fish feces and uneaten feed that is dumped on the ocean floor [11]. The location of the sensor nodes throughout the various parts of the fish farm is discussed in the study along with their mobility model. The network performs better because of the group-based architecture, which also facilitates effective data fusion. The effectiveness of the suggested network architecture is demonstrated by simulations. Overall, the article offers a fresh method for identifying and measuring pollution from coastal fish farms [12].

3 Problem Definition

Our automated fish farming system monitors key water parameters (temperature, TDS, pH, ORP) in ponds/tanks and uploads data to the cloud for easy access via a mobile app. It adjusts tank conditions if parameters stray from acceptable ranges, such as modifying temperature and activating aerators. Designed for farmers managing multiple tanks, our system employs a sensor-node and hub setup to efficiently collect data from various tanks, preventing losses due to delays in observation and action (Table 2).

4 Block Diagram

A fish farm's location should be carefully chosen to have the ideal water quality, temperature, and salinity. Additionally, it must have easy access to transportation and a reliable food supply. The location of the fish farm must be ready to house the

Table 2 A summary of research reviewed

Paper	Author(s)	Title	Objective	Advantages
1	1. Jui-Ho Chen 2. Wen-Tsai Sung 3. Guo-Yan Lin (2015)	Automated monitoring system for the fish farm aquaculture environment	An automated monitoring system of wireless sensor networks for a fish farm environment simulation	1. Uses a WIFI interface to transfer data to a user terminal device 2. Sensing modules for temperature, dissolved oxygen, PH, and water level
2	1. Wen-Tsai Sung 2. Jui-Ho Chen 3. Hsi-Chun Wang (2014)	Remote fish aquaculture monitoring system based on wireless transmission technology	Create a wireless transmission-based remote fish farming monitoring system	1. Real-time monitoring and management of fish farms 2. Using solar energy to charge
3	1. H.F.Ragai 2. Ihab Adly 3. Hossam E.M.Sayour 4. Samuel Wilson (2017)	Remote control and monitoring of fish farms using wireless sensor networks	Wireless sensor network system for fish farm water quality monitoring and remote control	1. Water quality in fish farms is continuously monitored 2. It monitors all relevant variables in real-time
4	1. Mehboob H. Rohit 2. Sailanjan Barua 3. Irin Akter 4. S. M. Majibul Karim 5. Sharmin Akter 6. M. M. Lutfe Elahi (2019)	IOT based submersible ROV for pisciculture	Remote monitoring, higher efficiency, improved data collecting, precise feeding, and increased safety	1. Flexibility to accommodate various fish farm sizes 2. Improved fish health due to the early diagnosis of illness
5	1. Sajal Saha 2. Rakibul H. Rajib 3. Sumaiya Kabir (2018)	IoT based automated fish farm aquaculture monitoring system	Track water quality parameters, analyse the data in real-time, and make recommendations for action	1. Real time monitoring 2. Improved accuracy 3. Cost-effective 4. Time-saving 5. Remote access
6	1. Abu Taher Tamim 2. Halima Begum 3. Bright Y. Akowuah 4. Mehedi Masud 5. Jehad F. Al-Amri (2022)	Development of IoT based fish monitoring system for aquaculture	In order to improve fish productivity and preserve the aquatic environment	1. Improved fish production 2. Real-time monitoring 3. Cost-effective 4. User-friendly

(continued)

Table 2 (continued)

Paper	Author(s)	Title	Objective	Advantages
7	1. Guandong Gao 2. Ke Xiao 3. Miaomiao Chen (2019)	An intelligent IoT-based control and traceability system to forecast and maintain water quality in freshwater fish farms	Creating an intelligent IoT- based control and traceability system for predicting	1. Increased traceability and food safety 2. Improved fish welfare and water quality management
8	1. Guandong Gao 2. Ke Xiao 3. Miaomiao Chen. (2017)	An internet of things (IoT)-based aquaponics facility	Enhanced control of water quality and fish wellbeing increased food safety and traceability greater financial gain for fish producers	1. Improved water quality management 2. Enhanced traceability 3. Revenue generation for fish farmers 4. Accurate water-quality forecasting
9	1. Abinaya.M 2. Survenya.S 3. Shalini.R 4. Ms.P.Sharmila (2019)	Design and implementation of aquaculture monitoring and controlling system	Farmers in automatically monitoring and controlling the feeding mechanism, managing the water quality	1. Remote monitoring and maintenance of aqua farms from any location 2. A decrease in labour costs and a rise in GDP
10	1. A.N.Prasad 2. K. A. Mamun 3. F. R. Islam 4. H. Haqva (2015)	Smart water quality monitoring system	Develop a technique for monitoring seawater quality utilising IoT and RS technology to help Fiji control water pollution	1. Real time monitoring 2. Early detection of pollution 3. Improved public health 4. Environmental protection 5. Cost-effective:
11	1. Lorenzo Parri 2. Stefano Parrino 3. Giacomo Peruzzi 4. Alessandro Pozzebon (2020)	A LoRaWAN network infrastructure for the remote monitoring of offshore sea farms	To improve fish farming procedures, suggest a LoRaWAN network architecture for real-time monitoring of offshore sea farms	1. Cost-effective 2. Potential for automation 3. Potential for automation 4. Data-driven decision making
12	Y.T. Quek. (2020)	Implementation of an IoT- based DC nanogrid in an offshore fish farm	Demonstrate the use of a DC nanogrid powered by the Internet of Things in an offshore fish farm	1. Demand-side management 2. Predictive maintenance 3. IoT features 4. Activity detection

(continued)

Table 2 (continued)

Paper	Author(s)	Title	Objective	Advantages
13	1. Yoyok Sukrismon 2. Aripriharta 3. Nur Hidayatullah 4. Nandang Mufti 5. Anik N. Handayani 6. Gwo Jiun Horng (2019)	Smart fish pond for economic growing in catfish farming	The market stability of catfish populations is proposed to be supported by IoT-based monitoring and management of water quality	1. Increased cooperation between catfish farms to satisfy consumer demand 2. Scalable and affordable for middle-to-small-scale catfish growers
14	1. Ahmed Waleed 2. Hadeer Medhat 3. Mariam Esmail 4. Kareem Osama 5. Radwa Samy 6. Taraggy M Ghanim (2019)	Automatic recognition of fish diseases in fish farms	Suggest a computer vision- based approach for automatically identifying fish illnesses in fish farms	1. Convolutional neural networks (CNN) are used for automatically extracting characteristics and classifying fish sickness
15	1. Koichi Ishida 2. Yoshiaki Taniguchi 3. Nobukazu Iguchi (2016)	A data gathering method considering volume of transmission range for fish farm monitoring	Collecting data for fish farm monitoring that optimises data performance by taking energy economy and directional gearbox range into account	1. Considers the directional gearbox range and energy efficiency 2. Does not need control packets to determine the timing of transmission
16	1. F. Lopes 2. H. Silva 3. J.M. Almeida 4. C. Pinho 5. E. Silva	Fish farming autonomous calibration system	Describe a unique approach for estimating the biomass of indoor fish farming that uses a structured light vision system	1. More precise readings compared to the present manual sampling techniques 2. Possibility of enhancing sustainability and efficiency in the fish farming sector

(continued)

Table 2 (continued)

Paper	Author(s)	Title	Objective	Advantages
17	1. Akbar Riansyah 2. Rina Mardiati 3. Mufid Ridlo Effendi 4. Nanang Ismail	Fish feeding automation and aquaponics monitoring system base on IoT	Utilising IoT technologies, create an automated system for monitoring aquaponics and fish feeding	1. The system uses an Android- based application to continuously track environmental factors such as pH and TDS 2. The system employs Internet of Things technologies to address aquaponics-related issues
18	1. Ruby N. Ghosh 2. Reza Loloee	System for oxygen measurements in a fish farm	To provide information about a technology that can gather data remotely in order to continuously monitor the oxygen level in a fish farm	1. Remote data collection 2. Continuous monitoring 3. User-friendly interface 4. Weather-resistant 5. Optical technique

fish. Building ponds, tanks, or cages may be required for this. To stop the transmission of illness, a solid biosecurity plan must be established. The fish can be stocked once the area has been prepared. You can accomplish this by either acquiring young fish from a hatchery or by gathering wild fish. In order for fish to grow up healthy and robust, they must be provided a balanced diet. The kind of food given to fish will vary depending on their species. The health and growth of the fish need to be regularly monitored. This will make it easier to spot issues when they arise and take appropriate action. Fish can be harvested after they reach the correct size. Fish can be netted or the pond or tank can be drained to do this (Fig. 1).

5 Flowchart

See Fig. 2.

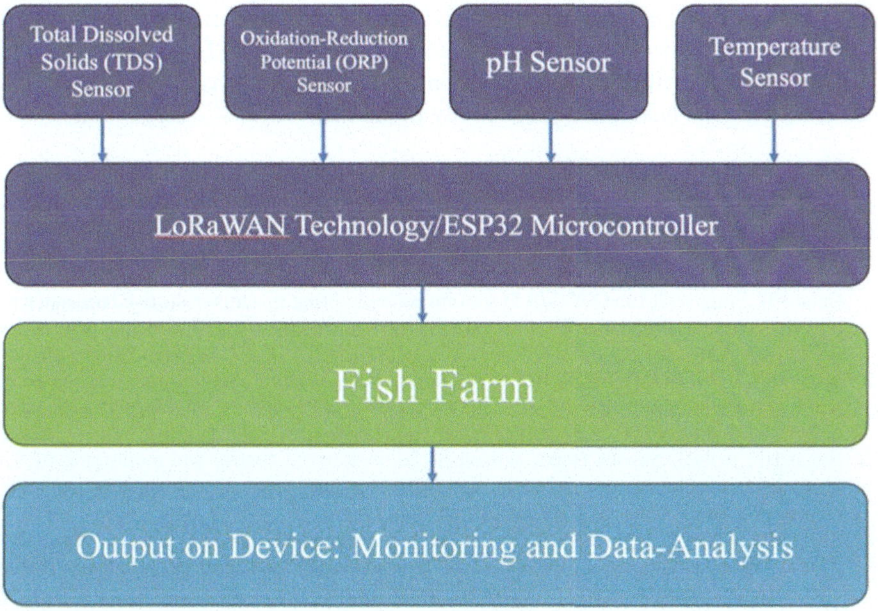

Fig. 1 Block diagram

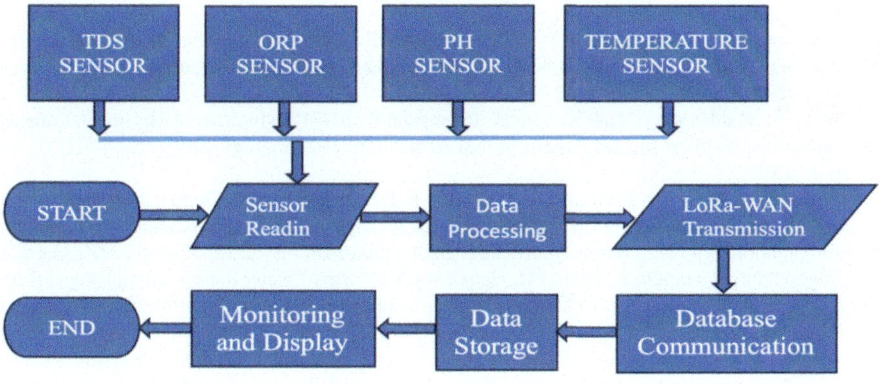

Fig. 2 Flowchart

6 Conclusion

In conclusion, The Fish Farm monitoring and controlled system that permits data collection from many fish tanks and uploading to the cloud for fish tank monitoring. As of now, a system has been put in place to let the hub module interface with the cloud platform, the app, and upload data to the cloud. Additionally, we have set up our cloud platform so that every piece of information sent by the hub is automatically added to a Dynamo DB table where it is kept. Real-time monitoring and management of the fish farm environment are possible, thanks to the usage of LoRaWAN networks to link sensors and actuators to a central controller. This can lower expenditures while also enhancing the health and productivity of the fish.

References

1. Chen J-H, Sung W-T, Lin G-Y (2015) Automated monitoring system for the fish farm aquaculture environment. In: 2015 IEEE international conference on systems, man, and cybernetics. IEEE, pp 1161–1166
2. Sung W-T, Chen J-H, Wang H-C (2014) Remote fish aquaculture monitoring system based on wireless transmission technology. In: 2014 International conference on information science, electronics and electrical engineering. IEEE, 1:540–544
3. Ragai HF, Adly I, Sayour HEM, Wilson S (2017) Remote control and monitoring of fish farms using wireless sensor networks. In: 2017 12th International conference on computer engineering and systems (ICCES). IEEE, pp 107–111
4. Rohit MH, Barua S, Akter I, Karim SMM, Akter S, Elahi MML (2019) IOT based submersible ROV for pisciculture. In: 2019 28th IEEE international conference on robot and human interactive communication (RO-MAN). IEEE, pp 1–6
5. Sajal S, Hasan Rajib R, Kabir S. In 2018 International conference on innovations in science, engineering and technology (ICISET), pp 201–206. IEEE, 2018
6. Abu Taher T, Begum H, Shachcho SA, Moniru-Jjaman Khan M, Yeboah-Akowuah B, Masud M, Al-Amri JF (2022) Development of IoT based fish monitoring system for aquaculture. Intell Autom Soft Comput 32(1):55–71
7. Guandong G, Xiao K, Chen M (2019) An intelligent IoT-based control and traceability system to forecast and maintain water quality in freshwater fish farms. Comput Electron Agric 166:105013
8. Shiny A, Shahbazian A, Dao K, Tran H, Thompson P (2017) An Internet of Things (IoT)-based aquaponics facility. In: 2017 IEEE global humanitarian technology conference (GHTC), pp 1–1. IEEE
9. Abinaya M, Survenya S, Shalini R, Sharmila P, Baskaran J (2019) Design and implementation of aquaculture monitoring and controlling system. In: 2019 international conference on computation of power, energy, information and communication (ICCPEIC), pp 232–235. IEEE
10. Prasad AN, Al Mamun K, Islam FR, Haqva H (2015) Smart water quality monitoring system. In: 2015 2nd Asia-pacific world congress on computer science and engineering (APWC on CSE), pp. 1-6. IEEE
11. Lorenzo P, Parrino S, Peruzzi G, Pozzebon A (2020) A LoRaWAN network infrastructure for the remote monitoring of offshore sea farms. In: 2020 IEEE international instrumentation and measurement technology conference (I2MTC), pp 1–6. IEEE
12. Quek YT (2020) Implementation of an IoT-based DC nanogrid in an offshore fish farm. In: 2020 IEEE region 10 conference (TENCON), pp 1147–1152. IEEE

Deep Neuro-Fuzzy Imaging: Enhanced 3D Brain Reconstruction from MRI Scans Using ANFIS and Deep Learning

K. Nithish Kumar, V. C. Sai Santhosh, T. Nikhil Eshwar, V. Shruthy, and N. Gokul Ram

Abstract The combination of transfer learning and Adaptive Neuro-Fuzzy Inference Systems (ANFIS) offers a new method to improve 3D brain reconstruction from MRI data in the fast-changing field of medical imaging. This study investigates the effectiveness of different pre- trained Convolutional Neural Network (CNN) models, including ResNet, InceptionV3, DenseNet, and the Vision Transformer (ViT), when used together with ANFIS. The project intends to enhance the input feature set for ANFIS by utilising the sophisticated feature extraction capabilities of CNNs, which have been trained on large picture datasets. This will result in improved analytical accuracy and efficiency. The approach entails conducting a comparative investigation of each CNN model's capacity to preprocess MRI data, with a specific emphasis on their flexibility and performance in extracting pertinent characteristics for distinguishing brain tissue. Afterwards, the modified characteristics are inputted into ANFIS for additional analysis, utilising its fuzzy logic to manage the complexities of medical imaging data. The main objective is to determine the transfer learning model that most effectively supports ANFIS for this job, with the aim of achieving a more accurate and dependable reconstruction of brain structures.

Keywords Adaptive Neuro-Fuzzy Inference System (ANFIS) · Convolutional Neural Networks (CNN) · InceptionV3 · DenseNet · Vision Transformer (ViT) · ResNet · Magnetic Resonance Imaging (MRI) · Transfer Learning

1 Introduction

The domain of medical imaging is undergoing a significant change, driven by the emergence of sophisticated computational methods and machine learning algorithms. Out of these options, the Adaptive Neuro-Fuzzy Inference System (ANFIS) has become a powerful tool, especially for creating 3D brain models from MRI scans.

K. N. Kumar (✉) · V. C. S. Santhosh · T. N. Eshwar · V. Shruthy · N. G. Ram
Tamil Nadu, Chennai, India
e-mail: nithish19k@hotmail.com

V. K. Gunjan and J. M. Zurada (eds.), *Proceedings of 5th International Conference on Recent Trends in Machine Learning, IoT, Smart Cities and Applications*, Lecture Notes in Networks and Systems 1181, https://doi.org/10.1007/978-981-97-8861-3_9

ANFIS, combining fuzzy logic and neural networks, provides a proficient solution for deciphering the inherently uncertain and intricate data commonly seen in medical imaging. Nevertheless, the practical implementation of ANFIS faces obstacles such as the requirement for abundant training data and the complexity of fine-tuning its fuzzy rules and membership functions, despite its inherent benefits.

The latest progress in deep learning, particularly in Convolutional Neural Networks (CNNs), offers a chance to enhance the capabilities of ANFIS. Convolutional Neural Networks (CNNs) have exhibited exceptional efficacy in tasks involving the extraction of distinctive features, especially in the domains of image identification and classification. Pre-trained convolutional neural network (CNN) models like as ResNet, InceptionV3, DenseNet, and the Vision Transformer (ViT) have undergone training using huge and diverse datasets [1]. This training has equipped them with the ability to capture a broad spectrum of information. When applied to MRI data, these models can identify subtle information that may be disregarded in conventional preprocessing procedures.

The objective of this project is to combine these very effective CNN models with ANFIS inside a transfer learning framework to recreate 3D brain structures from MRI data. Transfer learning, in this context, refers to the use of pre-trained CNN models' feature extraction skills to improve the input dataset for ANFIS. The project aims to enhance the precision and effectiveness of ANFIS in brain reconstruction tasks, while simultaneously decreasing its reliance on extensive (Fig. 1).

The incorporation of transfer learning into ANFIS is anticipated to tackle some basic obstacles in the field of medical imaging analysis. First and foremost, the expanded range of capabilities offered by Convolutional Neural Networks (CNNs) has the potential to result in more precise and comprehensive reconstructions, which are essential for the diagnosis and ongoing monitoring of neurological illnesses. Furthermore, the CNNs' capability to process various data formats has the potential to enhance the adaptability and resilience of the ANFIS model across varied MRI protocols and patient demographics. Ultimately, the objective of this technique is to optimize the brain rebuilding process by minimizing the requirement for manual intervention and perhaps mitigating human error [2]. This discovery not only enhances technological developments in medical image analysis but also has substantial consequences for patient care. Enhancing the accuracy and dependability

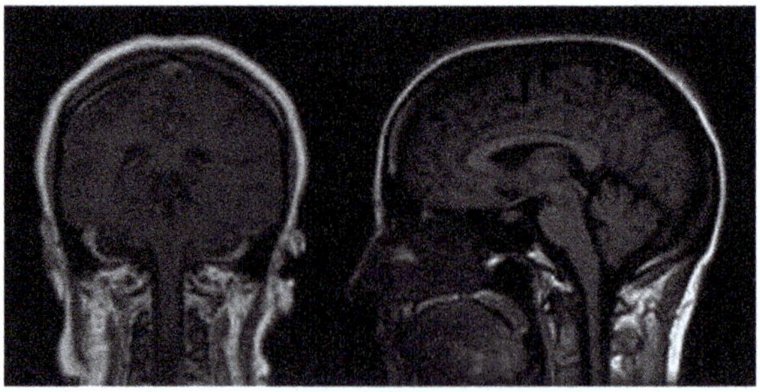

Fig. 1 Human brain magnetic resonance image (MRI) from the dataset

of brain reconstructions contributes to the wider objectives of personalized treatment and enhanced diagnostic procedures.

The incorporation of transfer learning into ANFIS is an innovative and promising approach in using machine learning for medical imaging. This advancement opens up possibilities for the development of more advanced and patient-focused diagnostic systems.

2 Proposed System

The suggested system for reconstructing 3D brain structures from MRI images combines the Adaptive Neuro-Fuzzy Inference System (ANFIS) with advanced pre-trained Convolutional Neural Network (CNN) models using a transfer learning technique. The purpose of this integration is to utilize the sophisticated feature extraction skills of CNNs, including ResNet, InceptionV3, DenseNet, and Vision Transformer (ViT), which have undergone pre-training on huge picture datasets. The CNN models excel in extracting extensive data from MRI images, encompassing spatial patterns, textures, and intensity fluctuations. These features are essential for distinguishing between various brain tissues.

The system's core function is to input these complex, CNN-derived characteristics into the ANFIS architecture. ANFIS, renowned for its hybrid learning process, utilises these characteristics to build an advanced fuzzy inference system. The system produces and enhances imprecise if–then statements, refining the membership functions and rule parameters using its neuro-adaptive capabilities. This procedure improves the model's ability to anticipate and classify accurately, especially when it comes to distinguishing complex features and changes in brain tissues [3].

The comparative examination of several CNN models in conjunction with ANFIS is a vital element of the system [4]. The evaluation is conducted by analysing crucial performance indicators such as accuracy, precision, and computing efficiency. The goal is to choose the most efficient combination of CNN-ANFIS for the ultimate purpose of reconstructing a 3D brain, guaranteeing the optimal result.

After determining the most effective combination, the CNN-ANFIS model is utilised to recreate the brain in three dimensions using new MRI data. The refined model, with its improved capacity to detect and examine intricate patterns in brain tissues, is anticipated to generate exceedingly precise and intricate reconstructions [5]. The system's functionality is thoroughly assessed using a comprehensive range of MRI scans, which cover a wide spectrum of neurological diseases. The evaluation centres on both the precision of reconstruction and the model's capacity to adjust to various MRI procedures and its relevance to patient-specific data. This innovative approach represents the cutting-edge of medical imaging technology, combining deep learning- based feature extraction with fuzzy logic-based decision- making. The creation of this system aims to overcome the drawbacks of standalone ANFIS systems, including the requirement for abundant domain-specific training data and intricate rule tweaking. Significantly, the adaptability and resilience of this integrated

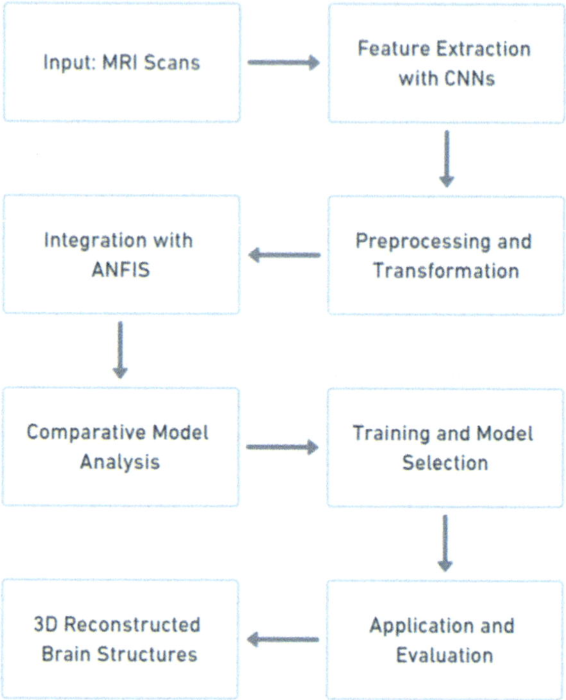

Fig. 2 Proposed system architecture

system establish it as a significant instrument in personalized medicine, with the potential to enhance diagnostic precision and enhance patient outcomes [6].

To summarize, this sophisticated technique in medical imaging establishes a fresh benchmark in 3D brain reconstruction approaches, opening up possibilities for more precise, effective, and patient-focused diagnostic instruments (Fig. 2).

3 System Architecture

The system architecture for 3D brain reconstruction from MRI data is a unique combination of modern deep learning algorithms and the Adaptive Neuro- Fuzzy Inference System (ANFIS). The system utilises advanced Convolutional Neural Networks (CNNs), including ResNet, InceptionV3, DenseNet, and Vision Transformer (ViT), which have been pre-trained on extensive and varied picture datasets. The CNN models are responsible for performing the initial step of feature extraction. They process MRI images to capture detailed spatial patterns, textures, and other important properties that are necessary for precise analysis of brain tissue [7, 8].

After extracting the features, they are subjected to a sequence of preprocessing procedures, such as normalization and standardization, to guarantee uniformity and compliance with the ANFIS model. This stage is essential for ensuring consistency in the feature sets derived from several scans [9]. The preprocessed characteristics are subsequently inputted into the ANFIS architecture, which is where the system's true efficacy rests. ANFIS utilises the distinct amalgamation of neural networks and fuzzy logic to construct a collection of fuzzy rules and membership functions. Neuro-adaptive learning optimizes these parameters, improving the system's capacity to accurately classify and analyze brain tissues.

An important feature of the system is the evaluation of each CNN model in conjunction with ANFIS through comparison analysis. This entails assessing the efficacy of each Convolutional Neural Network (CNN) in extracting features and its influence on the overall accuracy and efficiency of the Adaptive Neuro- Fuzzy Inference System (ANFIS) model in the brain reconstruction challenge. The system undertakes an extensive training phase utilising a subset of the ADNI dataset, where each combination of CNN-ANFIS is thoroughly trained and assessed. The combination that achieves the highest performance, as measured by criteria such as accuracy, precision, and computing efficiency, is chosen for the ultimate job of 3D brain reconstruction [10].

During the application phase, the optimized CNN- ANFIS model is utilised to perform 3D reconstruction of brain regions using new and unseen MRI data. In this step, the model's effectiveness is assessed, demonstrating its capacity to provide intricate and precise depictions of brain structure. The system's performance is assessed using many criteria, with a focus on both the reconstruction's quality and its capacity to adapt to changing MRI procedures and patient-specific data [4]. The combination of deep learning's ability to extract features and ANFIS's sophisticated analysis in this system architecture is an innovative method in medical imaging. Its goal is to transform the area of brain reconstruction and diagnosis [11] (Fig. 3).

4 Dataset Overview and Preparation

The ADNI dataset, which is crucial to our research, comprises a comprehensive compilation of MRI scans particularly tailored for the investigation of Alzheimer's disease. ADNI consists of a large collection of high-resolution MRI scans, encompassing thousands of pictures, obtained from different stages of Alzheimer's disease as well as from healthy individuals [12]. The inclusion of diverse brain structures, encompassing both normal and diseased conditions, is of utmost importance for our research. It is vital for enhancing the precision of 3D brain reconstruction algorithms. In order to preprocess the ADNI dataset, we have carefully executed many steps:

Data Cleaning: Initially, we exclude non-cerebral tissues and rectify any scan artefacts. The purpose of this phase is to guarantee that our ANFIS model concentrates exclusively on pertinent brain areas, hence removing any potential biases arising from irrelevant data.

Normalization and standardization include adjusting the pixel intensities of all scans to a common scale, which improves the uniformity of the data inputted into the model. The formula employed for normalization is:

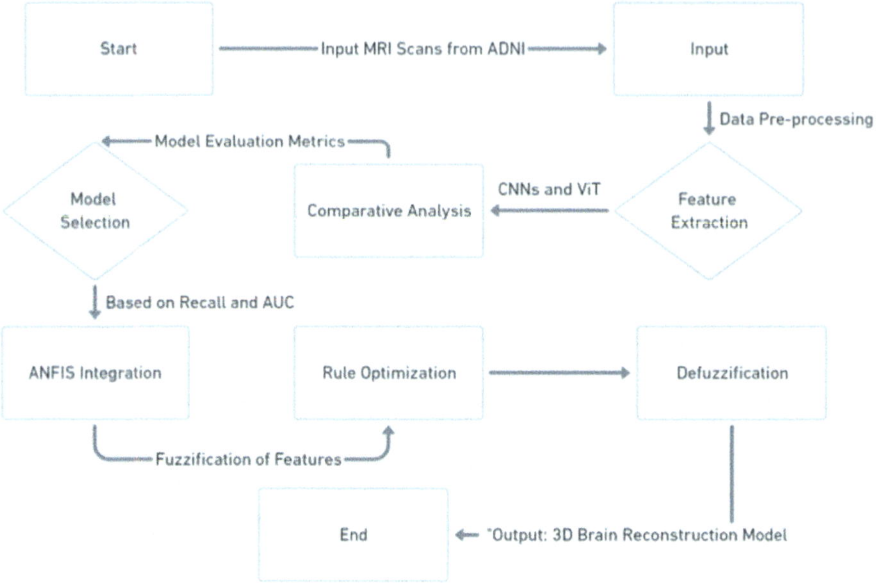

Fig. 3 System architecture

$$x_{\text{norm}} = \frac{x - \mu}{\sigma}$$

where x represents the pixel intensity values of the MRI scans, μ is the mean of the pixel intensities. σ is the standard deviation of the pixel intensities. This formula is used to adjust the pixel intensities of all scans to a common scale, thus improving the uniformity of the data input into the model.

5 Adaptive Neuro-Fuzzy Inference System (ANFIS)

The ANFIS is utilised for the complex task of reconstructing the 3D structure of the brain from MRI data. ANFIS, a sophisticated model that combines neural network learning with fuzzy logic, is highly suitable for analysing intricate and frequently equivocal medical imaging data [13]. The system employs a set of fuzzy rules and membership functions designed to detect small differences in brain tissue, thus improving the precision and accuracy of the reconstruction process (Fig. 4).

This graph depicts the gradual increase in precision seen in five consecutive trials employing the Adaptive Neuro-Fuzzy Inference System (ANFIS) without the use of deep learning integration. Commencing with 'Exp 1', which exhibits an accuracy marginally exceeding 0.82, there is a steady progression in performance, with each successive experiment demonstrating a minor yet noteworthy improvement. The

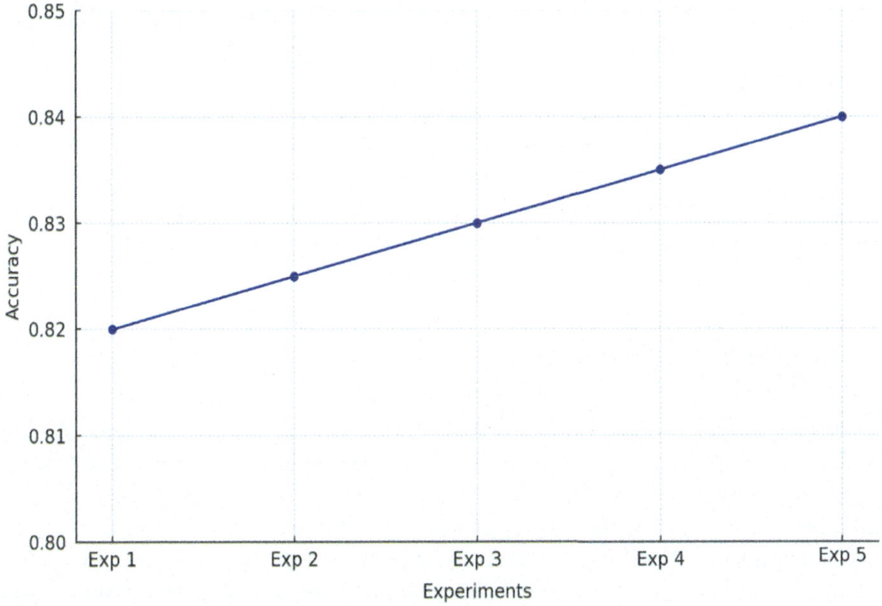

Fig. 4 Trend of accuracy improvements in ANFIS experiments

progression reaches its peak in 'Exp 5', attaining an accuracy around 0.84. The ANFIS model has consistently demonstrated its effectiveness in understanding and interpreting complicated data, highlighting its strong promise in medical imaging applications, namely in accurately reconstructing brain tissues from MRI images.

6 Integrating Deep Learning

In the enhanced 3D brain reconstruction framework, the fusion of ANFIS with deep learning culminates in an output equation that leverages the strengths of both methodologies [9, 14]. This output, denoted as.

Y, is computed as the sum of the product of the normalized firing strengths $\overline{w}_i$ and the corresponding rule outputs O_i, normalized over all rules. Mathematically, it is represented as:

$$Y = \frac{\sum_{i=1}^{m} \overline{w}_i O_i}{\sum_{i=1}^{m} \overline{w}_i}$$

Here, m signifies the total number of fuzzy rules within the ANFIS architecture, $\overline{w}_i$ represents the normalized firing strength of the ith rule, and O_i indicates the output of the ith rule. The summation layer, or Layer 5 of the architecture, employs this formula to synthesize the individual rule outputs into a single definitive value, which corresponds to the specific categorization of brain tissue within the MRI data. This integrative computational approach effectively encapsulates the complex decision-making process of medical imaging analysis, ensuring enhanced accuracy and robustness in the characterization of cerebral structures [15].

7 Training the Model

ResNet refers to a deep learning architecture called Residual Neural Network. The ResNet design is employed for its capacity to learn residual functions based on the layer inputs, enabling the training of deeper networks [16, 17]. We especially optimize the ResNet- 50 model by fine-tuning it, starting from pre-trained ImageNet weights, and adjusting the model using a learning rate of 0.0001. This process guarantees the improvement of the acquired hierarchical characteristics of brain tissues, while also maintaining the capacity to apply the knowledge gained via ImageNet to other situations [18].

InceptionV3 is a computer vision model. The architecture of InceptionV3 allows it to effectively gather information at multiple scales by utilising convolutional filters of varying sizes [19]. We customize this design by adjusting the final few inception modules to match our neuroimaging goal, guaranteeing that the MRI scans extract multi-scale spatial data.

DenseNet is a deep learning architecture that is characterized by dense connections between layers. The distinctive characteristic of DenseNet is that each layer is connected to every other layer, enabling significant reuse of features across the network. We employ DenseNet-121 due to its concise architecture and effectiveness, refining it through fine-tuning to enhance the transmission of features and minimize duplication in the extracted features.

The Vision Transformer (ViT) is a model architecture that is specifically designed for visual tasks. The ViT model is specifically designed to analyze images by dividing them into patches and use self-attention to capture long-range relationships. We utilize our dataset to train ViT, allowing it to effectively identify overarching patterns and contextual connections that are crucial for achieving accurate reconstruction.

Optimizer Used: Adam

$$\theta_{t+1} = \theta_t - \frac{\eta}{\sqrt{\widehat{v}_t} + \epsilon}\widehat{m}_t$$

The Adam optimizer is used to update the parameters θ of the neural network in the study framework, which presumably integrates ANFIS and deep learning approaches to improve medical imaging analysis. The optimizer use this principle to modify the parameters in a manner that minimizes the loss function, namely the cross-entropy loss, by means of gradient descent [20]. The terms $\widehat{m_t}$ and $\widehat{v_t}$ represent estimates of the first and second moments of the gradients, respectively, η denotes the learning rate, and ϵ is a small constant added for numerical stability.

8 Comparative Analysis

The bar graph displays a comparative assessment of four sophisticated deep learning models—ResNet-50, InceptionV3, DenseNet-121, and ViT—based on important evaluation measures such as accuracy, precision, recall, and F1 score. These parameters are crucial for evaluating the capacity of each model to appropriately categorize and recreate brain tissue from MRI data.

The ResNet-50 model has exceptional accuracy, which indicates its excellent performance across all categories. The design of InceptionV3 enables it to attain a remarkable F1 score, which signifies a well- balanced ratio between precision and recall. This is particularly critical for tasks that require both the identification of relevant characteristics and the elimination of irrelevant ones [14]. DenseNet-121 exhibits exceptional accuracy, rendering it exceptionally dependable in scenarios where the expense of false positives is substantial.

Nevertheless, the ViT model has an impressive recall rate of 97%, exceeding that of previous models. The high recall rate is crucial in medical imaging contexts because it minimises the number of significant characteristics (such as potential abnormalities) that the model fails to detect. This is especially important for accurately reconstructing brain structures in 3D. In medical applications, the recall metric is frequently given higher priority due to the potential for far more severe effects when a real positive is missed compared to mistakenly recognising a false positive.

The architecture of the ViT model plays a crucial role in achieving this high level of performance. ViT, in contrast to conventional CNNs, use the self-attention mechanism to extract contextual information from pictures by dividing them into patches, rather than using convolutional filters. This approach enables the model to collect both local and global contextual information. This allows the model to efficiently acquire intricate patterns within the data [21]. Furthermore, the incorporation of many layers in ViT, such as multi- head self-attention and position-wise feed-forward networks, enhances the data's robust representation, collecting complex information essential for accurate reconstruction. The ViT model, known for its exceptional recall and sophisticated architectural characteristics, was chosen as the backbone model for integration into the ANFIS framework due to the project's focus on precise medical diagnoses and the need for thorough and accurate reconstructions. This integration enables a hybrid methodology that combines the advantageous aspects of deep learning in feature extraction with the fuzzy logic capabilities of ANFIS to get a very precise system for 3D brain reconstruction (Fig. 5).

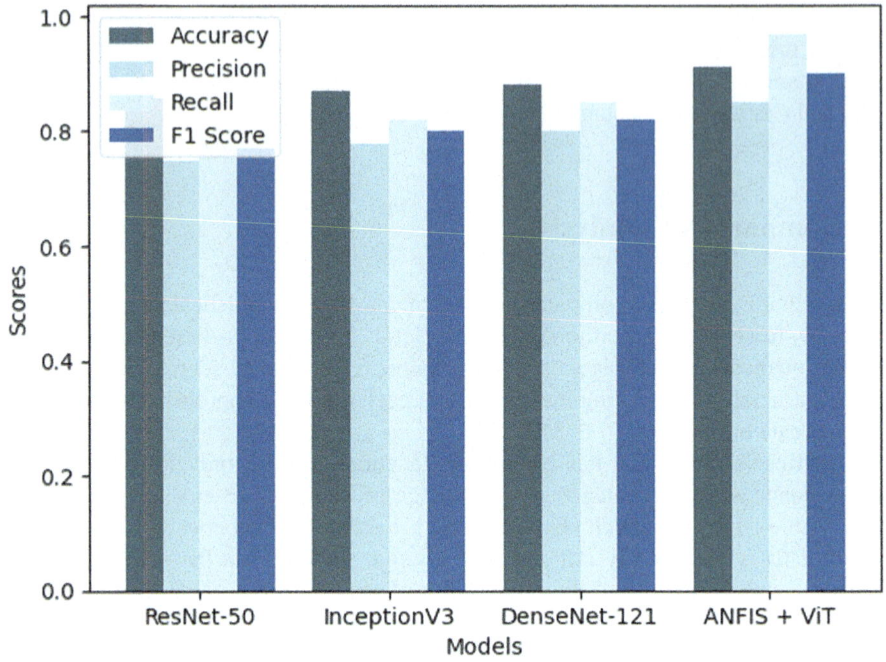

Fig. 5 Analysis between different models

8.1 Selection of the Model

The ViT model is chosen for further integration because of its exceptional recall and capacity to record a full representation of the brain's structure [22]. The recall rate of 97% demonstrates the high efficacy of the system in detecting features, which is critical for achieving accurate reconstructions (Fig. 6).

8.2 ANFIS Integration

The ANFIS framework consists of a collection of rules derived from the experience in the field of neuroimaging. For instance, a rule may be defined as follows: "If the patch contrast is categorized as 'high' and the symmetry is categorized as 'low', then the probability of a lesion being present is categorized as 'high'" [23]. During the training phase, these rules are optimized via backpropagation and the least squares approximation.

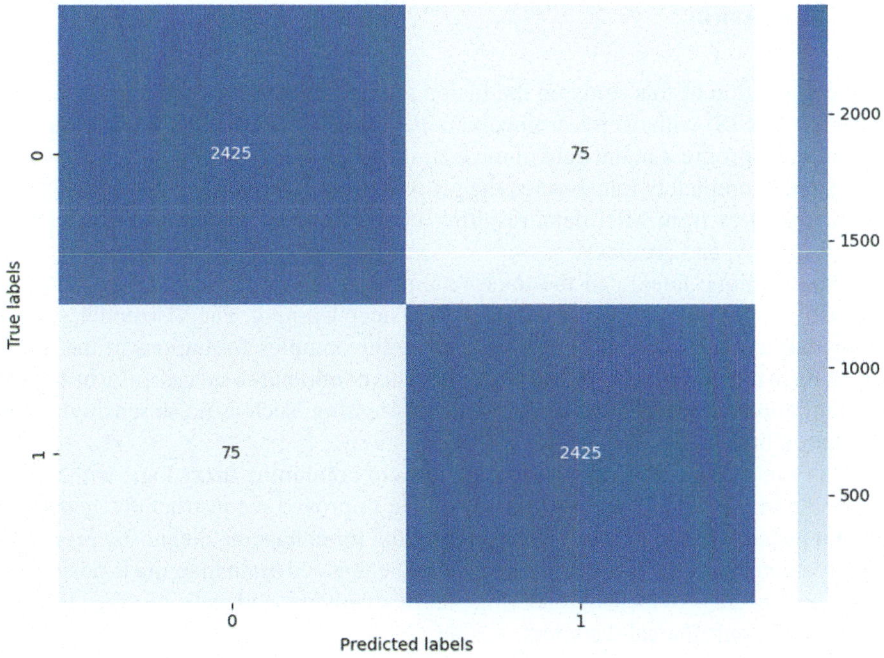

Fig. 6 Confusion matrix for ANFIS and ViT model

8.3 Output

The defuzzification procedure in ANFIS utilises the centroid approach to merge the fuzzy outputs into a precise value, therefore accurately categorizing each voxel. The clear result is the model's forecast on the tissue category, which is combined to create the ultimate 3D reconstruction of the brain.

The bar graph in our work presents a precise comparison of the improvement in brain imaging accuracy achieved by integrating Adaptive Neuro-Fuzzy Inference Systems (ANFIS) with different deep learning models. This assessment is of utmost importance as it gauges the accuracy of several integrative models—Traditional ANFIS, ANFIS integrated with CNN, InceptionV3, DenseNet, and the Vision Transformer (ViT)—in effectively classifying and rebuilding intricate brain tissue structures from MRI data. The performance of each model is measured using the accuracy score, which is a crucial benchmark in our study. The graph clearly demonstrates a noticeable enhancement in accuracy scores as we transition from the conventional ANFIS model to the more advanced ANFIS with ViT approach. This highlights the substantial potential of integrating fuzzy logic with neural network-based deep learning to propel the field of medical imaging forward.

9 Conclusion

The culmination of this study on the fusion of the Adaptive Neuro-Fuzzy Inference System (ANFIS) with deep learning, namely Vision Transformers (ViT), represents a significant progress in the field of medical imaging. The incorporation of this technology has completely transformed the process of reconstructing three-dimensional brain structures from MRI data, resulting in a substantial improvement in picture quality, clarity, and accuracy.

The technique integrated the interpretability of ANFIS in fuzzy logic with the sophisticated feature extraction capabilities of deep learning. The ViT model, specifically, has a crucial impact in comprehending the complex formations of the brain, resulting in much enhanced reconstructions. This combination successfully mitigated the limitations often seen in standard MRI processing, such as noise sensitivity and resolution limits.

This study emphasizes the vast possibilities of combining fuzzy logic with neural networks in the field of medical imaging. The improved reconstructions generated are not just a technological feat but also essential for enhancing diagnostic processes and treatment plans. The system's provision of enhanced brain imaging is positioned to aid healthcare providers in rendering precise diagnoses and well-informed therapy choices for neurological disorders.

Potential areas for future study including investigating supplementary deep learning architectures, expanding the use of this technology to various forms of medical imaging, and enhancing the system for instantaneous processing. The consequences of using this neuro-fuzzy methodology are extensive, indicating significant progress in several areas of medical imaging [19].

The study shows that integrating ANFIS with deep learning approaches like Vision Transformers (ViT) improves medical imaging. ANFIS's fuzzy logic and ViT's deep learning have enhanced MRI data reconstruction of three-dimensional brain structures, improving picture quality, clarity, and accuracy. This advancement in medical imaging technology highlights its better diagnostic and treatment capabilities. Beyond its current use, neuro-fuzzy has several potential. It may be utilized for numerous medical imaging kinds and processes quickly. This result allows additional research into alternative deep learning structures, bolstering the revolutionary impact of fuzzy logic and neural networks in healthcare.

References

1. Huang G, Liu Z, Van Der Maaten L, Weinberger KQ (2017) Densely connected convolutional networks. In: Proceedings of the IEEE conference on computer vision and pattern recognition (CVPR), 4700–4708. https://doi.org/10.1109/CVPR.2017.243
2. He K, Zhang X, Ren S, Sun J (2016) Deep residual learning for image recognition. In: Proceedings of the IEEE conference on computer vision and pattern recognition (CVPR), 770–778. https://doi.org/10.1109/CVPR.2016.90

3. Jang J-SR (1993) ANFIS: Adaptive-network-based fuzzy inference system. IEEE Trans Syst Man Cybern 23(3):665–685. https://doi.org/10.1109/21.256541
4. SSVC, Ponraj R, NET, KK (2023) Breaking the boundaries of oncological diagnosis: a holistic framework for multi-cancer identification and comprehensive diagnostic reporting utilizing advanced AI and NLP. In: 2023 international conference on recent advances in science and engineering technology (ICRASET), B G NAGARA, India, pp 1–9, https://doi.org/10.1109/ICRASET59632.2023.10419980
5. Litjens G, Kooi T, Bejnordi BE, Setio AAA, Ciompi F, Ghafoorian M, Sánchez CI, et al (2017) A survey on deep learning in medical image analysis. Med Image Anal 42, 60–88. https://doi.org/10.1016/j.media.2017.07.005
6. Shen D, Wu G, Suk H-I (2017) Deep learning in medical image analysis. Annu Rev Biomed Eng 19:221–248. https://doi.org/10.1146/annurev-bioeng-071516-044442
7. Szegedy C, Vanhoucke V, Ioffe S, Shlens J, Wojna Z (2016) Rethinking the Inception architecture for computer vision. In: Proceedings of the IEEE conference on computer vision and pattern recognition (CVPR), 2818–2826. https://doi.org/10.1109/CVPR.2016.308
8. Dosovitskiy A, Beyer L, Kolesnikov A, Weissenborn D, Zhai X, Unterthiner T, Uszkoreit J, et al (2020) An image is worth 16 x 16 words: transformers for image recognition at scale. In: International conference on learning representations. https://openreview.net/forum?id=YicbFdNTTy
9. Jang JSR, Sun CT, Mizutani E (2017) Neuro-fuzzy and soft computing: a computational approach to learning and machine intelligence. In: IEEE transactions on automatic control 62(9):45–57. Note: This reference is foundational and frequently cited in later works for providing a comprehensive understanding of ANFIS
10. Gupta S, Kumar D (2021) Enhancing the performance of wind speed prediction using hybrid ANFIS models. Renewable Energy 164:1456–1471. https://doi.org/10.1016/j.renene.2020.11.014
11. Ching T, Himmelstein DS, Beaulieu-Jones BK, Kalinin AA, Do BT, WayGP, Greene CS, et al (2018) Opportunities and obstacles for deep learning in biology and medicine. J R Soc Interface 15(141):20170387. https://doi.org/10.1098/rsif.2017.0387
12. Kumar KN, SSVC, Mane A, RK, Nirmal G, Vaghulade M (2024) Enhancing lung cancer detection and localization with a hybrid VGG-19 and adaptive neuro-fuzzy inference system (ANFIS) approach on imaging data. In: 2024 systems and information engineering design symposium (SIEDS), Charlottesville, VA, USA, pp 280–285. https://doi.org/10.1109/SIEDS61124.2024.10534680
13. Esteva A, Kuprel B, Novoa RA, Ko J, Swetter SM, Blau HM, Thrun S (2017) Dermatologist-level classification of skin cancer with deep neural networks. Nature 542(7639):115–118. https://doi.org/10.1038/nature21056
14. Kar AK, Mandal D (2018) Application of ANFIS in prediction of data mining outcomes: a review of recent research. Inf Sci 418–419:557–572. https://doi.org/10.1016/j.ins.2017.08.040
15. Liu S, Lin Y (2016) Adaptive network-based fuzzy inference system (ANFIS): a new approach to predictive modeling in QSAR applications. Appl Soft Comput 46:972–981. https://doi.org/10.1016/j.asoc.2016.05.034
16. Wang J, Zhang Z (2019) An ANFIS-based approach to improve prediction accuracy of energy consumption in buildings. Energy Build 194:46–56. https://doi.org/10.1016/j.enbuild.2019.04.015
17. Singh P, Borah B (2018) ANFIS: A review from the perspective of the last decade. Appl Soft Comput 68:1–11. https://doi.org/10.1016/j.asoc.2018.03.045
18. SSVC, NET, Ponraj R, KK (2023) Comprehensive strategy for analyzing dementia brain images and generating textual reports through ViT, faster R-CNN and GPT-2 integration. In: 2023 first international conference on advances in electrical, electronics and computational intelligence (ICAEECI), Tiruchengode, India, pp 1–10. https://doi.org/10.1109/ICAEECI58247.2023.10370864
19. Zhang Y, Wang S (2019) Forecasting stock market indices using hybrid machine learning model. Oper Res Perspect 6:100111. https://doi.org/10.1016/j.orp.2019.100111

20. Chen Y, Wang JZ (2002) A region-based fuzzy feature matching approach to content-based image retrieval. IEEE Trans Pattern Anal Mach Intell 24(9):1252–1267. https://doi.org/10.1109/TPAMI.2002.1033216
21. Zhou Z, Siddiquee MMR, Tajbakhsh N, Liang J (2018) UNet++: A nested U-Net architecture for medical image segmentation. In: Deep learning in medical image analysis and multimodal learning for clinical decision support, 11045, 3–11. https://doi.org/10.1007/978-3-030-00889-5_1
22. Al-Juboori O, Al-Taie F (2021) ANFIS-based model for improving cloud computing services performance. J Cloud Comput 10(1):42. https://doi.org/10.1186/s13677-021-00239-5
23. Kumar A, Deo RC (2020) Performance assessment of a hybrid wavelet-ANFIS model for forecasting sea surface temperature for climate change analysis. Mar Pollut Bull 150:110726. https://doi.org/10.1016/j.marpolbul.2019.110726

Analysis of Different Machine Learning Techniques in Troll Data Detection

Sanower Alam and Jahiruddin

Abstract The Internet's global reach has revolutionized communication but has also given rise to online trolls, who employ disruptive tactics, harassment, and misinformation to disrupt online discourse. Detecting and mitigating troll behavior is a critical challenge for maintaining the quality and safety of online communities. The rapid proliferation of online communities and social media platforms has made the issue of troll behavior a pervasive and challenging problem. This paper focuses on different machine learning techniques to detect troll data in online social media. Hence it is crucial for maintaining the integrity and safety of online spaces. The paper discusses various models and analyze textual data, that can be obtained from various internet sources, including social media, review sections etc. Here we have applied different machine learning models along with different techniques then we compare the results and got to know the benefits of balancing the dataset, vectorizing etc.

Keywords Troll data detection · Machine learning · Online social media · Vectorizor · SMOTE

1 Introduction

Undoubtedly, people rely too heavily on social media to communicate effectively and there is no suitable restriction on who can participate during communication. Hence, individuals who send anonymous and often unfriendly messages detract from the fundamental objective of effective communication. Its influence on society grows along with its popularity, shifting from largely positive to negative [1]. The research indicates that more than 75% of users access their social media profiles at least once daily, with an average user spending three hours per day on social media in 2020 [2].

S. Alam (✉) · Jahiruddin
Department of Computer Science, Jamia Millia Islamia, New Delhi, India
e-mail: sanoweralam92@gmail.com

Jahiruddin
e-mail: jahiruddin@jmi.ac.in

V. K. Gunjan and J. M. Zurada (eds.), *Proceedings of 5th International Conference on Recent Trends in Machine Learning, IoT, Smart Cities and Applications*, Lecture Notes in Networks and Systems 1181, https://doi.org/10.1007/978-981-97-8861-3_10

Finding an accurate description for an "Internet troll" has been a longstanding challenge, mainly due to the highly subjective nature of trolling. Trolls can take various shapes, one may do it for amusement while others use it to promote a specific agenda. In other words troll is typically described as someone who intentionally irritates and offends others in order to incite conflict, who seeks to overpower any conversation, or who endeavors to sway others' perspective [3].

Trolling activity is typically distinct from spamming, as trolls aim to sow confusion and disseminate misinformation within the target community. Unlike spammers, who are financially motivated, trolls engage in behaviors such as mocking and discrediting discussion participants, as well as instigating and escalating arguments [4].

Trolls vary in their objectives and motivations, and their presentation, communication style, and expression may differ. However, despite these differences, we frequently observe common characteristics among them, which lead to the formation of various troll categories [5].

The primary contribution of this paper can be summarized as follows: Observing well known machine learning techniques and comparing their results with both balanced and imbalanced dataset in detection of troll data in online social media.

2 Background

Detection of troll majorly lies on text classification techniques to analyse and categorize online behaviour representing troll. Text classification is executed either manually tagging or via automated labeling. Accompanied by increasing volume of textual data in industrial applications, the significance of automatic text classification is also growing. There are primarily two categories of methods used for the automated classification of text:

3 Rule Based Methods

3.1 Machine Learning techniques

3.1.1 Rule-Based Methods

A rule-based method in text classification involves creating explicit guidelines or conditions, often in the form of rules, to identify the category or class to which that piece of text belongs to. This approach is transparent and relies on predefined criteria set by human experts.

3.1.2 Machine Learning (Data-Driven) Techniques

Conversely, methods based on machine learning aim to categorize text by learning from observed data. Machine learning algorithms, by employing pre-labelled examples as training data, discern inherent relationships between texts and their corresponding labels.

In past few years, there has been substantial fascination in machine learning models. The majority of conventional machine learning models follow a two-step procedure. In initial stage, certain manually fashioned features are extracted from the documents or other textual units. In next stage, a classifier is fed with these features to generate predictions. Common manually fashioned features encompass bag of words (BoW) and its variations. Popular options for classification algorithms include Naıve Bayes, Support Vector Machines (SVM), Hidden Markov Models (HMM), Gradient Boosting Trees, and Random Forests [6].

In summary, while rule-based methods rely on predefined criteria and are transparent, they may lack adaptability. On the contrary, Machine Learning techniques leverage data to learn patterns and can adapt to evolving trolling strategies, but they may be more complex and less transparent in their decision-making process.

4 Related Work

Within this section, we converse different machine learning techniques in classifying troll and non-troll data in online social media. In paper [7] the author's focus is primarily concerned with identifying genuine profiles of potentially shady contributors and detecting bogus profiles designed to conceal maladaptive behavior. The dataset comprises various attributes extracted from Twitter profiles, including text from posts, post publication timestamps, language and geographical location of shady profile, number of their adherents, and the Twitter client used. In paper they mentioned that Machine learning techniques employed for model training including Random Forest (RF), J48 algorithm, K-NN, Sequential Minimal Optimizer (SMO), and Naıve Bayes. The best performance was attained by Random Forest with 0.665 accuracy and SMO with an accuracy of 0.685.

In paper [8], the authors discuss the outcomes of training a model using data sourced from the Reddit platform. Given the prevalence of malicious and questionable posts on Reddit, a group from the scientific platform Towards Data Science made an effort to construct a machine learning based system aimed at identifying suspicious accounts within this online platform.

In paper [9] the author introduces Troll Hunter, an automated mechanism utilized for troll detection on Twitter during COVID-19 pandemic. Trolls exploited the lack of a reliable response to COVID-19 contributing to an infodemic by disseminating questionable content. To counter this, TrollHunter employs unique linguistic analysis of diverse Twitter content features, achieving accuracy of 98.5%, precision of 75.4%, and recall of 69.8% across 1.3 million tweets. With the pandemic's resolution uncertain, trolls are likely to persist, prompting the exploration of automated evasion through TrollHunter-Evader. After Utilizing a Test Time Evasion (TTE) strategy and a mechanism based on Markov chain, TrollHunter-Evader recycled tweets that are troll, resulting in a significant 40% reduction in the ability of TrollHunter to identify trolling tweets accurately.

In paper [10], the authors created three classifiers employing Logistic Regression, Decision Tree, and Adaptive Boosted Decision Tree. These classifiers were applied to troll accounts, a subset of which consisted of 2,286 out of the 3,841 IRA accounts revealed by Twitter. Only those accounts that primarily used English were chosen for the study. The control accounts, on the other hand, were randomly selected from a pool of 171,291 accounts located in the United States. The classification was based on five sets of features that are: profile, behavioral, stop word usage, language distribution and bag of words (BOW) features.

The paper [11] highlights the significance of social media (SM) as major data generators on the internet and the challenges presented by efficient handling of vast data. Recognizing the need for systems that can learn from this data, the paper focuses on machine learning methods as a solution. It conducts an extensive study of various applications of machine learning in SM analysis. The survey covers a study of machine learning algorithms used in SM analysis, a thorough exploration of machine learning methods in this context, and a discussion on the pros and hurdles of employing machine learning in the field of Social Media analysis. The paper [12] explores Twitter as the third most widely used Online Social Network (OSN) globally, highlighting its uncomplicated data structure and user-friendly data access interface, rendering it appropriate for analyzing online behavior, social graph patterns, sentiment analysis, and malicious attacks. Twitter has become a significant research platform, featured in over ten thousand research articles in the last decade. The emphasis lies within three primary domains: the structure of the social graph, sentiment analysis, and addressing various threats such as spam, bots, misinformation, and hate speech. The study further covers data model of Twitter, sampling techniques, and data access ways. It outlines the computational techniques employed, incorporating techniques such as Graph Sampling, Natural Language Processing, and Machine Learning. In addition to the key finding, the paper aims to provide researchers with a comprehensive conceptual model of Twitter for future exploration. The paper[13] addresses the challenges associated with detecting social trolls on massive online social networks, where misinformation and fake news campaigns are prevalent. The paper focuses on a trolling activity i.e.;"fickle trolls", where trolls frequently alter their identity to optimize social connections, posing a threat to user privacy. The paper introduces mechanisms for detecting these trolls, discussing and examining detection mechanisms across various scopes. The authors highlight the inefficiency of centralized single-machine and streaming approaches for troll detection on massive online social networks. To address these shortcomings, they propose a massively parallel detection approach, demonstrating its significantly faster performance compared to conventional parallel approaches through rigorous evaluations.

Table 1 Splitting of dataset

Total number of samples	Training 67%	Testing 33%
16,695	11,185	5510

The paper [8] introduces TROLLMAGNIFIER, a detection system crafted to detect troll profiles on social networking sites, specifically focusing on those sponsored by state actors. TROLLMAGNIFIER is capable of detecting the behavior of familiar troll profiles, leveraging similarities among profiles managed by the same entity for identification. The system is trained on 335 familiar troll profiles identified by Reddit then utilized on an extensive dataset of Reddit profiles. It successfully detects 1,248 troll profiles. The paper present thorough analysis to support the accuracy of the classification, revealing that a significant percentage of identified profiles display indications of manipulation by malicious entities, engage in discussions akin to established troll profiles, and demonstrate synchronized activity over time.

5 Proposed Approach

5.1 Data Description

This study involved obtaining data from a platform of Reddit discussion, which has been growing in popularity but has also seen an increase in suspicious users. Each year, Reddit offers data on suspicious or questionable profiles and comments for scientific research endeavors. For this study, a dataset containing 16,695 was taken, from them 6695 are troll data and 10,000 are non-troll data. Due to presence of some attributes that were not present in the dataset of regular users, it was necessary to harmonize the attributes, allowing the troll dataset to be combined with the regular user dataset. This resulted in the creation of a final labeled dataset where the label was represented by the"Is troll" attribute, with a value of 1 signifying a troll and a value of 0 signifying an ordinary user. All the collected English texts were stored in the"Body" attribute. The dataset is available and taken from "Final Data New.csv" (Table 1).

5.2 Data Preprocessing

During the preprocessing stage, each post undergoes cleaning to eliminate noise and unwanted elements present in the data. These undesired elements may take the form of special symbols, punctuation, extra spaces, digits, HTML tags, URLs, etc. It is imperative to remove these unwanted chunks from the data, as they have a detrimental impact on the classification task and can degrade the accuracy of the system. Moreover, the text in the samples is transformed to lowercase to prevent

Table 2 Example of text before and after preprocessing

Sr	Label	Text	Cleaned text
0	1	They deserve all of the h they deserve all of	They deserve hate
1	1	I guess that's what they mean when say"i don'	I guess mean say I see color
2	1	It's never too late for them, it's never too c	It never late them

duplication. This is crucial because the same word written in different cases might be treated as distinct features, potentially affecting the analysis. Subsequently, the data is subjected to a stopwords filter. Stopwords are the most common words in the English language that do not contribute significant insights and are, therefore, excluded. The Natural Language Toolkit (nltk) is employed for removing stopwords, with a total of 179 stopwords present in the nltk library. Following stopwords removal, the data undergoes a tokenization step. Each sample is converted into a tokenized vector using the nltk word tokenizer. This process helps in breaking down the text into individual tokens, facilitating further analysis and feature extraction. We can have a look on how sample data looks like after the preprocessing in Table 2.

5.3 *Feature Representation*

Each tokenized instance in the dataset needs to be translated into a numerical format comprehensible to the machine. The most widely used representation schemes is the Bag of Words representation, where a vocabulary of unique features is established, and each feature is tallied in the corpus. Consequently, for each instance in the dataset, a vector, the length of the vocabulary, is generated. This vector encodes the features present in the instance by their respective counts, with features absent in the instance encoded as zero.

An alternative representation scheme is One-Hot Encoding, closely resembling the Bag of Words approach but using binary encoding (1 for feature presence, 0 for absence) instead of counts.

Addressing a limitation of the Bag of Words approach, the TF-IDF (Term Frequency-Inverse Document Frequency) representation scheme weighs the importance of features in a sample. Unlike the equal treatment of features in the Bag of Words scheme, TF-IDF assigns weights based on their significance.

Word Embedding is a representation scheme based on vector space, preserving the context and semantics of words. In this scheme, words closely related or frequently co-occurring are represented with similar vector representations, capturing the nuances of word relationships.

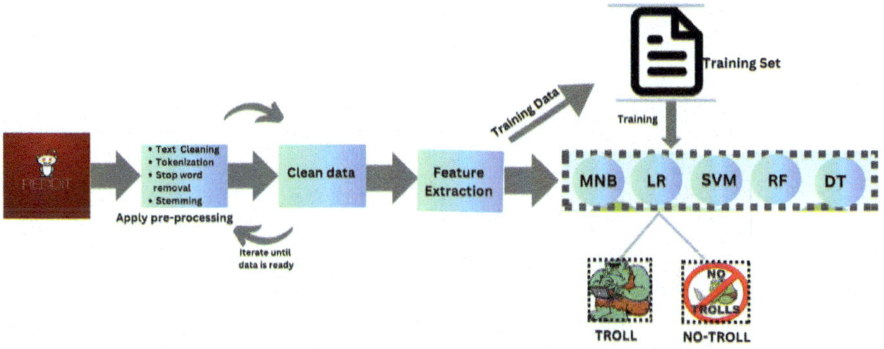

Fig. 1 Workflow of methodology used in troll detection

5.4 Used Methods

For our work we have choosen different machine learning approaches: Multinomial Naıve Bayes (MNB), Logistic Regression (LR), Support Vector Machine (SVM), Random forest (RF) and Decision Tree (DT). These models have allowed us to compare the outcomes of using the bag of words: count vectorizer and the Tfidf vectorizer with a balanced dataset and imbalanced dataset. After calculating the suitable accuracy, precision, recall, and F1 score measures, we managed to acquire outcomes. Thus, we can evaluate and understand the distinctions and advantages of various machine learning techniques and algorithm, such as vectorization, and how having a balanced dataset affects the results (Fig. 1).

6 Experiment and Results

The Python code was executed using the online Jupyter Notebook-based service, Google Research Collaboratory. **With imbalanced dataset and Count vectorizer**: The results of all five models i.e., MNB, LR, SVM, RF and DT four models on imbalanced dataset and using bag of words have been briefed in Table 3. The performance of all of the models has been measured on different parameters like accuracy, precision, recall/sensitivity, and F1 score. The metrics used to assess the proposed method are formally described in Eqs. 1, 2, 3, and 4. In these equations, True Positives (TP), False Positives (FP), and False Negatives (FN) are defined. TP represents the count of troll text correctly identified as troll text, FP represents the count of non-troll text incorrectly identified as troll text, and FN represents the count of troll text incorrectly identified as non-troll text.

$$Accuracy = \frac{TP + TN}{TP + FP + TN + FN} * 100 \quad (1)$$

$$Precision(\pi) = \frac{TP}{FP + TP} \tag{2}$$

$$Recall(p) = \frac{TP}{FN + TP} \tag{3}$$

$$F - score(F1) = \frac{2 * \pi * p}{\pi + p} \tag{4}$$

With Imbalanced Data and Tfidf Vectorization: Now we have imbalanced data but now we are vectorizing the data with Tfidf vectorizer and the results have been summarized in Table 4.

With balanced dataset and Bag of words as well as Tfidf vectorizer: Here we have balanced the dataset with oversampling technique SMOTE and used both Tfidf vectorizer and bag of words vectorizer one by one and both the results came out to be same. The results have been summarized in Tables 5 and 6.

Table 3 Test results of MNB, LR, SVM, RF and decision tree trained on imbalanced text data with bag-of-words representation

Methods	Accuracy	Precision	Recall	F1 Score
MNB	86.82	0.79	0.92	0.85
LR	95.0	0.95	0.92	0.93
SVM	95.0	0.99	0.88	0.93
RF	96.0	0.99	0.91	0.95
DT	96.0	0.99	0.91	0.95

Table 4 Test results of MNB, LR, SVM and RF trained on imbalanced text data with Tfidf representation

Methods	Accuracy	Precision	Recall	F1 Score
MNB	82.0	1.00	0.54	0.70
LR	94.0	0.97	0.87	0.92
SVM	96.0	0.98	0.92	0.95
RF	98.0	0.99	0.97	0.98
DT	98.0	0.99	0.97	0.98

Table 5 Test results of MNB, LR, SVM, RF AND decision tree trained on balanced text data and bag of words vectorizer

Methods	Accuracy	Precision	Recall	F1 Score
MNB	90.22	0.99	0.76	0.86
LR	93.0	0.92	0.90	0.91
SVM	97.0	0.98	0.93	0.96
RF	99.0	0.99	0.98	0.99
DT	99.0	0.99	0.98	0.99

Table 6 Test results of MNB, LR, SVM, RF and decision tree trained on balanced text data and Tfidf vectorizer

Methods	Accuracy	Precision	Recall	F1 Score
MNB	90.22	0.99	0.76	0.86
LR	93.0	0.92	0.90	0.91
SVM	97.0	0.98	0.93	0.96
RF	99.0	0.99	0.98	0.99
DT	99.0	0.99	0.98	0.99

7 Conclusion

The study presents methodologies for detecting trolls within online communities. The suggested method relies on employing machine learning techniques (MNB, LR, SVM, RF AND DT) using fundamental properties extracted from the text of troll posts. The models derived from these approaches underwent testing, with the most favourable outcomes achieved by Random Forest and Decision Tree in both scenario when the dataset is balanced as well as when dataset is imbalanced. But one thing is very clear that the accuracy is more when the dataset is balanced.

In the future, there are various avenues through which the research could be expanded like we can use different deep learning model to see the changes in results and for better Troll detection in real time environment. We can also use multimodal data in future for troll detection and compare the results of various models.

References

1. Khan U, Khan S, Rizwan A, Atteia G, Jamjoom MM, Samee NA (2022) Aggression detection in social media from textual data using deep learning models. Appl Sci 12(10):5083
2. Shanahan T, Tran TP, Taylor EC (2019) Getting to know you: Social media personalization as a means of enhancing brand loyalty and perceived quality. J Retail Consum Serv 47:57–65
3. Tomaiuolo M, Lombardo G, Mordonini M, Cagnoni S, Poggi A (2020) A survey on troll detection. Future internet 12(2):31

4. Tsantarliotis P, Pitoura E, Tsaparas P (2017) Defining and predicting troll vulnerability in online social media. Soc Netw Anal Min 7:1–15
5. Berghel H, Berleant D (2018) The online trolling ecosystem. Computer 51(08):44–51
6. Minaee S, Kalchbrenner N, Cambria E, Nikzad N, Chenaghlu M, Gao J (2021) Deep learning–based text classification: a comprehensive review. ACM computing surveys (CSUR) 54(3):1–40
7. Galán-García P, Puerta JGdl, Gímez CL, Santos I, Bringas PG (2016) Supervised machine learning for the detection of troll profiles in twitter social network: Application to a real case of cyberbullying. Oxford University Press
8. Saeed MH, Ali S, Blackburn J, De Cristofaro E, Zannettou S, Stringhini G (2022) Trollmagnifier: Detecting state-sponsored troll accounts on reddit. In: 2022 IEEE symposium on security and privacy (SP), pp 2161–2175. IEEE
9. Jachim P, Sharevski F, Treebridge P (2020) Trollhunter [evader]: Automated detection [evasion] of twitter trolls during the covid-19 pandemic. In: New security paradigms workshop 2020, pp 59–75
10. Im J, Chandrasekharan E, Sargent J, Lighthammer P, Denby T, Bhargava A, Hemphill L, Jurgens D, Gilbert E (2020) Still out there: modeling and identifying russian troll accounts on twitter. In: Proceedings of the 12th ACM conference on web science, pp 1–10
11. Balaji T, Annavarapu CSR, Bablani A (2021) Machine learning algorithms for social media analysis: a survey. Comput Sci Rev 40:100395
12. Antonakaki D, Fragopoulou P, Ioannidis S (2021) A survey of twitter research: Data model, graph structure, sentiment analysis and attacks. Expert Syst Appl 164:114006
13. Shafiei H, Dadlani A (2022) Detection of fickle trolls in large-scale online social networks. J Big Data 9(1):1–21

Comparative Analysis of Machine Learning Algorithms for Prostate Cancer

Bharti Thakur and Abhinav

Abstract Worldwide, second deadly cancer encompassed by males is Prostate cancer. With the outcome of artificial intelligence, diagnosis along with analysis of prostate cancer has become more accurate. Artificial intelligence methods can rapidly examine big amount of data which includes images and tissue specimens, to recognize sequence and based on sequence prognosis can be performed related to prostate cancer. This paper includes a dataset containing 100 patient observations from prostate biopsies. Data pre-processing is performed to clean the data, including label encoding and Min–Max scaling. Three machine learning models, Support Vector Machine, Random Forest, and Logistic Regression are evaluated. Logistic regression achieved the highest accuracy of 0.86667 when compared to the other two machine-learning algorithms. The study's findings can benefit medical practitioners and researchers for the enhancement of prostate cancer treatment.

Keywords Artificial intelligence · Machine learning · Support vector machine · Random forest · Cross validation

1 Introduction

Prostate cancer originates in the prostate gland. In males, a walnut-shaped gland that produces seminal fluid, which transports sperm is known as prostate [1]. In 2020, United States witnessed 191,930 individuals being diagnosed with prostate cancer, leading to an estimated 33,330 fatalities [1]. Prostate cancer is among the regular cancers. Prostate cancer was diagnosed in about 1 in 8 men at some point in their lifetime. About 6 out of 10 men who are 65 or older are diagnosed, and it is less common in men who are under 40. The average diagnostic age for men is 66. Age, caste, inherited genetic factors, and family medical history are risk factors

B. Thakur (✉) · Abhinav
Yogananda School of AI, Computers and Data Science, Shoolini University, Solan, Himachal Pradesh, India
e-mail: bharthithakur.thakur@gmail.com

V. K. Gunjan and J. M. Zurada (eds.), *Proceedings of 5th International Conference on Recent Trends in Machine Learning, IoT, Smart Cities and Applications*, Lecture Notes in Networks and Systems 1181, https://doi.org/10.1007/978-981-97-8861-3_11

for the disease. The rate of death is higher in African men compared to Caucasian men partly due to their greater prostate cancer development risk [2]. In the prostate gland, most prostate cancers grow slowly and are restricted. Although some prostate cancers develop at a slow pace, some develop fast and are aggressive. Early-stage symptoms of prostate cancer are rare, although some symptoms, such as bone pain, spinal compression sensations, painful urination, and blood in the urine or semen can help in the diagnosis.

Measuring prostate-specific antigen, a protein generated by cells of the prostate gland is one very popular screening technique [3]. Diet and physical activity are great contributors towards prostate cancer development. [1].

With the assistance of machine-learning techniques a lot of research has been performed on cancer diagnosis, which has shown promising results in improving the accuracy and efficiency of detection. Machine learning algorithms inspect medical records, imaging scans, to pick out patterns so that predictions can be carried out. These techniques can aid in early detection, risk assessment, and personalized treatment planning for prostate cancer patients. Identification of attributes can be grouped into two types one is by the hand and make use of information present within the image. Secondly attributes are automatically acquired from machine learning algorithms.

1.1 Risk Factors of Prostate Cancer

- At the age of 50 the risk of cancer increases. Men at the age of 65 or above are recognized with 60% of prostate cancer. Men recognized with prostate cancer suffer with distinctive problems like painful ejaculation, blood in the urine and back discomfort [2].
- Familial prostate cancer is the term for prostate cancer that runs in families. Shared genes, environmental variables, and lifestyle choices all contribute to the development of this kind of prostate cancer [4].
- According to chronological investigations, geographic variances, and descriptive epidemiologic studies of migrants, prostate cancer incidence and death are strongly positively correlated with fat consumption, particularly polyunsaturated fat [5].

1.2 Methodologies for Prostate Cancer Treatment

1.2.1 Digital Rectal Examination

Physical examination in which expert inserts a gloved, lubricated finger into the rectum to assess size, shape, texture of the prostate. DRE can discover irregularity like lumps on the other hand hard areas that might indicate prostate cancer.

1.2.2 Prostate-Specific Antigen Test

Blood screening calculating amount of PSA, a protein produced by the prostate. Elevated PSA levels can indicate various prostate conditions, including cancer. PSA testing is a common screening tool [3]. Elevated PSA levels may prompt further investigations, such as biopsy.

1.2.3 Multiparametric Magnetic Resonance Imaging

It combines multiple imaging sequences to provide detailed images of the prostate. Radiologists assess the images for abnormalities suggestive of cancer. mpMRI is used for prostate cancer detection, localization, and staging. It can help guide biopsy procedures.

1.2.4 Transrectal Ultrasound

A procedure in which a probe is pushed towards rectum to generate ultrasound images. TRUS is used in combination with biopsy procedures to guide the needle to suspicious areas for sample collection.

1.2.5 Prostate Biopsy

In this method compact specimen of prostate tissues collected using a needle. These samples are then tested by using microscope for the existence of cancer cells. Biopsy is the definitive practice for diagnosing prostate cancer. It is often prompted by abnormal PSA levels, DRE findings, or suspicious imaging results [5].

1.2.6 PCA3 Test

The Prostate Cancer Antigen 3 test measures levels in reference to a specific RNA molecule in urine. Elevated PCA3 levels may indicate the presence of prostate cancer.

1.2.7 Prostate Imaging-Reporting and Data System

A standardized system deals with interpretation and report mpMRI findings. It assigns scores to lesions based on their likelihood of being cancerous [6]. PI-RADS helps standardize the interpretation of imaging results, aiding in treatment planning.

1.2.8 Genomic Testing

Molecular tests analyze the expression of certain genes or genomic markers associated with prostate cancer. Examples include the Oncotype DX Prostate Cancer Assay. Genomic testing can help predict the aggressiveness of prostate cancer and guide treatment decisions [17].

1.2.9 Liquid Biopsy

Examination of blood-borne tumor cells in the blood to detect genetic mutations associated with prostate cancer [18]. Liquid biopsy is an area of active research for non-invasive cancer detection and monitoring.

1.3 Machine Learning

Machine learning, a section of artificial intelligence. It educates machines how to handle data more competently. It includes programming computers to use free data sources. The primary motivation behind AI is to investigate and create predictions that can draw on historical data and generate desires on fresh information sources [6]. Utilizing modern algorithms to evaluate patient data and find patterns that help forecast the likelihood of prostate cancer. Machine learning models acquire the ability to predict the existence of cancer. This strategy improves medical judgment, which eventually results in better patient outcomes and more efficient care of prostate cancer [7]. Machine learning is applied in support to cure Prostate cancer through different approaches.

1.3.1 Early Detection and Diagnosis

- Image Classification: Machine learning models, especially deep learning algorithms, applied on medical imaging data to contribute in favour of early detection. These models can identify suspicious regions in the images that may require further examination.
- Biomedical Information Analysis: ML models can analyze complex biomedical data, including patient age, prostate weight, cancer volume and prostate-specific antigen (PSA) levels. These models help in predicting the likelihood of prostate cancer and determining its severity [13].

1.3.2 Prognosis and Risk Assessment

- Survival Prediction: Machine learning models are used to predict the survival time or outcomes for patients with prostate cancer. These models consider various clinical and pathological factors to provide more personalized and accurate prognostic information.
- Risk Stratification: Machine learning algorithms can assist in stratifying patients into different risk groups based on their likelihood of disease progression. This helps in tailoring treatment plans and interventions according to individual risk profile [14].

1.3.3 Treatment Planning

Personalized Treatment Recommendations: Machine learning models analyze patient data to recommend personalized treatment plans based on factors such as genetic markers, imaging results, and patient history. This helps in optimizing treatment strategies for better outcomes.

1.3.4 Research and Biomarker Discovery

- Identification of Biomarkers: Machine learning algorithms analyze genomic and proteomic details to identify potential biomarkers connected with prostate cancer. These biomarkers can function like indicators of disease presence, progression, or response to treatment [16].
- Data Integration: ML techniques help integrate diverse datasets, including clinical, molecular, and imaging data, facilitating a more comprehensive understanding of prostate cancer and its underlying mechanisms.

1.3.5 Decision Support Systems

Clinical Decision Support: Machine learning models can serve as decision support systems for healthcare professionals by providing insights into diagnosis, treatment planning, and risk assessment. This aids clinicians in making informed decisions based on a broader range of data.

1.3.6 Automated Histopathological Analysis

Digital Pathology: Machine learning is applied to digitized pathology slides to assist pathologists in analysing tissue samples. Automated detection and classification of cancerous regions contribute to faster and more accurate diagnostics.

2 Literature Review

To find crucial information, the researchers used a variety of machine learning algorithms. Muktevi Srivenkatesh in the year 2020, used machine learning techniques on the Kaggle set and made prediction of prostate cancer. The accuracy outcomes from the Support Vector Machines, Linear Regression, Naive Bayes and Random Forest algorithms fluctuated between 70 and 90%. The Linear Regression together with Random Forest algorithms produced the results with the greatest accuracy (90%) when compared to other Machine-learning Algorithm [8].

P. Mohamed Shakeel and Gunasekaran Manogaran in the year 2020, collected prostate cancer information from the DBCR dataset that manages person's age, prostate weight, Gleason grade. When the study's accuracy rates were examined of pro, it was observed that the Radial trained extreme learning neural network (RTELNN) is the most effective classifier with an accuracy of 99.3%, SVM(95.90%), MLP(98.15%), and Neural Networks(97.12%) [9].

Yixuan Yuan in the year 2019 used deep learning-based multi-parametric transfer learning model and performed the classification of prostate cancer on MRI images. The model achieved an accuracy state cancer classification (accuracy of 86.92%) [10].

Bejoy Abraham and Madhu S. Nair in the year 2018 used a unique approach to confirm level set in prostate cancer using poly-parametric MRI biomarkers. In this method, they used a deep network of Stacked Sparse Autoencoders to extract high-level features applying SoftMax classifier with 47.3% accuracy [11].

Islam Reda in the year 2018, suggested a completely automated CAD model for prostate cancer early detection based on convolutional neural networks. They used DWI datasets that are gathered at seven distinct b-values from 45 patients (25 malignant, 20 benign) and achieved accuracy rates of 97.8%, obtained from distinct b-values through CNN based classifier [12].

Srđan Jović in their research, applied machine learning approach to forecast the probability of prostate cancer. Three methods, namely Extreme Learning Machine, Artificial Neural Networks and Genetic Programming were used on the same dataset. The outcome showed that the coefficients of determination for ELM, ANN, and GP were 0.9976, 0.9647, and 0.9204, respectively. Based on these findings, the ELM model demonstrated the highest accuracy in predicting prostate cancer probability. Therefore, it can be concluded that the ELM model provides the most accurate predictions for prostate cancer probability in this study [13].

Huaiyu wen implemented Support Vector Machine, Naïve Bayes and Neural Network on prostate cancer dataset to categorize prostate cancer patients' survival time between two classes, less than 60 months and higher than 60 months. Findings demonstrate neural networks are most accurate (85.64%) for estimating the chances of survival [14]. Using data from population-based health surveys, Roffman et al. constructed a multiparameterized ANN to anticipate the uncertainty of prostate cancer using prebiopsy clinical and demographic factors.

With a high level of accuracy (89.4%) the trained ANN demonstrated encouraging results. The results indicate that ANNs may be used as a noninvasive method for estimating the risk of prostate cancer based on individual health data. Analysis emphasizes on machine learning approaches, in the direction of early identification and risk assessment of prostate cancer, potentially enhancing clinical judgment and patient-specific treatment planning [15].

In their study, Wang et al. successfully created predictive models utilizing machine learning approaches to identify cases of prostate cancer based on prebiopsy data. The models performed incredibly well in identifying major prostate cancer cases. He utilized four machine learning methods SVM, Least Squares Support Vector Machine ANN, and RF. All methods showed good performance in detecting significant prostate cancer cases, with Artificial Neural Network achieving highest correctness of 95.27% [16].

Distinct machine learning algorithms used to forecast prostate cancer, with varying degrees of accuracy. Notably, studies using the Kaggle dataset in 2020 found accuracy ranging from 70 to 90%, demonstrating the efficiency of techniques like Support Vector Machines and Random Forests. However, these studies have limitations. Some, such as the study by Laabidi and Aissout, only managed to attain 81.3% accuracy.

Furthermore, relying solely on accuracy may fail to consider false negatives or limited sensitivity. Also, using specific datasets might not cover the broader population, impacting the generalizability of the results. Radial trained extreme learning neural network (RTELNN) was identified as the most effective classifier highest validity of 99.3% in one study. Various researchers have used diverse approaches which includes deep learning on MRI images, convolutional neural networks (CNN) on DWI datasets, and stacked autoencoders on multi-parametric MRI biomarkers. Neural Networks demonstrated efficacy in predicting survival time for prostate cancer patients, with an accuracy of 85.64%. Models like Extreme Learning Machine (ELM) showed high accuracy in predicting the probability of prostate cancer, outperforming Artificial Neural Networks and Genetic Programming Multiparameterzed artificial neural network model achieved the highest accuracy (89.4%) for predicting threat of prostate cancer derived from prebiopsy clinical and demographic factors. Machine learning approaches, when applied to prebiopsy clinical and demographic factors, may serve as noninvasive methods for estimating the risk of prostate cancer, aiding in early identification and risk assessment. Literature survey highlights that machine learning shows promise in prostate cancer research, ongoing efforts are needed to address challenges and limitations, including the generalizability of results, interpretability of models, and the incorporation of these techniques into practical clinical settings (Table 1).

Table 1 Summary of Dataset and Model used in research paper with outcomes

References	Dataset	Machine learning classifier/technique	Accuracy (%)
[8]	Kaggle	SVM, KNN, LR, NB, RF	90
[7]	Kaggle	RNN, NB, RF, SVM, KNN, DT, LR, XG Boost	81.3
[9]	DBCR	RTELNN, SVM, MLP, NN	99.3
[10]	mp-MRI	Multi-parametric transfer learning	86.92
[11]	PROSTATEx-2 2017 challenge	SSAE	47.3
[12]	DWI	CNN	97.8
[13]	Prostate cancer data acquired by measurement	ELM, ANN, GP	99.76, 96.47, 92.04
[14]	SEER prostate cancer dataset	SVM, C4.5, NB, NN	85.47, 84.97, 71, 85.64
[15]	National Health Interview Survey (NHIS)	ANN	89.4
[16]	The real-world prostate cancer dataset of 1652 Chinese men	SVM, RF, LSSVM, ANN	95.27

3 Additional Information Required from Authors

3.1 Dataset

Information set comprises of 100 observations and 10 variables. The goal is to build a predictive model that can accurately classify patients into two classes: malignant or benign based on the provided attributes. The 'diagnosis result' column serves as the target variable or label, while the other columns serve as the input features. By using this dataset, researchers or practitioners can develop and evaluate predictive models for prostate cancer diagnosis and potentially assist in early detection and treatment planning.

3.2 Methodology

3.2.1 Pre-processing

Pre-processing is used to spot duplicates and missing values. Label encoding technique is used to convert categorical variables into numerical format. In label encoding, each unique label is assigned an integer, and the categorical variable is represented by these integers. The Min–Max scaler is used for feature scaling and the range lies between 0 and 1. Min–max scale is used to avoid the dominance of features.

3.2.2 Training and Testing Data

On the Prostate cancer dataset, training and testing carried out to examine the model's accuracy. There are two sections to the dataset. The first portion is used for testing, while the second is used for training. Evaluation is based on 70 and 30% split.

3.2.3 Machine Learning Algorithms

We have used (1) SVM (2) random forest (3) logistic regression.

- SVM applied because of its strong classification capability, making it ideal for predicting prostate cancer. Its proficiency in handling complex data patterns and high-dimensional sets is particularly valuable [18]. The SVM model is then built, it may classify recent instances [14].
- Effectiveness of the RF classifier employed in several medical investigations showed that this classifier could be useful for expressing knowledge useful in medicine [19, 20]. Random Forest Classifier model, trains it on the training data, evaluates its performance using various metrics, and visualizes the confusion matrix. This process helps assess the model's predictive accuracy and overall effectiveness for prostate cancer prediction.
- All variants of logistic regression are commonly employed in medical applications [21, 22].

3.2.4 Cross-Validation

On the same dataset, the ninefold cross-validation is executed to see if the sliced dataset has changed in order to improve accuracy. The model is trained and evaluated 9 times.

3.2.5 Confusion Matrix

Interpretation of a machine learning model's performance on a test data is summed up by a confusion matrix. Confusion matrix evaluates the effectiveness of machine learning models.

3.3 Result

In this research, we train SVM, Random Forest and Logistic Regression. Evaluation metrics calculated the accuracy, f1_socre, TPR, and FPR. Table 2 represents the result of the evaluation metrics. We divided the dataset into 70 and 30% split. According to findings, logistic regression attained the highest accuracy of 0.86667.

In comparison to Random Forest, Logistic Regression and SVM show higher precision. SVM and logistic regression had a high precision score of 0.825 and 0.85520 but random forest model had a precision score of 0.76666. Precision values represent the percentage of accurately predicted positive cases among all projected positive instances.

Recall score for the Logistic Regression was 0.87559,

F1_score strikes a balance between the two metrics. Logistic Regression had a high F1_score score of 0.86111.SVM performed better than other models and displayed a higher TPR of 0.89473. Logistic regression, which has a low rate of false positives of 0.09090, has the lowest FPR (Table 3).

Table 2 Represents the parameter used for ML classifier

Models	Parameters	Parameters uses
Random forest	No of estimator Criteria follow Random state	20 Gini 100
Logistic regression	Random state Maximum iteration	42 100
SVM	Kernel	Linear

Table 3 Represent the result of the models

Models	Accuracy	Precision	Recall	F1_score	TPR (True positive rate)	FPR (False positive rate)	Mean square error
Random forest	0.76667	0.76666	0.8	0.76	0.7	0.1	0.23334
SVM	0.83334	0.825	0.81100	0.81684	0.89473	0.27272	0.16666
Logistic regression	0.86667	0.85520	0.87559	0.86111	0.84210	0.09090	0.13333

3.4 *Discussion*

This paper shows how different machine-learning approaches impact prostate cancer prediction. The purpose intends to create a predictive model that can correctly divide patients into two groups: benign (cancer-negative) or malignant (cancer-positive).

Our study has some limitations.

- The accuracy and amount of the dataset used substantially influence the study's findings. The effectiveness and generalizability of the model may suffer if the dataset is too small or inadequately represents the target population.
- Prostate cancer prediction datasets may suffer from class imbalance. This imbalance can bias the model towards the majority class and affect performance metrics.

4 Conclusion

The study aimed to compare the performance of three different models, Random Forest, Support Vector Machine and Logistic Regression, in predicting prostate cancer. Logistic regression has the best accuracy 0.86667. Our research demonstrated the potential of machine learning methods, especially Logistic Regression, predicting prostate cancer using the given features. It emphasized the importance of careful model selection and evaluation in medical applications. It is essential to recognize that the success of machine learning models in clinical settings relies not only on high accuracy but also on practical implementation and ethical considerations. Future research explores these models on more diverse datasets to ensure their healthiness true to life situation.

References

1. Mucci LA, Wilson KM, Giovannucci EL (2017) Epidemiology of prostate cancer. In: Loda M, Mucci L, Mittelstadt M, Van Hemelrijck M, Cotter M (eds) Pathology and epidemiology of cancer. Springer, Cham. https://doi.org/10.1007/978-3-319-35153-7_9
2. Panigrahi GK, Praharaj PP, Kittaka H, Mridha AR, Black OM, Singh R, Deep G (2019) Exosome proteomic analyses identify inflammatory phenotype and novel biomarkers in African American prostate cancer patients. Cancer Med 8(3):1110–1123
3. Barlow H, Mao S, Khushi M (2019) Predicting high-risk prostate cancer using machine learning methods. Data 4(3):129. https://doi.org/10.3390/data4030129
4. Patel AR, Klein EA (2009) Risk factors for prostate cancer. Nat Clin Pract Urol 6(2):87–95
5. Bostwick DG, Burke HB, Djakiew D, Euling S, Ho SM, Landolph J, Timms B (2004) Human prostate cancer risk factors. Cancer Interdiscip Int J Am Cancer Soc 101(S10):2371–2490
6. Mahesh B (2020) Machine learning algorithms-a review. Int J Sci Res (IJSR) 9(1):381–386
7. Laabidi A, Aissaoui M (2020) Performance analysis of Machine learning classifiers for predicting diabetes and prostate cancer. In: 2020 1st international conference on innovative research in applied science, engineering and technology (IRASET), IEEE, pp 1–6

8. Srivenkatesh M (2020) Prediction of prostate cancer using machine learning algorithms. Int J Recent Technol Eng 8(5):5353–5362
9. Shakeel PM, Manogaran G (2020) Prostate cancer classification from prostate biomedical data using ant rough set algorithm with radial trained extreme learning neural network. Heal Technol 10:157–165
10. Yuan Y, Qin W, Buyyounouski M, Ibragimov B, Hancock S, Han B, Xing L (2019) Prostate cancer classification with multiparametric MRI transfer learning model. Med Phys 46(2):756–765
11. Abraham B, Nair MS (2018) Computer-aided classification of prostate cancer grade groups from MRI images using texture features and stacked sparse autoencoder. Comput Med Imaging Graph 69:60–68
12. Reda I, Ayinde BO, Elmogy M, Shalaby A, El-Melegy M, Abou El-Ghar M, El-Baz A (2018) A new CNN-based system for early diagnosis of prostate cancer. In: 2018 IEEE 15th international symposium on biomedical imaging (ISBI 2018), IEEE, pp 207–210
13. Jović S, Miljković M, Ivanović M, Šaranović M, Arsić M (2017) Prostate cancer probability prediction by machine learning technique. Cancer Invest 35(10):647–651
14. Wen H, Li S, Li W, Li J, Yin C (2018). Comparision of four machine learning techniques for the prediction of prostate cancer survivability. In: 2018 15th international computer conference on wavelet active media technology and information processing (ICCWAMTIP), IEEE, pp 112–116
15. Roffman DA, Hart GR, Leapman MS, Yu JB, Guo FL, Ali I, Deng J (2018) Development and validation of a multiparameterized artificial neural network for prostate cancer risk prediction and stratification. JCO Clin Cancer Inform 2:1–10
16. Wang G, Teoh JYC, Choi KS (2018) Diagnosis of prostate cancer in a Chinese population by using machine learning methods. In: 2018 40th annual international conference of the IEEE engineering in medicine and biology society (EMBC), IEEE, pp 1–4
17. Thakur B, Gupta G, Kumar N (2022) Hybrid genetic model with ANOVA for predicting breast neoplasm using METABRIC gene data. Mater Today Proc 56:1847–1852
18. Thakur B, Kumar N, Gupta G (2022) Machine learning techniques with ANOVA for the prediction of breast cancer. Int J Adv Technol Eng Explor 9(87):232 Alickovic E, Subasi A (2016). Medical decision support system for diagnosis of heart arrhythmia using DWT and random forests classifier. J Med Syst 40(4):108
19. Jin Z, Shang J, Zhu Q, Ling C, Xie W, Qiang B (2020) RFRSF: Employee turnover prediction based on random forests and survival analysis. In: Web information systems engineering–WISE 2020: 21st international conference, Amsterdam, the Netherlands, Springer International Publishing, Part II 21, pp 503–515
20. Harrell FE Jr, Lee KL, Califf RM, Pryor DB, Rosati RA (1984) Regression modelling strategies for improved prognostic prediction. Stat Med 3(2):143–152
21. Ludbrook J (1991) Practical statistics for medical research. Aust N Z J Surg 61(12):963–964
22. Dreiseitl S, Ohno-Machado L, Kittler H, Vinterbo S, Billhardt H, Binder M (2001) A comparison of machine learning methods for the diagnosis of pigmented skin lesions. J Biomed Inform 34(1):28–36

Breaking Barriers in Communication Using Long Short-Term Memory Networks for Sign Language Recognition

Pratik R. Gopale, Sarang L. Joshi, and Hrushikesh B. Kulkarni

Abstract Sign language recognition is pivotal in enhancing communication and accessibility for individuals with hearing impairments. This paper introduces an innovative system leveraging Long Short-Term Memory (LSTM) networks, a form of recurrent neural network renowned for capturing temporal dependencies in sequential data. The objective is to proficiently interpret and classify sign language gestures extracted from real-time video or image data. The proposed system employs a comprehensive methodology encompassing dataset selection, preprocessing, feature extraction, LSTM model architecture design, training, and evaluation. A diverse, well-annotated dataset is utilized, covering various sign language gestures with corresponding labels. The LSTM model is meticulously configured with appropriate architecture parameters, including the number of LSTM layers and memory cells. Experimental results underscore the robustness and efficiency of the proposed sign language recognition system. Real-time capabilities are achieved by integrating the trained LSTM model into a user-friendly interface for live video or image data processing. The system undergoes performance and usability evaluation through user studies, providing insights for future enhancements and applications. This LSTM-based sign language recognition system exhibits promising potential in enhancing accessibility and communication for individuals with hearing and speech impairments. Deep learning techniques and temporal modeling contribute to accurate interpretation and classification of sign language gestures, fostering improved inclusion and interaction within the deaf community.

Keywords Sign language recognition · Long Short-Term Memory (LSTM) Networks · Accessibility · Deep learning · Inclusive communication

---

P. R. Gopale · S. L. Joshi (✉) · H. B. Kulkarni
School of Mechatronics Engineering, Symbiosis Skills and Professional University, Pune, Maharashtra, India
e-mail: jsarang70@gmail.com

V. K. Gunjan and J. M. Zurada (eds.), *Proceedings of 5th International Conference on Recent Trends in Machine Learning, IoT, Smart Cities and Applications*, Lecture Notes in Networks and Systems 1181, https://doi.org/10.1007/978-981-97-8861-3_12

1 Introduction

Sign language detection typically refers to the process of identifying and interpreting sign language gestures or expressions performed by individuals. In the context of technology and artificial intelligence, sign language detection often involves the use of computer vision and machine learning techniques to recognize and understand signs made by individuals using their hands, facial expressions, and body movements [1, 2].

Sign language detection has applications in various fields, including assistive technology for individuals with hearing impairments, human-computer interaction, and accessibility in communication. Developing effective sign language detection systems involves creating robust datasets, implementing suitable preprocessing techniques, and training models to accurately recognize and interpret the intricate movements and expressions inherent in sign languages.

The use of deep learning methods can help to reduce communication barriers for persons with impaired speaking or hearing abilities [3]. Kothadiya et al. proposed a deep learning-based model that detects and recognizes the words from a person's gestures, achieving around 97% accuracy over 11 different signs. Similarly, Urmee et al. [4] developed a convolutional neural network (CNN) based model using Xception architecture that achieves 98.93% accuracy over the test-set for real-time Bangla Sign language (BdSL) detection. This indicates the potential of deep learning methods in improving sign language detection and recognition. Borg et al. [5] proposed a multi-layer RNN for sign language detection, leveraging dynamic information of hand motion during detection. Their system uses features extracted automatically from a 2-stream convolutional neural network (CNN) that takes video image data and motion data as input. The authors compare their system against the state-of-the-art, and attain an improvement of around 18%, indicating the effectiveness of their network in sign language detection. Bhadra et al. [6] proposed a deep multi-layered convolution neural network for the automatic detection of sign language from hand gesture images. Their experimental results demonstrate the efficacy of the proposed methodology in sign language detection task, highlighting the potential of deep learning methods in this area. While there are advancements in sign language detection using deep learning methods, Jantunen et al. [7] discuss the complexities and challenges in developing automated translation technology for sign languages. They argue that a more diverse and critical perspective towards these issues is needed in order to avoid generalizations and bias that is often manifested within dominant research paradigms particularly in the fields of spoken language research and speech community. In conclusion, the use of deep learning methods has shown promising results in sign language detection and recognition, with high accuracy rates achieved in various studies [3–7]. However, the development of automated translation technology for sign languages presents complex challenges that require a diverse and critical perspective [7].

The study of sign language recognition involves the intersection of many fields and disciplines. At present, the two mainstream research directions of sign language recognition are data gloves and visual sign language recognition. The former uses the data collected by the sensor for sign language recognition and translation, while the latter uses the camera to capture the user's hand characteristics for sign language recognition and translation [8]. In recent years, there has been a growing interest in developing sign language recognition systems based on deep learning techniques, particularly Long Short-Term Memory (LSTM) networks. One study presented a sign language recognition system that combines an improved convolutional neural network (CNN) and LSTM neural network, which is designed to not only recognize and translate sign language but also generate sign language [8]. The system achieved a sign language recognition rate of 95.52% and a sign language accuracy of 90.3%. Another study proposed a real-time American Sign Language alphabet translation system based on the YOLO network and LSTM, achieving a mean average precision of 81.74% and a processing rate of 61.35 frames-per-second (Rivera-Acosta et al., [9]. Additionally, a framework based on a CNN with LSTM layers for recognizing hand gestures from a video stream in real-time achieved an accuracy of 98.46% on the test dataset [10]. Furthermore, a study proposed an American Sign Language recognition system for dynamic signs in real-time using the Media Pipe framework and an LSTM network, achieving a test accuracy of 97.2% [11]. Additionally, a study on Kurdish sign language recognition using the Kinect V2 sensor proposed a feature engineering approach and compared the performance of LSTM and echo system network (ESN) classifiers, with both achieving high accuracy, but the ESN exhibiting significantly lower training time [12]. These studies demonstrate the potential of LSTM networks in sign language recognition systems, showing high accuracy rates and real-time performance. The combination of CNN and LSTM networks has shown promise in recognizing and translating sign language, as well as generating sign language, providing a valuable tool for communication for the deaf-mute community. Further research in this area could focus on improving the robustness and generalization of these systems to recognize a wider range of sign language gestures and improve real-time performance.

2 Objective of Present Work

The main objective of present work is to develop a real-time hand sign language recognition system for the Marathi language using computer vision and deep learning techniques. The system should accurately detect and interpret hand gestures representing individual Marathi characters and phrases, providing a means of communication for individuals with speech and hearing impairments. The project aims to bridge the communication gap and promote inclusivity by enabling seamless interaction between hearing-impaired individuals and the wider Marathi-speaking community.

3 LSTM (Long Short-Term Memory) for Sign Language Detection (SLD)

3.1 LSTM Over CNN

While CNNs can be applied to sign language recognition tasks with some modifications, LSTM is more naturally suited for this project. LSTM's ability to handle sequential data, capture long-term dependencies, and process variable length sequences align well with the requirements of sign language recognition, where the temporal aspect of gestures is essential for accurate classification. However, it's worth noting that the choice of the model ultimately depends on the specific characteristics of the dataset and the complexity of the recognition task.

3.2 Methodology of SLD Using LSTM

- **Data Collection-** In this sign language recognition for Marathi hand signs, we have collected a diverse and representative dataset to train the recognition model. The first step was to select the set of hand signs that needed to be recognized, which included all the Marathi alphabet letters and several commonly used words and phrases in Marathi sign language (Fig. 1).
- **Data Preprocessing-** Preparation of the collected dataset to be suitable for training the recognition model. The data processing steps involved various transformations and manipulations to ensure that the data was in the appropriate format and represented the gestures effectively.
- **Hand Detection and Tracking-** In MediaPipe Holistic, hand detection and tracking is performed using a combination of two separate models: the "Palm Detection" model and the "Hand Landmark" model.
- **Feature Extraction-** In the context of hand gesture recognition, feature extraction refers to the process of transforming raw input data, such as images or video frames of hands, into meaningful representations that capture relevant information for classification or analysis. The goal of feature extraction is to highlight the distinctive characteristics of hand gestures that can be used by machine learning models to recognize different gestures accurately (Fig. 2).
- **Real-Time Implementation-** Real-time performance in sign language recognition systems is a critical aspect, particularly for applications where immediate and accurate recognition of sign gestures is required. The term "real-time" refers to the system's ability to process and analyze incoming video frames at a speed that matches the video's capture rate, typically 25 to 30 frames per second (fps).

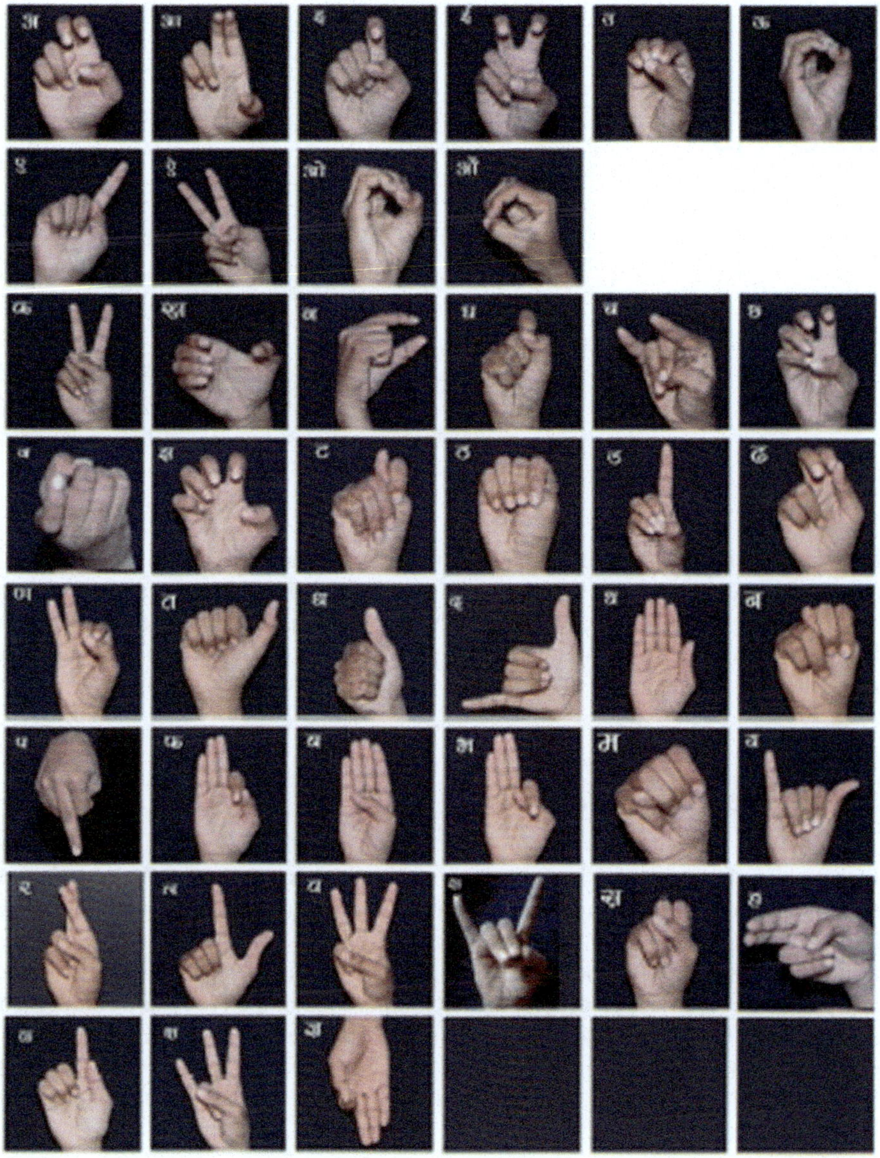

Fig. 1 Sign Symbols for Marathi language

3.3 LSTM Layers in the Program

LSTM layers are the backbone of the sign language recognition model, enabling the effective processing and understanding of the temporal dynamics in the sequences of

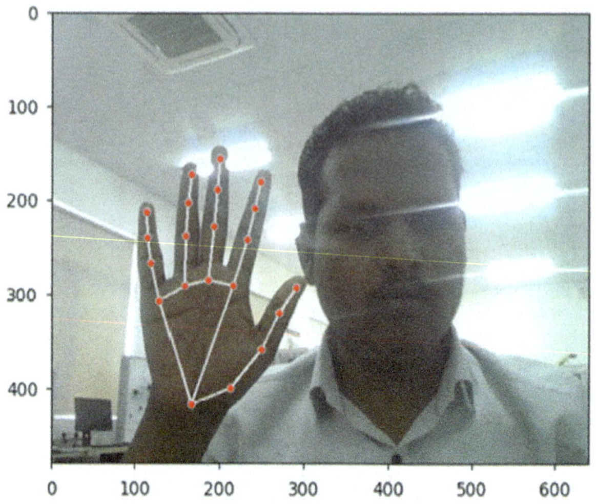

Fig. 2 Image of extracting key points

hand landmarks. Their ability to handle and model sequences over time is crucial in accurately interpreting sign language gestures. Through this detailed understanding, we've gained insight into how LSTM layers contribute to the success of the sign language recognition system, making it a powerful tool in the domain of gesture and sign language recognition.

3.4 Application of MediaPipe

In the realm of computer vision and machine learning, gesture recognition has gained significant traction due to its potential in various applications, including sign language interpretation. Recognizing hand gestures is a critical aspect of sign language recognition systems. One of the powerful tools for hand gesture recognition is MediaPipe, an open-source framework developed by Google. MediaPipe provides a comprehensive suite of tools and pre-trained models to facilitate hand landmark detection, pose estimation, and facial recognition.

4 Results and Performance Evaluation

In the hand gesture recognition project for Marathi sign language, the trained model achieved an impressive accuracy of 93.75% on the test dataset. This high accuracy indicates that the model is proficient in recognizing a vast majority of hand gestures with great precision. The success of the model was assessed using the confusion

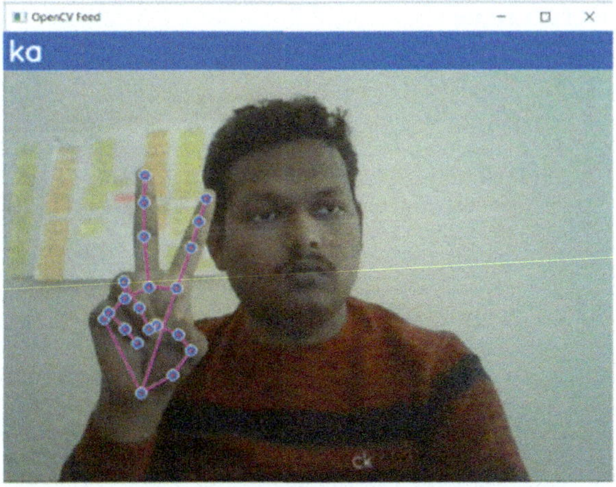

Fig. 3 Sign language recognition for Marathi letters

matrix, which provided a comprehensive view of its performance across different classes (hand gestures) (Fig. 3).

The confusion matrix revealed that the model accurately classified a significant number of hand gestures as true positives (TP). It demonstrated a remarkable ability to correctly identify the intended hand signs for most of the Marathi alphabet letters and common words in the sign language. Moreover, the model exhibited minimal false negatives (FN), indicating that it rarely misclassified a hand gesture as another class.

5 Conclusion

Overall, the hand gesture recognition system exhibits robustness and effectiveness, making it a valuable tool for empowering the hearing-impaired individuals to communicate effectively through hand gestures. The successful implementation of this technology holds the potential to revolutionize accessibility and communication in various domains, opening doors for greater inclusivity and understanding.

As the project moves forward, the team remains dedicated to enhancing the model's performance, exploring advanced techniques, and continuously refining the recognition system. The ultimate goal is to make the system even more accurate, versatile, and user-friendly, empowering individuals with hearing impairments to express themselves confidently and facilitating seamless interactions through the richness of Marathi sign language.

References

1. Patil S, Joshi S, Kulkarni HB, Hagawane P, Shinde P (2022) Marathi sign language hand gesture recognition using accelerometer and 3D printed gloves. In: 2022 14th international conference on computational intelligence and communication networks (CICN), IEEE, pp 72–77
2. Mhatre S, Joshi S, Kulkarni HB (2022) Sign language detection using LSTM. In: 2022 IEEE international conference on current development in engineering and technology (CCET), IEEE, pp 1–6
3. Kothadiya D, Bhatt C, Sapariya K, Patel K, Gil-González AB, Corchado JM (2022) Deepsign: sign language detection and recognition using deep learning. Electronics 11(11):1780
4. Urmee PP, Al Mashud MA, Akter J, Jameel ASMM, Islam S (2019) Real-time bangla sign language detection using xception model with augmented dataset. In: 2019 IEEE international WIE conference on electrical and computer engineering (WIECON-ECE), IEEE, pp 1–5
5. Borg M, Camilleri KP (2019) Sign language detection "in the wild" with recurrent neural networks. In: ICASSP 2019–2019 IEEE international conference on acoustics, speech and signal processing (ICASSP), IEEE, pp 1637–1641
6. Bhadra R, Kar S (2021) Sign language detection from hand gesture images using deep multi-layered convolution neural network. In: 2021 IEEE second international conference on control, measurement and instrumentation (CMI), IEEE, pp 196–200
7. Jantunen T, Rousi R, Rainò P, Turunen M, Moeen Valipoor M, García N (2021) Is there any hope for developing automated translation technology for sign languages. Multilingual Facilitation 1:61
8. Li T, Yan Y, Du W (2022) Sign language recognition based on computer vision. In: 2022 IEEE international conference on artificial intelligence and computer applications (ICAICA), IEEE, pp 927–931
9. Rivera-Acosta M, Ruiz-Varela JM, Ortega-Cisneros S, Rivera J, Parra-Michel R, Mejia-Alvarez P (2021) Spelling correction real-time American sign language alphabet translation system based on yolo network and LSTM. Electronics 10(9):1035
10. Siriak R, Skarga-Bandurova I, Boltov Y (2019) Deep convolutional network with long short-term memory layers for dynamic gesture recognition. In: 2019 10th IEEE international conference on intelligent data acquisition and advanced computing systems: technology and applications (IDAACS), IEEE, 1:158–162
11. Farhan Y, Madi AA (2022) Real-time dynamic sign recognition using MediaPipe. In: 2022 IEEE 3rd international conference on electronics, control, optimization and computer science (ICECOCS), IEEE, pp 1–7
12. Mirza SF, Al-Talabani AK (2021) Efficient kinect sensor-based Kurdish sign language recognition using echo system network. ARO- Sci J Koya Univ 9(2):1–9

Data Transmission and Reception with Error Detection Circuitry

Salma S. Shahapur, Pallavi P. Gaonkar, Nikita M. Sambrekar, Shravani S. Sawant, H. P. Rajani, Sahana Devali, and Kartik Solanki

Abstract We, Human Beings are social creatures and communication is the most important aspect of our daily lives that we use to convey information, ideas and thoughts with each other. With the advent of digital electronics, a faster and more efficient communication is now possible. But this merit comes with an uncertainty of the received data being error-free. Communication channel in which the data gets transmitted in, is very much likely or always has a chance of inducing an error in the data being transmitted. This is because the channel can be affected by various factors such as noise, interference, attenuation, distortion, crosstalk, and jitter. These factors can cause errors in the signal being sent. The goal of this project is not just transmitting and receiving the data, but to transmit signal such that, the person receiving this data at the receiving end is able to confirm its reliability. The aim is to bring about this reliable transmission of data with this technique known as Cyclic Redundancy Check or CRC for short. To sum it all up we will be using CRCs to detect any degree of error in a message signal with the ultimate goal of making digital communication always reliable.

Keywords CRC · Data · Noise · Communication channel · Error

1 Introduction

Here we have implemented error detection-based project to apply the knowledge and skills and gain a deeper understanding of a fundamental concept that is essential to our future careers. We were eager to tackle a challenge that would help us acquire critical skills and knowledge, and were convinced that error detection was a field

S. S. Shahapur (✉) · P. P. Gaonkar (✉) · N. M. Sambrekar · S. S. Sawant · H. P. Rajani · S. Devali · K. Solanki
Jain College of Engineering, VTU, Belagavi, Karnataka, India
e-mail: salma.jce@gmail.com

P. P. Gaonkar
e-mail: pallavimpg10@gmail.com

V. K. Gunjan and J. M. Zurada (eds.), *Proceedings of 5th International Conference on Recent Trends in Machine Learning, IoT, Smart Cities and Applications*, Lecture Notes in Networks and Systems 1181, https://doi.org/10.1007/978-981-97-8861-3_13

where we could make a meaningful contribution. Our motivation stemmed from our ambition to develop and learn as professionals, and we viewed this project as the ideal chance to accomplish our objectives.

The objective of this project is not just transmitting and receiving the data, but to transmit data in which the person receiving this data at the receiving end is able to confirm its reliability. The aim is to bring about this reliable data transmission using the above technique. Here we have implemented error detection-based project to apply the knowledge and skills and gain a deeper understanding of a fundamental concept that is essential to our future careers. We were eager to tackle a challenge that would help us acquire critical skills and knowledge, and were convinced that error detection was a field where we could make a meaningful contribution. Our motivation stemmed from our ambition to develop and learn as professionals, and we viewed this project as the ideal chance to accomplish our objectives.

2 Literature Survey

The error detection and correction put some extra data to a signal, which receivers use to check flow of the delivered message & to get back the data that has been found to be manipulated [1].

Error detection & correction techniques can be either in order or out of order. In an ordered scheme, transmitter transmits the original data & joints discrete amount of bit, obtained from the data bit by some encoding algorithm in CRC [2].

If error correction is needed, a receiver can apply the obtained algorithm to the received data bits and compare its output with the obtained bits; if the values differ that means an error has occurred at a particular point during the transmission of data. If error corrections are needed then the receiver may apply the decoding algorithm to obtained data bit and the obtained bits to recover the original error-free data. In the system which uses out of order technique, the original message is converted into encoded message which carries the same info and that has at least similar bits when compared to the original message.

3 Methodology

Figure 1 shows block & diagram of transmitter and receiver. The overall design consists of three important parts, i.e., Transmitter, Receiver and Channel. The transmitter is programmed to send the data along with the error detecting redundant code [3]. The receiver is programmed to receive and display the data with added circuitry of error detection if found any. The channel consists of three lines, one for data being transmitted, another for common ground and the third is the clock to synchronize both sides. The channel consists of a break switch to induce some errors in the message being sent [4].

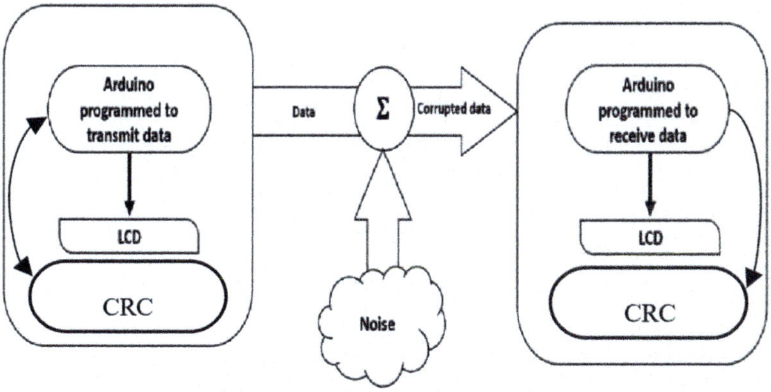

Fig. 1 Block diagram of transmitter and receiver when noise is applied in the communication

3.1 Crc Circuit

- Figure 2 shows that CRC technique of detecting changes in the communication channel. Cyclic redundancy check takes Generator polynomial available to both transmitter & receiver [5].
- The polynomial is of the form like x^3-x –1. This generates key 1001. Another example is x^2 –1 that shows key 110.
- N- Represents no of bits of data to be sent from transmitter.
- K-Represents no of bits in the key obtained from polynomial generator [6].
- Sender Side: The binary source is first multiplied by adding k-1 zeros at last of the signal using modulo2 binary division to divide source by the key and stores remainder.
- Receiver Side: Performs modulo-2 division and if the remainder is zero, then no error.
- This polynomial can be expressed in hexadecimal in the form 0xd265.
- This generator polynomial has a hamming distance of 4.
- This hamming distance is valid for message length lower than 32,751, this means that any 4-bit errors in this bit range can be detected with 100% certainty.
- The way this operates is that, the CRC circuit consists of D type flip-flops and XOR gates which simulate the division of the message signal by the generator polynomial.
- The output of this division can be visualised as a reminder [7].
- Once the crc of the original message signal is generated, it is read out from the d flipflops by shifting it with the help of 16 zeroes and is transmitted by the transmitter along with the original message [8].
- This means that we are adding the original binary message and the reminder.
- The receiver end also has a similar crc calculation circuit with the generator polynomial known [9].

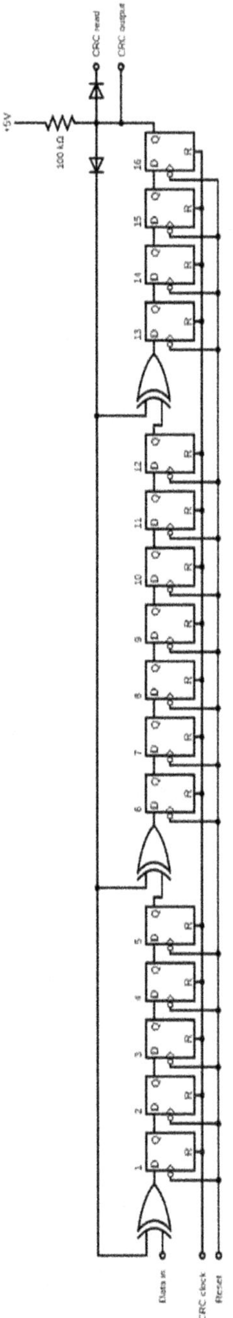

Fig. 2 CRC circuit

- When the receiver calculates CRC, it finds that the polynomial is completely divisible by the key if there is no error occurred in the message and if any of the key used for this project is famously known as CRC Modem is a 16-bit CRC key and of the form $x^{16}–x^{12}–x^5–1$ [10].
- Flip-flops are non-zero then it means there is some error in signal
- Error is shown on the LCD [11].

3.2 System Working Algorithm

- Step 1: Start the transmitter and receiver Arduino UNO [12].
- Step 2: Power up both Arduino UNO's.
- Step 3: At the transmitter side give the message to be transmitted and convert it into binary code and calculate CRC.
- Step 4: Transmit the binary coded data along with CRC. Step 5: When the data is transmitted via channel the noise will be entered into data.
- Step 6: Through the channel data reaches the receiver end and receiver receives the data.
- Step 7: CRC is calculated subsequently from received data at the receiver side.
- Step 8: If in the receiver end CRC of the signal is zero, then the received data is reliable and the error detection process stops.
- Step 9: If in the receiver end CRC of data is not zero, then the CRC of the received data is not zero then again data is sent to transmitter for calculating CRC.
- Step 10: This process continuous until CRC of received data & receiver becomes zero.

3.3 System Working Flowchart

See Fig. 3.

4 Result

4.1 Module Testing

See Table 1.

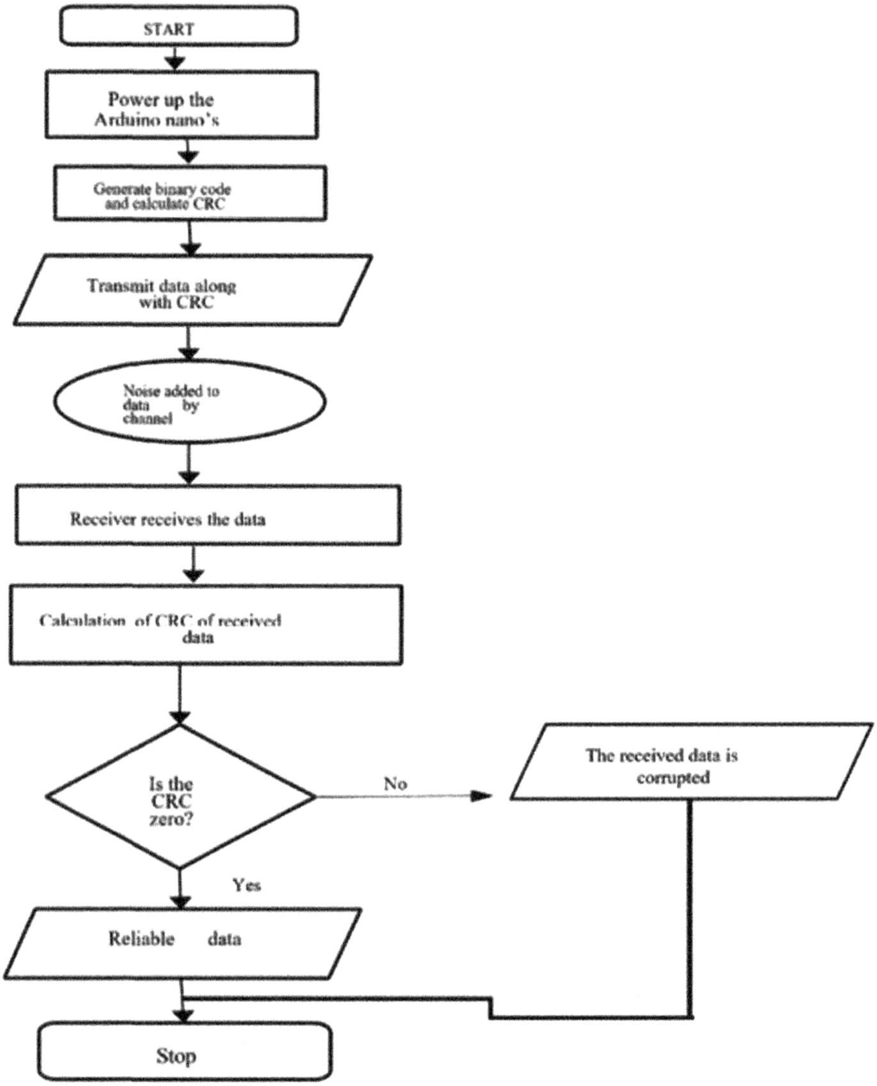

Fig. 3 Flow chart

Table 1 Power testing

Test	Expected	Actual
Connect the SMPS adaptor to power supply and measure the output voltages by using digital multimeter	Digital voltmeter should show the output as + 5VDC	Output obtained on digital voltmeter is + 4.93 VDC

Table 2 Power on led test

Test	Expected	Actual
Apply + 5VDC power to Arduino board	Power ON LED should glow	Power ON LED was glowing

Table 3 Arduino pin testing

Test	Expected	Actual
Apply + 5VDC power to Arduino board	LEDs should glow by writing a LEDs blinking program	LEDs were glowing

Table 4 LCD screen testing

Test	Expected	Actual
Apply power to LCD and run a standard test code to print "HELLO!" on the screen using Arduino	LCD should turn ON and display the message "Hello!"	The LCD screen successfully powered up and displayed the "Hello!" message

4.2 Arduino Board Testing

See Tables 2, 3 and 4.

4.3 Transmitter Testing

While transmitting the message "I'm errorless:)", the CRC circuit should calculate the CRC to be (1101 0000 0100 1011) and this should be visible on the LCD as well as the output of D Flip-Flops (Fig. 4).

4.4 Receiver Testing

When receiving the message "I'm errorless:)" along with the correct CRC and given that no error is induced in the channel, the CRC calculation circuit should display the CRC to be zero on the output of the D Flip-Flops (Fig. 5).

Fig. 4 Transmitter testing

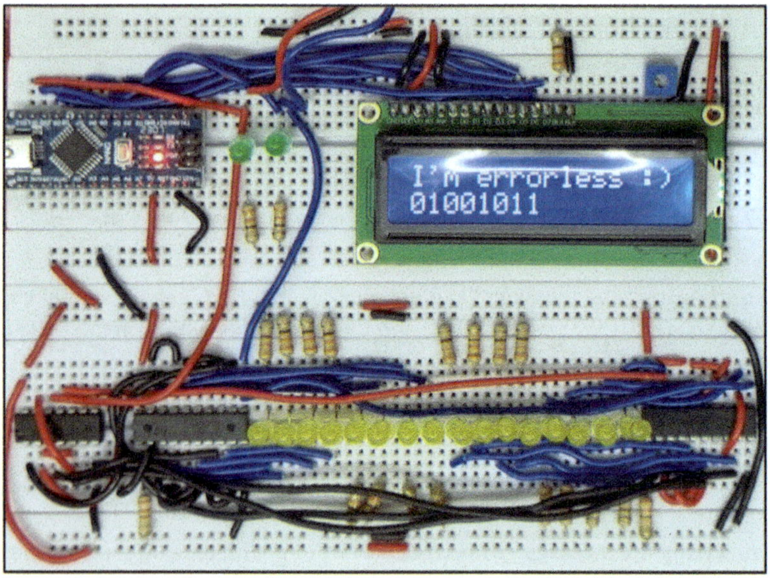

Fig. 5 Receiver Testing

4.5 Transmitter Code

```
#include<LiquidCrystal.h>

// Transmit rate in bps

#define TX_RATE 5
#define TX_CLOCK 2
#define TX_DATA 3
#define LCD_D4 4   #define
LCD_D5 5
#define LCD_D6 6
#define LCD_D7 7   #define
LCD_RS 8
#define LCD_EN 9

const char *message = "Hello, world!";   void setup() {
pinMode(TX_CLOCK, OUTPUT);   pinMode(TX_DATA,
OUTPUT);

 // Initialize the LCD screen
 LiquidCrystal lcd(LCD_RS, LCD_EN, LCD_D4,
LCD_D5, LCD_D6, LCD_D7);   lcd.begin(16,
2);  lcd.setCursor(0, 0);  lcd.print(message)   for (int byte_idx
= 0; byte_idx < strlen(message); byte_idx++)
{    char tx_byte = message[byte_idx];

   // Clear the second line of the display     lcd.noCursor();
lcd.setCursor(0, 1);    lcd.print("        ");
lcd.setCursor(byte_idx, 0);    lcd.cursor();      for (int bit_idx
= 0;  bit_idx < 8; bit_idx++) {      bool tx_bit  = tx_byte &
(0x80 >> bit_idx);      digitalWrite(TX_DATA, tx_bit);
delay((1000 / TX_RATE) / 2);

     // Update the LCD       lcd.noCursor();
lcd.setCursor(bit_idx, 1);      lcd.print(tx_bit ? "1" : "0");
lcd.setCursor(byte_idx, 0);
lcd.cursor();

     // Pulse clock       digitalWrite(TX_CLOCK, HIGH);
delay((1000 / TX_RATE) / 2);      digitalWrite(TX_CLOCK,
LOW);
   }  }
digitalWrite(TX_DATA, LOW);
}  void loop()
{
 // put your main code here, to run
repeatedly: }
```

4.6 Receiver Code

```
#include <LiquidCrystal.h>

// Pin assignments   #define RX_CLOCK 2
#define RX_DATA 3

#define LCD_D4 4
#define LCD_D5 5
#define LCD_D6 6
#define LCD_D7 7
#define LCD_RS 8
#define LCD_EN 9

LiquidCrystal lcd(LCD_RS, LCD_EN, LCD_D4,
LCD_D5, LCD_D6, LCD_D7);   char message[16];
volatile byte rx_byte = 0; volatile int bit_position = 0;
volatile bool update_lcd = true;    void setup()
pinMode(RX_DATA, INPUT);   strcpy(message, "");
lcd.begin(16, 2);
  attachInterrupt(digitalPinToInterrupt(RX_CLOCK
)
, onClockPulse, RISING);    }  void
onClockPulse() {    bool rx_bit =
digitalRead(RX_DATA);    if
(bit_position == 8) {    rx_byte = 0;
bit_position = 0;
 }   if (rx_bit) {    rx_byte |=   (0x80 >> bit_position);
 }    bit_position +=
1;     if (bit_position
== 8)
{    strncat(message, (const char
*)&rx_byte,
1);  }   update_lcd = true;  }
void loop() {  if
(update_lcd)  {    update_lcd = false;
lcd.noCursor();    lcd.setCursor(0, 0);
lcd.print(message);    lcd.setCursor(0, 1);
   for (int i = 0; i < 8; i += 1) {      if (i < bit_position) {
lcd.print(rx_byte & (0x80 >> i) ? "1" :
"0");      } else {       lcd.print(" ");
}
   lcd.setCursor(strlen(message), 0);
lcd. cursor();}}
```

5 Conclusion

The conclusion of data transmission and reception with error detection provides a reliable mechanism for ensuring the integrity of data being transferred between systems or devices.

Error detection techniques involve adding extra bits or information to the transmitted data, which can be used to check for errors upon reception.

Upon receiving the data, the receiver performs the error detection algorithm using the additional information. If the calculated result matches the received error detection code, it indicates that the data has been received correctly.

If there is a mismatch, it signifies that errors may have occurred during transmission, and further action can be taken, such as requesting retransmission or notifying the sender. The use of error detection techniques helps to ensure data integrity, reducing the chances of undetected errors and enabling reliable communication between systems. It is particularly crucial in applications where accuracy and reliability are essential, such as telecommunications, networking, file transfers, and digital storage systems. In conclusion, incorporating error detection mechanisms in data transmission and reception processes provides a valuable means of verifying data integrity and increasing the reliability of communication systems.

References

1. Ilievska N (2019) Towards the fixed length redundancy code. In: 2019 27th telecommunications forum (TELFOR). https://doi.org/10.1109/telfor48224.2019.8971113
2. Zeh A, Bezzateev S (2012) Describing a cyclic code by another cyclic code. In: 2012 IEEE international symposium on information theory proceedings https://doi.org/10.1109/isit.2012.6284054
3. Fujimoto K (2018) Checksum and cyclic redundancy check mechanism. Encycl Database Syst 423–424. https://doi.org/10.1007/978-1-4614-8265-9_1474
4. Turdiev O, Khomonenko A, Gofman MV (2022) Comparison of probable code number (PNC) and cyclic redundancy code (CRC). In: Vestnik of Russian New University. Series: Complex Systems: Models, Analysis, Management, 4, 119–131. https://doi.org/10.18137/rnu.v9187.21.04.p.119
5. Bhure RD, Manjunathachari K (2020) Quad code sequence generation using cyclic redundancy technique to enhance target detection in doppler rader. J Electr Eng Technol 15(6):2745–2757. https://doi.org/10.1007/s42835-020-00519-1
6. Zhilin I, Kreschuk A (2016) Generalized concatenated code constructions with low overhead for optical channels and NAND-flash memory. In: 2016 XV international symposium problems of redundancy in information and control systems (REDUNDANCY). https://doi.org/10.1109/red.2016.7779358
7. Yacoub EB (2023) Trapping and absorbing set enumerators for protograph-based generalized lowdensity parity-check code ensembles. In: 2023 XVIII International Symposium Problems of Redundancy in Information and Control Systems (REDUNDANCY) https://doi.org/10.1109/redundancy59964.2023.10330177

8. Deundyak VM, Kosolapov Yu V (2019) On the strength of asymmetric code cryptosystems based on the merging of generating matrices of linear codes. In: 2019 XVI international symposium "problems of redundancy in information and control systems" (REDUNDANCY). https://doi.org/10.1109/redundancy48165.2019.9003319
9. Sais M, Rafalia N, Abouchabaka J (2023) DNA technology for big data storage and error detection solutions hamming code vs cyclic redundancy check (CRC) E3S web of conferences 412, 01090. https://doi.org/10.1051/e3sconf/202341201090
10. Ferreira DC, van der Linden MG, de Oliveira LC, Onuchic JN, Pereira de Araújo AF (2016) Information and redundancy in the burial folding code of globular proteins within a wide range of shapes and sizes. Proteins Struct, Funct, Bioinform 84(4):515–531, Portico. https://doi.org/10.1002/prot.24998
11. Guarnieri G, Albani L, Ramponi G (2008) Minimum-error splitting algorithm for a dual layer LCD display—Part I, background and theory. J Disp Technol 4(4):383–390. https://doi.org/10.1109/jdt.2008.2001159
12. Abilash V, Kumar JPC (2017) Ardunio controlled landmine detection robot. In: 2017 third international conference on science technology engineering & management (ICONSTEM). https://doi.org/10.1109/iconstem.2017.8261366

An Amphibious Surface Cleaning Robot with GVM Wireless Control and Automatic Waste Segregation

Salma S. Shahapur, Megha M. Bagodi, Rakshita M. Patil, Komal Wajantri, Keshav Patil, Pooja Kanabur, and Abhishek Hadimani

Abstract The driving force behind this initiative is the environmental threat that non-biodegradable waste causes to terrestrial and marine life, is the main reason behind this. In addition to emphasizing the design and construction of amphibious surface cleaning mechanisms and the segregation of waste, this project's primary goal is to cut down on labor and time requirements, which will increase the efficiency of the machinery for river surface cleaning, land surface cleaning, waste separation, and implementable water quality monitoring. The term "GVM" (gesture, voice, manually) refers to the hand operated framework that is controlled manually, meaning that the robot can be directed with the hand. Similarly, the voice controlled system allows the robot to be directed with the voice, and the gesture controlled allows system for directing the robot with gesture. Robot is designed so that waste or debris is collected by the front escalator net, which then delivers the collected surface waste to the waste tank where waste is segregated according to its type. For example, if the waste is dry waste, that is if the robot is operation on the ground, it will be situated in one chamber; if it is wet waste, biodegradable waste, such as marine litter, it will be positioned in a different tank, where a coded Arduino board and an ultrasonic sensor will activate an automated setup to sort it. Further we can also add the additional values such as, Water quality meter, on which we will affix a sensor to measure certain river water parameters like PH and turbidity so we are able to keep an eye on everything.

Keywords Amphibious · GVM (gesture voice manually) · Waste tank · Front escalator net · Segregation · Automated setup · Water parameters

---

S. S. Shahapur (✉) · M. M. Bagodi · R. M. Patil · K. Wajantri · K. Patil · P. Kanabur · A. Hadimani
Jain College of Engineering, VTU, Belagavi, Karnataka, India
e-mail: salma.jce@gmail.com

V. K. Gunjan and J. M. Zurada (eds.), *Proceedings of 5th International Conference on Recent Trends in Machine Learning, IoT, Smart Cities and Applications*, Lecture Notes in Networks and Systems 1181, https://doi.org/10.1007/978-981-97-8861-3_14

1 Introduction

We've developed an innovative solution to tackle the escalating issue of polluted water bodies and landscapes—a cutting-edge robotic system known as the "Amphibious Surface Cleaning Robot with GVM Wireless Control and Automatic Waste Segregation." With nearly two million tons of plastic reaching oceans [1], causing harm to marine life, this technology aims to revolutionize environmental cleanup. The GVM wireless control, featuring Gesture, Voice, and Manual modes, ensures effortless operation, enhancing user experience. This versatile robot moves seamlessly on land and water, offering a comprehensive cleaning solution. Construction kits contain servo motors, touch, distance or light sensors [2]. Its Automatic Waste Segregation function, powered by advanced sensors and algorithms, streamlines waste separation during the cleaning process, promoting proper disposal and environmental sustainability. As we embrace the era of smart automation, this inventive cleaning robot sets the standard for autonomous solutions, emphasizing efficiency, user-friendly design, and ecological consciousness.

A. Novelties brought to this project.

1. Cleaning of Versatile Surfaces: This robot is very adaptable, cleaning a variety of surfaces both on land and in the water with ease and fitting in with the surroundings.
2. Capability to Amphibious: Because of its amphibious characteristics, the robot can move between land and water surfaces with ease, making it possible to thoroughly clean an array of locations, including ponds, lakeshores, and urban landscapes.
3. Gesture Control via Wireless: With its simple wireless gesture control technology, users can effortlessly direct the robot's motions and activities with simple hand gestures, offering a hands-free and engaging control experience.
4. Voice Manual Control: By utilizing voice instructions, users may operate the robot, improving accessibility and convenience. This capability comes in very handy when it's necessary to operate hands-free, including while cleaning up dangerous garbage or in places where manual control is difficult.
5. Monitoring in real time: With its real-time monitoring features, the robot can update customers on its current state of operation, cleaning progress, and any obstructions in its route.
6. Use of Renewable Energy Utilizing: renewable energy sources, such as solar panels or rechargeable batteries, the robot reduces its environmental footprint and facilitates extended cleaning sessions.
7. Disk for Automatic Waste Segregation: The robot has a customized disk for automated trash segregation that is driven by a servo motor. It can gather and arrange different kinds of waste into designated sections.
8. Compliance with Environmental Standards: The robot's construction complies with environmental laws and guidelines, guaranteeing that its activities support environmentally friendly waste management techniques and promote a healthier, cleaner environment.

9. Modular Design for Upgradability: The robot's modular design makes it easy for add new features and improvements, enabling it to keep up with changing user needs and technology developments.
10. Remote Monitoring and Control: Through a specialized interface, users may remotely monitor and control the robot's actions, enabling real-time modifications and intervention as needed.

2 Literature Survey

In [3] the authors proposed an unmanned ship Ro-Boat, which can detect, collect and remove debris, chemical sewage present on the surface autonomously. But still, there is no provision for the security of bots in water. Gao and Fu [4] proposed a miniature water surface garbage cleaning bot. STC12C5A60S2 controller is used along with WiFi for wireless control. A mobile app is developed to send instructions to the controller. A lightweight prototype is developed, but is restricted to a range 20 meters and is not waterproofed. However, the high cost of the sensors precludes the commercial development of efficient floor cleaning robots. At this moment, most of the commercially available mobile robots are based upon random path-planning algorithms [5] with very few sensors as magnetic encoders and contact or non-contact collision detection, sensors of the mobile robot and the previous or acquired information of the area where the robot operates [6]. Considering this problem, a system has been introduced through this paper based on dual-band Global System for Mobile Communications (GSM) module to get the seamless connectivity between waste segregation machine and monitoring room or person without considering the distance. The 8051 microcontroller is used, wherein the model consists of a stationary unit that collects waste into a funnel shaped collector. Sensors are placed on this funnel [7]. The aim of integration of remote hands on experiments into control of robotics systems is to minimize the gap between the robot theory and applications to teach students the use of remote robot control and to decrease the education costs [8].

3 Methodology

The three primary components of the constructed prototype are the flight controller, the control system, and the automatic waste segregation feature. The filter conveyor contains pores, which allows our robot to collect waste and release excess water when it is operating on the river's surface. The surplus water is then returned to the distributor, where it is distributed in accordance with that.

In Fig. 1's block diagram, the control mechanism comprises an Arduino Uno, HC-05 Bluetooth, L298N motor driver, and two DC motors. User commands, transmitted via the controller, are received by a Bluetooth module, then processed by

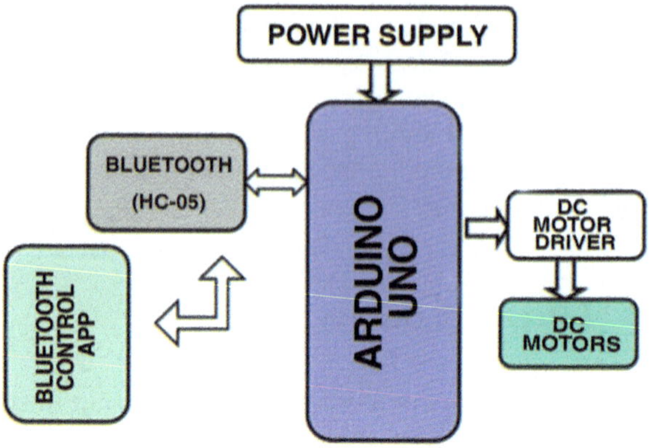

Fig. 1 Block diagram of controller

the Arduino for motor actions like forward, backward, left, right, and stop. The GVM system allows control through gesture, voice, or manual inputs, translating analog signals into electrical commands for seamless user access. This efficient design enables effective robot control through various means, emphasizing accessibility and versatility. We examine Fig. 2 for the automatic waste segregation process. Three ultrasonic sensors, three LED, moisture sensor, LCD displays, and one servo motor are all connected to the Arduino, it is the brains behind the operation. When a dry waste bin is full, an LCD display will show the range and the message "Dry bin is full." Similarly, when a wet bin is full, an LCD will display the range and the message "Wet bin is full." These two ultrasonic detectors are used to track the amount of waste in the bins they are installed within. In Fig. 2, the waste segregation system connects three ultrasonic sensors, three LEDs, a moisture sensor, LCD displays, and a servo motor to the Arduino, functioning as the core. When the dry or wet waste bins are full, the LCD displays the range with corresponding messages. Ultrasonic detectors, installed within the bins, monitor waste levels, ensuring efficient waste management.

Furthermore, the RED led will flash at that moment if both bins are filled. Now for the fun part: after the flight conveyor delivers the waste to the distributor, if the waste falls within the ultrasonic sensor's defined range, the sensor will detect it and feed the data to the Arduino; if the sensor detects moisture, the servo motor will rotate to an angle of 290; once the work is completed, the servo will return to its neutral state that is angle of 0, and the blue led will blink. Following this, the LCD will display the detected waste as "it's a wet garbage," with the volume of both the dry and wet bins, labeled "DRY WET." Likewise, all of these steps will be repeated for the dry waste, with the exception that the servo motor will now rotate to an angle of 70 before coming to a stop and flashing the yellow light. In case when both bins are full, the red light will blink and the LCD will show "both bins are full," alerting us to the requirement to clean both bins. The LCD will not display "DRY garbage detected" or the range of the bins. This is how it will function within the automatic mode.

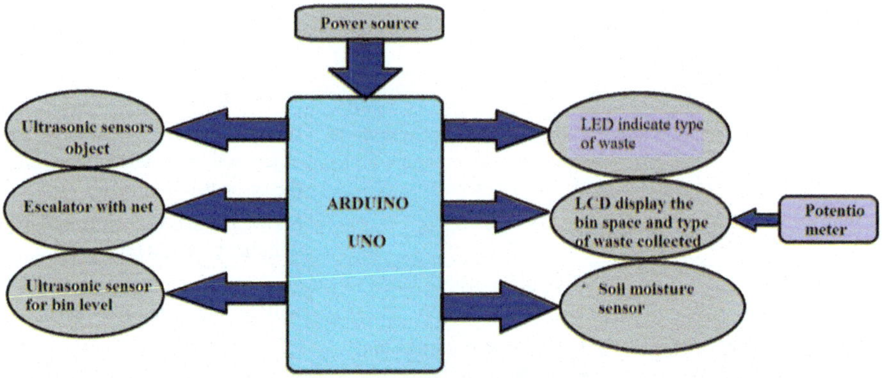

Fig. 2 Block diagram of automatic waste segregation mechanism

Since the robot is amphibious, it can operate on both land and water surfaces. This means that collected waste may mix, becoming both wet and dry. To avoid this, we have developed automatic waste segregation, which has many functions of its own as previously mentioned. If the garbage is not there then the LCD will be displaying "No Garbage".

4 System Design

The design portion of our suggested model is now available for viewing. The 3D model can be seen as in Fig. 3.

Notice how the wheel is simply fastened to the propeller, enabling the robot to function as an amphibious one. The front portion, which we refer to as the conveyor,

Fig. 3 3D view of developed model

Table 1 Sensors and their working range

Sensors and their working range		
Components	Range	Working
Ultrasonic sensor which are in the bin	We have two of them both of which have the range < 5	If object follows in that range the object will be detected meaning the respective bin is full
Liqud crystal display	3.3–5 V	Shows the message in accordance with the received command
Moisture sensor	Sense the moisture when value is < 1000	Once the moisture is detected it will be considered to be wet
Object detection through ultrasonic sensor	Detect value of object that should be lying under the range 10	Object detection is the main work that is carried out by this
Servo motor	For wet waste 290 Dry waste 70	The primary element in the process of separating the trash is the servo motor

is what will gather waste from the water's surface and the land's surface, where it drags the waste upside using the externally attached motors rotating in an anti-clockwise manner. This waste will then be delivered to the distributer, where the presence of garbage is detected by ultrasonic sensor that is visible in the model; once detected, it will examine what kind of garbage it is, causing the LED light to blink. Simultaneously, the moisture sensor will also be operating, which we have precisely placed on the distributer disk. Following all of these actions, the LCD will display the information.

This is how the suggested model will appear from the exterior alone. Next, for Buoyancy, we have connected the bottle or another material with the reduced density to ensure that the robot would float when working on the river surface. The additional front wheel is attached just for the balance.

Now let us know that in what range our sensors will be working by looking at the below table (Table 1).

5 Control System

To comprehend anything, we must first consult the circuit diagram. It will enable us to understand how the system is truly operated. Let's first examine how the control mechanism functions.

Here in Fig. 4, the Arduino Uno has a connection to the Bluetooth. Two motors can be operated by one motor driver, as demonstrated by the fact that when we give a command using voice, gesture, or manual input, the command is translated and received by the receiver. The robot will then be directed in accordance with the code we provide, moreover, because this is radio frequency based and no need to

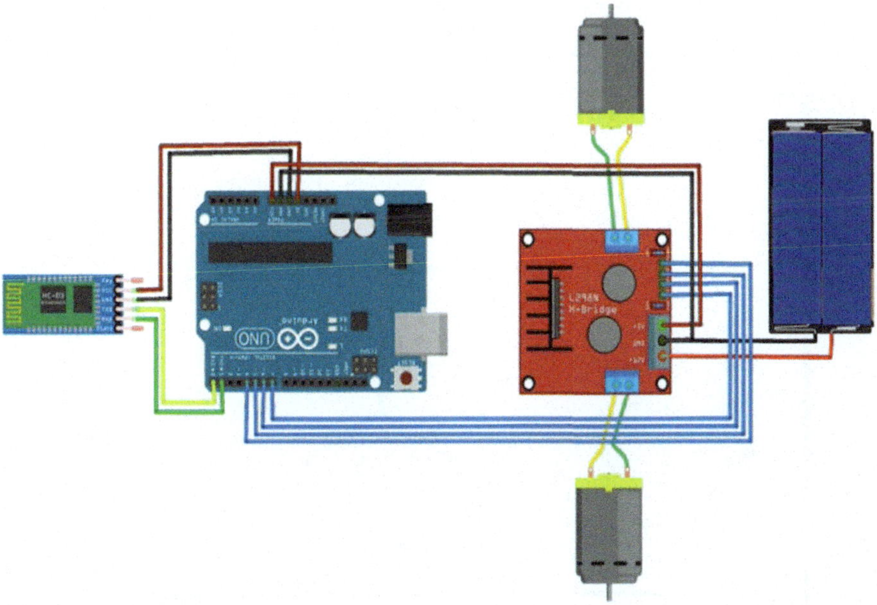

Fig. 4 Circuit diagram of the control mechanism

connect the controller with a wire-it can be wireless. The paper [9] explain about the implementation and design of gesture controlled robot. This constructs the communication link between technology and mankind. The target of this work is to upgrade the complete security to the robot and to simplify the controlling mechanism [10]. The voice directions are handled, utilizing an advanced mobile phone.

Now let us see the algorithm on which we have coded the Arduino to work as the Gesture, voice, manually controlled control system as in Fig. 5.

6 Waste Segregation Mechanism

Garbage segregation is the process by which garbage is divided into dry and wet categories. Because it is automated, it can execute all of these tasks on its own, making it operate more efficiently. It segregates the waste into wet waste and dry waste. It utilizes capacitive sensors to perceive wet and dry waste and parallel resonant impedance identifying component to perceive metallic things [11].

As in Fig. 6, we shall now examine the operational element one by one as follows. The mixed waste is sorted based on the following methods at the industrial level [12].

1. When the conveyor passes waste via the distributer disk, the sensor will detect any thing that falls within the designated range of the distributer disk's detection;

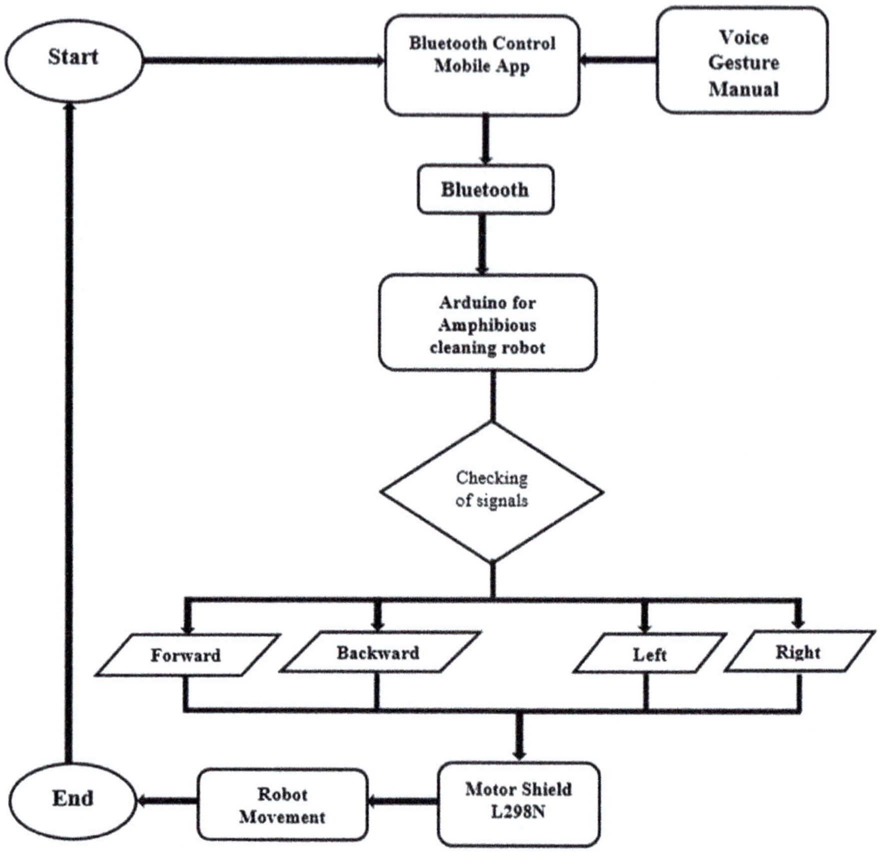

Fig. 5 Flowchart of control mechanism

nevertheless, the flat surface situated at the top of the servo motor itself is the distributer disk.

2. As it is a prototype, the ultrasonic sensor we have provided has a range of 10, but we can extend it for huge scaling and object identification as well.
3. Next, we have a moisture sensor that we have linked to the distributor's surface so that every time waste falls, it will pass through it and be measured for moisture.
4. Currently, There is a servo that rotates the waste distributor disk at a precise angle, directing the waste to fall into a certain bin according to its moisture content.
5. Ultrasonic sensor has been put in the bin to detect when it is full and will alert us when it is time to empty it. This sensor senses the quantity of waste fed into the bin.
6. Because everything is enclosed in a box that is hidden from view, our LCD display allows us to monitor everything and make judgments quickly.
7. As soon as a container is filled, the LEDs here will flicker to indicate what kind of rubbish is identified.

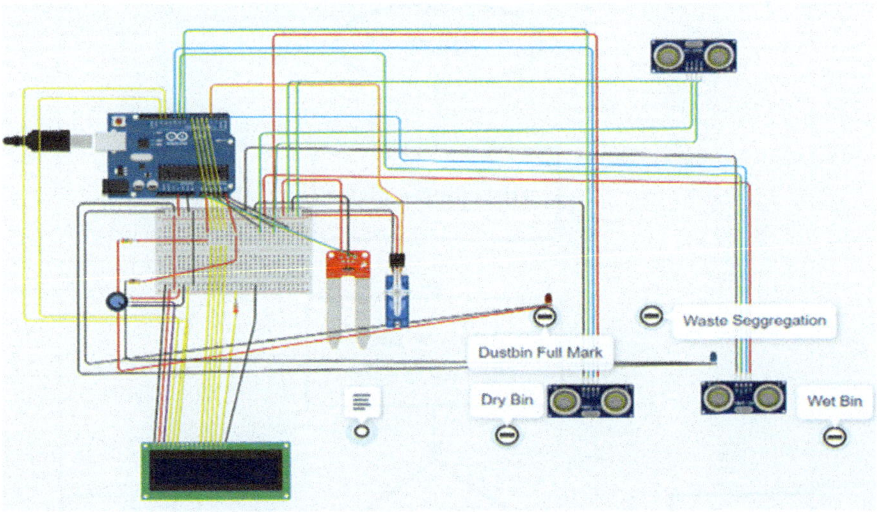

Fig. 6 Circuit diagram of automatic waste segregation

Let's examine the algorithm that powers this system's operation using Fig. 7.

7 Result

Our constructed prototype, with some of the functional components indicated, is depicted in the Fig. 8.

In the Fig. 9 the LCD display is showing "No Garbage Found" This indicates that there is nothing there that an ultrasonic sensor can detect.

When ultrasonic sensor detects the garbage it checks for the moisture content if the moisture is not there then the servo will tilt right, further LCD will display that and for detected dry garbage green LED will blink as per Fig. 10.

Similarly, as we have observed in Fig. 10, here in Fig. 11 Moisture sensor detects the wet garbage and tilt left bin followed by blinking of blue LED with LCD displayed message.

Here in Fig. 12 the values can be seen, where below DRY, the bin volume is present. Similarly, for WET also.

Ultrasonic sensors which are placed inside the bins will detect the garbage, if the respective bin is full LCD that will display it as shown in the Fig. 13. When both bins are full that time red light will blink and LCD will display as "Both Bins are Full".

In the Fig. 14 the top photo is about controlling the robot through Gesture, in the bottom left we have Voice command and in right we are having manual command which stands for GVM.

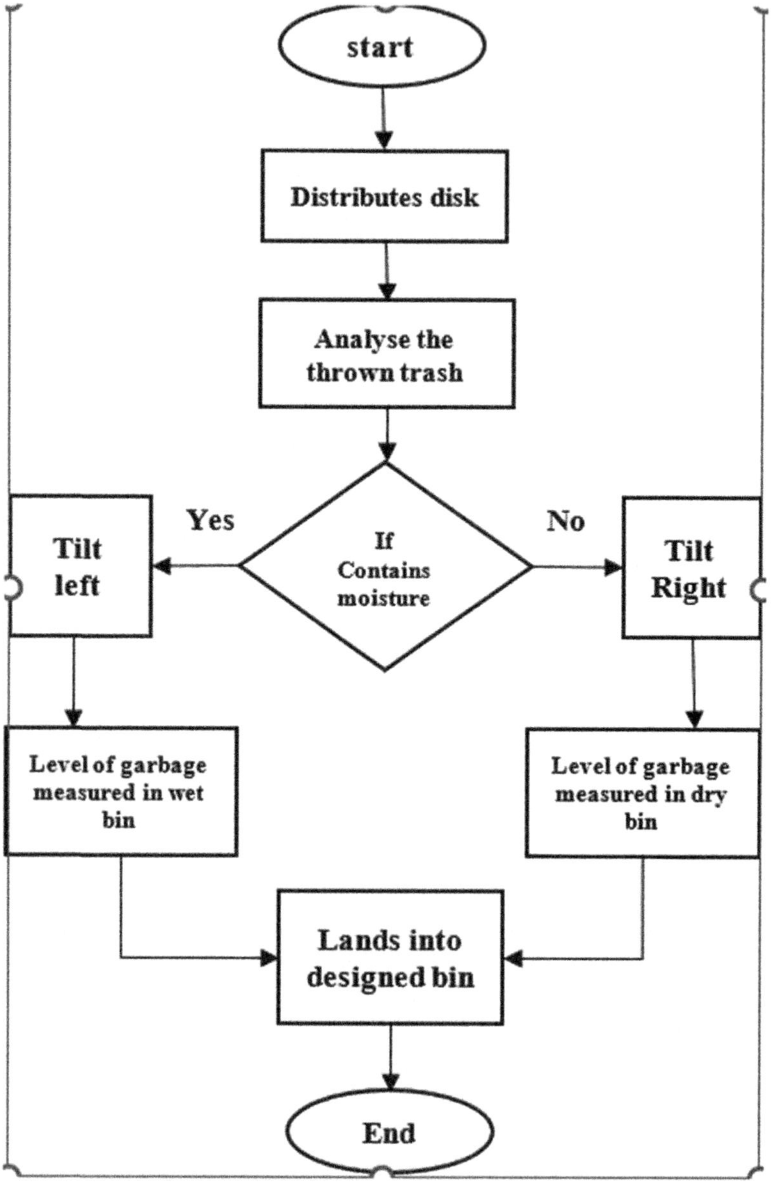

Fig. 7 Algorithm of automatic waste segregation

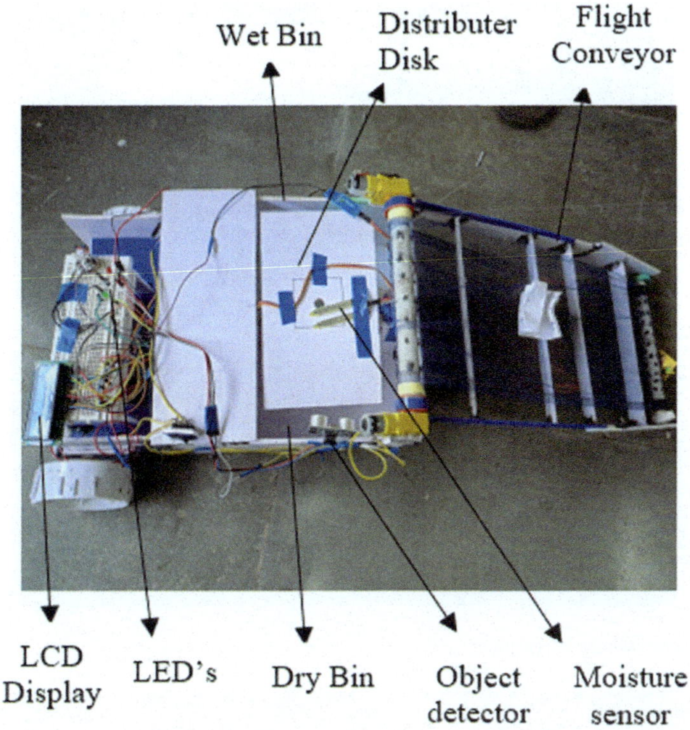

Fig. 8 Top view of the model

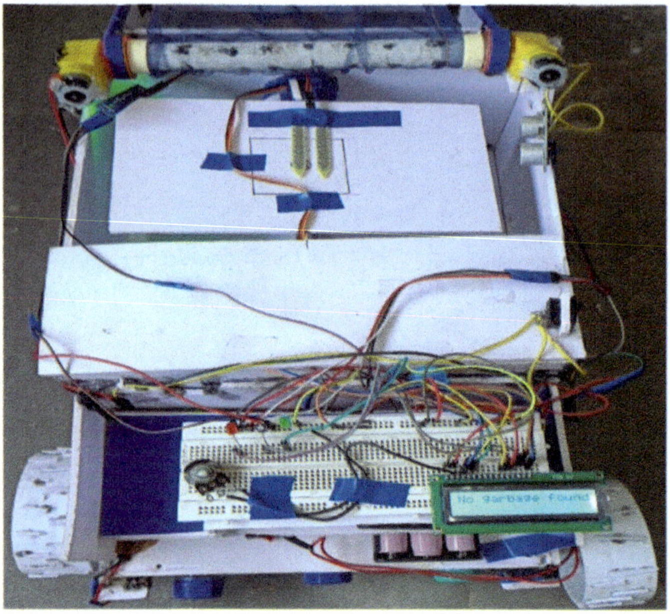

Fig. 9 Scenario where no garbage

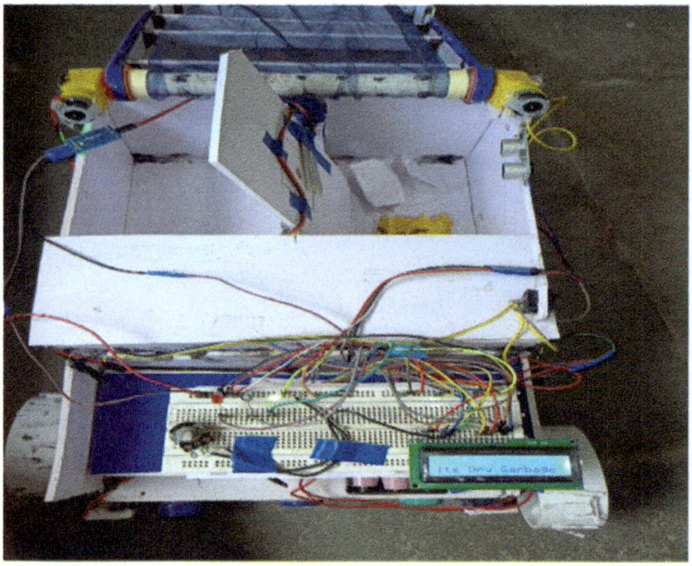

Fig. 10 It is dry garbage

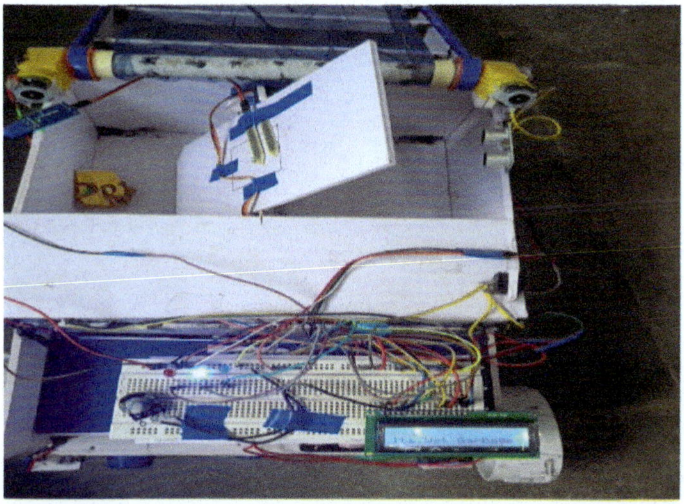

Fig. 11 It is Wet Garbage

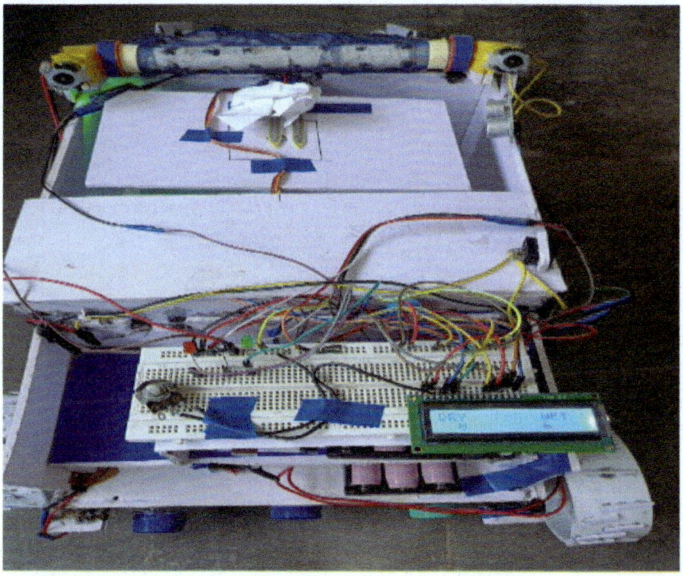

Fig. 12 Volume of bins when garbage is present

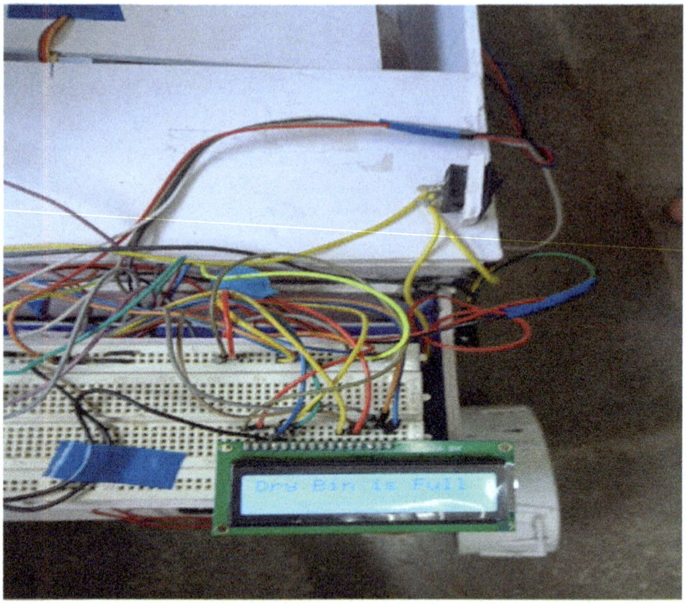

Fig. 13 Indicator of bins

Fig. 14 Control mechanism

8 Conclusion

For waste management and the environment, the innovative combination of an automated waste segregation system and an amphibious surface cleaning robot with wireless controls is groundbreaking. Sorting waste is made easy with this innovative approach, which also handles a variety of cleaning difficulties. While manual control gives versatility for specialized cleaning needs, voice commands and wireless gestures enable the robot to navigate various surfaces with ease. With the use of intelligent sensors, its automated waste segregation system can easily differentiate between dry and wet trash, resulting in effective separation for recycling and appropriate disposal. In addition to clearing up the area, this lessens the damaging effects that incorrect waste management has on the ecosystem. This all-encompassing strategy encourages cleaner, healthier, and more sustainable communities and is in line with the worldwide movement for eco-friendly behaviors.

References

1. Srivastava A, Prathna TC (2021) Urban water resource management: experience from the revival of Rajokri lake in Delhi. AIMS Environ Sci 8(5):421–434. https://doi.org/10.3934/environsci.2021027
2. Dogruel L, Joeckel S, Bowman ND (2014) Choosing the right app: an exploratory perspective on heuristic decision processes for smartphone app selection. Mob Media Amp Commun 3(1):125–144. https://doi.org/10.1177/2050157914557509
3. Kumar IN, Banker A (2023) Street cleaning machinery study–a review of existing tools and techniques. Int J Sci Res (IJSR) 12(10):150–156. https://doi.org/10.21275/sr23207150105
4. Gao X, Fu X (2020) Miniature water surface garbage cleaning robot. In: 2020 International conference on computer engineering and application (ICCEA). https://doi.org/10.1109/iccea50009.2020.00176
5. Swathi R, Ananya R, Kshama KM, Sreenivas Rao, DMV (2023) Design and implementation of a bluetooth controlled floor cleaning robot. IJARCCE 12(10). https://doi.org/10.17148/ijarcce.2023.121024
6. Choset H, Acar EU, Zhang SM (2001) Sensor-based coverage for demining: exact cellular decompositions and uncertain detectors. Dyn Syst Control. https://doi.org/10.1115/imece2001/dsc-24637
7. Nagarajapandian M, Prasanth UR, Kumar GS, Selvan ST (2015) Automatic irrigation system on sensing soil moisture content. IJIREEICE 96–98. https://doi.org/10.17148/ijireeice.2015.3120
8. Carusi F, Casini M, Prattichizzo D, Vicino A (2004) Distance learning in robotics and automation by remote control of Lego mobile robots. In: IEEE international conference on robotics and automation. ICRA 04. 2004. https://doi.org/10.1109/robot.2004.1308088
9. Ashish A, Gautam G, Kumar A (2014) Robot controlled by voice using mobile phone. J Adv Res Electr Amp Electron Eng 1(1):14–19. https://doi.org/10.53555/nneee.v1i1.267. ISSN: 2208-2395
10. Meghana M, Usha Kumari Ch, Sthuthi PJ, Mrinal P, Abhinav VSK, Reddy PS, Vikranth K, Kumar ST, Kumar PA (2020) Hand gesture recognition and voice controlled robot. Mater Today Proc 33:4121–4123. https://doi.org/10.1016/j.matpr.2020.06.553

11. Chandramohan A, Mendonca J, Shankar NR, Baheti NU, Krishnan NK, Suma MS (2014) Automated waste segregator. In: 2014 Texas instruments India educators conference (TIIEC). https://doi.org/10.1109/tiiec.2014.009
12. Manoiu V-M (2018) Waste collection, sorting and storage and food waste management on disney cruise line ships. European Sci J. https://doi.org/10.19044/esj.2018.c4p1

Automatic Speech Recognition (ASR) Elevator

C. M. Amar, Salma S. Shahapur, C. M. Abhishek, K. Uday Kiran, Arya Pulakeshi Talukar, Sahana Devali, and Krupa Rasane

Abstract Voice Recognition Technology has become a game-changer in the modern world of elevator systems, offering unique accessibility, hygienic conditions, and easier access. By using advanced speech recognition algorithms to identify and understand user commands, the voice recognizing elevator system help people to avoid physical touch with push buttons. To seamless integration of this technology, users including those with disabilities can navigate buildings with ease and get past conventional barriers and limitations. With voice recognition technology, the system can identify and understand spoken commands from the user. The elevator can then take the user inevitably to the desired floor, saving them from having to push any buttons. This is great way to mitigate the spread of germs and increase demand for disability people by minimizing physical contact with elevator buttons. The foundation is made to be user friendly inclusive of individuals with disabilities, and intuitive. It functions by recognizing and grasping spoken commands from the users through speech recognition software. This is a fantastic contribution of how technology can develop convenience and ease of our lives.

Keywords Speech recognition module · Inevitably · Viruses · Microcontroller · Electromyograph

C. M. Amar (✉) · S. S. Shahapur · C. M. Abhishek · K. Uday Kiran · A. P. Talukar · S. Devali · K. Rasane
Jain College of Engineering, VTU, Belagavi, Karnataka, India

C. M. Abhishek
e-mail: abhicm14@gmail.com

A. P. Talukar
e-mail: drrajanihp@jainbgm.in

V. K. Gunjan and J. M. Zurada (eds.), *Proceedings of 5th International Conference on Recent Trends in Machine Learning, IoT, Smart Cities and Applications*, Lecture Notes in Networks and Systems 1181, https://doi.org/10.1007/978-981-97-8861-3_15

1 Introduction

Elevators are essential to daily life whilst so many people count on them, especially when visiting offices, hospitals, airports, and hotels. Nevertheless, they are additionally believed to be among those spots where by viruses are most susceptible to spread, notably the ones that transmit by contact. The motive of this research is to tackle the problem of contact being the main means of transmission of viruses with a view to offer an adequate and secure transition environment especially at this stage of COVID-19 epidemic [1–3]. A researcher pertains to voice-driven elevators to their attempt to avert the virus from proliferating. These elevators are designed to perform by imparting conventional keypad operated elevators to voice-controlled ones. The researchers' subject of preference was how to retain people secured from the touch based virus in elevators by avoiding direct physical contact with the device [4]. One of the foremost significant parameters in the paper is the voice recognition model, which identifies human speech and sends it to the microcontroller for conversion of C programming language. In this way, the elevator is moved to the floor number that the user has entered, which is recognized by the microcontroller, and the screen displays that the elevator's floor number are the two most noticeable outputs. A person might want to construct an electrical elevator using voice recognition and a microcontroller for a few different reasons.

- **Accessibility**: The elevator is simpler to use for people with disabilities, such as blind or those with impaired mobility;
- **Convenience**: Voice recognition technology makes it possible to operate the elevator without physically touching buttons or control panels, which can be more hygienic;
- **Personalization**: The elevator is able to recognize different users to provide a more customized experience such as calling the elevator to a specific floor in response to a voice command.
- **Customization**: Design and functionality of the elevator is made more adaptable and unique by using a microcontroller. It is easily programmable and also can be updated or modified to carry out range of tasks.

Employing voice recognition in a microcontroller-based electrical elevator system has a number of benefits.

The Benefits Include

- **Enhanced Availability**: People with disabilities or mobility issues can use elevators more easily because voice recognition technology eliminates the need for them to physically push buttons.
- **Increased effectiveness**: Voice recognition technology can expedite the elevators arrival at its destination by allowing users to speak the floor they wish to go to instead of pressing a button

- **Increased safety**: Voice recognition provides yet another level of protection by restricting access to specific floors or sections of a building to only authorized person[5]

2 Literature Review

In [6] the Smart Stair Lift for Disabled, which has an IOT-based design, a separately fitted mobile chair, and the capacity to carry a wheelchair user. Additionally, it is less expensive than the civil installation and maintenance cost of an elevator is less. The stairway can be used with family members, other individuals, or an assistant. It is simple to put together and take apart, and it works by figuring out how much weight is on the chair. For stair lifts that are motorized and have the capacity to hold a stable individual in a seat, stair railings provide the essential means of navigation. There is no way to operate manually. One person at a time may use either forward or backward. In an emergency, evacuation can be difficult. Requires a backup battery in order to work. This system is limited to usage on straight staircases; curved stairs are not compatible with it.

In [7] integrated staircase lifts are recommended for home access. This staircase has a classic appearance and can be used as a platform lift at either end. Any wheelchair user or not, can access the top or bottom staircase platform so there is never a need to "call" or wait for the platform. The decision to use the stairs or an elevator, Ready to go on any level, self-powered models are offered along with possible aesthetic advantages for integrating the device into an already-existing house or in event of power outage. This is a fascinating study and invention that will be helpful for grounded independent homes, like villas, bungalows, hospitals, and outdoor entrances but not for multi-story residential complexes. The method is electrohydraulics, primarily an entry door solution for a meter-high home.

In [8] the authors offer a framework, method, and block diagram for a speech recognition system that uses a microcontroller and MATLAB software to be used in an elevator with no technical details regarding the motor, drive parameters, or capacity of the elevator. The results of this research project don't live up to the expectations.

In [9] for people with disabilities, Victorovna et al. (2017) recommend automating electromechanical lifts with mobile controls. It has something to do with the lift control system which uses Z-wave and open remote technologies. The wheelchair user can be elevated to a height of up to three meters by utilizing the cradles that come with the mobile control lift, which has a maximum load capacity of 300 kg. This is a mathematical model; no prototype is shown; the proposal cannot be guaranteed in the absence of a platform or a space that has been specially designed.

In [10] the authors proposed the Electromyography (EMG) Sensor-based mobility assistance robot for people with disabilities. This is a standing platform equipped with a Mobility Assistance Robot (MAR). Neutron motion is tracked during the electromyography process, and with the assistance of an Android mobile application and its own muscles, the robotic vehicle can be moved in any direction. With the help of this very expensive mobility aid—which the patient must own—people with walking disabilities can go outside, interact with others, dine out, and maintain their independent lifestyle. Its public use model, which is token-based, can function in luxurious outdoor environments.

In [11] the author introduces the implementation and analysis of an ARM 7 processor- based speech recognition system. The author discusses the MFCC and VQ algorithm in MATLAB 7.7 the control circuitry in Keil vision3, and the accompanying hardware setup diagram. Regularity of Mel Positive outcomes are obtained when the wrapping technology is applied.

In [12] the block diagram for the creation of the database and voice command training is presented. The MFCC from the database is compared with the MAT lab database that was previously created.

3 Complete Setup and Working

Here Figs. 1, 2, and 3 show the Front view, Back view of the prototype on how the intended speech-recognition electrical elevator would function in the real world, which were implemented using a microcontroller. When compared to traditional electric elevators, the proposed project in experimental results showed remarkable performance.

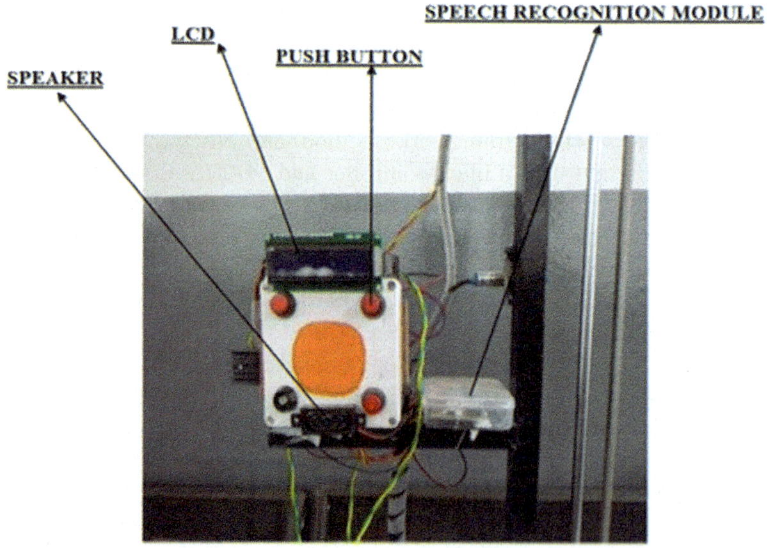

Fig. 1 Front veiw

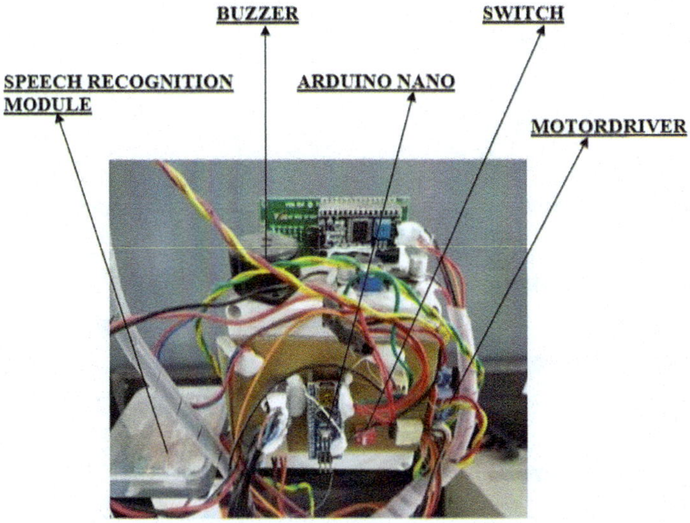

Fig. 2 Back view

Fig. 3 Back view of prototype

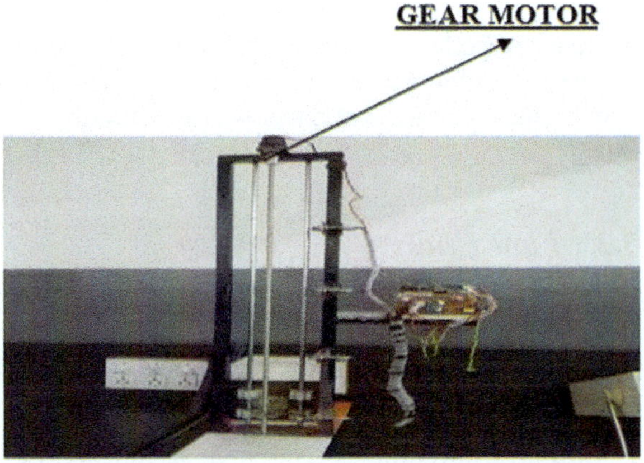

3.1 Block Diagram

In the Fig. 4 a speech recognition module is a piece of hardware designed to recognize, understand, and translate spoken language or commands into digital signals that a computer or microcontroller can process. It is a crucial component of voice recognition systems, allowing gadgets to comprehend spoken commands and react appropriately, as well as to communicate with users via voice commands. It is made up of a signal processing unit that evaluates the audio data that has been recorded and a microphone or array of microphones for capturing audio input. These modules are frequently combined with additional hardware elements, like embedded systems

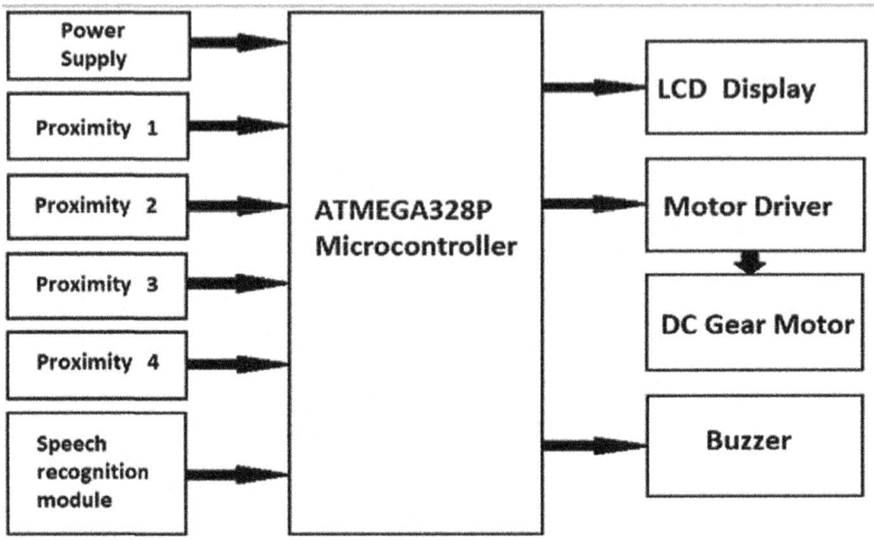

Fig. 4 Block diagram

or microcontrollers, to create a full voice recognition system. They improve accessibility for people with disabilities, allow hands-free operation, and offer simple and human-machine interaction that occurs naturally.

3.2 Flow Chart

Figure 5 shows the process of the flowchart for operating the elevator system with a floor display and button control. It shows how to operate an elevator by entering commands into a keyboard and pressing buttons in a specific order.

- **Initialization**: The system first establishes its internal state to ensure that everything is ready of use.
- **Ground Floor Check**: When the system reaches the ground floor, it uses additional sensors or limit switches to pinpoint its precise location.
- **Ground Level Exhibit**: When the elevator is on the ground floor, the system updates the ground floor display to reflect the current location. Handling serial input and buttons is done
- **Button test**: The system continually verifies that the button has been pressed for each floor option that is accessible.
- **Check for Serial Input**: The system keeps looking for signals coming from external devices or controllers

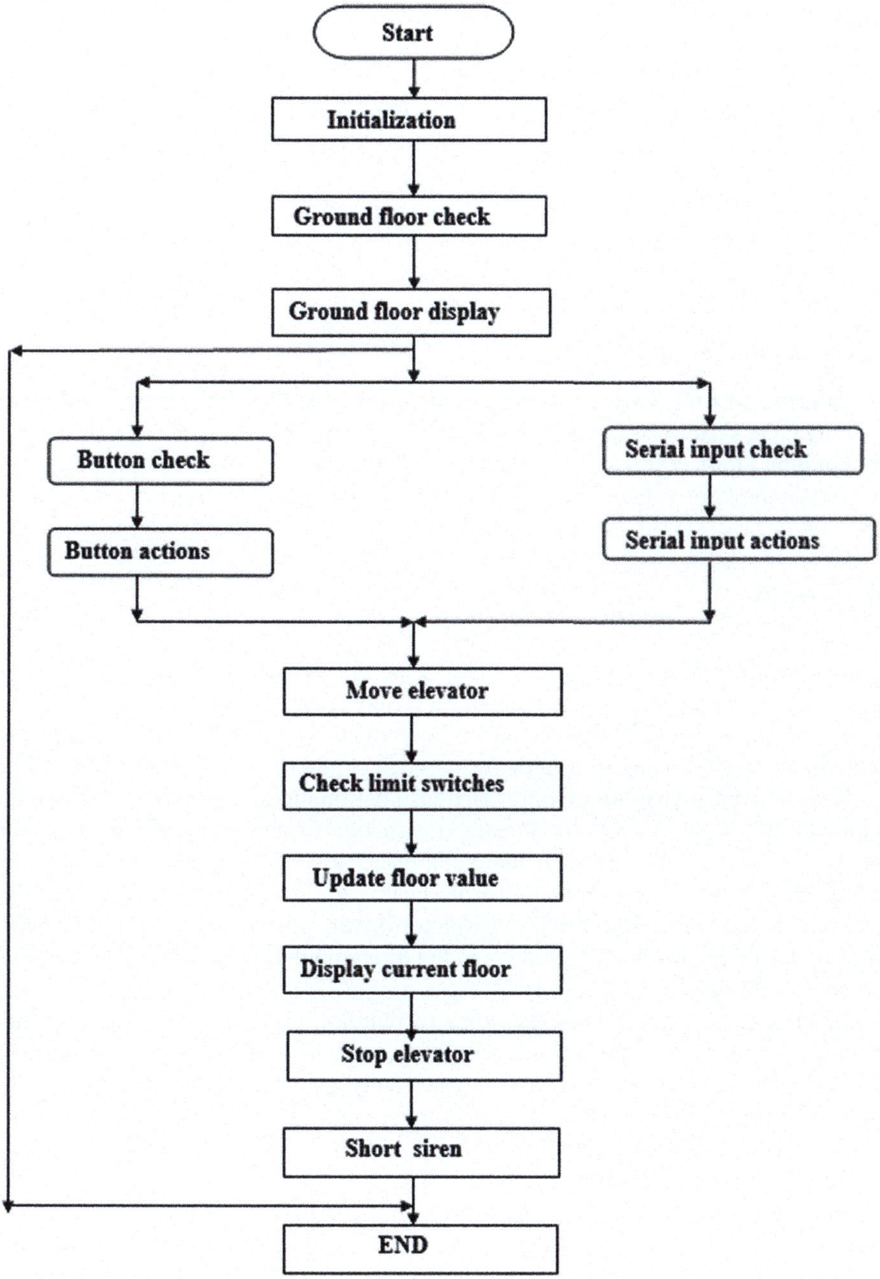

Fig. 5 Flow chart

On/Off Buttons

Locate the button and press it to reveal a particular bottom:

- **Elevator Stop**: The elevator comes to a stop at its present location.
- **Update the floor position**: The chosen floor is reflected in the elevator's internal floor value.
- **Launch the lift**: The chosen floor is reached by the elevator. Sequencing methodologies are proposed depending on how the input signals are sequenced.
- **Move Elevator**: To pinpoint the exact floor to which the elevator to be moved, a variety of inputs are used, limited switches and floor display updates.

While in the Elevator

- **Checking Limit Switch**: To ensure that elevator stays within its allowed travel range, the system keeps a close eye on the limit switch.
- **Update Floor Value**: The system uses the current location of the elevator to update the floor value in the absence of limit switches.

4 Result

The main controller used in this project is the microcontroller. The L289N motor driver is in charge gear motor that displays the elevator's location on a screen. Proximity sensors locate the elevator and a microcontroller receives voice commands from the user after being translated by a speech recognition module.

It initially ensures to figure out if any push button input has been detected on the ground floor, if not, it waits for an audio instruction from the user. After the speech recognition module identifies an audio command, it sends a signal to the microcontroller. The signal is first converted from analog to digital by the microcontroller, and then it was directed to the L289N motor driver, which in turn controls the DC gear motor that helps the lift to move in the direction that is intended (clockwise or counterclockwise).

Modern technology is used by voice recognition elevators to translate spoken commands into a hands-free experience. When a user approaches the elevator and say ground floor, the system knows what they said and uses speech recognition algorithms to translate spoken words into text. After processing the text, pre-programmed floor selection commands are compared to it.

The elevator system starts the procedures required to move the elevator to the ground floor as shown in the Fig. 6. By eliminating the need to physically press buttons, this seamless integration of voice recognition technology improves user experience and provides a practical and effective way to navigate between floors. In Fig. 7 we come to know that the position of the Elevator is displayed.

The elevator arrives at the first floor when the user speaks the command, "go to first floor". Figures 8 and 9 illustrate that the elevator is positioned in the first floor.

Fig. 6 Elevator at ground floor

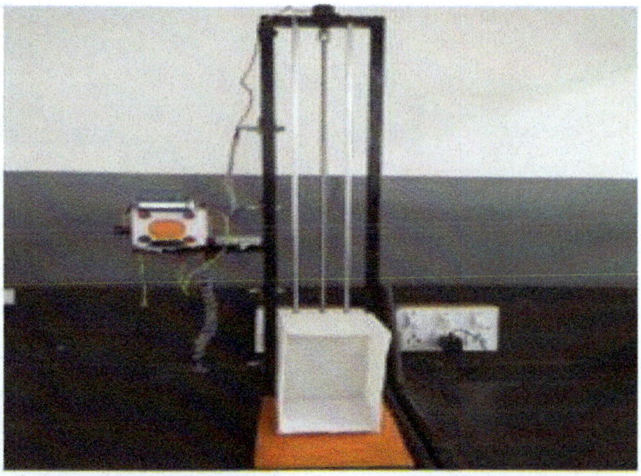

Fig. 7 LCD showing position of elevator

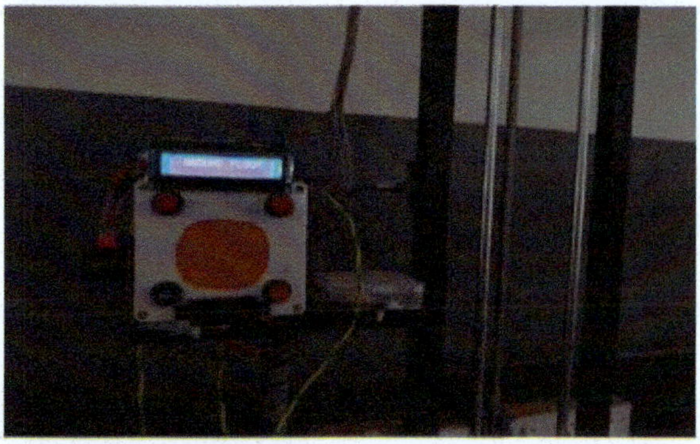

Fig. 8 Elevator at first floor

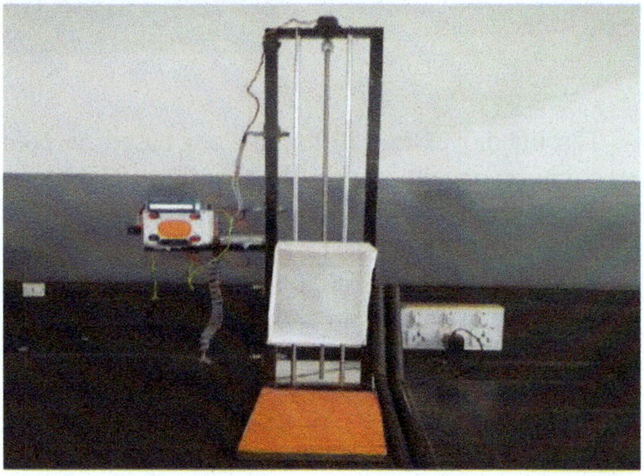

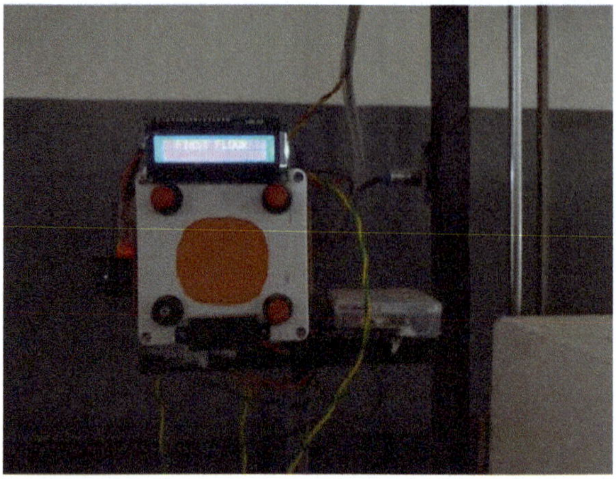

Fig. 9 LCD showing first floor

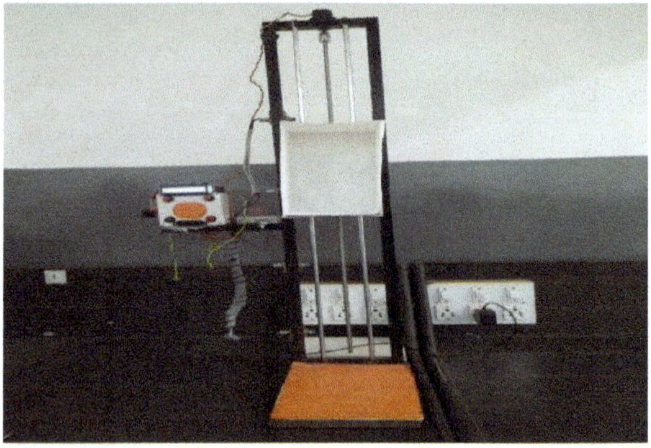

Fig. 10 Elevator at first floor

The lift moves to a different floor based on commands entered into the voice recognition, once it reaches the first floor. The lift travels to the second floor, as in Fig. 10 and stops with the help of proximity sensors before providing information on the screen, as in Fig. 11.

As indicated by Figs. 12 and 13 the elevator ultimately reaches the third floor in accordance with the command with the assistance of a voice recognition system.

4.1 Pseudo Code

Setup

Initialize the serial communication Initialize the LCD display

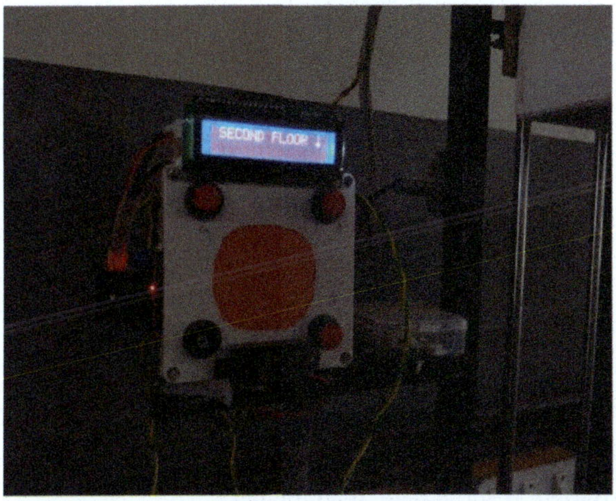

Fig. 11 LCD showing second floor

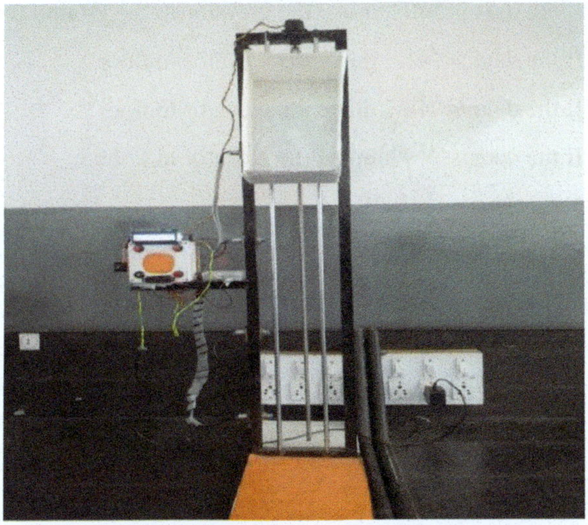

Fig. 12 Elevator at Third floor

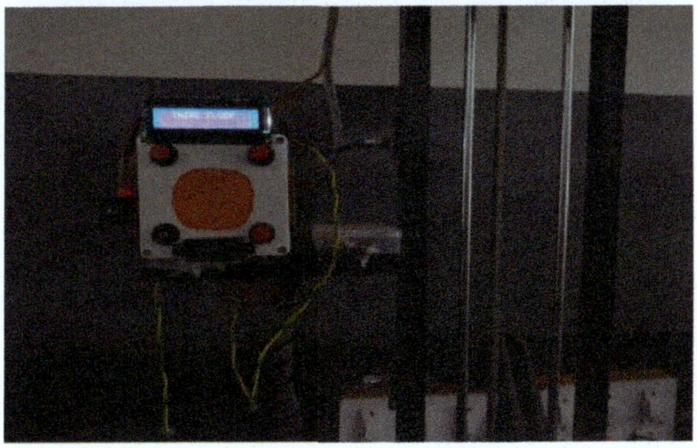

Fig. 13 LCD showing Third floor

Set pin modes for buttons, limit switches, motor pins, and buzzer Display a message on the LCD indicating initialization is complete Move the elevator to the ground floor

Loop

Check if any of the buttons are pressed

If button 1 is pressed, move the elevator to the ground floor

If button 2 is pressed, move the elevator to floor 1

If 3 is pressed, move the elevator to floor 2

If 4 is pressed, move the elevator to floor 3

Check if any serial data is available

If there is serial data, read the data

If the data is "40", move the elevator to the ground floor

If the data is "41", move the elevator to floor 1

If the data is "48", move the elevator to floor 2

If the data is "49", move the elevator to floor 3

Functions

gotogroundfloor():Move the elevator to the ground floor

gotofloor(int floornum):Move the elevator to the specified floor shortsiren():Play a short siren sound

motorclock():Turn on the motor in clockwise direction motoranticlock():Turn on the motor in anticlockwise direction motorstop():Stop the motor

5 Conclusion

An electric elevator with five floors that responds to sound commands was designed and put into service using the microcontroller. Sound commands are converted by the microcontroller into electrical signals that are sent between floors. The project was put to the test, executed practically, and performed admirably in a number of areas, the most important of which are arguably the security and service aspects. The elevator operates only when it has explicit instructions stored in it in order to ensure the safety of people, and it provided exceptional assistance to patients and individuals with special needs. There are several contexts in which this project can be implemented, such as homes, business, and medical facilities. It is also used to lessen the transmission of COVID-19 through touch, particularly in elevators.

References

1. Bhalerao M, Ohol S (2022) Smart vertical lift for disabled person. In: Proceedings of the international conference on industrial engineering and operations management. https://doi.org/10.46254/in02.20220470
2. Salim MH, Jaitapkar S, Ziyaan F, Ibrahim K (2022) Voice based lift control. Int J Res Appl Sci Eng Technol 10(5):3526–3529. https://doi.org/10.22214/ijraset.2022.42962
3. Hussain SM, Al Saadi EASN (2022) Voice enabled elevator system. J Stud Res. https://www.jsr.org/index.php/path/article/view/1490
4. Patil V (2021) Review on automated elevator-an attentive elevator to elevate using speech recognition. J Adv Res Power Electron Power Syst 08(1 and 2):20–26. https://doi.org/10.24321/2456.1401.202102
5. Fakhruldeen HF, Meri AA, Sa'id AH, Makttoof AN, Kadhim AM, Al-Asady HAJ (2023) An Arduino-based voice-recognition elevator for special purposes. Indones J Electr Eng Comput Sci 31(2):828. https://doi.org/10.11591/ijeecs.v31.i2.pp828-834
6. Navya K, Kumar BP, Mounika GH, Vineeth B, Rao KP, Pasha IA (2018) Smart stair lift for disabled and elderly. Int J Pure Appl Math 120(6):4647–4660
7. Mattie JL, Borisoff JF, Leland D, Miller WC (2015) Development of an integrated staircase lift for home access. J Rehabil Assist Technol Eng 2:205566831559407. https://doi.org/10.1177/2055668315594076
8. Gatane A, Dalvi A, Arkal P, Jagdale SM (2016) Using speech recognition create smart elevator controlling. Int Res J Eng Technol (IRJET) 3(03):1075–1077
9. Zudilova TV, Ivanov SE, Ivanova LN (2017) The automation of electromechanical lift for disabled people with control from a mobile device. In: 2017 computing conference. https://doi.org/10.1109/sai.2017.8252167
10. Ahmed SF, Ali A, Joyo MK, Rehan M, Siddiqui FA, Bhatti JA, Liaquat A, Dezfouli MMS (2018) Mobility assistance robot for disabled persons using electromyography (EMG) sensor. In: 2018 IEEE international conference on innovative research and development (ICIRD). https://doi.org/10.1109/icird.2018.8376304
11. Vijayan MN (2014) Analysis and implementation of speech recognition system using ARM7 processor. J Inf Eng Appl ISSN: 2224-5782. www.iiste.org
12. Thomas Mohan, Amrutha K, Anjana Anilkumar, Helen Johnson, Silsha K (2018) Voice operated intelligent lift. Int Res J Eng Technol (IRJET) e-ISSN: 2395–0056

Integrated Smart Water Management System for Urban Sustainability

Tejas K. Rayangoudar, Shahak Patil, Salma S. Shahapur, Chandan P. Kangralkar, Dayanand V. Daddi, Aditya A. Dhamanekar, and Balayya Walaginamath

Abstract TheWaterlogging and flooding. This study proposes an innovative prototype combining electronic devices, sensors, tanks, a natural sump, and "instant water filter" for efficient rainwater recycling. The system employs multiple tanks for water filteration, beginning with rainsensor activating the entire system upon rainfall. Rainwater is collected in the first tank, where an ultrasonic sensor monitors water levels. In cases of rain without a corresponding rise in tank water levels, a GSM module alerts the municipality, signaling potential blockages in the entry path. Manual cleaning addresses surface impurities, while colloidal impurities are treated with potash alum. The water undergoes sedimentation before passing through a sump filter, consisting of glass wool, activated charcoal, ceramic rocks, purified sand, and cellulose, effectively removing maximum impurities. Purified water accumulates in the second tank, equipped with turbidity and LDR sensor.. The turbidity sensor assesses water hardness and softness, while the LDR sensor checks water clarity. An Instant filter, boasting 98% efficiency, is employed for instant water filtering at the second tank's outlet valve, ensuring removal of remaining colloidal and soluble impurities, making the water fit for drinking. The treated water is suitable for drinking, irrigation, and various other purposes, contributing to urban water management. In addition to rainwater recycling, the study addresses sewage treatment, emphasizing three stages: primary, secondary (Activated Sludge Process), and tertiary treatment. Primary treatment involves screening, filtration, sedimentation, and coagulation for the removal of coarse, suspended, and floating solids. Secondary treatment focuses on organic matter removal through biological treatment. The sludge formed is removed, and the secondary effluent is treated with chlorine or UV radiation before discharge into running water or reuse for watering plants. Tertiary treatment is initiated for residual water with high concentrations of phosphates, heavy metal ions, colloidal impurities, and non-degradable organic matter. This involves the removal of phosphate,

---

T. K. Rayangoudar · S. Patil · S. S. Shahapur (✉) · C. P. Kangralkar · D. V. Daddi · A. A. Dhamanekar · B. Walaginamath
Jain College of Engineering, VTU, Belagavi, Karnataka, India
e-mail: salma.jce@gmail.com

C. P. Kangralkar
e-mail: chandankangralkar28@gmail.com

V. K. Gunjan and J. M. Zurada (eds.), *Proceedings of 5th International Conference on Recent Trends in Machine Learning, IoT, Smart Cities and Applications*, Lecture Notes in Networks and Systems 1181, https://doi.org/10.1007/978-981-97-8861-3_16

heavy metal ions, organic matter, and colloidal impurities through lime treatment, sulphide ions, activated charcoal addition, and alum, respectively. This integrated approach offers a promising solution for mitigating water-related challenges in urban environments and contributes to sustainable water management practices.

Keywords Waterlogging · LDR sensor · Ultrasonic sensor Turbidity sensor · Sump water filter · Rain water sensor

1 Introduction

The challenges of water scarcity loom large for developing nations like India, driven by factors such as population explosion, rapid industrialization, and global warming [1]. With a population exceeding 1.3 billion, India grapples with a mere 4% share of the world's fresh water resources, underscoring the pressing need for sustainable water management [2]. Research reveals that in India, chemical and microbial contamination affects over 70% of surface water, while more than 33% of rural groundwater sources face pollution, highlighting the critical importance of addressing water quality issues [3]. In urban centers such as Bangalore, the rapid urbanization and intense, short-duration rainfall events pose significant challenges, leading to waterlogging and flooding [4]. Rapid urbanization, population increase and environmental changes make communities more vulnerable to floods and waterlogging, which is becoming worse everyday, which is main reason for waterlogging in Rajshahi city, Bangladesh [5]. Beijing "7.21" rainstorm event and New York hurricane "Sandy" caused serious urban waterlogging, leading to significant causalities and property losses [6]. This paper addresses the urgent need for innovative water management solutions in the face of such environmental issues. One hour of rain is sufficient to trigger waterlogging and flooding in cities like Bangalore, necessitating immediate attention to sustainable water management practices.. Recognizing this, our research focuses on the development and implementation of a prototype designed to mitigate the adverse effects of rainwater inundation through the integration of electronic devices, sensors, tanks, a natural sump, and an Instant filter. The primary goal of this research is to introduce a comprehensive rainwater recycling system that combines advanced electronic technologies and traditional water treatment methods to alleviate the impact of rain-induced waterlogging and flooding. Our prototype involves a multi-step process, beginning with the deployment of a rain sensor outside the first tank to detect rainfall in the specific area. Once rain is detected, the entire system is activated, directing the rainwater into the first tank. An Ultrasonic sensor monitors the water level, and if anomalies arise, such as a lack of water level increase during rainfall, the system communicates the issue to the municipality via a GSM module, prompting action for potential blockage clearance. The rainwater undergoes a meticulous filtration process, starting with manual cleaning of surface impurities and treatment of colloidal impurities using potash alum. The water is then left for sedimentation, followed by passage through a sump filter containing glass wool, activated charcoal, ceramic rocks, purified sand, and cellulose. This process ensures the removal of maximum impurities, resulting in purified water that is subsequently accumulated in a second tank. The second tank is equipped with turbidity and LDR sensors. The turbidity sensor assesses water hardness and softness, while the LDR sensor checks water clarity. Based on predefined threshold values, the water is categorized as clean or impure. In both scenarios, the water in the second tank is compelled to pass through an Instant filter, an instant water filtering device situated at the outlet valve of the second tank. With an impressive 98% efficiency, the instant water filter clears any remaining colloidal and soluble impurities, rendering the water suitable for drinking. The treated water is now suitable for a variety of purposes, including drinking, irrigation, and general consumption, contributing to sustainable urban water management practices.

In addition to rainwater recycling, this study delves into sewage treatment, emphasizing three crucial stages: primary, secondary (Activated Sludge Process), and tertiary treatment. The primary and secondary treatments focus on the removal of solids and organic matter, respectively, while the tertiary treatment addresses residual water with concentrated contaminants. This research offers an integrated and sustainable approach to address the challenges posed by rain-induced waterlogging in urban environments, providing a blueprint for effective water management systems that can be applied to cities facing similar issues globally.

2 Literature Survey

The challenge of rapid urbanization leading to waterlogging and floods in cities during short-duration rainfall is a pervasive issue globally. Many researchers have explored innovative solutions to address this concern, and several themes emerge in the existing literature.

2.1 Rain Water Harvesting and Recycling

Two rainwater harvesting methods: surface runoff (collected in underground tank) and rooftop (collected from roof catchment). Harvested rainwater is pure and can be consumed directly. Quality varies; rooftop systems yield drinking water, while ground catchments may need filtration [7].

2.2 Smart Urban Water Management System

The utilization of electronic devices, sensors, and tanks for rainwater recycling has gained attention. A similar rainwater harvesting system was implemented, focusing on effective utilization of rain sensors and Ultrasonic sensors for efficient water accumulation [8]. The incorporation of GSM modules for real-time communication with municipal authorities in case of blockages aligns with the concept of smart urban water management. This paper highlights the importance of incorporating intelligent systems for timely issue resolution in urban environment.

2.3 Filtration Technologies

The sump filtration system, consists of glass wool, activated charcoal, ceramic rocks, purified sand, and cellulose. Cellulose has the potential to be used as an adsorbent in natural/modified or as a precursor for activated carbon for adsorbing substances from water [9]. Porous ceramics consist of a high level of porosity that results in properties such as high specific area, high permeability and high tortuosity. These factors make porous filters effective by physically (non-chemical process) removing suspended solids, filamentous bacteria and protozoa from drinking water [10].

2.4 Sensor Based Water Quality Monitoring

The use of turbidity and LDR sensors in the second tank for assessing water hardness, softness, and clarity corresponds to the evolving field of sensor-based water quality monitoring [11]. We explored the application of similar sensors for real-time water quality assessment, emphasizing their role in ensuring the safety of water for various purposes.

2.5 Instant Water Filtering Device

The introduction of the instant water filter with 98% efficiency reflects advancements in instant water filtering technology. We demonstrated a comparable device and its success in removing remaining colloidal and soluble impurities, making water fit for consumption. Regarding sewage treatment, the existing literature emphasizes the importance of the three-stage approach—primary, secondary, and tertiary treatment.

According to the pH value of sewage water, a particular amount of alum and chlorine is added into the water. Alum helps to settle the sludge at the bottom and chlorine is used to kill the bacteria in the water [12].

3 Methodology

3.1 Hardware Design

See Fig. 1.

3.2 Rainwater Harvesting and Filtration System for Urban Water Management

A. Rainwater Collection

Objective: Efficiently manage rainwater in urban areas prone to waterlogging. Figure 1 shows the hardware block diagram of our module. Procedure: Installation of rain sensor to detect rainfall. Activation of system upon rainfall detection.

B. Water Level Monitoring and Blockage Detection

Objective: Ensure proper functioning of the rainwater collection system. Procedure: Ultrasonic sensor in the first tank monitors water levels. GSM module alerts municipality if water levels are not increasing during rainfall, suggesting a blockage. Manual cleaning of surface impurities and treatment of colloidal impurities with potash alum.

C. Sedimentation and Sump Filtration

Objective: To remove suspended impurities and facilitate further filtration. Procedure: Sedimentation process for settling suspended particles. Passage of water through a sump filter with layers of glass wool, activated charcoal, ceramic rocks, purified sand, and cellulose.

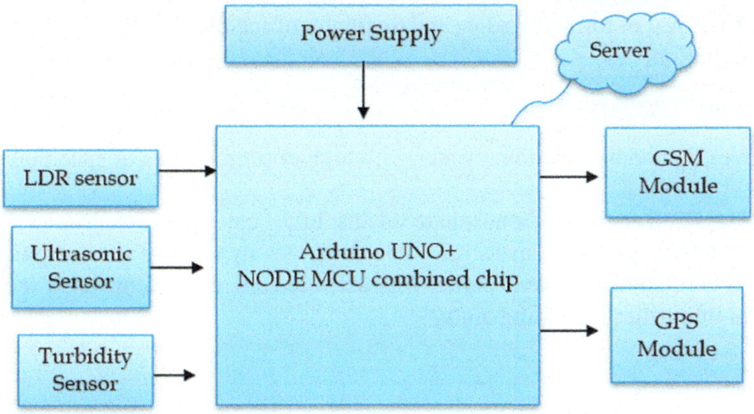

Fig. 1 Block diagram

D. Second Tank and Quality Monitoring

Objective: accumulate filtered water and monitor its quality. Procedure: Transfer of purified water to the second tank. Inclusion of turbidity and LDR sensor to asses water hardness, softness and clarity. Instant water filter usage for instant water filtering with 98% efficiency.

E. Final Water Supply

Objective: To provide clean water for various purpose in urban areas. Procedure: instant water filter ensures removal of remaining impurities. Supply oftreated water for drinking, irrigation and other purpose in urban areas.

3.3 Sewage Treatment

Sewage/waste water treatment involves breakdown of complex organic compounds in the wastewater into simpler compounds that are stable and nuisance-free, either physio-chemically and or by using micro-organisms [13].

A. Primary Treatment

Objective: Removal of coarse, suspended and floating solids. Procedure: Fig. 2. Shows the primary treatment of sewage water. Screening for large suspended or floating matter. Filtration through conventional sand beds. Sedimentation to remove suspended impurities. Coagulation with potash alum or ferrous sulphate for colloidal particle removal.

B. Secondary Treatment (Activated Sludge Process)

Objective: removal of organic matter. Procedure: Fig. 3. Shows the secondary treatment of sewage water Biological treatment in large tanks with activated sludge containing microorganisms. Vigorous oxygenation for aerobic oxidation of organic matter. Removal of formed sludge, and treatment of secondary effluent with chlorine or UV radiation.

C. Tertiary Treatment

Objective: T0maddress residual water with high concentrations of specific contaminants.

Procedure: Removal of phosphate as insoluble calcium phosphate with lime. Removal of heavy metal ions as insoluble sulphides by treating with sulphide ions. Removal of remaining organic matter with activated charcoal. Removal of remaining colloidal impurities by adding alum.

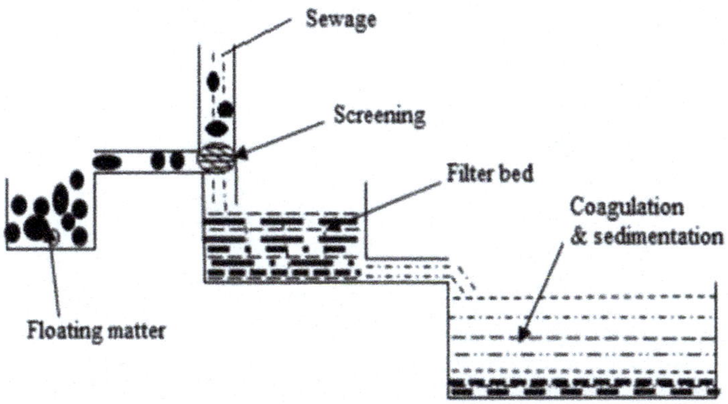

Fig. 2 Primary treatment

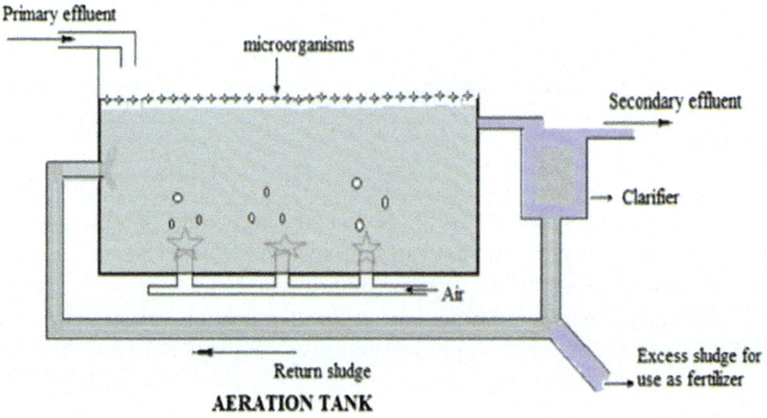

Fig. 3 Secondary treatment

4 Programming: Pseudocode

```
void loop()
{ // Check if it's raining
 int raindropValue = analogRead(raindropPin);
 if (raindropValue < 500) {
  // Read LDR value
 int ldrValue = analogRead(ldrPin);
  // Read turbidity value
 int turbidityValue = analogRead(turbidityPin);
 // Calculate water level
 float distance = getWaterLevel();
 // Check if water level is not increasing
 if (distance <= previousDistance) {
 // Send GPS location through GSM module
    sendGPSLocation();}
```

5 Result

Rainwater Harvesting system: Fig. 4. Shows the top view of our model which comprises of sensors. Rain sensor activates the system foe water accumulation. Ultrasonic senors monitor water levels, alerting authorities if blockage occur. Manual and chemical cleaning precedes sedimentation. Sump filtration (glass wool, activated charcoal, ceramic rocks, sand, cellulose) removes impurities. Figure 5. Shows the overview of sump filter. Purified water stored in a second tank with turbidity and LDR sensors.

Water Quality Assessment: Turbidity sensor detects hardness/softness. LDR sensor assesses water clarity second tank water sent through "instant water filter" (98% efficiency); the resultant water deemed fit for drinking, irrigation, and general use. Figure 6. Shows the combined view of the model. Figure 7. Shows the output observed of impure water in the serial monitor before filtration, and Fig. 8. Shows the output observed of impure water after passing through our model thereby leading to filtered output.

Sewage Treatment:

Three-stage process (primary, secondary, tertiary).

Primary: screening, filtration, sedimentation, cogulation.

Secondary: activated sludge process for organic matter removal. Effluent treated with chlorine/UV for bacterial removal.

Tertiary: Removal of phosphates, heavy metal ions, organic matter and colloidal impurities.

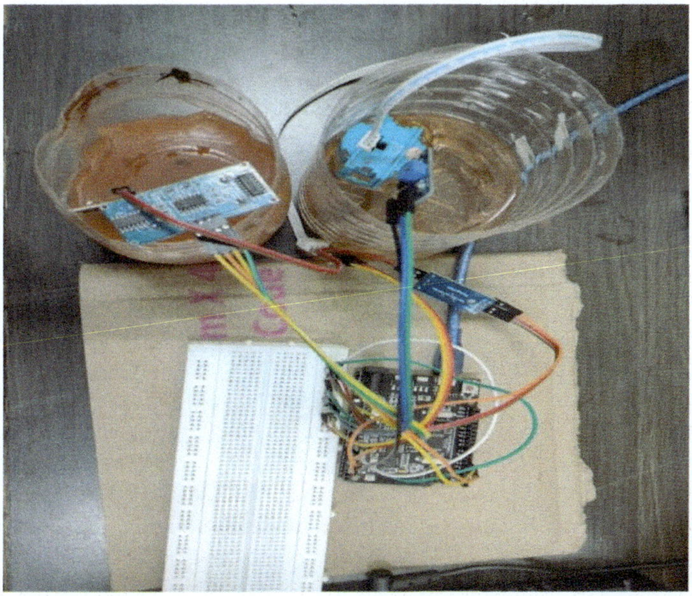

Fig. 4 Top view of the prototype

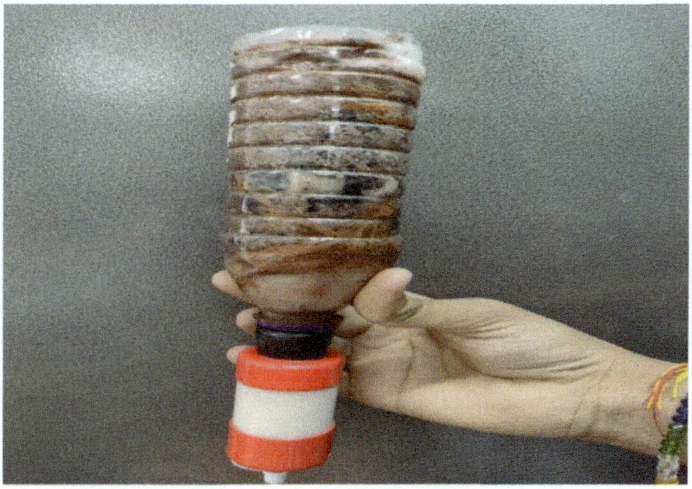

Fig. 5 Sump filter

These processes collectively address urban waterlogging, floods, and sewage contamination issues, providing a comprehensive solution for sustainable water management.

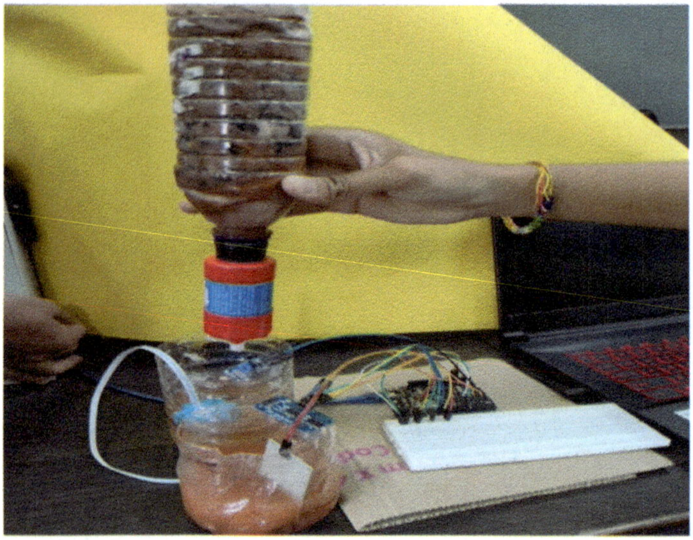

Fig. 6 Front view of the prototype

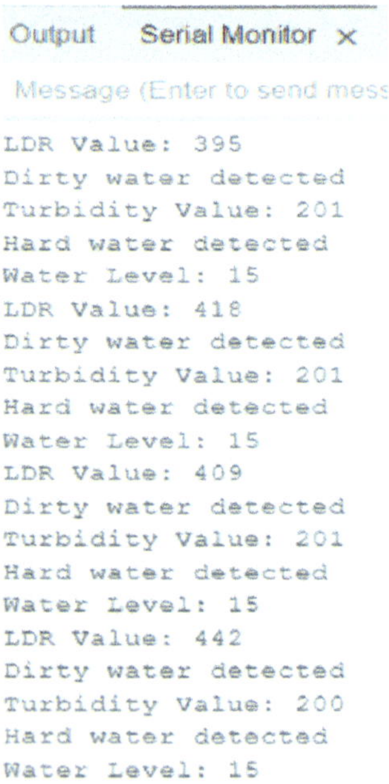

Fig. 7 Observed output of impure water

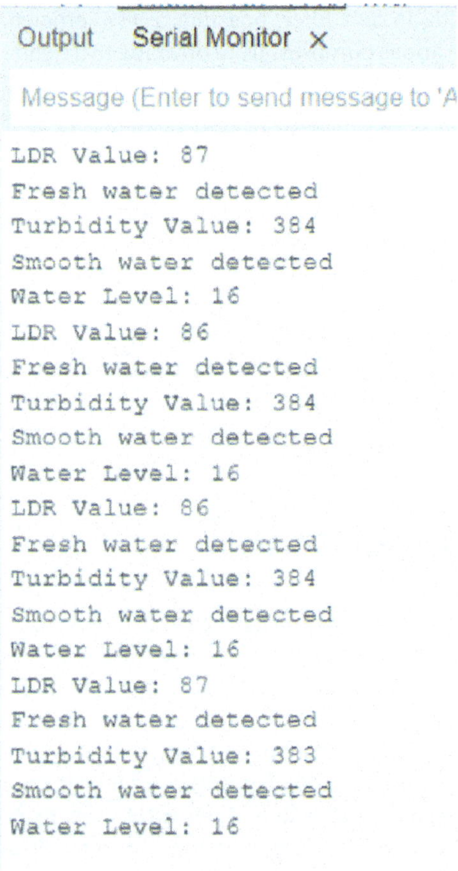

Fig. 8 Observed output of filtered water

6 Discussion

The proposed water recycling system addresses the pressing issue of waterlogging in densely populated cities during heavy rainfall. By integrating electronic devices, sensors, and a multi-tank filtration system, the prototype effectively captures, treats, and recycles rainwater. The rain sensor initiates the system, accumulating water in the first tank.

An Ultrasonic sensor monitors water levels, promoting alerts if blockage occur.

Manual and chemical methods include potash alum, manage surface and colloidal impurities, respectively. Sedimentation and sump filtration further refine the water, with a diverse filter composition including glass wool, activated charcoal, ceramic rocks, purified sand, and cellulose. This ensures maximum impurity removal, yielding purified water stored in a second tank.

Turbidity and LDR sensors in the second tank assess water quality. If water fails clarity or hardness standards, it undergoes additional purification through the Instant filter, boasting a 98% efficiency rate. The resultant clean water serves various purposes, from drinking to irrigation, contributing to sustainable urban water management.

In parallel, the sewage treatment process follows a three-stage approach: primary, secondary, and tertiary treatment. Methods such as screening, filtration, sedimentation, and coagulation efficiently remove contaminants. Secondary treatment involves biological processes like activated sludge, while tertiary treatment targets specific impurities with lime, sulphide ions, activated charcoal, and alum.

The combined water recycling and sewage treatment methodologies present a comprehensive approach to address water-related challenges in urban environments. Tha combined water recycling and sewage treatment methodologies present a comprehensive approach to address innovative system contributes to water conservation, pollution reduction, and sustainable water reuse practices, crucial for mitigating the impact of rapid urbanization on water resources.

7 Conclusion

In summary, our devised water recycling system, equipped with advanced sensors and a meticulous multi-tank filtration process, offers an effective solution to combat waterlogging issues in densely populated cities during heavy rainfall. The integration of electronic devices, rain sensor, and purification techniques ensures the efficient capture, treatment, and reuse of rainwater.

The developed prototype demonstrates practicality in urban settings, with an automated system that alerts authorities to potential blockages, manual and chemical processes for impurity removal, and a diverse sump filtration system for comprehensive water refinement. The resulting purified water, stored in a secondary tank, meets stringent quality standards.

Furthermore, our sewage treatment methodology, encompassing primary, secondary, and tertiary stages, showcases a holistic approach to contaminant removal. The incorporation of biological treatments, specific ion treatments, and advanced filtration techniques ensures the effective purification of sewage water.

In conclusion, our innovative water management system stands as a viable sustainable solution to the challenges posed by rapid urbanization. By addressing waterlogging concerns and implementing advanced purification process, our system contributes to the conservation of the water resources, reduction of pollution, and the establishment of sustainable practices for urban water reuse.

References

1. Ram SA, Irfan ZB (2021) Application of System Thinking Causal Loop Modelling in understanding water Crisis in India: A case for sustainable Integrated Water resources management across sectors. HydroResearch 4:1–10. https://doi.org/10.1016/j.hydres.2021.02.001
2. Bindal S, Singh CK (2019) Predicting groundwater arsenic contamination: regions at risk in highest populated state of India. Water Res 159:65–76. https://doi.org/10.1016/j.watres.2019.04.054
3. Shaji E, Santosh M, Sarath KV, Prakash P, Deepchand V, Divya BV (2021) Arsenic contamination of groundwater: a global synopsis with focus on the Indian Peninsula. Geosci Front 12(3):101079. https://doi.org/10.1016/j.gsf.2020.08.015
4. SA (2014) Flood related disasters: concerned to urban flooding in Bangalore, India. Int J Res Eng Technol 03(28):76–83. https://doi.org/10.15623/ijret.2014.0328013
5. Sarker D, Islam M (2023) Exploring the reasons of waterlogging in Rajshahi City and develop community based proposals to mitigate the increasing waterlogging problem through participatory approach. SSRN
6. Zheng W, Li X, Lam N, Wang D, Yin L, Yin Z (2017) Impact of land use on urban water-logging disaster: a case study of Beijing and New York cities. Environ Eng Manag J 16(5):1211–1216. https://doi.org/10.30638/eemj.2017.127
7. Pradhan R, Sahoo J (2020) Smart rainwater management: new technologies and innovation. Smart Urban Dev. https://doi.org/10.5772/intechopen.86336
8. Sharma V (2020) Arduino based smart water management. Int J Eng Res V9(08). https://doi.org/10.17577/ijertv9is080239
9. Suhas Gupta VK, Carrott PJM, Singh R, Chaudhary M, Kushwaha S (2016) Cellulose: a review as natural, modified and activated carbon adsorbent. Biores Technol 216:1066–1076. https://doi.org/10.1016/j.biortech.2016.05.106
10. Simonis JJ, Basson AK (2012) Manufacturing a low-cost ceramic water filter and filter system for the elimination of common pathogenic bacteria. Phys Chem Earth, Parts A/B/C 50–52:269–276. https://doi.org/10.1016/j.pce.2012.05.001
11. Mukta M, Islam S, Barman SD, Reza AW, Hossain Khan MS (2019) IoT based smart water quality monitoring system. In: 2019 IEEE 4th international conference on computer and communication systems (ICCCS). https://doi.org/10.1109/ccoms.2019.8821742
12. Sanamdikar ST, Harne KR (2012) Advanced method for sewage water treatment. Int J Adv Technol Civ Eng 175–179. https://doi.org/10.47893/ijatce.2012.1026
13. Topare NS, Attar SJ, Manfe MM (2011) Sewage/wastewater treatment technologies: a review. Sci Revs Chem Commun 1(1):18–24

A Simplified Deep Learning Approach for Diabetic Macular Edema Detection

Tejal M. Shirodkar, G. K. Aishwarya, Nandita Kottargi, Satish Chikkamath, Kaushik Mallibhat, and S. R. Nirmala

Abstract Diabetes is a chronic disorder that affects the health of an individual by increasing blood sugar levels. Adverse effects of diabetes can cause severe eye-related problems like Diabetic Macular Edema (DME), leading to vision loss. DME is characterized by the formation of hard exudates in the macular region, causing damage to the retinal microvasculature. Early detection is, therefore, required to prevent vision loss, as DME has no permanent cure. A deep learning framework is utilized in the present work to detect DME. This is followed by preprocessing of the input image, such as cropping and resizing the image. The images of DME are obtained from the DME VQA dataset built from IDRiD and eOphta datasets. The model was trained and tested in the ratio 4:1. The accuracy obtained is 86.5% from the Dense-net model employed in the present work. **Motivation**: The clinical diagnosis of DME may cause human error as different clinicians have a chance to interpret symptoms differently, leading to variations in results in diagnosis, and limited information from a patient's medical history can obstruct accurate results. The clinical diagnosis may be time-consuming, leading to patient treatment delays. Therefore, CNN models are used in detecting DME as they can analyze fundus images effectively by capturing abnormalities that indicate the presence of DME. CNN models help healthcare clinicians in many ways, including early detection of DME, providing accurate results with high efficiency, and allowing them to focus on patients' further treatment planning and other complex tasks.

T. M. Shirodkar (✉) · G. K. Aishwarya · N. Kottargi · S. Chikkamath · K. Mallibhat · S. R. Nirmala
Electronics and Communication Department, KLE Technological University, Hubballi, Karnataka 580021, India
e-mail: tejalshirodkar0510@gmail.com

S. Chikkamath
e-mail: chikkamath@kletech.ac.in

K. Mallibhat
e-mail: kaushik@kletech.ac.in

S. R. Nirmala
e-mail: nirmala.s@kletech.ac.in

V. K. Gunjan and J. M. Zurada (eds.), *Proceedings of 5th International Conference on Recent Trends in Machine Learning, IoT, Smart Cities and Applications*, Lecture Notes in Networks and Systems 1181, https://doi.org/10.1007/978-981-97-8861-3_17

Keywords Diabetes · Diabetic macular edema · Hard exudates · Macula · Fundus images · DenseNet architecture

1 Introduction

Diabetes is a long-lasting disorder caused by the increase in sugar content in the blood. It is estimated that nearly 9.3% of the worldwide population is affected by diabetes, and this percentage would be increased by 25% by 2030 and 51% by 2045 [1]. Long-term diabetes affects blood vessels, leading to cardiovascular diseases and problems with Kidneys, Eyes, feet, and nerves. Diabetes can cause vision loss through Diabetic Retinopathy, Diabetic Macular Edema (DME), Cataracts, and other eye problems. Diabetic Macular Edema (DME) is a complication of diabetes that is closely related to diabetic retinopathy (DR). DME is identified by the accumulation of fluid content in the extracellular space within the macular region. Left untreated, it causes blurred vision, leading to vision loss due to damage to the photoreceptors present in the fovea, which is responsible for sharp and central vision [2]. The risk factor of having DME in diabetic patients is around 8% [1] (Fig. 1 and Table 1).

CNN models can be trained using Retinal fundus images of the eye or OCT images of the retina. Early detection is essential for diagnosing DME at early stages [3] to reduce vision loss, as DME has no cure. Still, regular treatment can help improve the condition by preventing vision loss. Screening of DME is recommended after 1–2 years for DME patients [2] (Figs. 2 and 3).

The paper discusses the utilization of CNNs as the primary architecture for image processing in the context of Diabetic Macular Edema. CNNs are specifically designed for tasks involving image analysis because they can learn and extract features directly from images.

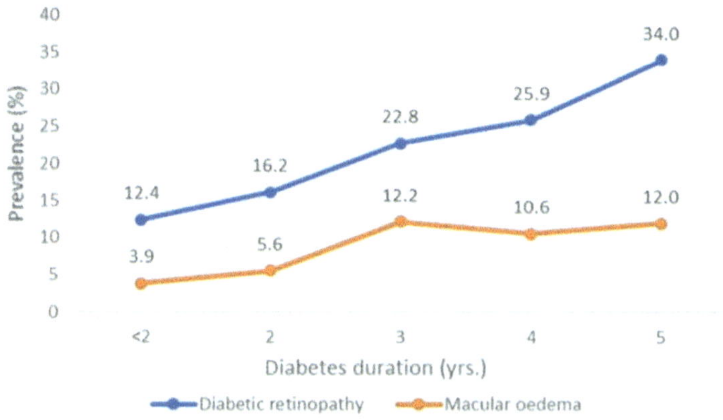

Fig.1 Rate of increase of diabetic retinopathy and diabetic macular edema per year

Table 1 Evolution of methods for DME detection

Evolution of methods for DME detection	Explanation	Limitations	Advantages
Colour fundus photography	Consist of fundus camera that records colour images of the interior surface of the eye	1. Variability of pigmentation 2. Drusen appearance 3. Lack of depth resolution of fundus 4. Lack of detailed quantitative information	To identify abnormalities within the eye and monitor their progression over time
Machine learning technique	Machine learning is a branch of artificial intelligence that concentrates on leveraging data and algorithms to resemble the way that humans learn, gradually progressing its accuracy	1. Data acquisition 2. Time and resources 3. Results interpretations 4. High error chances	1. Recognizing images 2. Detecting fraud 3. Medical diagnoses 4. Speech recognition
Deep learning technique	Deep learning is a method in artificial intelligence that helps computers to recognize complex patterns in pictures, texts, and other data to produce accurate results	1. Data requirements 2. Computational resources 3. Overfitting 4. Lack of contextual understanding	1. Automatic feature learning 2. Handling huge and complicated datasets 3. Enhanced performance

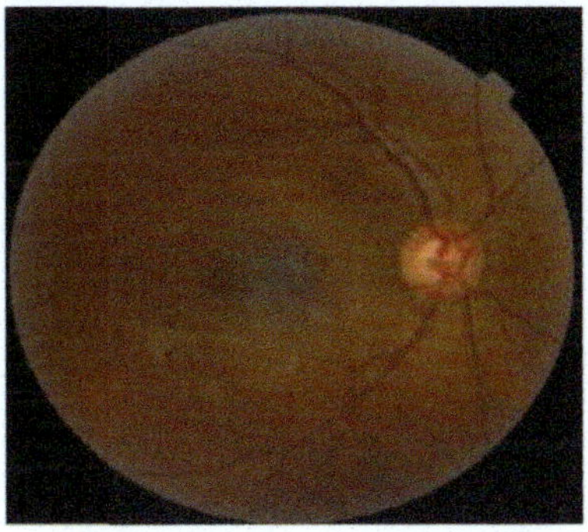

Fig. 2 Retina Fundus image [1]

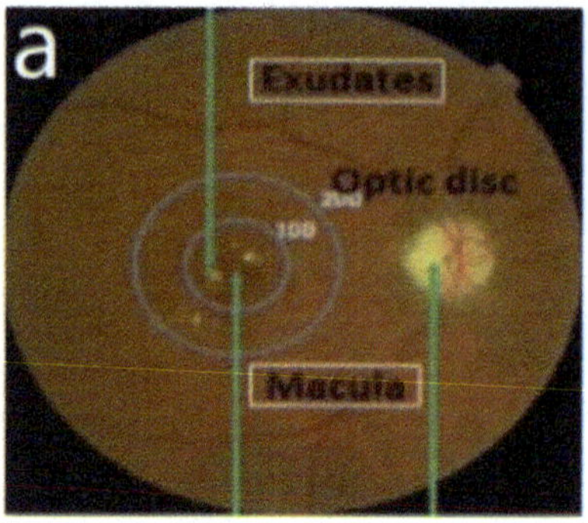

Fig. 3 Fundus image affected by DME [4]

The CNNs are trained on a dataset that includes fundus images affected by DME conditions. The model is expected to classify the presence and absence of DME in a given image. A Convolution Neural Network (ConvNet/CNN) accepts an input image, assigns learnable weights and biases, performs specific operations, and gives the desired output [1]. CNN network consists of convolutional layers that help in feature extraction as they contain a set of convolutional operators that convolute the input image and reduce the high dimensionality of images without losing their information.

In detecting DME using convolutional neural networks (CNNs), max-pooling is also a critical component of the CNN architecture. Max pooling is used in CNNs for object recognition tasks because it helps identify an object's most distinct features, such as its edges and corners, and even for downscaling an image. ReLU is added as an activation layer to prevent the exponential growth that helps operate the neural network, followed by a flattening layer for converting the resulting feature maps into one-dimensional vectors. Fully connected layers help preprocess the image, followed by the SoftMax layer, as it alters the output of the last layer into an essential probability distribution.

1.1 Objectives

The objectives of the study include

- Perform a comprehensive literature review to investigate what preprocessing, feature selection, feature extraction, and machine learning algorithms are used to detect Diabetic Macular Edema using retinal fundus images.
- Propose a machine/deep learning algorithm to detect and classify the given retinal fundus images into Diabetic Macular Edema (DME) and non-DME classes.

2 Literature Survey

This table provides an overview of different machine learning algorithms and deep learning techniques applied for DME detection. The reported accuracies suggest that these methods are helpful in automated detection and grading of DME, depending on the dataset used (Table 2).

Challenges faced for DME detection

- Generally, the input image taken is of the eye area, which contains many details and information that affects the identification of the DME region, so the DME region needs to be focused [5].
- The optic disc generally appears the same as that of the DME region due to its same color values [5].
- To avoid confusion, the area of the DME region must be highlighted using the threshold limit to differentiate the region of DME from the rest of the image [5].
- In detecting different types of DME, data for training and testing must be significant in quantity to provide accurate results [5].

2.1 Proposed Pipeline

The proposed pipeline describes the flow of events that occur to achieve accurate results. It involves taking a retinal fundus image as input and undergoes preprocessing focusing on the macular region for feature extraction, i.e., hard exudates (DME) extraction. The DME dataset is divided into a ratio of 4:1 to train and test the model using DenseNet architecture to give results in binary classification, i.e., standard fundus image or DME-infected fundus image (Fig. 4).

Dataset—The dataset which we considered for our study was obtained from the VQA dataset that is built on IDRiD and eOptha datasets that consisted of 433 images for training, 112 for validation and 134 for testing the model.

3 Proposed Architecture

Preprocessing: The images are pre-processed by resizing and cropping them so they are accessible for feature extraction and detecting hard exudates in the macular region, i.e. DME. The proposed architecture uses a Convolutional Neural Network that detects Diabetic Macular Edema using binary classification. The dataset consists of the images used to train and test the model in the ratio of 4:1.

In this paper, we are using DenseNet architecture to train the model. It is a deep learning architecture, which contains dense blocks that connect each layer to the other layer, and it consists of a thick layer function that adds the batch normalization layer. ReLu activation layer, and 3×3 convolution layer, the output of which is

Table 2 Use of different machine learning algorithms to check its accuracy

Machine learning algorithms	Techniques	Steps	Datasets	Performance (accuracy) (%)	References
Learning based algorithms	SVM classifier	The images are pre-processed using normalization, segmentation of exudates using SVM classifier, Detection of macula using histogram equalization, Grading of Diabetic Macular Edema according to region Surrounding the macula	DIARE TDB1	92.11	[6]
			MESSI DOR	90	
	SVM classifier	Pre-processing using contrast- limited histogram equalization (CLAHE), GLCM Features i.e. extraction of statistical texture features from the images, feature selection using ANOVA, classification of DME using SVM classifier	IR images	100	[10]
	Naïve Bayes classifier	Image preprocessing using Histogram equalization, ILM and Choroid Layer Segmentation, Features Extraction using Segmentation based Fractal Texture Analysis, Feature Set Formulation, Classification using Naïve Bayes Classifier	200 OCT images	98	[12]

(continued)

Table 2 (continued)

Machine learning algorithms	Techniques	Steps	Datasets	Performance (accuracy) (%)	References
Non-learning based algorithm	Marker controlled watershed transformation. algorithm	Exudates detection using contrast-limited adaptive histogram equalization (CLAHE), Fovea localization, classification using ETDRS grading scale	MESSI DOR	85.2	[1]
	Particle Swarm Optimization	Image preprocessing using Median filtering, PSO based Exudates Segmentation, Optic disc removal, Fovea localization and Macular region extraction, classification	MESSIDO R	93	[8]
	Computer Assisted grading technique	Image preprocessing using adaptive histogram Equalization, Detection of hard exudates, Suppression of Optic. disc, Grading of diabetic macular edema	MESSI DOR, Diaretdb1 and DMED	98.80	[7]

(continued)

Table 2 (continued)

Machine learning algorithms	Techniques	Steps	Datasets	Performance (accuracy) (%)	References
	KNN classifier, and one Random Forest clas sifier	Cropping and resizing of OCT images, Training is done using stochastic gradient descent algorithm, A pre-trained VGG-16 CNN model was used for DME detection	The Singapore Eye Research Institute (SERI) database	93.75	[9]
	ELM classifier	Image preprocessing using Adaptive Histogram Equalization, feature extraction	DRIVE	98	[11]
		Using ELM classifier, similarity based severity			

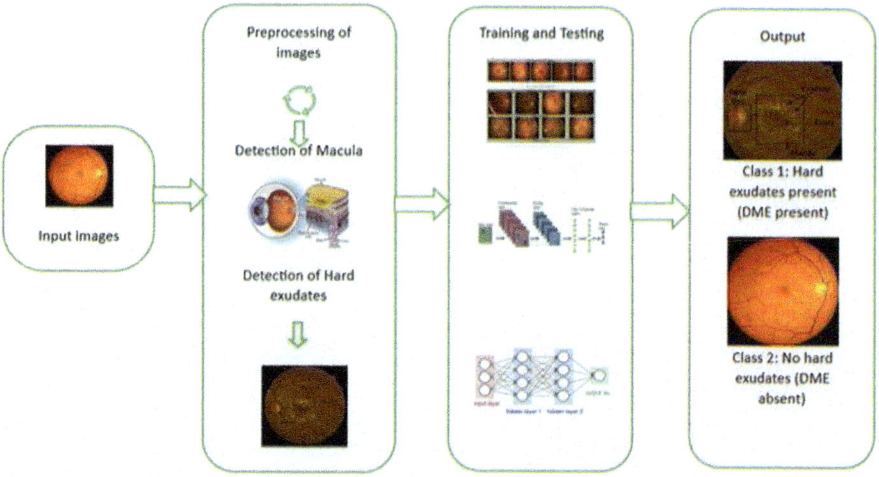

Fig.4 Proposed pipeline

concatenated with the input. The transition block function reduces feature map size and dimensionality by using 1 × 1 convolution and 2 × 2 for max pooling. The model starts with an initial convolutional layer with a filter size of 7 × 7, and Max pooling is applied with a 3 × 3 pool size with a stride of 2 to reduce spatial dimensions further. A Dense layer with SoftMax activation is used for classification with the specified number of output classes (Fig. 5).

4 Results

The model was trained using nvidia Rtx A6000 GPU available at makerspace of the university [13]. The necessary packages for creating the model are Keras, Matplotlib, os, NumPy, pandas, cv2, sklearn, and other needed ones. The model is tested with accuracy of 86.5% using DenseNet architecture. The comparison of accuracies of three CNN models has been done and following are the results shown in Table 3. Among the three CNN models the better accuracy was obtained using DenseNet.

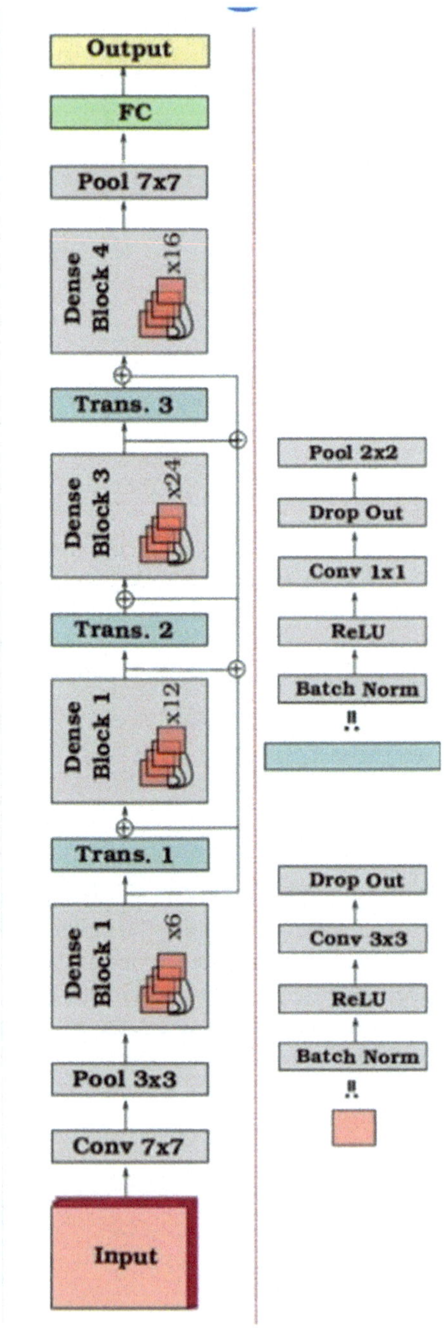

Fig. 5 DenseNet architecture

Table 3 Accuracies of different CNN models

CNN model types	Accuracy (%)
DenseNet	86.5
LeNet	79.4
AlexNet	68.4

5 Conclusion

In conclusion, using a deep learning framework specific to DenseNet proves effective in the binary classification of DME from preprocessed images. The model is trained and tested in a ratio of 4:1, giving an accuracy of 86.5%. The methodology holds good for timely intervention and preventive measures associated with Diabetic Macular Edema (DME).

References

1. Gadde SS, Kiran KVD (2022) DMERCNET: a quirky deep learning classifier for diabetic retinopathy and macular edema risk using CNN. In: 2022 IEEE Delhi section conference (DELCON). IEEE, pp 1–8
2. Bressler I et al (2022) Autonomous screening for diabetic macular edema using deep learning processing of retinal images. medRxiv 2022-08
3. Nasir N et al (2021) A transfer learning approach for diabetic retinopathy and diabetic macular edema severity grading. Rev d'Intelligence Artif 35(6):497–502
4. Bilal A, Sun G, Mazhar S (2021) Survey on recent developments in automatic detection of diabetic retinopathy. J Fr Ophtalmol 44(3):420–440
5. Hussein ZJ, Al Saadi EH (2022) Recent developments used for early detection of diabetic macular edema by retinal images: a survey. In: 2022 Iraqi international conference on communication and information technologies (IICCIT). IEEE, pp 335–341
6. Rekhi RS, Issac A, Dutta MK (2017) Automated detection and grading of diabetic macular edema from digital colour fundus images. In: 2017 4th IEEE Uttar Pradesh section international conference on electrical, computer and electronics (UPCON). IEEE, pp 428–486
7. Sundaresan V et al (2015) Computer-assisted grading of diabetic macular edema on retinal color fundus images. In: 2015 37th annual international conference of the IEEE engineering in medicine and biology society (EMBC). IEEE, pp 4330–4333
8. Li F et al (2022) Deep learning-based automated detection for diabetic retinopathy and diabetic macular oedema in retinal fundus photographs. Eye 36(7):1433–1441
9. Zhang Q et al (2020) Identifying diabetic macular edema and other retinal diseases by optical coherence tomography image and multiscale deep learning. Diabetes Metablc Syndr Obes 4787–4800
10. Wu J et al (2021) Diabetic macular edema grading based on improved faster R-CNN and MD-ResNet. Signal Image Video Process 15:743–751
11. Kumar SJJ, Ravichandran CG (2017) Macular Edema severity detection in colour fundus images based on ELM classifier. In: 2017 international conference on I-SMAC (IoT in social, mobile, analytics and cloud) (I-SMAC). IEEE, pp 926–933
12. Ariya A, Asha S (2020) Automated detection of macular Edema using a machine learning algorithm. In: AIP conference proceedings, vol 2222, no. 1. AIP Publishing
13. Chikkamath S et al (2020) Effective utilization of maker space for facilitating product realization course. J Eng Educ Transform 33(3):37–42

Innovative Rumor Detection on Social Media Text: A Comprehensive Study of Dual Co-Attention Ensemble Based Voting Approach

Anjali Gangone, G. J. Bharat Kumar, and Swapna Mandu

Abstract This study delves into the crucial task of rumor detection amidst the rapid spread of information online, focusing on the efficacy of advanced dual co-attention ensemble models for ensuring digital communication's reliability. By integrating Gated Recurrent Unit (GRU) Representation, Self-Attention (SA) Models, and Bidirectional Encoder Representation from Transformers (BERT) Sentiment Models, alongside various Natural Language Processing (NLP) techniques, the research examines the potential of these models to improve rumor detection accuracy significantly. The investigation encompasses eight distinct models: the GRU Representation Model, the SA Model, the BERT Sentiment Model, and various dual co-attention ensemble combinations thereof (GRU + SA, SA + BERT, BERT + GRU, and GRU + SA + BERT), employing a sophisticated voting mechanism for rumor classification. This comprehensive evaluation across key performance metrics Accuracy, Precision, Recall, and F1 Score revealed the substantial impact of leveraging dual co-attention ensemble model strengths, particularly through dual co-attention and a strategic voting mechanism. Among the dual co-attention ensemble models, the SA + GRU + BERT configuration emerged as a superior performer, achieving remarkable results with an accuracy of 93.5%, a precision of 92%, a perfect recall of 99%, and an F1 score of 95%. These findings underscore the dual co-attention ensemble models' exceptional capability to address rumor detection challenges effectively. The study not only showcases the promising prospects of utilizing cutting-edge AI for rumor detection but also provides insights for refining these models further. It suggests avenues for future research, including model finetuning, ethical considerations, and

A. Gangone (✉)
Department of Computer Science and Engineering, CMR Institute of Technology, Hyderabad, India
e-mail: anjali.gangone@gmail.com

G. J. Bharat Kumar · Swapna M
Department of Computer Science and Engineering, Sreyas Institute of Engineering and Technology, Hyderabad, Telangana, India
e-mail: dr.bharatkumar@sreyas.ac.in

Swapna M
e-mail: drswapnam@sreyas.ac.in

V. K. Gunjan and J. M. Zurada (eds.), *Proceedings of 5th International Conference on Recent Trends in Machine Learning, IoT, Smart Cities and Applications*, Lecture Notes in Networks and Systems 1181, https://doi.org/10.1007/978-981-97-8861-3_18

the necessity for real-time monitoring. This research underscores the transformative potential of AI in enhancing the accuracy of online information, paving the way for future advancements in rumor detection.

Keywords Rumor detection · Dual co-attention · GRU · BERT · Sentiment analysis · NLP · Ensemble models

1 Introduction

The advent of social media has markedly accelerated the flow of information, heralding a new era where the majority of the global population now encounters breaking news first on these platforms, rather than through traditional media outlets such as television or newspapers. This shift has been both a boon and a bane, as the convenience and immediacy of social media also predispose users to the rapid dissemination of rumors, which can have deleterious effects on both society and governance structures [1, 2]. Given the potential for harm, the importance of rumor detection and mitigation cannot be overstated [3].

Rumor detection is fundamentally a binary classification task, where information is categorized into "Rumor" or "Non-Rumor" classes. This process relies on a labeled dataset to train classifiers, employing supervised learning techniques—a category of machine learning where the model is trained on a pre-labeled dataset to make predictions. Various algorithms, including decision trees, neural networks, Bayesian classifiers, and Support Vector Machines (SVM), are utilized in the development of binary classifiers [4].

The concept of a rumor is nuanced, with multiple definitions existing within scholarly literature. Generally, a rumor is identified as misinformation, whether spread maliciously or unintentionally, characterized by its unverified status and tendency to propagate through social networks [5]. The Oxford Dictionary defines a rumor as a story or report of uncertain veracity that circulates among people, gaining credibility not through direct evidence but through widespread belief.

Understanding the psychology behind rumors is a critical first step in developing effective detection systems. Rumors can be categorized based on their truthfulness (true, false, or unverified) and their credibility (low or high). The literature identifies three primary types of rumors: pipe-dream rumors, which are characterized by wishful thinking; bogy rumors, which stoke fear or anxiety; and wedge-driving rumors, which aim to sow discord or hatred among groups [9]. By analyzing these characteristics and the psychological underpinnings of rumors, researchers can devise sophisticated systems capable of distinguishing between harmful rumors and factual information, thereby safeguarding the integrity of information disseminated on social media platforms.

2 Related Works

Here, the most related works that are dealing with the rumor detection with different datasets, methodologies and the metrics. Reference [12] is the baseline model for this research, as it is done on Afan Oromo.

In a comparative study of machine learning models for rumor detection across diverse datasets and languages, several key findings were reported. The Passive Aggressive model applied to a custom Facebook dataset in Amharic achieved an accuracy of 86%, precision of 89%, recall of 83%, and an F1 score of 86% [6]. A CNN model on the same dataset significantly improved performance with an accuracy of 93%, precision of 93%, recall of 95%, and an F1 score of 94% [7]. For a combined custom Weibo and Twitter dataset in Chinese, the Bi-GCN model reported 88% accuracy, 84% precision, 86% recall, and a 93% F1 score [8]. A remarkable performance was observed with a Bi-LSTM model on a custom English Twitter dataset, achieving 97% accuracy, 96% precision, 98% recall, and a 97% F1 score, marking the highest reported efficacy [9].

Further studies include a multi-BiLSTM model on a custom Weibo dataset in Chinese, attaining 96% accuracy, 97% precision, 95% recall, and a 96% F1 score [10], and a Bi-LSTM model on a custom Facebook dataset in Afaan Oromo with 80% accuracy [11]. A novel approach using self-attention-based dual co-attention ensemble models on custom Facebook and Twitter datasets in Afaan Oromo achieved 89% accuracy, 87% precision, 97% recall, and a 92% F1 score [12]. The KZWANG Framework on custom Twitter and Weibo datasets in Chinese reported 95% accuracy, 94% precision, 95% recall, and a 94% F1 score [13].

Additional contributions include a BiGRU model on a custom Weibo dataset in Chinese with 95% accuracy, 92% precision, 93% recall, and a 93% F1 score [14], a BERT-based Dual-Co-Attention model on the same platform achieving 95% accuracy, 94% precision, 96% recall, and a 95% F1 score [15], and a Self-Attention based Multi View model with 96% accuracy, 94% precision, 98% recall, and a 96% F1 score [16]. These studies underscore the effectiveness and potential of various machine learning models in identifying and classifying rumors across different languages and social media contexts.

3 Methodology

3.1 Data Collection and Preparation

Collecting data from Facebook required a structured approach. I used Facebook's Graph API, a developer tool provided by the platform, to access publicly available posts and comments in Afaan Oromo and also and for Facebook Pages Scraper is used. The API allows users to query public content within certain limitations and constraints set by Facebook. To extract data from Twitter, we used an automated

	text	label
0	Ayyaana Irreechaa booru Hora Finfinneetti ifta...	rumor
1	Abbaa manaa yoo taate Yihowaan haadha manaakee...	rumor
2	Baatii Onkoloolessa bara waajjirrii ABO Yeroo...	rumor
3	Nageenyii koo isheen onnee koo ishee akka male...	rumor
4	Ajaaiba garu gaafa calqaba nagaa wal gaafannee...	rumor

Fig. 1 Sample dataset

web scraping tool Octo parse which allows you to pull all the information you see on the targeted website our queries were designed to retrieve posts and comments that were written in Afaan Oromo or contained a significant portion of Afaan Oromo text. These posts and comments were then stored in our dataset for further analysis.

A total amount of 10,000 Afaan Oromo corpus is collected for both Facebook and Twitter.

The labeling process involved human annotators who were proficient in Afaan Oromo. They reviewed the content of each instance and determined whether it contained information that could be classified as a rumor. Rumors were defined as unverified or potentially false claims or pieces of information. Instances that did not meet this criterion were labeled as "non-rumor." The calculation of Cohen's Kappa involved comparing the labels assigned by the two annotators to the same set of instances. The high inter-annotator Kappa value of 98% suggests an exceptionally strong level of agreement between the annotators. This high agreement indicates that the labeling process was conducted rigorously and with a high degree of consistency. The dataset size collected was 9804. Figure 1 shows the sample dataset.

3.2 Data Cleaning and Preprocessing

Once the dataset was ready, 196 Duplicates were removed, then in total, our dataset comprises 9804 instances, with 4978 instances labeled as "rumor" and 4826 instances labeled as "nonrumor." Then HTML Tags and URLs were removed. Then replaced number with its names and abbreviations with full forms. Then Punctuations and special characters were removed. Then Stop words were removed and the data was stemmed and lemmatized by using custom function with custom Afan Oromo specific stop word, word suffix and prefix corpus.

3.3 Label Encoding

Here, we delve into the crucial steps of encoding for our study, it encompasses the transformation of the categorical labels, specifically 'rumor' and 'non-rumor,' into numerical values that enable our deep learning model to effectively process and analyze the data.

The labeling process involved a categorical assignment where instances were designated as 'rumor' and 'non-rumor' based on their content. To make the data compatible with the model, these categorical labels were encoded into numerical values. 'Rumor' instances were assigned the label '1,' while 'non-rumor' instances received the label '0'.

The goal of our study is to develop a neural network model that can classify social media texts into two categories: rumor and nonrumor. To achieve this, I proposed a dual co-attention ensemble multi model architecture, as shown in Fig. 2, which utilizes three different models: self-attention model, GRU model and BERT Model. To prepare the text data for our model, I tokenized both the text. Additionally, I used pre-trained word embeddings for Afaan Oromo words and phrases, which were 200–300 dimensional vectors, to initialize our embedding layer. To implement and train our model, I utilized TensorFlow, which is popular deep learning frameworks. By using this approach, I will able to achieve accurate results in automatically detecting rumors on social media.

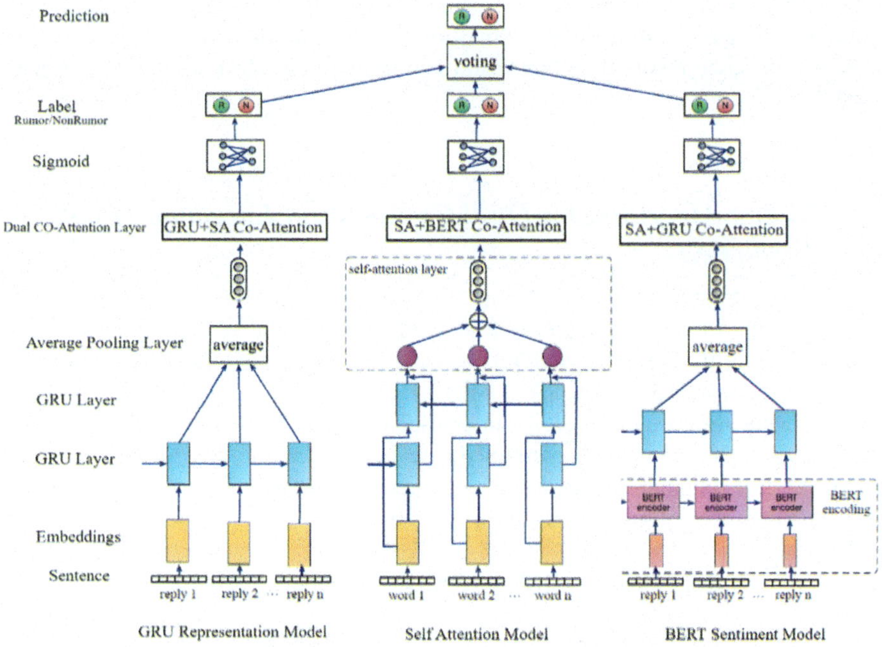

Fig. 2 Proposed dual co-attention ensemble co-attention based rumor detection model

3.4 Proposed Model

The dual co-attention ensemble multi model proposed for classifying social media texts into rumor and nonrumor has a framework consisting of three models—GRU representation model, Self-Attention model, and BERT sentiment Model. The GRU Model employs GRU layer to generate representations for social media text. The self-attention model uses bidirectional GRU and self-attention layer to assess the veracity of the social media text. The BERT Model extracts sentiment embeddings using fine-tuned BERT encoders. The vote layer combines the outcomes of all three views and makes a prediction by majority voting. This framework is expected to enhance the accuracy of the model in classifying rumor and nonrumor events on social media. The model is implemented and trained using TensorFlow.

The proposed dual co-attention ensemble multi model includes three sub-networks, namely the GRU representation model, Self-Attention model, and BERT sentiment Model. The GRU Model sub-network is comprised of an embedding layer, a bidirectional GRU network, and a self-attention layer. It processes the social media text by representing it as a vector of word embeddings and uses GRU to extract lexical and temporal features. The self-attention layer helps to assign weights to words in the text based on their significance. Similarly, for the GRU Model, I generated a post embedding by averaging the word embeddings and feed it to the GRU network. The BERT sentiment model uses a fine-tuned BERT model to generate sentiment embeddings for each post and sends them to GRU. Each model is followed by a sigmoid layer that predicts whether a post is a rumor or nonrumor. Finally, the results of the three models are combined using a majority voting classifier. If two or more views produce the same result, then that result is considered as the final prediction of the model.

3.5 Voting Algorithm for Dual Co-Attention Ensemble Model Predictions

After building the GRU Representation Model (GRU), Self-Attention Model (SA) and BERT Sentiment Model (BERT), for predicting the output of the dual co-attention ensemble models GRU + SA, SA + BERT, BERT + GRU and GRU + SA + BERT, the following algorithm is used, it can be used for dual co-attention ensemble model with any number of models combined, the main principle behind this algorithm design is, if a statement needs to be categorized as rumor, it must be voted by the maximum number of models as rumor.

$$\text{Hybrid\_Model} = \{\text{Model}_1, \text{Model}_2, \text{Model}_3, \ldots \text{Model}_N\}$$

$$f(x) = \text{Predicted\_Voted}_{\text{output}}(\text{Hybrid\_Model})$$

$$= \sum_{i=1}^{n} [\text{Prediction}(\text{Hybrid\_Model})]_i - \frac{|\text{Hybrid\_Model}|}{2}$$

$$f(x) = \begin{cases} 0\,(\text{nonrumor}), x < 0 \\ 1\,(\text{rumor}), x \geq 0 \end{cases}$$

where,

| Dual co-attention ensemble_Model| is the length of the dual co-attention ensemble_ Model

Prediction (Dual co-attention ensemble_Model) i is actual predicted output of each individual model i.

4 Results Discussion

All 8 models were implemented using Anaconda Python Distribution using Jupyter Notebook python editor and the hardware environment use was an Intel Core i3 processor Laptop with 8 GB of RAM.

4.1 Comparing with Baseline Model

The accuracy comparison of the baseline model [12] and the proposed model for rumor class are shown in Table 1 below. The proposed model outperforms the baseline, achieving higher accuracy (0.93 vs. 0.89), indicating its superior ability to identify rumor and nonrumor instances. With a precision of 0.92 (over the baseline's 0.87), it demonstrates fewer false positives and greater reliability in rumor detection. Its recall of 0.99 surpasses the baseline's 0.97, showing it misses fewer true rumors. The F1 Score of 0.95, compared to the baseline's 0.92, signifies improved precision and recall balance, enhancing overall model performance while reducing both false positives and negatives.

Table 1 Accuracy comparison of baseline and proposed model

Model	Accuracy	Precision	Recall	F1 score
Baseline model				
GRU + SA + BERT	0.89	0.87	0.97	0.92
Proposed model				
GRU + SA + BERT	0.93	0.92	0.99	0.95

Table 2 Experimental results

Model	Class	Accuracy	Precision	Recall	F1 Score
GRU	nonrumour (0)	0.7487	0.75	**0.89**	0.74
	rumor (1)		0.79	0.72	0.75
SA	nonrumour (0)	0.8000	0.82	0.75	0.73
	rumor (1)		0.79	0.90	0.84
BERT	nonrumour (0)	0.7574	0.73	0.71	0.56
	rumor (1)		0.75	0.85	0.77
GRU + SA	nonrumour (0)	0.7897	0.81	0.71	0.71
	rumor (1)		0.83	0.90	0.83
SA + BERT	nonrumour (0)	0.8125	0.92	0.85	0.45
	rumor (1)		0.84	0.91	0.80
BERT + GRU	nonrumour (0)	0.8584	0.89	0.81	0.56
	rumor (1)		0.90	0.97	0.81
GRU + SA + BERT	nonrumour (0)	**0.9312**	**0.97**	0.81	**0.88**
	rumor (1)		**0.92**	**0.99**	**0.95**

Bold denotes the highest values of each metric

4.2 Results of Proposed Model

Each model was trained for 30 epochs and the data set has only two labels, I used Binary_crossentropy loss function and Adam optimizer and with accuracy, precision, recall and F1score metrics for each model for each class. The experimental results with highest achieved value of each metric is denoted in bold are shown in the Table 2. The presented table offers an in-depth analysis of various models' performance on a binary classification task, where "nonrumour" and "rumor" are the target classes. Performance metrics encompass accuracy, precision, recall, and F1 score. The following sections provide a detailed discussion of the findings.

4.3 Result Analysis

The results presented above detail the performance of various models, including standalone Gated Recurrent Unit (GRU), Self-Attention (SA), Bidirectional Encoder Representations from Transformers (BERT), and their ensemble combinations (GRU + SA, SA + BERT, BERT + GRU, and GRU + SA + BERT), in the context of rumor detection classified into two categories: nonrumor (0) and rumor (1). Performance metrics such as Accuracy, Precision, Recall, and F1 Score are used to evaluate each model's effectiveness.

The results presented above detail the performance of various models, including standalone Gated Recurrent Unit (GRU), Self-Attention (SA), Bidirectional Encoder Representations from Transformers (BERT), and their ensemble combinations (GRU + SA, SA + BERT, BERT + GRU, and GRU + SA + BERT), in the context of rumor detection classified into two categories: nonrumor (0) and rumor (1). Performance metrics such as Accuracy, Precision, Recall, and F1 Score are used to evaluate each model's effectiveness.

4.4 Standalone Models Performance

The GRU model shows a moderate performance with a slight preference towards identifying nonrumor instances over rumor instances, evidenced by a higher recall for nonrumor (0.89) compared to rumor (0.75).

The SA model demonstrates a balanced performance, with notably higher precision and recall for rumor detection (0.90 and 0.84, respectively) than the GRU model, indicating its better capability at identifying rumors.

The BERT model presents a closer performance between nonrumor and rumor classes but with lower F1 scores compared to the GRU and SA models, suggesting it might be more conservative in classifying rumors, potentially leading to higher false negatives.

4.5 Ensemble Models Performance

The combination of GRU + SA improves upon the standalone models in terms of accuracy and F1 scores for rumor detection, indicating that the ensemble approach benefits from the strengths of both models.

SA + BERT shows a significant improvement in precision for nonrumor detection (0.92) and maintains high precision for rumor detection (0.91), suggesting this combination is particularly effective at correctly identifying true instances of each class.

BERT + GRU achieves the highest accuracy (0.8584) among the dual combinations, with remarkable precision in rumor detection (0.97), highlighting the synergistic effect of combining BERT's contextual understanding with GRU's sequential data processing.

The triple ensemble model (GRU + SA + BERT) outperforms all other models across almost all metrics for both classes, achieving the highest accuracy (0.9312) and exceptional precision and recall for rumor detection (0.99 and 0.95, respectively), indicating a highly effective combination for rumor detection tasks.

4.6 Discussion

The superior performance of the ensemble models, particularly the GRU + SA + BERT combination, underscores the value of integrating multiple approaches to capture the nuanced characteristics of rumors. This model leverages GRU's ability to handle sequential data, SA's focus on the importance of different parts of the input, and BERT's deep contextual understanding, resulting in a robust model capable of high precision and recall. The variance in F1 scores, especially in models like SA + BERT for nonrumor detection (0.45), suggests a trade-off between precision and recall in some models. This indicates the importance of choosing the right model based on the specific requirements of the application, whether that's minimizing false positives or false negatives. Overall, the results highlight the effectiveness of dual co-attention ensemble models in rumor detection, with the combined strengths of GRU, SA, and BERT models providing a comprehensive approach to accurately identifying and classifying rumors on social media platforms. Future work could explore further optimization of these ensemble models, the integration of additional data sources, or the application of these models in real-time rumor detection systems.

5 Conclusion and Future Works

This study highlights the effectiveness of dual co-attention ensemble models, particularly the GRU + SA + BERT combination, in detecting rumors on social media. By integrating multiple AI techniques, this approach achieves high accuracy and excels in precision and recall, significantly reducing both false positives and negatives. The findings suggest such models are invaluable in combating online misinformation and enhancing digital communication integrity.

To further enhance rumor detection, it is recommended to refine the ensemble models through varied AI technique combinations and configurations. Additionally, developing real-time detection systems, diversifying datasets to include multiple languages and cultural contexts, and integrating user feedback mechanisms can improve the models' effectiveness and applicability.

Future research could explore incorporating more machine learning models into the ensemble framework, extend analyses across different social media platforms, address ethical concerns related to automated detection, and assess the models' real-world impact on misinformation spread and public discourse.

References

1. Gereme FB, Zhu W (2019) Early detection of fake news before it flies high. In: ACM international conference proceeding series, pp 142–148. https://doi.org/10.1145/3358528.3358567

2. Ahmed H, Traore I, Saad S (2017) Detection of online fake news using N-gram analysis and machine learning techniques. In: Lecture notes in computer science (including subseries lecture notes in artificial intelligence and lecture notes in bioinformatics), 10618 LNCS (October), pp 127–138. https://doi.org/10.1007/978-3-319-69155-8_9
3. Zhang J, Dong B, Yu PS (2020) FakeDetector: effective fake news detection with deep diffusive neural network. In: Proceedings—international conference on data engineering, 2020-April, pp 1826–1829. https://doi.org/10.1109/ICDE48307.2020.00180
4. Girgis S, Amer E, Gadallah M (2019) Deep learning algorithms for detecting fake news in online text. In: Proceedings—2018 13th international conference on computer engineering and systems, ICCES 2018, pp 93–97. https://doi.org/10.1109/ICCES.2018.8639198
5. Thota A, Tilak P, Ahluwalia S, Lohia N, Ahluwalia S, Lohia N (2018) Fake news detection: a deep learning approach. SMU Data Sci Rev 1(3):10. http://digitalrepository.smu.edu. Available at: https://scholar.smu.edu/datasciencereview/vol1/iss3/10
6. Jimalo KM, Babu PR, Assabie Y (2017) Afaan Oromo news text categorization using decision tree classifier and support vector machine: a machine learning approach. Int J Comput Trends Technol 47(1):29–41. https://doi.org/10.14445/22312803/ijctt-v47p104
7. Bahad P, Saxena P, Kamal R (2020) ScienceDirect ScienceDirect fake news detection using Bi-directional LSTM-recurrent neural fake news detection using Bi-directional LSTM-recurrent neural network. Procedia Comput Sci 165(2019):74–82. https://doi.org/10.1016/j.procs.2020.01.072
8. Wubetu BD, Ayodeji OS (2022) Detection of fake news and hate speech for Ethiopian languages: a systematic review of the approaches. Demilie Salau J Big Data 9(66):1–17
9. Kedir AL (2022) Classification and detection of Amharic language fake news on social media using machine learning approach. Electr Sci Eng 4(1):1–6
10. Ajeet RP, Aditee M, Keshav S, Aishwarya P, Anusha N (2019) Analysis of techniques for rumor detection in social media. In: International conference on computational intelligence and data
11. Neji I (2022) Lexicon-stance based Amharic fake news detection. Addis Ababa Institute of Technology, ADDIS ABABA
12. Anjali G, Bharat KGJ, Swapna G, Tujube M (2023) Innovative rumor detection on social media text: a comprehensive study of dual co-attention ensemble attention based voting approach. In: IEEE Publishers as a Book Chapter in Intelligent computing, machine learning, artificial intelligence, robotics, sensors with ISSN 2195-3988, ISSN 2195-3996 (electronic), ISBN 978-981-19-2357-9 and ISBN 978-981-19-2358-6
13. Hailemichael EN (2021) Fake news detection for Amharic language using deep learning. Adama Science and Technology University, ADAMA
14. Arega KL (2020) Social media fake account detection for Afan Oromo language using machine learning. New Media Mass Commun (IISTE) 90(2020):1–6
15. Tian B, Xi X, Tingyang X, Peilin Z, Wenbing H, Yu R, Junzhou H (2020)Rumor detection on social media with Bi-directional graph convolutional networks. In: Association for the advancement of artificial intelligence (www.aaai.org) (AAAI-20)
16. Sushila S, Vahida A (2022) Rumor detection in social network based on user, content and lexical features. Multimedia Tools Appl 81:17347–17368

Artificial Intelligence for Glaucoma and Diabetic Retinopathy: A Review

Karanbir Singh Grover and Nitika Kapoor

Abstract This article dives into the utilization of machine learning (ML) and deep learning (DL) algorithms for the diagnosis of glaucoma and diabetic retinopathy, significant ocular conditions. Glaucoma, marked by optic nerve deterioration, often stems from elevated intraocular pressure, while diabetic retinopathy results in damage to retinal blood vessels, causing vision impairments. The diagnostic process encompasses stages like data acquisition, preprocessing, data augmentation, feature extraction, and model selection. A diverse range of models, including convolutional neural networks (CNNs) and traditional ML algorithms, are deployed for disease classification. Evaluation metrics like AUC, Accuracy, F1 score, and DiceOC assess model performance. The article also addresses challenges such as data quality and ethical considerations, emphasizing the imperative of data privacy in healthcare AI research. Ultimately, this research endeavors to contribute to healthcare advancement by developing precise and efficient diagnostic tools for glaucoma and diabetic retinopathy.

Keywords Machine learning · Deep learning · Glaucoma · Diabetic retinopathy · Fundus images

K. S. Grover (✉) · N. Kapoor
Department of CSE, Chandigarh University, Patiala, Punjab 147001, India
e-mail: kbir156@gmail.com

N. Kapoor
e-mail: nitika.cse@cumail.in

V. K. Gunjan and J. M. Zurada (eds.), *Proceedings of 5th International Conference on Recent Trends in Machine Learning, IoT, Smart Cities and Applications*, Lecture Notes in Networks and Systems 1181, https://doi.org/10.1007/978-981-97-8861-3_19

1 Introduction

1.1 Glaucoma

An eye condition called glaucoma causes damage to the optic nerve, transmitting signals to the brain from the eye. As millions of nerve fibers within the optic nerve deteriorate, they lose their ability to transmit clear signals, ultimately leading to vision impairment and potential blindness.

Elevated intraocular pressure (IOP), stemming from increased pressure within the eye, represents the primary catalyst for glaucoma. This pressure originates from the fluid circulating within the eye. While this fluid is continually generated and drained, any obstruction or malfunction within the drainage system can trigger fluid buildup, elevating intraocular pressure. The escalation in intraocular pressure can inflict damage upon the optic nerve through two distinct mechanisms. First, it may compress the nerve fibers, prompting their demise. Second, it can curtail blood flow to the nerve, inducing further damage to the fibers.

Glaucoma manifests in four principal types:

- Open-angle glaucoma, the most prevalent form, arises from a gradual constriction of the eye's drainage system.
- Angle-closure glaucoma, less common, results from a swift blockage of the drainage system, causing rapid escalation of eye pressure.
- Congenital glaucoma is present from birth, attributed to an anomaly in the eye's drainage system.
- Secondary glaucoma emerges due to an underlying eye condition like injury, inflammation, or a tumor.

1.1.1 Facts

Globally, 2.2 billion people are suffering from vision impairment diseases. According to WHO, half of these cases could have been prevented. Low and middle income regions has approximately 4 times the cases than in higher income regions. Worldwide there were around the 80 million patients who are suffering from Glaucoma in the year 2020 [1]. The cases are expected to increase to 111 million by the year 2040 [2]. In America, more than 3 million people are affected with Glaucoma. According to an estimation from National Eye Institute, by the year 2050 in the number can increase to more than 6 million patients. Among these patients, the study suggests that 50% will belong to Hispanic descent.

Risk factors for glaucoma include:

- Age: People beyond the age limit of 40 are more likely to get glaucoma [3].
- Family history: Family heritage of glaucoma increases a person's risk of developing the condition [3].

- Diabetes: Glaucoma is more common in people with diabetes.
- Eye injuries: People who have had eye injuries are more likely to develop glaucoma.
- Some medications: Some medications, such as steroids, can increase your risk of glaucoma.

1.2 Diabetic Retinopathy (DR)

The blood vessels in the retina are harmed by the diabetes condition known as diabetic retinopathy. High levels of sugar in the blood have the potential to harm the blood arteries in the retina and begin to leak or bleed. This can lead to a number of vision problems, including blurred vision, blind spots, and even blindness [4].

The two primary forms of DR are:

Non proliferative Diabetic Retinopathy (NPDR):This is the beginning phase of DR. Small bulges in the retina's blood vessels are its defining feature. These bulges can leak fluid or bleed, but they usually do not cause any noticeable vision problems.

Proliferative Diabetic Retinopathy (PDR). This stage of diabetic retinopathy is more severe. The creation of new arteries in the retina is one of its defining features. These young blood arteries are delicate, feeble, and prone to bleeding. Bleeding from these new blood vessels can cause a sudden loss of vision [5].

1.3 Facts

According to the International Diabetes Federation globally 463 million were suffering from diabetes mellitus, which is common development due to diabetic retinopathy, leading to partial and complete vision loss worldwide. They estimated that by the year 2045, cases will increase to more than 700 million.

According to the data in year 2020 total number of people suffering from diabetic retinopathy was around 103 million, out of which the life threatening diabetic retinopathy (VTDR) was 28.54 million globally. According to the studies done by year 2045 number of cases of people suffering from DR will increase to 160.5 million of which people suffering from VTDR will be 44.82 million [6, 7].

Risk factors for diabetic retinopathy include:

- Genetic factors: Some ethnic groups may be more genetically predisposed to developing diabetic retinopathy.
- Socioeconomic factors: People from lower socioeconomic groups may be less likely to have access to regular eye exams and may have difficulty controlling their blood sugar levels.

- Differences in health care: People from certain ethnic groups may have less access to quality health care, including eye care.

2 Artificial Intelligence

Artificial intelligence (AI) powered machines have the capacity to acquire knowledge and tackle problems, mimicking the approaches typically employed by human intelligence. The field of AI research has demonstrated remarkable achievements in developing robust techniques for addressing a wide range of challenges, spanning from gaming strategies to medical diagnostics.

2.1 Machine Learning

ML, a subset of AI, enables computers to acquire knowledge without direct programming. Through training on data, machine learning algorithms gain the ability to make predictions or decisions based on that information.

How does ML work?

ML algorithms work by finding patterns in data. They do this by analyzing the data and looking for relationships between the different features. For instance, a system using machine learning that has been taught to categorize pictures of dogs and cats may search for patterns in the size, shape, and fur's color.

Applications The following are some of the most typical machine learning applications:

- Image recognition: This is the ability to identify objects in images.
- Natural Language Processing: This is the ability to understand and process human language.
- Speech recognition: This is the ability to understand and transcribe human speech. Machine learning is used in speech recognition systems for voice assistants, dictation software, and call centers.
- Fraud detection: This is the ability to identify fraudulent transactions. Machine learning is used in fraud detection systems for credit card companies, banks, and insurance companies.
- Recommendation systems: This is the ability to recommend products or services to users. Machine learning is used in recommendation systems for e-commerce websites, streaming services, and social media platforms [4].

2.1.1 Types

ML is categorized into four main types:

- Supervised Learning: Machine learning, a subset of artificial intelligence (AI), enables computers to acquire knowledge without direct programming. Through training on data, machine learning algorithms gain the ability to make predictions or decisions based on that information.
- Unsupervised Learning: Unsupervised learning operates without labeled data. The algorithm discovers patterns within the data independently.
- Semi-Supervised Learning: A blend of labelled and unlabelled input is used by the system as it learns through semi-supervised learning. It employs labeled data to enhance predictions on the unlabeled data.
- Reinforcement Learning: Reinforcement learning is driven by trial and error. The algorithm receives rewards and penalties and learns to take actions that maximize rewards and minimize penalties.

2.2 *Deep Learning*

In order to extract insights from data, deep learning, a subset of machine learning, uses artificial neural networks. These networks, inspired by the human brain, find application in diverse tasks like speech recognition, image identification, and natural language processing.

Key features of deep learning

- ANN: Neural networks, which are used in deep learning models, are modelled after the way the human brain is organized. Each node in a neural network performs fundamental calculations and is connected to the others through connections. These nodes are organized in layers, and the greater the depth of the network, the more intricate patterns it can comprehend.
- Training on large amounts of data: Deep learning models require a lot of data to learn from. This data can be labeled, which means that each data point has a known label. For example, if the model is being trained to classify images of dogs and cats, the training data might include a set of images of dogs that are labeled "dog" and a set of images of cats that are labeled "cat".
- Backpropagation: In order to train deep learning models, backpropagation is used. Backpropagation is a technique for modifying the neural network's weights to improve prediction accuracy. The weights are modified in accordance with the prediction error, which is determined using a loss function.
- Optimization algorithms: In order to train deep learning models, algorithms for optimization are essential. They are employed to determine the optimal set of weights for the neural network.

Deep learning model types:

- Convolutional neural networks (CNNs): Image identification and other activities requiring the processing of grid-like input are performed with CNNs. CNNs employ filters to extract features from the input data.

- Recurrent neural networks (RNNs): For applications involving sequential data, such speech recognition and natural language processing, RNNs are employed. RNNs function by transferring data from one layer to the next's input, which enables them to learn long-term dependencies.
- Generative adversarial networks (GANs): GANs are used to generate fresh data for tasks like creating graphics, texts, and music. Two neural networks are trained against one another to create GANs. The goal of the generator network is to generate new data that might be taken for real data. The discriminator network, the opposing network, aims to discern between authentic and fraudulent data.
- Autoencoders: Autoencoders are used for tasks that involve compressing and reconstructing data. Before being handled by autoencoders, the input data is firstly transcribed into a lower-dimensional form. The data is then rebuilt using the lower-dimensional form.
- Deep reinforcement learning: Deep reinforcement learning is a type of reinforcement learning that uses deep learning to train agents. Deep reinforcement learning can be utilized to provide groundbreaking results in a range of activities, including playing video games and operating robots.

Challenges of ML and Deep learning:

- Data requirements: Deep learning models require a lot of data to learn from. This data can be difficult and expensive to collect, especially for some tasks, such as medical diagnosis.
- Overfitting: This is a problem that occurs when the deep learning model learns the training data too well. This can lead to the model making poor predictions on new data.
- Explainability: It can be difficult to explain how a deep learning model makes its predictions. This can make it difficult to trust the model and to use it in safety–critical applications.

3 Scope of AI in Healthcare

AI-driven health interventions can be categorized into four main areas:

- Diagnosis: A number of illnesses, such as TB, malaria, dengue, and cervical cancer, may be diagnosed using AI. For example, machine learning algorithms have been used to identify tuberculosis cases from digital chest radiographs, and to predict the risk of dengue fever severity [5].
- Morbidity or mortality risk assessment: AI can also be used to assess the risk of a patient developing a disease or dying. This information can be used to identify patients who need early intervention, or to prioritize patients for treatment [5].
- Disease outbreak prediction and surveillance: AI can be used to track the spread of diseases and to predict when and where outbreaks may occur. This information can be used to prevent or contain outbreaks, and to protect public health [5].

- Health policy and planning: Decisions on health planning and policy can be informed by AI. For instance, resource allocation in hospitals and patient duration of stay have both been predicted using machine learning models [5].

Challenges and limitations of AI in healthcare

The application of AI to medical care is currently in its infancy, and a number of difficulties and restrictions still need to be resolved. These include:

- Data availability and quality: AI algorithms require large amounts of data to train and test. However, healthcare data is often siloed and not easily accessible. This can make it difficult to develop and deploy AI solutions.
- Bias: If artificial intelligence systems are educated on data that is not representative of the population, they may be prejudiced. This can result in unreliable or biased outcomes.
- Interpretability: Understanding how AI systems make judgements might be challenging. It may be challenging to trust and employ these algorithms as a result.
- Regulation: There are few regulations governing the use of AI in healthcare. This can make it difficult for healthcare organizations to adopt and use AI solutions.

AI has the ability to transform healthcare and raise the standard of patient care despite these obstacles. In the years to come, we may anticipate seeing even more ground-breaking and inventive uses of AI in healthcare as the technology advances [8].

3.1 *Scope of Machine Learning in Classifying Glaucoma and Diabetic Retinopathy*

The diagnosis of glaucoma and diabetic retinopathy has seen significant progress through the application of machine learning (ML) and deep learning (DL) models. Researchers have leveraged various image features, including optic disc characteristics, retinal nerve fiber layer thickness, and cup-to-disc ratio, extracted from retinal images using diverse image processing techniques to develop accurate classification models.

One notable study employed a support vector machine (SVM) algorithm to classify glaucoma based on optic disc features, achieving an impressive 85.7% accuracy (Takagi et al. 2019) [4]. In another breakthrough, a deep convolutional neural network (CNN) was employed to classify diabetic retinopathy using retinal fundus images, resulting in an accuracy of 96.6% (Gulshan et al. 2016) [9]. Ensemble methods, such as random forests and gradient boosting, have also been explored to further enhance disease classification accuracy (Lee et al. 2019) [4].

Building upon these achievements, Ghosh et al. utilized a CNN model to not only detect diabetic retinopathy but also classify its different stages, yielding an impressive 95% accuracy [4]. Similarly, Shaban et al. deployed a 21-layer neural network to identify various stages of diabetic retinopathy, although with a slightly lower accuracy of 89%. Kumari et al. took a different approach, utilizing a pretrained VGG16 model to detect and classify diabetic retinopathy stages with the APTOS dataset [10, 11]. Some authors have combined neural networks with machine learning classifiers to achieve exceptional results. Gayathri et al. applied the M-CNN model alongside J48 as a classifier, reaching an accuracy of approximately 98% [12].

In addition to these approaches, some researchers have integrated deep learning models with specific pre-processing and segmentation techniques. Soomro et al. utilized a filtering scheme in combination with Adaptive Wiener morphological filtering to attain a commendable accuracy of 94% [4]. Sonali Chaudharry et al. employed the Grey Level Co-occurrence Matrix (GLCM) and CNN to classify and detect diabetic retinopathy into different stages, conducting experiments on the Stare and DIARETDB 0 & 1 databases.

Chen et al. applied a CNN-based model to the ORIGA and SCES datasets, achieving notable AUC values of 0.832 and 0.887, respectively. Similarly, Alghamdi et al. utilized a CNN-based model on the PAMDI and HAPIEE datasets [13]. Another innovative approach was taken by Orlando et al., who employed the VGG-S model for glaucoma detection and enhanced images using a contrast-limited histogram pre-processing technique, yielding an AUC value of 0.6655 on the Drishti dataset.

Moreover, some researchers have adopted a multi-model, multi-dataset strategy. Sertan et al. employed ResNet-50, ResNet-150, and GoogLeNet on five datasets, revealing that the best accuracy was achieved with the ResNet-150 model on the RIM-ONE dataset [14].

3.2 Diagnosing Glaucoma and Diabetic Retinopathy with ML and DL Algorithms

The process of diagnosing glaucoma and diabetic retinopathy using ML and DL involves several key stages, from data acquisition to final evaluation. This comprehensive approach aims to understand generalize model's effectiveness.

Data Acquisition: Data acquisition is the foundational step in building any disease diagnosis model. Researchers often obtain data from specialized medical departments or clinics dedicated to the specific disease under investigation. Alternatively, publicly available datasets, such as those on UCI and Kaggle, are utilized. Selection of the data source must consider data quality, quantity, and ethical and legal considerations regarding patient data privacy.

Data acquired from medical departments offers high quality but can be costly and time-consuming to collect. Additionally, privacy concerns arise due to personal patient information. Public datasets provide cost-effective access to diverse data but may vary in quality and lack direct patient access, potentially affecting results.

Preprocessing: Preprocessing is vital for refining data quality by eliminating noise and unwanted information. Techniques like filtering, segmentation, and normalization are applied based on data quality. For instance, in glaucoma datasets, methods like contrast enhancement, Gaussian blur, and adaptive histogram equalization are employed. Diabetic retinopathy datasets often use StandardScaler, CLAHE (Contrast Limited Adaptive Histogram Equalization), and green channel transformation for normalization [8, 15].

Data Augmentation: Data augmentation expands the dataset by generating various image variations through transformations like rotation, scaling, flipping, and shearing [14]. This technique enhances model quality, prevents overfitting.

Feature Extraction and Feature Selection: Feature extraction and selection are critical in medical image analysis. Feature selection involves picking the most relevant features for classification, while feature extraction transforms original features into a more useful set. Techniques such as PCA, ICA, wavelet transform, and LCBO algorithm are used to choose optimal features [8, 16]. Feature selection strategies like CFS, MIFS, and WFS are also employed.

These steps ensure informative feature sets for accurate disease classification. Classification and Evaluation: After data preparation, ML and DL models are utilized for image classification, with performance evaluation using metrics like AUC, Accuracy, F1 score, DiceOC, and DiceOD. Various models, from CNN architectures like Inception and VGG16 [17] to traditional ML algorithms such as Random Forest and SVM, are applied. Some studies also explore hybrid models combining different techniques for improved accuracy.

Each model has its advantages and drawbacks. Traditional ML algorithms are simple and interpretable but may struggle with complex data. CNNs excel at capturing image patterns but require large datasets and computational power. Hybrid models leverage both traditional ML and deep learning for enhanced performance but may need more expertise (Table 1).

In the above Table 2 we discussed the Drishti-GS dataset and several publications which outperformed the researchers on this particular dataset. We try to understand their models and detect progression. Similarly we analysed the top works from

Table 1 Highest performing models for Drishti-GS Dataset

Dataset	Model	DiceOC	DiceOD	Paper
Drishti-GS	EfficientNet + U-Net	0.9381	0.9784	Kamble et al. [18]
	ET-Net	0.9314	0.9752	Zhang et al. [19]
	pOSAL	0.901	0.974	Wang et al. [15]
	U-Net	0.8806	0.9643	Olaf et al. [20]

Table 2 Highest performing models for Drive dataset

Dataset	Model	AUC	F1 score	Accuracy	Paper
DRIVE	FR-UNet	0.9889	0.8316	0.97	Liu et al. [21]
	RV-GAN	0.9887	0.869	0.979	Kamran et al. [22]
	SA-UNet	0.9864	0.8263	0.9698	Guo et al. [23]
	UNet	0.9855		0.9712	Uysal et al. [14]
	VGN	0.9802	0.8263		Shin et al. [24]

different authors on public available dataset such as DRIVE(Diabetic retinopathy dataset) dataset and listed the top performing models and respected metrics.

4 Conclusion

In conclusion, early diagnosis of vision-threatening eye diseases like glaucoma and diabetic retinopathy is crucial to prevent vision loss or blindness. Unfortunately, there is a shortage of specialized doctors who can accurately diagnose these diseases. Machine learning and deep learning models have shown great potential in assisting doctors with the diagnosis of these diseases by providing a pre-diagnosis with a certain confidence score, saving time and increasing accuracy. Our review of various techniques used to classify these diseases highlights the potential of these models in improving the quality of eye disease diagnosis. Further research and development in this field could lead to more accurate and efficient diagnosis of eye diseases, ultimately improving the quality of life for millions of people around the world.

References

1. Gupta ST (2022) Optimized hybrid machine learning approach for smartphone based diabetic retinopathy detection. Multimedia Tools Appl
2. Allison K, Patel D, Alabi O (2020) Epidemiology of glaucoma: the past, present, and predictions for the future. Cureus 12(11)
3. Schuster AK, Erb C, Hoffmann EM, Dietlein T, Pfeiffer N (2020) The diagnosis and treatment of glaucoma. Deutsches Arzteblatt Int 117(13):225–234. https://doi.org/10.3238/arztebl.2020.0225
4. Lee J, Shin J, Kim J, Kim J, Kim K (2019) Ensemble deep learning for glaucoma detection from color fundus photographs
5. Schwalbe N, Wahl B (2020) Artificial intelligence and the future of global health. The Lancet 395(10236):1579–1586
6. Teo ZL, Tham Y-C, Yu M, Chee ML, Rim TH, Cheung N, Bikbov MM, Wang YX, Tang Y, Lu Y, Wong IY, Ting DSW, Tan GSW, Jonas JB, Sabanayagam C, Wong TY, Cheng C-Y (2021) Global prevalence of diabetic retinopathy and projection of burden through 2045: systematic review and meta-analysis. Ophthalmology 128(11):1580–1591. ISSN 0161-6420. https://doi.org/10.1016/j.ophtha.2021.04.027

7. Quigley AT (2006) The number of people with glaucoma worldwide in 2010 and 2020. Br Ophthalmol 90:253–254
8. Takagi S, Sugimoto M, Nakamura Y, Hasegawa A, Sawada A (2019) A support vector machine algorithm for detecting glaucoma in fundus images
9. Gulshan V, Peng L, Coram M, Stumpe MC, Wu D, Narayanaswamy A, Venugopalan S, Widner K, Madams T, Cuadros J, Kim R, Raman R, Nelson PC, Mega JL, Webster DR (2016) Development and validation of a deep learning algorithm for detection of diabetic retinopathy in retinal fundus photographs. JAMA 316(22):2402–2410. https://doi.org/10.1001/jama.2016.17216
10. Kumari P, Saxena P (2023) Automated detection and multi-stage classification of diabetic retinopathy through CNN. AIP Conf Proc 2705(1): 020016. https://doi.org/10.1063/5.0133784
11. Islam MR, Abdulrazak LF, Nahiduzzaman M, Goni MOF, Anower MS, Ahsan M, Haider J, Kowalski M (2022) Applying supervised contrastive learning for the detection of diabetic retinopathy and its severity levels from fundus images. Comput Biol Med 146:105602. https://doi.org/10.1016/j.compbiomed.2022.105602
12. Gayathri S, Gopi VP, Palanisamy P (2021) Diabetic retinopathy classification based on multipath CNN and machine learning classifiers. Phys Eng Sci Med 44(3):639–653
13. Alghamdi HS, Tang HL, Waheeb SA, Peto T (2017) Automatic optic disc abnormality detection in fundus images: a deep learning approach, pp 17–24. https://doi.org/10.17077/OMIA.1042
14. Uysal E, Bilici M, Zaza B, Özgenc M, Boyar O (2021) Exploring the limits of data augmentation for retinal vessel segmentation
15. Wang S, Yu L, Yang X, Fu C-W, Heng P-A (2019) Patch-based output space adversarial learning for joint optic disc and cup segmentation. IEEE Trans Med Imaging 38(11):2485–2495. https://doi.org/10.1109/TMI.2019.2899910
16. Gadekallu TR, Khare N, Bhattacharya S, Singh S, Maddikunta PK, Ra I-H, Alazab M (2020) Early detection of diabetic retinopathy using PCA-firefly based deep learning model. MDPI (Deep Neural Netw Appl) 9:274
17. Ahmed MT, Ahmed I, Rakin RA, Akter MT, Jahan N (2023) An effective deep learning network for detecting and classifying glaucomatous eye. Int J Electr Comput Eng (IJECE) 13(5):5305–5313
18. Kamble R, Samanta P, Singhal N (2020) Optic disc, cup and fovea detection from retinal images using U-Net++ with EfficientNet encoder. In: Fu H, Garvin MK, MacGillivray T, Xu Y, Zheng Y (eds) Ophthalmic medical image analysis. OMIA 2020. Lecture notes in computer science, vol 12069. Springer, Cham. https://doi.org/10.1007/978-3-030-63419-310
19. Zhang Z, Fu H, Dai H, Shen J, Pang Y, Shao L (2019) ET-Net: a generic edge-attention guidance network for medical image segmentation. In: Shen D et al. (eds) Medical image computing and computer assisted intervention—MICCAI 2019. MICCAI 2019. Lecture notes in computer science, vol 11764. Springer, Cham. https://doi.org/10.1007/978-3-030-32239-749
20. Ronneberger O, Fischer P, Brox T (2015) U-Net: convolutional networks for biomedical image segmentation. In: Navab N, Hornegger J, Wells W, Frangi A (eds) Medical image computing and computer-assisted intervention—MICCAI 2015. MICCAI 2015. Lecture notes in computer science, vol 9351. Springer, Cham. https://doi.org/10.1007/978-3-319-24574-428
21. Liu W et al (2022) Full-resolution network and dual-threshold iteration for retinal vessel and coronary angiograph segmentation. IEEE J Biomed Health Inform 26(9):4623–4634. https://doi.org/10.1109/JBHI.2022.3188710
22. Kamran SA, Hossain KF, Tavakkoli A, Zuckerbrod SL, Sanders KM, Baker SA (2021) RV-GAN: segmenting retinal vascular structure in fundus photographs using a novel multi-scale generative adversarial network. Lecture notes in computer science, vol 12908. Springer, Cham. https://doi.org/10.1007/978-3-030-87237-34
23. Guo C, Szemenyei M, Yi Y, Wang W, Chen B, Fan C (2021) SA-UNet: spatial attention U-Net for retinal vessel segmentation, pp 1236–1242. https://doi.org/10.1109/ICPR48806.2021.9413346
24. Shin SY, Lee S, Yun ID, Lee KM (2019) Deep vessel segmentation by learning graphical connectivity. Med Image Anal 58:101556. ISSN 1361-8415. https://doi.org/10.1016/j.media.2019.101556

Query-Based Text Summarization: A Comparative Investigation and TextRank Implementation

Varad Unhale, Suyash Udchan, Mohak Shah, and Bhakti D. Kadam

Abstract Owing to the proliferation of internet and multimedia applications, there has been a huge increase in digital content in the form of text, images, audio and video worldwide. The various applications of text summarization has attracted the computer vision researchers to generate the optimal text summaries. Several methodologies have been reported in literature to generate text summaries. However, the key challenge is incorporating user's preference as text summarization is a subjective task. User generated queries act as a guiding beacon in the summarization process. Instead of producing generic text summaries, query-based text summarization techniques offer user preferred responses. This paper begins by discussing the fundamental concepts and objectives of text summarization, emphasizing on the role of user queries in the summarization process. It proceeds to categorize existing query-based summarization techniques into distinct paradigms, including extractive, abstractive, and hybrid approaches, highlighting the advantages and limitations of each. This paper presents the evolution of query-based text summarization with developed techniques, available datasets, evaluation metrics, and performance comparison. After conducting a survey on query-based text summarization, a text summarization using the TextRank algorithm is implemented and Recall-Oriented Understudy for Gisting Evaluation (ROUGE) scores are calculated for the generated summaries.

Keywords Query-based Text Summarization · Extractive Summarization · Abstractive Summarization · Deep Learning · Evaluation Metrics

---

V. Unhale · S. Udchan · M. Shah (✉) · B. D. Kadam
Department of Electronics and Telecommunication Engineering, SCTR's Pune Institute of Computer Technology, Pune, India
e-mail: shahmohak2311@gmail.com

V. Unhale
e-mail: varadunhale.vu@gmail.com

S. Udchan
e-mail: suyash.udchan@gmail.com

B. D. Kadam
e-mail: bdkadam@pict.edu

V. K. Gunjan and J. M. Zurada (eds.), *Proceedings of 5th International Conference on Recent Trends in Machine Learning, IoT, Smart Cities and Applications*, Lecture Notes in Networks and Systems 1181, https://doi.org/10.1007/978-981-97-8861-3_20

1 Introduction

Text summarization is an essential task in a world where information is abundant but time and attention are limited. The ability to distill the most crucial information from lengthy documents or vast datasets enables users to quickly access the information that matters most to them. However, generic summarization methods may fall short of addressing the specific information needs of users. This is where query-based text summarization comes into play. Query- based text summarization, also known as topic-based, user-focused, or query-focused summarization, integrates user-provided query information to craft the text summaries. In contrast to generic text summarization, which aims to provide broad document summaries, query-based summarization extracts or generates text that handles query-specific points and condenses the documents. In essence, text summarization can be seen as a higher-level category comprising three forms: generic summarization, extractive query-based summarization, and abstractive query-based summarization. It often results in a summary of content related to the user's question.

There have been surveys, like the one conducted by Afantenos et al. [1], that addressed text summarization in the context of medical documents. Nevertheless, these surveys limited their focus to medical-related queries. In 2010, Damova et al. [2] conducted a survey on query-based text summarization, but the field has evolved significantly in the following decade, with the introduction of technologies such as Word2Vec, end-to-end sequence generation models, and various attention mechanisms. The challenge lies in developing an efficient and effective query-based text summarization method that can distill essential information from extensive textual datasets in response to user-generated queries, considering the diversity of summarization paradigms and their unique advantages and limitations. Therefore, this survey aims to provide a comprehensive overview of the specific techniques in query-based text summarization and bridge the gap in the literature, considering the recent developments in this field.

This paper presents a survey of generic and query-based text summarization techniques including existing techniques, performance comparison, limitations and challenges in the same. The paper also discusses an implementation of Tex-tRank algorithm with obtained results.

2 Problem Statement

Given an input text, the aim of the query-based text summarization model is to generate a meaningful text summary focusing on user inputted query. Figure 1 shows the systematic block diagram of query-based text summarization. Text pre-processing, Text extraction, Query pre-processing, training of the deep learning model and generating a text summary are the major steps in query-focused text summarization. From the perspective of text summarization, query-based text

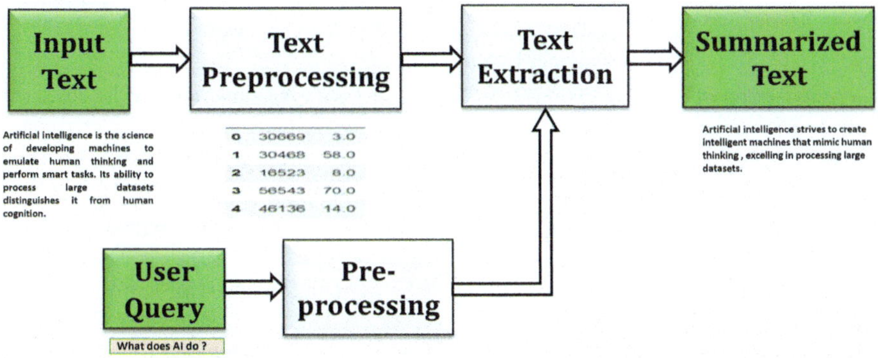

Fig. 1 Block diagram of query-based text summarization

summarization can be categorized into two main approaches: query-based extractive and query-based abstractive text summarization. This type of summarization takes into account user-provided queries, which can be in the form of individual words or complete sentences. Additionally, research in this area can be divided into Single-Document Summarization (SDS) and Multi-Document Summarization (MDS). Figure 2 presents the broad categorization of Text Summarization techniques.

The survey's structure overview begins by examining a key division: extractive techniques and abstractive techniques in text summarization. These two approaches differ fundamentally, with extractive techniques focusing on selecting prominent sections from a document and directly extracting them. Consequently, query-based extractive summarization primarily evaluates the relevance of the content concerning the query. In contrast, abstractive text summarization aims to identify the essential information in the text and generate new text that encapsulates the identified information, which is not present in the original document. While most research in text summarization revolves around either extractive or abstractive techniques, there are also efforts to combine these two approaches using hybrid methods.

3 Existing Techniques

A crucial element in the development of a successful deep learning or machine learning algorithm is the establishment of a validated evaluation process that guarantees the generalization of the developed method. Evaluation comprises two key components: evaluation datasets and evaluation metrics. Choosing the appropriate evaluation datasets and metrics is task-dependent and essential for accurate generalization. This section provides a brief overview of existing approaches reported in the literature for both generic and query-based text summarization. Table 1 presents a comparison of existing text summarization techniques in terms of methodology used, datasets experimented, achieved results and limitations.

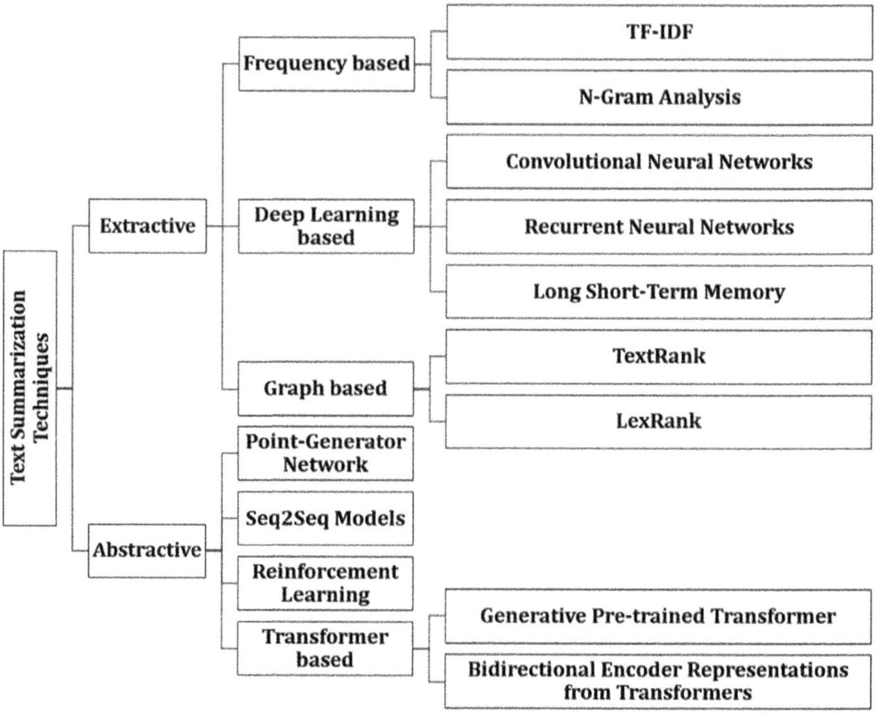

Fig. 2 Categorization of text summarization techniques

In the paper authored by Aakash Sinha et al. [3], feedforward neural net- works are utilized in conjunction with Word2Vec and Fasttext for word and sentence vectorization, respectively, all while employing the softmax activation function. Their work is evaluated using the DUC 2002 dataset and ROUGE metrics, demonstrating that optimal performance is achieved when processing 40 sentences at a time. The paper authored by Angel Herandez-Castenda et al. [4] employs clustering, Genetic Algorithm, LDA, TF-IDF, and N-grams techniques. They have evaluated their method using the DUC02 and TAC11 datasets, achieving an impressive 86% accuracy rate for their generated summaries and suggesting its potential.

The research work presented in [5] utilizes cosine similarity, word-order similarity, semantic similarity, a hybrid similarity measure, and clustering algorithms for query-based summarization. The authors Ahmed A. Mohamed et al. introduced multiple summarization methods in [6], with the Q-Summarizer method standing out as particularly effective for generating query-based summaries with high relevance to the query. The paper authored by Samridhi Murarka et al. [7] presents a hybrid approach for query-relevant text summarization. Their methodology involves tokenization, normalization, stop-word removal, part- of-speech tagging, lemmatization, and context modeling using Latent Semantic Indexing. They also incorporate a Text Rank Algorithm.

Table 1 Comparative study of video activity classification techniques

Research work	Methodology	Dataset	Results	Limitations
[3]	Feedforward neural networks, Word2Vec for word vectorization Fasttext for sentence, vectorization, Softmax activation function	DUC 2002	The best performance in terms of ROUGE scores was observed when using a 'page-len' parameter set to 40	Enhancing the model's capability to generate summaries exceeding the 'page-len'
[10]	BERT, a bidirectional transformer model, boosts NLP tasks through pre-training and efficient fine-tuning, achieving top results	BooksCorpus and English Wikipedia	Unsupervised pre-training improves language understanding, benefits low-resource tasks	Current language models are unidirectional, limiting context and pre-training options, impacting their performance
[5]	Cosine similarity, word-order similarity, semantic similarity, a hybrid similarity measure, and clustering algorithms	Various online websites of Amrita School of Engineering	These results suggest that setting the 'page-len' parameter to 65	Summarization methods can excel or struggle in specific domains due to specialized terminology
[6]	DGS-Summarizer, Q-Summarizer & QInc-Summarizer	DUC02 dataset	Effective in generating query-based summaries	Further experiments on larger text corpora are needed to evaluate the performance of the method
[4]	A text summarization method utilizing Doc2Vec, LDA, and a genetic algorithm to cluster sentences based on semantic and lexical features	DUC02 dataset	Excellent results, outperforming prior methods on DUC02, displaying competitiveness on TAC11	The paper lacks explicit limitations but potential concerns include scalability and GA parameter tuning

(continued)

Table 1 (continued)

Research work	Methodology	Dataset	Results	Limitations
[11]	Term Sense, Extraction Specificity Power, Informativeness Power, Sentence Selection, Maximal Marginal Relevance (MMR)	News corpus built	Optimal summarization results were achieved with a compression rate of 20 for documents with many closely related sub-topics	Further experiments on larger text corpora are needed to evaluate the performance of the method
[9]	Extractive text summarization using a statistical novel approach based on sentence ranking	DUC 2002	The proposed model improves the accuracy compared to traditional approaches	The need for labeled data, anaphora and cataphora problems, and the ongoing
[11]	A text summarization model is built using a sequence-to-sequence architecture with attention and a pointer mechanism	The dataset used in the paper is created from CNN and Daily Mail news articles	The experiment results show that the model performs better when queries are incorporated, as indicated by ROUGE scores	The model's performance relies on queries, posing a limitation when they are absent or less informative
[12]	MeanSum is a model for summarizing multiple documents without human-labeled data. It tests different architectures and pre-trained models to enhance summarization quality	The document mentions using Yelp dataset	The MeanSum model performs well in evaluations, surpassing extractive models in ROUGE scores and sentiment accuracy	Designed for multi-document summarization and doesn't work well for single-document summarization due to the lack of redundancy cues
[13]	A combination of neural language models and different encoders for context capture. The model is trained using negative log-likelihood on input-summary pairs, and beam search decoding is utilized for efficient summary generation	DUC-2003 and DUC-2004, The training data is sourced from the Gigaword dataset, with training pairs	ABS and MOSES + models outperformed TOPIARY	Pretrained language models like BERT can inherit biases from the internet data they are trained on, which can result in biased or objectionable outputs

Table 2 Evaluation metrics for text summarization

Metric	Information	Papers cited in
ROUGE	Recall-Oriented Understudy for Gisting Evaluation (ROUGE) is a widely-used metric for checking how well text summaries perform. It measures the overlap between machine-generated summaries and reference summaries, considering precision, recall, and the F1 score. It's a tool that helps us assess how accurately our summaries capture the essence of the original text	[3, 5, 7, 8, 11, 12, 14–19]
F1-Score	The F1-Score is a metric that combines precision and recall into a single value, making it useful for summarization evaluation	[11, 19]
MMR-MD Parameter	Maximal Marginal Relevance for Multi-Domain (MMR-MD) parameter is used to control redundancy in multi-document summaries. It affects the trade-off between removing redundancy and maintaining relevance in summaries	[20]

A neural attention model is designed by the authors in [8] to generate concise summaries from input sentences. Their approach combines a neural language model with various encoders, with a focus on minimizing the negative log-likelihood of input-summary pairs during training, utilizing stochastic gradient descent for parameter estimation. In a separate study by Johan Hasselqvist et al. [9], they present a sequence-to-sequence model with attention and a pointer mechanism tailored for query-based abstractive summarization. This model considers both a document and a query as input, processing them using bidirectional Recurrent Neural Network (RNN) encoders. It strategically focuses on relevant segments of the document by using attention mechanism and employs a generator mechanism to craft summaries. Information about the evaluation metrics used in text summarization is provided in Table 2.

4 Implementation

In this research, we dive into two main ways of summarizing text: extraction- based and abstraction-based approaches. The spotlight is on extraction-based summarization, which involves pulling out keyphrases from source documents without changing the original text. This method proves particularly handy for handling diverse types of text resources like books, news articles, blog posts, research papers, emails, and tweets. This work is essentially driven by an algorithm called TextRank, which takes its cues from the PageRank algorithm, a concept initially explored by Jinghua Wang et al. [21]. You might be familiar with PageRank—it's the algorithm commonly used to rank web pages in online search results. In our work, we applied TextRank to develop a text summarization system. What TextRank does is identify similarities between sentences in articles, helping us generate concise and informative summaries. The proposed approach included preprocessing a dataset containing articles, including attributes such as article ID, title, text, and source. We imported the data into a Pandas dataframe, extracted the article text, and performed necessary preprocessing steps. It included converting to lowercase, removing punctuation, numbers, and special characters, as well as eliminating stop words. These clean sentences were then represented as vectors using GloVe word vectors which is one of the various word embeddings provided by Jeffrey Pennington et al. [22]. In the implementation phase, a systematic process is followed as:

1. **Sentence Representation**: Sentences were represented as vectors using word embeddings.
2. **Similarity Matrix Preparation**: A similarity matrix was constructed based on sentence similarities.
3. **Graph/Network Conversion**: The similarity matrix was transformed into a graph/network.
4. **TextRank Algorithm Application**: The TextRank algorithm was applied to rank sentences.
5. **Summary Extraction**: A summary was extracted by selecting the top N sentences based on their rankings.

4.1 Results

Figure 3 shows the result of text summarization using TextRank algorithm in the form of Generated Summary and ROUGE scores. Table 3 presents the comparison of calculated ROUGE sores between GPT 3.5 and Bard. This demonstrates the effectiveness of TextRank-based summarization system in generating bullet-point summaries from multiple articles. This work contributes to the broader field of NLP and addresses the practical need for efficient information condensation in the era of information overload.

5 Conclusion

This comprehensive survey provides as a definitive exploration of the intricacies surrounding query-based text summarization. Through the meticulous categorization of research papers, it has illuminated four distinct taxonomies, artfully differentiating between machine learning methods, namely supervised and unsupervised, and the summary types, whether extractive or abstractive. By doing so, this survey not only highlights the remarkable developments in query-based text summarization but also underscores the notable gaps and scarcities in certain taxonomies, offering a concise summary of the current state of research. Furthermore, it emphasizes the

```
GENERATED SUMMARY:
Automatic summarization is the process of reducing a text document with a computer program in order to create a summary that retai
ns the most important points of the original document.
ROUGE Scores: [{'rouge-1': {'r': 0.7058823529411765, 'p': 1.0, 'f': 0.8275862020451843}, 'rouge-2': {'r': 0.6744186046511628, 'p':
 1.0, 'f': 0.8055555507445988}, 'rouge-l': {'r': 0.7058823529411765, 'p': 1.0, 'f': 0.8275862020451843}}]
```

Fig. 3 Result of text summarization using TextRank algorithm in the form of Generated Summary and ROUGE scores

Table 3 Comparison of ROUGE scores

Parameters	GPT 3.5	Bard
Rouge-1	0.8889	0.8275
Rouge-2	0.8824	0.8055
Rouge-L	0.8889	0.8275

potential for cross-pollination of methodologies from the broader domain of generic text summarization into the specialized field of query-based summarization. A text summarization based on TextRank algorithm is implemented and results are reported for the same.

References

1. Afantenos S, Karkaletsis V, Stamatopoulos P (2005) Summarization from medical documents: a survey. Artif Intell Med 33(2):157–177. https://doi.org/10.1016/j.artmed.2004.07.017. PMID: 15811783
2. Damova M, Koychev I (2010) Query-based summarization: a survey
3. Sinha A, Yadav A, Gahlot A (2018) Extractive text summarization using neural networks. arXiv:1802.10137
4. Herandez-Castenda A, Garcia-Hernandez RA, Ledenza Y, Millan-Hernandez CE (2020) Extractive automatic text summarization based on lexical-semantic keywords. IEEE
5. Venu G, Chandu M (2020) Extractive approach for query based text summarization. IEEE
6. Mohamed AA, Rajasekaran S (2006) Improving Query-based summarization using document graphs. IEEE
7. Murarka S, Singhal A (2020) Query-based single document summarization using hybrid semantic and graph-based approach. IEEE
8. Afantenos S, Karkaletsis V, Stamatopoulos P (2005) Summarization from medical documents: a survey. Artif Intell Med 33(2):157–177
9. Hasselqvist J, Helmertz N, Kageb¨ack M (2017) Query based abstractive summarization using neural networks. arXiv:1712.06100
10. Devlin J, Chang M-W, Lee K, Toutanova K, Ye X, Wei H (2008) Query-based summarization for search lists. IEEE
11. Kleinberg JM (1999) Authoritative sources in a hyperlinked environment. J ACM (JACM) 46(5):604–632
12. Chu E, Liu P (2019) MeanSum: a neural model for unsupervised multi- document abstractive summarization. In: International conference on machine learning. PMLR, pp 1223–1232
13. Rahman N, Borah B (2015)A survey on existing extractive techniques for query-based text summarization. In: 2015 international symposium on advanced computing and communication (ISACC)
14. Du Y, Huo H (2020) News text summarization based on multi-feature and fuzzy Logic. IEEE
15. Gupta VS, T.J (2012) Multi-document summarization using sentence clustering. In: 4th international conference on intelligent human computer interaction: advancing technology for humanity. IHCI, pp 1–5. https://doi.org/10.1109/IHCI.2012.6481826
16. Meena YK, Jai A, Gopalani D (2014) Survey on graph and cluster based approaches in multi-document text summarization. In: IEEE international conference on recent advances and innovations in engineering (ICRAIE-2014). IEEE, Jaipur, India
17. Bahdanau D, Cho K, Bengio Y (2014) Neural machine translation by jointly learning to align and translate. arXiv preprint arXiv:1409.0473

18. Conroy J, Schlesinger JD, O'leary DP (2006) Topic focused multi-document summarization using an approximate oracle score. In: Proceedings of the COLING/ACL 2006 main conference poster sessions. 2018. Bert: Pre-training of deep bidirectional transformers for language, pp 152–159
19. Rahman N, Borah B (2016) A survey on existing extractive techniques for query-based text summarization. IEEE
20. Mohamed AA (2018) Generating user focused content based summaries for multi-documents using document graphs. In: 5th IEEE symposium on signal processing and information technology (ISSPIT), pp 675–679. understanding. arXiv preprint arXiv:1810.04805
21. Wang J, Liu J, Wang C (2007) Keyword extraction based on PageRank. In: Zhou ZH, Li H, Yang Q (eds) Advances in knowledge discovery and data mining. PAKDD 2007. Lecture notes in computer science, vol 4426. Springer, Berlin, Heidelberg
22. Pennington J, Socher R, Manning C (2014) GloVe: global vectors for word representation. In: Proceedings of the 2014 conference on empirical methods in natural language processing (EMNLP), pp 1532–1543, Doha, Qatar. Association for Computational Linguistics

An Acoustic Method for Inspection of CNC Cutting Operations

Shilpa Sondkar and Kimaya Kerkar

Abstract CNC machining enables fine motion control of tools and workpieces simultaneously along multiple axes by a set of commands, simplifying manufacture of complex 3D components. Since the movements are not manually controlled, the operator does not have the tactile feedback based on which the cutting parameters can be fine-tuned to maintain and enhance the quality of cutting. One such aspect that is lost is the registration of the actual instant that the cutting tool makes contact with the workpiece. This can happen earlier or later than programmed due to variations in workpiece size and tool wear. This project uses commercially available sensors to analyse the change in sound and vibration from non-cutting movement to cutting operations, and applies MIR (Musical Information Retrieval) algorithms to determine the onset of cutting.

Keywords Sensor · MIR · CNC · Onset detection · Spectral flux · RNN

1 Introduction

Computer Numerical Control (CNC) is a machine control mechanism which uses preprogrammed commands to control the movement of a machine and its appendages. This process can be used to control a range of complex machinery, like grinders, lathes, mills and CNC routers. With CNC machining, three-dimensional cutting tasks can be accomplished in a single set of prompts, making complex, time consuming tasks simple. However, this also means that the intuition that an operator uses while operating the machine is lost. One can no longer feel the vibrations along the cutting process to determine the quality of the cut. Nowadays, sensors are being mounted on such CNC machines to record data and use both stochastic and probabilistic methods

S. Sondkar · K. Kerkar (✉)
Vishwakarma Institute of Technology, Pune, India
e-mail: ketan.kimaya21@vit.edu

S. Sondkar
e-mail: shilpa.sondkar@vit.edu

V. K. Gunjan and J. M. Zurada (eds.), *Proceedings of 5th International Conference on Recent Trends in Machine Learning, IoT, Smart Cities and Applications*, Lecture Notes in Networks and Systems 1181, https://doi.org/10.1007/978-981-97-8861-3_21

to analyse the cutting process, in both offline and online manners, incorporating these machines in a broader Industry 4.0 system [1]. Tool wear [2, 3], surface roughness [4], tool maintenance [5] and remaining useful life (RUL) can be predicted with the help of various sensors.

This can be extended to checking whether the machine is cutting according to the program or not. With variations in job blank sizes, tool wear and general wear and tear of the moving parts in the machine, the actual cutting might not always start at the programmed time. This paper presents a method to detect that instant when cutting begins so that one can use that to check whether the cutting has started at the right time. Since the machine generates unique sounds and vibrations, throughout the cutting process, these signals can be used to find that instant of time when the job makes contact with the cutting tool. Furthermore, techniques used for musical onset detection can be used to identify the change frequencies corresponding to the commencement of cutting operations. The algorithm would be used in the same manner, albeit with different thresholds for what the onset of a sound would be.

2 Literature Review

Onset detection is an actively researched problem in tasks for Musical Information Retrieval (MIR). A musical onset is the start of a note or a sound in an audio file. It is the first step for many other MIR tasks like beat detection, tempo estimation and music transcription. Instead of finding onsets in music, this application aims to find onset of cutting by CNC machines using audio signals generated during the process.

The basic process for onset detection is as follows: signal preprocessing, followed by generation of the onset detection function (ODF) and a peak picking algorithm to find the peaks in the ODF (Fig. 1).

Traditional methods for onset detection are typically done in the frequency domain using spectral analysis [6]. Here the spectral flux, i.e., the change in energy of the spectrogram is computed. The peaks of the spectral flux correspond to the onsets. This is similar to edge detection algorithms in image processing. Other techniques include pitch-based onset detection [7] and phase-based onset detection [8].

Neural networks have also been utilised for onset detection, as seen in [9]. Marolt et al. [10] use neural networks to improve peak picking on a hand-crafted onset detection function Schlüter & Böck[11] have trained a CNN model to generate an ODF using spectrograms of the audio in question. Eyben et al. [12] train a bidirectional

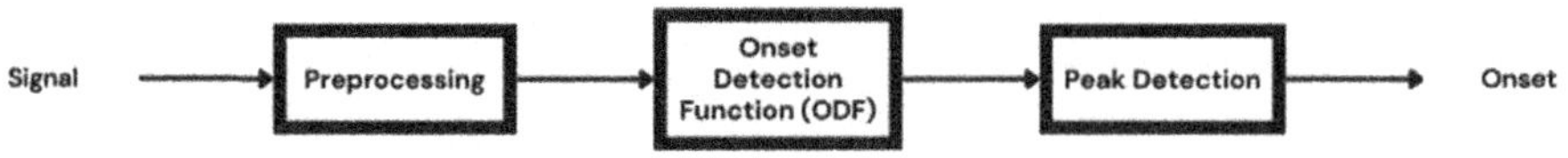

Fig. 1 General algorithm for onset detection

(RNN) on Mel-scaled magnitude spectrograms. Böck et al. [13] refine this model in several steps, defining the current state of the art in onset detection.

However, since neural networks require a lot of training, labelled data and information of the past, present and future sounds to detect onsets at a given instant, they do not lend themselves to online, or real time applications of onset detection. By modifying the peak picking algorithm, a spectral flux-based method can be applied for online onset detection, as proposed by Böck et al. [14]. Unidirectional RNNs have also been used by Böck et al. [13] for online onset detection by extracting information from three parallel Short-Time Fourier Transform (STFT)s and using that as input for the RNN. Both methods described above work well for online applications, while the RNN onset detector is more robust towards changes in the volume of signals.

3 Methodology

3.1 Algorithm

The offline spectral flux algorithm was selected for this application due to its simplicity and ease of use. The spectral flux is defined as the sum (or mean) of the first order differences of the magnitude spectrogram of an audio file [6]. This takes a STFT of an audio signal, with a certain number of frames per second, and finds the change in energy along the frames by subtracting them. The resultant is a signal with peaks at the intervals when the energy changes, i.e., the onsets of sounds. Only peaks above a certain threshold are selected to avoid false positives and accommodate for any noise that is present in the audio signal.

An online implementation of spectral flux, proposed by Böck et al. [14], can be done by reading the audio in framed chunks. Each frame would be 2048 samples, sampled at a rate of 44.1 Hz. The spectrogram of each frame is found and filtered by applying a Mel filter. Using the logarithmic magnitude of the spectrogram was found to have better results than a linear representation [12], so it can be implemented after pre-multiplying the function with a constant compression factor. Finally, it can be compared to the previous frame to generate the ODF as a function of the frames. The peaks are found by checking the ODF at a frame against 3 conditions-

1. The ODF at that frame is greater than its previous frames.
2. The ODF is greater than the mean of the ODFs at the previous frames.
3. The difference between the current frame and the last detected onset is greater than 0.

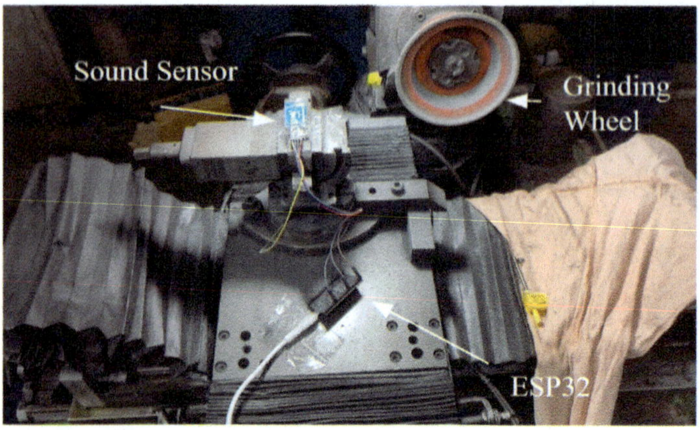

Fig. 2 Setup of the data acquisition system on the2 axis CNC grinding machine

3.2 Experiment

A sound sensor [15] was used to collect data from the CNC machine. The sensor data was collected with the help of an ESP32-WROOM microcontroller module. This has a fast clock speed of 80 MHz, which allows its Analog to Digital Convertor (ADC), to sample the audio data at high rates. Sampling at a frequency less than 1 kHz would lead to far too much data loss, given that the base cutting frequency is about 800 Hz. So, a sampling frequency of 2 kHz was used to ensure that the necessary features were successfully captured (Fig. 2).

3.3 Testing

To test the offline algorithm, a large number of signals were recorded using the setup described above. This data was saved in multiple. CSV files to avoid dealing with audio codecs and simplifying the process. Two types of cutting actions were recorded—plunging, where the grinding wheel makes a notch in the job and facing—where the grinding wheel rubs material off the job like sandpaper. Vibrations generated during both grinding actions were recorded, to ensure that there was no difference between the type of onsets generated by each of them.

Once recorded, a high pass filter with a cutoff of 200 Hz was applied on the signals to remove the frequencies caused by the spindle and movement of the axes. This filtered signal was used to obtain a spectrogram, which was then logarithmically scaled. The spectral difference between frames across the spectrogram was used as the ODF, and the peaks along the ODF were found to find the onsets. All the calculations and transformations were done using a python library for MIR applications called madmom [16].

4 Results and Discussions

The offline algorithm was tested on the cutting audios that were recorded and saved in .CSV files. Here, the spectral flux method would accurately be able to recognise the peaks in the ODF, generated by the spectral flux method. The threshold of the peak picking algorithm was altered and set to 7, to ensure only the required points were marked as onsets. The peaks found by this method were represented using a spectrogram in Figs. 3 and 4 shows a 5 sec snippet of the audio signal with an onset present. The offline method can successfully detect onsets in cutting with an accuracy of 89.6%. The peaks detected correspond to frequencies within the range of 1.5 kHz to 1.9 kHz.

The online algorithm was also implemented on recordings of the cutting process. Here, since the signal needed to be divided into frames, a 2 kHz sampling rate would not be sufficient for capturing all the auditory information. So, the audio was recorded in a .MP3 format and played so that the PC microphone could capture it and detect onsets in it. The PC microphone could be sampled at a rate of 44.1 kHz, which allows for frames of 2048 samples each and a total frame rate of 100 fps. The spectral flux method was used to generate the ODF for each frame in the framed signal. Some peaks were detected, after modifying the threshold of the peak picking algorithm. However, there are many erroneous peaks being detected. This is possibly due to the large amount of noise the PC microscope is subject to. More parameters could be

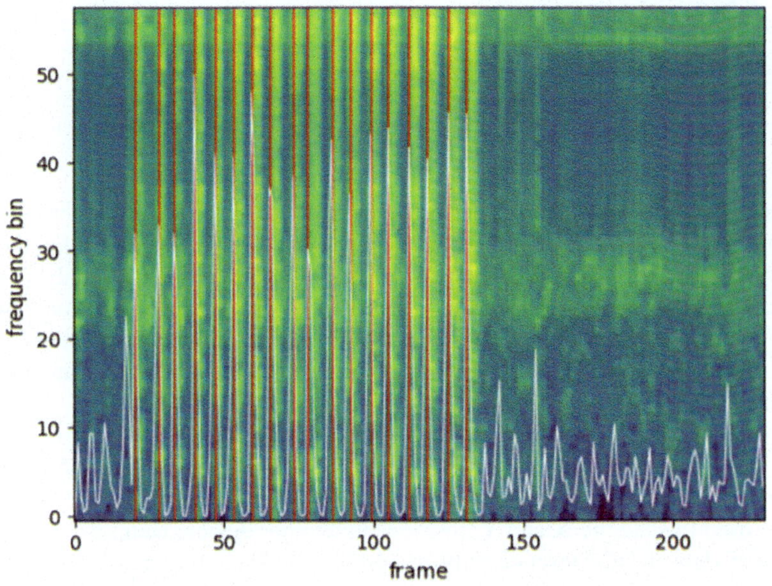

Fig. 3 The logarithmic spectrogram of a signal with the onsets marked in red vertical lines at the peaks of the spectral flux in white

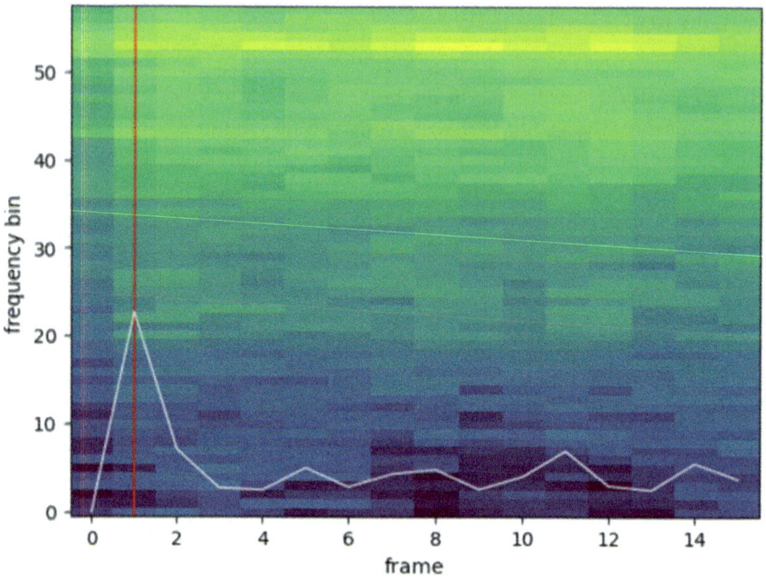

Fig. 4 The spectrogram of a 5 s audio signal with the spectral flux in white and onset of cutting in red

tuned to get better results in online peak picking and the same sound sensor could be used at a higher sampling rate, to get localised audio recording, without much noise.

5 Future Scope

So far, only the offline algorithm has successfully detected most of the onsets of cutting operations. The online variant needs more tuning, especially for the peak picking parameters. The online method will allow for real time audio analysis, which would then allow the onset of cutting to be detected as the cutting goes on, allowing the process to be immediately corrected if it is cutting at the wrong time, i.e., too early or too late. For this, the data will need to be sampled at a higher rate, i.e., 44.1 kHz instead of 2 kHz.

Once the online method is successfully running, the positional data from the CNC machine can be extracted to analyse whether cutting is happening at the right time. The machine described in the paper uses a Mach3 controller. A MODBUS connection can be used to send the X and Y positions to another device. Here, we can use that positional data and the onset of cutting to predict whether cutting is happening at the right time or not.

6 Conclusion

Using the methods and algorithms described above it is possible to create a system that will analyse the change in sound to determine onset of cutting. The proof of concept has been shown in offline mode and the validity of this algorithm running in online mode has also been established. Once the online algorithm has been fully developed, it has high potential to reliably automate the setup and operation of complex CNC processes.

References

1. Soori M, Arezoo B, Dastres R (2023) Machine learning and artificial intelligence in CNC machine tools, a review. Sustain Manuf Serv Econ pp 100009. https://doi.org/10.1016/j.smse.2023.100009
2. Zhang X, Han C, Luo M, Zhang D (2020) Tool wear monitoring for complex part milling based on deep learning. Appl Sci 10(19):6916. https://doi.org/10.3390/app10196916
3. Rizal M, Ghani JA, Nuawi MZ, Haron CH (2013) Online tool wear prediction system in the turning process using an adaptive neuro-fuzzy inference system. Appl Soft Comput 13(4):1960–1968. https://doi.org/10.1016/j.asoc.2012.11.043
4. Moreira LC, Li WD, Lu X, Fitzpatrick ME (2019) Supervision controller for real-time surface quality assurance in CNC machining using artificial intelligence. Comput Ind Eng 127:158–168. https://doi.org/10.1016/j.cie.2018.12.016
5. Bahr B, Motavalli S, Arfi T (1997) Sensor fusion for monitoring machine tool conditions. Int J Comput Integr Manuf 10(5):314–323. https://doi.org/10.1080/095119297131066
6. Bello JP, Daudet L, Abdallah S, Duxbury C, Davies M, Sandler M (2005) A tutorial on onset detection in music signals. IEEE Trans Speech Audio Process 13(5):1035–1047
7. Collins N (2005) Using a pitch detector for onset detection. In: Proceeding of the 6th international conference on music information retrieval (ISMIR)
8. Bello JP, Duxbury C, Davies M, Sandler M (2004) On the use of phase and energy for musical onset detection in the complex domain. IEEE Signal Process Lett 11(6):553–556
9. Schindler A, Lidy T, Böck S (2020) Deep learning for MIR tutorial. https://doi.org/10.48550/arXiv.2001.05266
10. Marolt M, Kavcic A, Privosnik M (2002) Neural networks for note onset detection in piano music. In: Proceedings of the international computer music conference (ICMC), Göteborg, Sweden, 2002
11. Schlüter J, Böck S (2014) Improved musical onset detection with convolutional neural networks. In: 2014 IEEE international conference on acoustics, speech and signal processing (ICASSP), pp 6979–6983. https://doi.org/10.1109/ICASSP.2014.6854953
12. Eyben F, Böck S, Schuller B, Graves A (2010) Universal onset detection with bidirectional long short-term memory neural networks. In: Proceeding of the 11th international society for music information retrieval conference (ISMIR), Utrecht, Netherlands
13. Böck S, Arzt A, Krebs F (2012) Online real-time onset detection with recurrent neural networks
14. Bock S, Krebs F, Schedl M (2012) Evaluating the online capabilities of onset detection methods
15. Sound sensor Sound sensor–Waveshare Wiki. https://www.waveshare.com/wiki/Sound_Sensor. Accessed 29 Nov 2023
16. Böck S, Korzeniowski F, Schlüter J, Krebs F, Widmer G (2023) Madmom: a new Python audio and music signal processing library. http://arxiv.org/abs/1605.07008. Accessed 29 Nov 2023

Enhancing Question Answer Generation from PDFs: A Fusion of BERT, RAKE, T5 and DistilBERT with RQUGE Evaluation

A. C. Akshitha Mary, Prachi Acharya, R. Rakshinee, Stefani Jeyaseelan, Thanvitha Nandakumar, V. Sakthivel, and P. Prakash

Abstract In educational institutes worldwide, the process of formulating questions for assessments and educational materials poses a recurring challenge. Traditional methods often involve manual question creation or keyword-based searches, leading to time-consuming and often inefficient processes, especially when dealing with extensive PDF documents. Addressing this prevalent issue, the research introduces a comprehensive approach to question–answer generation (QAG) from PDF documents. The system uses the most recent Natural Language Processing (NLP) models to extract text from PDFs using pdfplumber, create document summaries using BERT, extract keywords with RAKE, and create questions using a retrained T5 model. DistilBERT is then used to generate the answers. The RQUGE metric is used to assess the system's performance. The outcomes indicate how well the suggested method generates pertinent queries and precise responses, highlighting its potential for use in information retrieval and document comprehension.

A. C. A. Mary · P. Acharya · R. Rakshinee (✉) · S. Jeyaseelan · T. Nandakumar · V. Sakthivel · P. Prakash
School of Computer Science and Engineering, Vellore Institute of Technology, Vandalur-Kelambakkam Road, Chennai, India
e-mail: rakshinee.r2021@vitstudent.ac.in

A. C. A. Mary
e-mail: akshithamary.ac2021@vitstudent.ac.in

P. Acharya
e-mail: prachi.acharya2021@vitstudent.ac.in

S. Jeyaseelan
e-mail: stefani.jeyaseelan2021@vitstudent.ac.in

T. Nandakumar
e-mail: thanvitha.nandakumar2021@vitstudent.ac.in

V. Sakthivel
e-mail: sakthivel.v@vit.ac.in

P. Prakash
e-mail: prakash.p@vit.ac.in

V. K. Gunjan and J. M. Zurada (eds.), *Proceedings of 5th International Conference on Recent Trends in Machine Learning, IoT, Smart Cities and Applications*, Lecture Notes in Networks and Systems 1181, https://doi.org/10.1007/978-981-97-8861-3_22

Keywords Question Answer Generation · BERT · T5 · DistilBERT · RAKE · RQUGE Metric · Natural Language Processing · Information Retrieval

1 Introduction

The explosive growth of digital information has led to an overwhelming volume of textual data stored in diverse formats, with Portable Document Format (PDF) being a prevalent choice for document sharing and dissemination. This surge in document digitization has necessitated the development of advancedNatural Language Processing (NLP) [1] techniques to extract meaningful information automatically. One crucial task in this domain is Question Answer Generation (QAG), which involves the automated creation of questions based on a given text and the subsequent generation of relevant answers.

The significance of QAG lies in its potential to enhance information accessibility. Traditional methods of extracting information from documents often rely on manual processes or keyword-based searches, which can be time-consuming and inefficient, especially when dealing with large datasets. QAG systems, on the other hand, automate this process by not only extracting key information but also formulating questions that prompt concise answers, mimicking the way humans seek and comprehend information. The motivation behind this research stems from the growing need for sophisticated QAG systems, especially in handling the vast amounts of information locked within PDF documents. While serving as a popular format for preserving the original structure and layout of content, these documents pose unique challenges for automated information extraction due to their varied layouts, styles and rich textual content. Conventional information retrieval approaches struggle to provide accurate and contextually relevant results, prompting the exploration of advanced NLP models and techniques. Moreover, as the demand for automated information extraction continues to rise, there is a concurrent need for systems that not only retrieve relevant details but also present them in a structured and meaningful manner. QAG systems address this need by not only summarizing information but also by generating questions that guide the user through the core concepts of the document. This approach aligns with the natural way humans seek information, making it a valuable asset in diverse domains, from education to research and beyond.

The primary objectives of this research are rooted in addressing the limitations of existing QAG systems and harnessing the potential of advanced NLP models. The key objectives include: designing and implementing a robust pipeline encompassing various stages of QAG, from text extraction to question-and-answer generation. utilizing state- of-the-art NLP models, such as BERT, T5, and DistilBERT, to improve the system's accuracy and contextual awareness. Employing the RQUGE metric to assess the caliber of questions and answers that are created, offering a numerical representation of the system's efficacy, providing details on the real-world uses of the suggested QAG system and showcasing its adaptability to a variety of situations and use cases.

As the proposed paper delves into the intricacies of QAG from PDF documents, it is essential to underscore the broader significance of this research in the realm of information extraction and accessibility. The ability to automatically generate meaningful questions and answers from PDFs not only accelerates the process of information retrieval but also opens avenues for innovation in education, research and information-driven decision-making.

In academic settings, students and researchers often encounter extensive literature in PDF format. The QAG system can serve as a valuable companion, swiftly summarizing key concepts and generating insightful questions for a deeper understanding of the material. In professional settings, where time is often of the essence, an efficient QAG system can empower individuals to extract pertinent information swiftly, enhancing productivity and informed decision- making. Furthermore, the implications of this research extend to fields where large volumes of documentation are prevalent, such as legal and medical domains. Legal professionals dealing with extensive case files can benefit from a QAG system that extracts critical details and formulates relevant questions for further investigation. Similarly, healthcare practitioners managing voluminous medical literature can streamline their information retrieval process with the aid of an automated QAG system.

In the context of emerging technologies, the integration of advanced NLP models in the proposed architecture showcases the potential for synergy between machine learning and document analysis. The adaptability of the system to different domains highlights its versatility and applicability, positioning it as a promising tool for various industries seeking efficient solutions for information extraction from unstructured data.

The structure of the research paper adheres to a sequence that begins with related works, followed by the presentation of the proposed architecture, a detailed description of the experimental setup, the unveiling of results, concluding remarks with insights into future work and a comprehensive list of references.

2 Related Works

Numerous studies on the Automated Question Answer Generation model have been conducted in the past, and some of the most notable research being conducted currently is cited. Kettip Kriangchaivech et al. [2] proposed a method that automatically generated questions from Wikipedia passages using transformers on SQUAD. The evaluation of the model primarily focuses on the Word Error Rate (WER) as the metric for comparing the model-generated questions with the original SQuAD questions. To learn more about the quality of the questions, the WER distribution is examined, with particular attention to various WER ranges. In addition, the evaluation takes into account how many words each question has and how frequently the model-generated questions start with a certain word in comparison to the SQuAD questions. 9.94% of the model-derived questions had a WER of less than or equal to 5, 56.38% had a WER of between 6 and 10, 26.41% had a WER of between 11 and

15, 5.81% had a WER of between 16 and 20, and 1.45% of the generated questions had a WER of more than 21. These results were obtained from the experiment.

An Automatic Question Generation System using the "Text-toText Transfer Transformer" (T5) model was proposed by Saichandra Pandraju et al. [3]. The model generates questions from texts and tables. The dataset creation process involves modifying the ToTTo dataset to include tables and their corresponding highlighted cells, along with questions generated from the descriptions of the tables. The project's goal is to develop a modified version of the ToTTo dataset called TabQGen, which will enable questions to be generated from tables. The learning rate was set to a linear schedule warmup using an AdamW optimizer of 1e−4. Using NIST, BLEU, ROUGE-L, and METEOR as metrics, the experiment's performance was evaluated.

An Automatic Question Generation Model [4] based on a Deep Learning Approach that generates Wh-questions with different difficulty levels from uploaded PDF documents. The system aims to assist in educational applications by automating the process of question generation from textual content. This model gets a BLEU-4 score of 11.3.

A Reinforcement Learning and encoder-decoder framework-based Answer-driven Deep Question Generation (ADDQG) model [5]. Deep question generation (DQG) uses reasoning across several publications to produce sophisticated queries. ADDQG beats state-of-the-art models in both automatic and human evaluations, according to extensive trials conducted on the HotpotQA dataset. Comparing the experiment's results to the SemQG model, the average improvement was 2.83, 1.27, and 1.15 points for BLEU-4, METEOR, and ROUGE-L, respectively. The main distinction between ASs2s-a and ADDQG appears to be that the model architecture is better suited for complex question generation. Both of them encode the answer and document separately, obtaining an average improvement of 4.95, 3.11, and 4.88 points in terms of BLEU-4, METEOR, and ROUGE-L, respectively.

Question Generation Pre-Training for Text Generation (QURIOUS) was developed by Shashi Narayan et al. [6].The main benefits of this approach were that: (i) it was simple to gather a large amount of data for question generation from community quality assurance platforms like Quora, Yahoo Answers, and Stack Overflow; and, more significantly, (ii) text generators that were trained to produce questions that could be answered from passages or documents would identify important terms or concepts in the input and would learn to summarize and synthesize the information. Evaluations of the trial were conducted both automatically and by humans. In terms of both automatic and human evaluations, the model produced summaries and questions that were more natural and informative when tested for answer-focused question creation and summarizing tasks.

Katherine Stasaski et al. [7] enhanced the existing causal extraction system, which extracts causes and effects from unstructured text using a set of syntactic rules, to create a pipeline to produce and assess cause-effect questions straight from a text. Two English datasets were used to assess the approach: the Textbook Question Answering (TQA) dataset, which included middle school science textbooks, and SQuAD Wikipedia articles. A transformer-based quality assurance model from the Huggingface Transformers library, which is BERT-large fine-tuned on SQuAD

2.0 with whole-word masking, was employed to evaluate the generated questions' quality. The TQA dataset's F1 score for the best QG-QA pair of clause types is 0.59, whereas the Squad Dataset's F1 score is 0.63.

To produce questions from a passage and questions that are answered from the passage, a Natural Question Generation (QG) model [8] was proposed. Using a Graph Neural Network at each stage of the decoding process, they create the Iterative Graph Network-based Decoder (IGND) to simulate the prior generation. Based on SQuAD and MARCO datasets, experimental results show that the model performs better than the state-of-the-art models for sentence-level QG tasks. With a score of 20.33, BLEU-4 is considered the primary evaluation metric for text creation. This model produces the best outcomes. Three criteria were used to evaluate human evaluation: answerability, relevance, and fluency. The results showed that the model produced output that was of a higher caliber than what was needed.

An ASQ: Automatically Generating Question-Answer Pairs using AMRs [9] automatically mine questions and answers from a sentence using the Abstract Meaning Representation (AMR). They used the AMR Annotation Release 2.0 Corpus (39,260 sentences) and the AMR annotated Little Prince Corpus (1,562 sentences) to run ASQ to generate a question-answer pair that carried out a qualitative assessment of the outcomes from AMR 2.0. Seventy phrases were chosen at random from the AMR 2.0 test split to serve as the test set for error analysis in human evaluation. This evaluation is based on the quality assessment and error propagation from AMR parsers.

AnswerQuest: A System for Generating Question-Answer Items from Multi-Paragraph Documents was proposed by Melissa Roemmele et al. [10]. This model extracts questions directly from the given reference text. For the model's encoder and decoder layers, they use the Transformer architecture, and they improve the decoder by adding a copy mechanism. One sentence serves as the encoder input, while the question serves as the decoder output. The input sentence includes the question's answer. Regarding ratings, the outcomes mirror the automated assessment: RULEMIMIC questions receive higher ratings than STANDARD questions, and AUGMENTED questions receive higher ratings than RULEMIMIC questions. Compared to the HUMAN questions, every set of produced questions has a substantially lower rating. Overall, these findings once more demonstrate the value of adding automatically created questions to the training set. As a result, the AUGMENTED model is currently used in their demo.

3 Proposed Architectures

This section discusses the proposed architecture of the model. The QAG model, as shown in Fig. 1, integrates pdfplumber, BERT, RAKE, re-trained T5 and DistilBERT in an efficient workflow. Starting with text extraction, BERT summarizes the document, RAKE identifies keywords and the T5 model generates questions. DistilBERT produces accurate answers, showcasing the synergy of advanced NLP models. Evaluated with the RQUGE score, the model ensures a comprehensive assessment of question-and-answer quality.

The proposed Question Answer Generation (QAG) architecture, as illustrated in Fig. 2, is crafted to seamlessly transform user-provided PDF files into meaningful questions and answers. Delving into open-source tools such as Hugging Face and Gradio, the paper outlines a streamlined pipeline that harnesses the power of cutting-edge NLP models. Below is a detailed overview of each step within the proposed architecture:

Fig. 1 Workflow of the proposed model

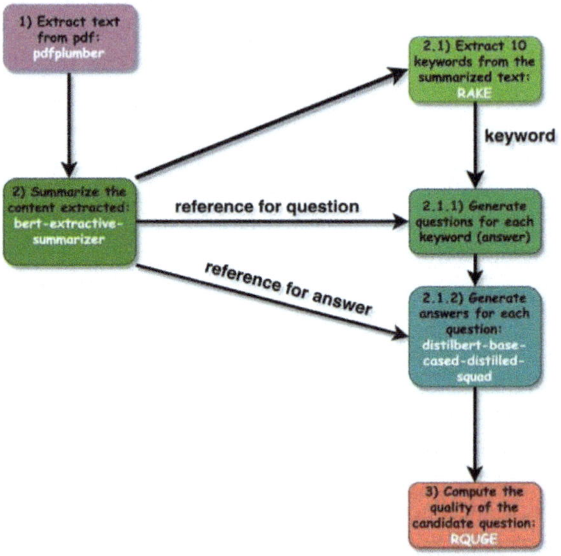

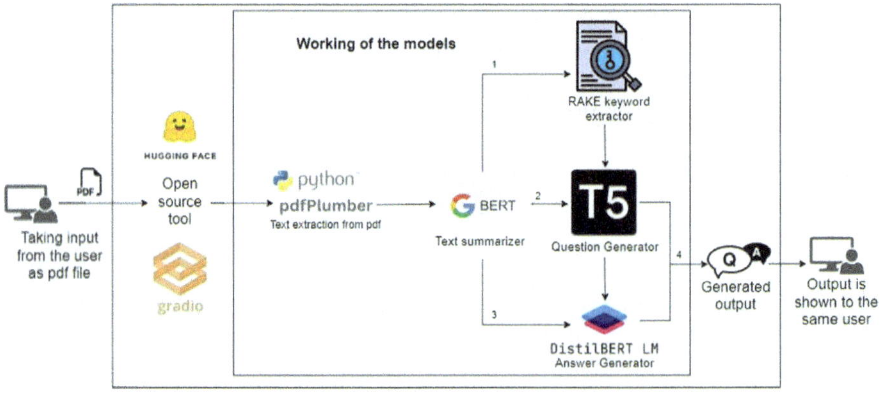

Fig. 2 The architecture of the proposed model

The process begins with the user providing a PDF file as input. This ensures the flexibility and user-friendliness of the system, allowing individuals to extract valuable insights from PDF documents effortlessly. Delving into the integration of state-of-the-art NLP models, two powerful open-source libraries, Hugging Face and Gradio, are leveraged to facilitate the architecture. Using the pdfplumber library, textual content is extracted from the user- provided PDF file. Pdfplumber is chosen for its reliability and efficiency in accurately converting the PDF content into machine-readable text. The extracted text undergoes processing through a pre-trained BERT (Bidirectional Encoder Representations from Transformers) model. BERT's contextual understanding enables the generation of concise and informative summaries of the input text, capturing essential information. Rapid Automatic Keyword Extraction (RAKE) is employed to identify key concepts and terms within the generated summary, enhancing the precision of subsequent question generation by pinpointing crucial information.

A re-trained T5 model is incorporated into the pipeline for question generation. The model is fine-tuned using a custom dataset and for each identified keyword, it generates questions related to the input text. This step ensures a focused and contextually relevant set of questions. Subsequently, the generated questions are paired with the original text and the DistilBERT model is employed for answer generation. DistilBERT, a more lightweight variant of BERT, efficiently produces contextually relevant answers. The final output consists of the questions generated for each keyword and their corresponding answers. This user-friendly output format ensures that individuals interacting with the system receive meaningful insights derived from the PDF document. Gradio, a user interface creation library, enhances the overall accessibility of the system. It facilitates a seamless interaction between the user and the QAG pipeline, providing a graphical interface for uploading PDF files and viewing the generated questions and answers.

4 Experimental Setups

4.1 Hyperparameter Tuning and RQUGE Evaluation

To systematically explore the impact of different hyperparameter configurations on the performance of the model for question-answering on the SQuAD [11] dataset, a series of experiments were conducted. The key hyperparameters under investigation were the learning rate (LR), batch size (BS) and the number of training epochs (Epochs). The primary evaluation metric for assessing model performance was the RQUGE score [12].

The following hyperparameter combinations were explored (Table 1).

Table 1 Hyperparameter combinations

Experiment	Learning rate	Batch size	Max epochs	RQUGE score
1	3e−4	4	1	4.27
2	5e−4	4	1	4.21
3	1e−4	8	1	3.79
4	1e−4	4	2	4.26
5	3e−4	4	1	4.32

4.2 Visualization

The results of the experiments are visualized in Fig. 3 below. Each bar represents a different hyperparameter configuration, with the x-axis indicating LR, BS and Epochs. The height of the bars corresponds to the achieved RQUGE score. Importantly, the blue bars represent different experimental configurations, while the green bar represents the configuration with the highest RQUGE score (Experiment 5).

The visualization clearly illustrates the comparative performance of each configuration in terms of the RQUGE metric. Experiment 5, indicated by the green bar, stands out as the most successful configuration with the highest RQUGE score among all experiments. These experiments and visualizations guide the understanding of the impact of hyperparameter choices on the model's performance, providing valuable insights for future iterations and improvements.

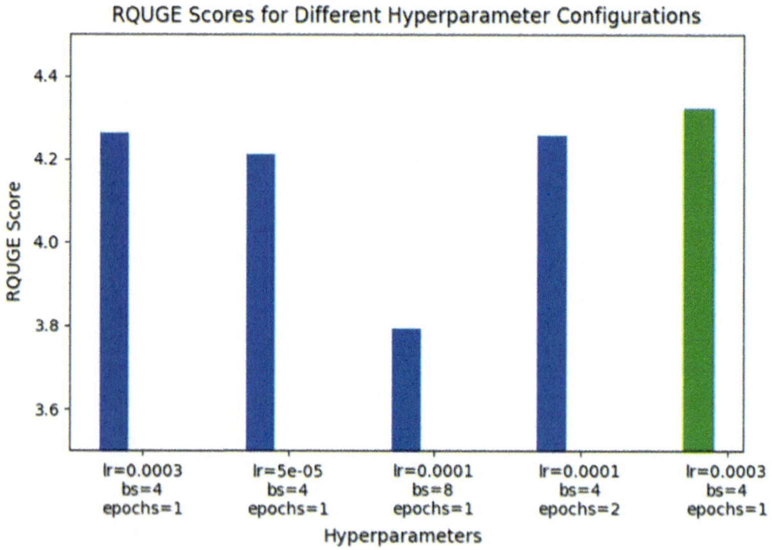

Fig. 3 RQUGE scores for different hyperparameter configurations

5 Results

5.1 Model Comparison

In this section, the comparative results of the proposed model against an existing baseline model on the task of question-answering using the SQuAD dataset are presented. The evaluation metric utilized for this comparison is the RQUGE score, a measure of the effectiveness of the model in generating relevant and high-quality responses.

In Fig. 4, the bar chart vividly illustrates the comparative RQUGE scores between the existing and the proposed models. Notably, the existing model obtained a commendable RQUGE score of 4.20, while the innovative approach propelled the proposed model to a significantly higher RQUGE score of 4.32.

This marked difference in RQUGE scores accentuates the prowess of the proposed model in generating responses that not only exhibit heightened accuracy but also contextual relevance. The model's performance stands out as superior when juxtaposed with the baseline model, showcasing its capacity to excel in the intricate task of question-answering. The observed uptick in the RQUGE score serves as a robust validation of the model's potential to elevate the efficiency and efficacy of question-answering tasks.

The experiments were meticulously conducted under consistent conditions, ensuring a fair evaluation of the models' capabilities. The RQUGE scores provided in the bar chart serve as a quantitative benchmark, offering insights into the models'

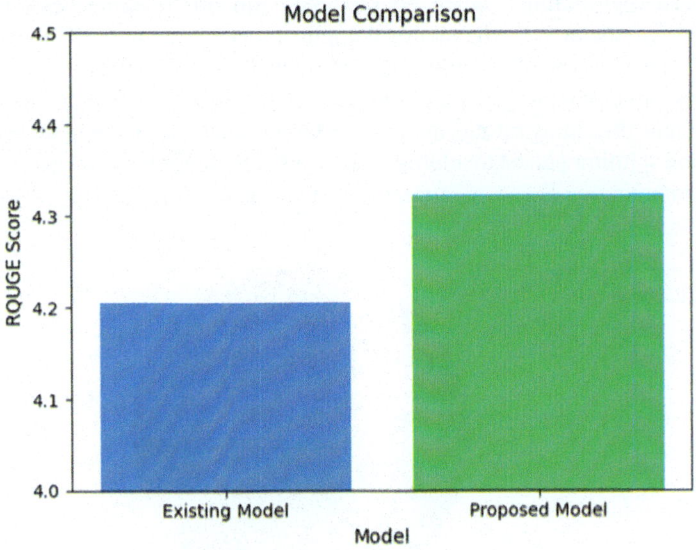

Fig. 4 RQUGE Scores

respective strengths. The higher RQUGE score achieved by the proposed model is indicative of its proficiency in generating responses that closely align with human-like comprehension. This alignment underscores the model's potential to contribute significantly to advancing question-answering systems.

The success of the model in surpassing the baseline establishes a compelling narrative for its prospects. The encouraging results invite further exploration and consideration for integration into real-world applications, particularly those demanding precision and relevance in responses to user queries. This achievement positions the proposed model as a promising solution for enhancing question-answering systems in diverse and practical settings.

5.2 Human Evaluation

In a crucial step toward assessing the practical efficacy of the question-answering model, a comprehensive human evaluation survey was initiated. Feedback was solicited from participants tasked with rating the quality of the generated answers on a 5-point scale. The outcomes of this survey as in Fig. 5 revealed an overwhelmingly positive response, signifying a unanimous vote of confidence from participants. An astonishing 46.7% of respondents bestowed the highest possible rating of 5, while an additional 43.3% expressed robust approval, assigning a rating of 4. This resounding endorsement underscores a shared consensus among users, affirming the model's capacity to deliver answers that are not only accurate but also contextually relevant, thus validating its effectiveness in real-world scenarios.

These survey findings, when coupled with quantitative metrics such as the RQUGE score, underscore the practical applicability of the model. The user-centric feedback gleaned from the survey provides invaluable insights into the model's performance from the perspective of those actively engaging with it. With nearly 90% of participants providing ratings indicative of high satisfaction, the model emerges as a promising solution poised to address real- world challenges in information retrieval and user interaction. These encouraging results further fortify the potential impact

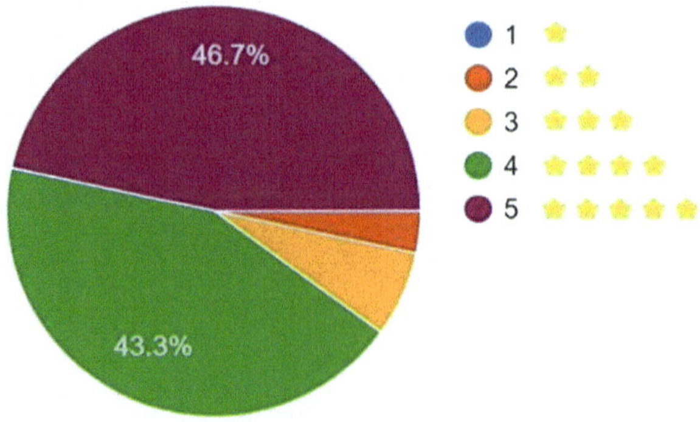

Fig. 5 Pie chart depicting human evaluation results

of the proposed model, positioning it as a catalyst for delivering heightened user experiences and reliable solutions in the realm of question-answering across diverse domains. The convergence of positive user sentiments and robust quantitative metrics underscores the model's promise and its ability to make substantive contributions to the field.

6 Conclusion and Future Works

In conclusion, the research introduces an innovative Question Answer Generation (QAG) architecture tailored for insightful information extraction from PDF documents. Open-source tools, including Hugging Face and Gradio, are utilized to seamlessly integrate advanced NLP models such as BERT, RAKE, T5 and DistilBERT into an intuitive pipeline. The user-friendly system facilitates seamless PDF uploads, providing concise summaries and delivering contextually relevant questions and answers.

The experimental evaluation, conducted on the SQuAD dataset, highlights the architecture's efficacy in accurately summarizing and extracting information from diverse PDF sources. Integration of RQUGE metrics showcases the system's capability to generate high-quality questions and answers. The user interface, powered by Gradio, enhances accessibility, ensuring the QAG system is both robust and user-friendly.

Looking ahead, the future work aims to enhance the versatility of the QAG model. The plan includes incorporating support for multi-document types to seamlessly handle diverse information sources. Additionally, expanding language capabilities to offer multi-lingual support is on the agenda, broadening the applicability of the system across global contexts. Furthermore, there are plans to diversify question types, incorporating multiple-choice questions (MCQs), filling in the blanks and exploring the complexity of questions to cater to varying educational and informational needs. Additionally, optimization efforts will be undertaken to enhance the efficiency and speed of the proposed QAG model, ensuring optimal performance across different scenarios.

References

1. https://www.udemy.com/course/question-generation-using-natural-language-processing/learn/lecture/23905916#overview
2. Kriangchaivech K, Wangperawong A. Question generation by transformers. U.S. Bank 1095 Avenue of the Americas New York, NY 10036
3. Pandraju S, Mahalingam SG (2021) Answer-aware question generation from tabular and textual data using T5. Int J Emerg Technol Learn 16(18)
4. Abdel-Galil H, Mokhtar M, Doma S (2021) Automatic question generation model based on deep learning approach. Int J Intell Comput Inform Sci

5. Wang L, Xu Z, Lin Z, Zheng H-T, Shen Y (2020) Answer-driven deep question generation based on reinforcement learning. In: 28th international conference on computational linguistics
6. Narayan S, Simoes G, Ma J, Craighead H, Mcdonald R (2020) QURIOUS: question generation pretraining for text generation
7. Stasaski K, Rathod M, Tu T, Xiao Y, Hearst MA (2021) Automatically generating cause- and-effect questions from passages. In: Proceedings of the 16th workshop on innovative use of NLP for building educational applications
8. Fei Z, Zhang Q, Zhou Y (2021) Iterative GNN-based decoder for question generation. In: Conference on empirical methods in natural language processing
9. Rakshit G, Flanigan J (2021) ASQ: automatically generating question-answer pairs using AMRs
10. Roemmele M, Sidhpura D, DeNeefe S, Tsou L (2021) AnswerQuest: a system for generating question-answer items from multi-paragraph documents. In: Proceedings of the 16th conference of the European chapter of the association for computational linguistics: system demonstrations
11. Dataset: https://huggingface.co/datasets/squad
12. Mohammadshahi A, Scialom T, Yazdani M, Yanki P, Fan A, Henderson J, Saeidi M (2022) RQUGE: reference-free metric for evaluating question generation by answering the question

Detecting Fundus Lesions for Diabetic Retinopathy (DR) Analysis

Soumya H. Hooli, Yashaswini S. Murgod, Kaveri S. Haramagatti, Satish Chikkamath, Kaushik Mallibhat, and S. R. Nirmala

Abstract Diabetic retinopathy (DR) is an eye condition caused by uncontrolled diabetes and can affect the retinal tissues and lead to vision impairment if left untreated for a long period. There are many computer-aided techniques and artificial intelligence algorithms to detect lesions. Detecting fundus lesions through these techniques is time-consuming, expensive, and more complex even for ophthalmologists. The contrast of image and noise is the challenge during DR diagnosis. Thus, the spot highlighting method and Contrast-Limited Adaptive Histogram Equalization (CLAHE) process are used to enhance the features of the eye fundus images towards detection of fundus lesions for diabetic retinopathy analysis. In addition to spot highlighting and CLAHE, data amalgamation techniques are used to overcome the challenges of over-fitting and noise.

Keywords Diabetic retinopathy · Microaneurysms · Hemorrhages · Exudates · Data amalgamation · CLAHE process · Spot highlighting process · GoogleNet

---

S. H. Hooli · Y. S. Murgod (✉) · K. S. Haramagatti · S. Chikkamath · K. Mallibhat · S. R. Nirmala
School of Electronics and Communication Engineering, KLE Technological University Hubballi, Hubbali, Karnataka, India
e-mail: 01fe21bei051@kletech.ac.in

S. H. Hooli
e-mail: 01fe21bei054@kletech.ac.in

K. S. Haramagatti
e-mail: 01fe21bei055@kletech.ac.in

S. Chikkamath
e-mail: chikkamath@kletech.ac.in

K. Mallibhat
e-mail: kaushik@kletech.ac.in

S. R. Nirmala
e-mail: nirmala.s@kletech.ac.in

V. K. Gunjan and J. M. Zurada (eds.), *Proceedings of 5th International Conference on Recent Trends in Machine Learning, IoT, Smart Cities and Applications*, Lecture Notes in Networks and Systems 1181, https://doi.org/10.1007/978-981-97-8861-3_23

1 Introduction

Diabetic retinopathy (DR) is an eye condition that has different levels of severity which mainly affects the eye's retina, that helps in processing vision. DR could damage major blood vessels in the eye fundus which is made up of the blood vessel, optic disc, fovea, and macula. It also diminishes the fundus light sensitivity causing vision loss and is one of the eye conditions caused by diabetes [1].

Diabetes is a body condition that is caused by the inability of the human body to control sugar levels [2]. There are two types of diabetes: Type I, which is inevitable and its cause is unknown. Type II, whose side effects are Microaneurysms, Hemorrhages, Exudates, cotton wool spots, etc., and it is treatable. If Type II diabetes goes without treatment for over five years it causes diabetic retinopathy.

There are mainly two stages in diabetic retinopathy: Proliferative Diabetic Retinopathy (PDR) and Non-proliferative Diabetic Retinopathy (NPDR) [3]. In PDR the chances of blindness are observed because of the uncontrolled growth and swelling of blood vessels and nerves. NPDR is the term used to describe the initial stage. Severe NPDR causes PDR.

DR can be detected by the changes in the retina which are Hemorrhages (HEM), Hard Exudates (HEX), Soft Exudates (SE), Microaneurysms (MAs), cotton wools, optic nerve head, and irregular growth in the blood vessels. Figure 1 explains the stages of DR from the initial stage (NPDR) to the final stage (PDR).

Identification of pigmented lesions in the eye fundus which specify the presence of DR is a challenge. If a patient with DR is diagnosed in the NPDR stage, the growth rate of the disease can be reduced as the condition of the retina will be at the initial stage. The major changes in the retina are Hemorrhages (HEMs), Microaneurysms (MAs), and Exudates (EXs), and the occurrence of these results in DR [4]. There are many computer-aided techniques in detecting fundus lesions such as CADe and CADx that work at the pixel level to locate fundus lesions manually [5]. The contrast of image

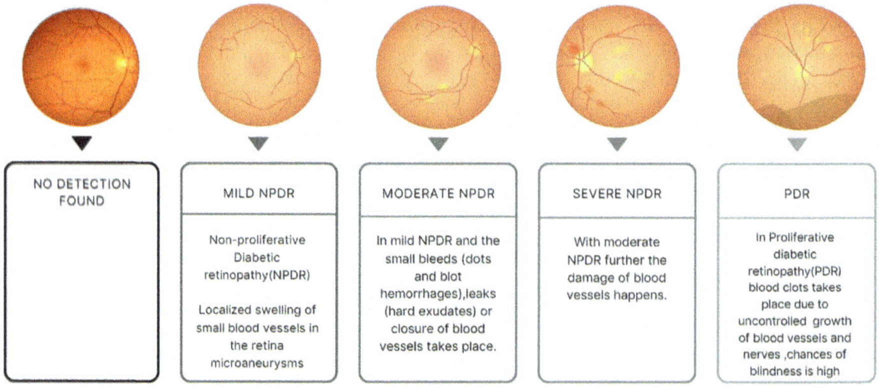

Fig. 1 Stages of DR

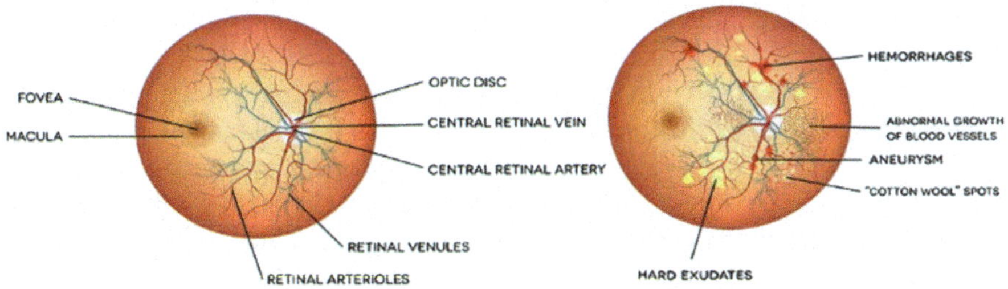

Fig. 2 Normal retina versus affected retina

and noise are some of the common challenges that we come across while detecting the DR due to which these techniques are time-consuming and more complex.

1.1 Objectives

The present work aims to:

1. Perform a comprehensive literature survey to arrive at the various pre-processing, feature extraction, and machine learning methods, and their performances toward the detection of lesions in Diabetic Retinopathy.
2. Investigate the impact of the amalgamation of datasets to avoid overfitting and arrive at an explicit technique for detecting the lesion in Diabetic Retinopathy.
3. Use the spot-highlighting process and Contrast-Limited Adaptive Histogram Equalization process to detect multiple hemorrhages and differentiate between normal blood clots and morphological changes caused by diabetic retinopathy.

The most widely used Deep learning (DL) architecture is Convolution Neural Network (CNN) in which overfitting is the challenge. We aim to overcome this challenge and detect the lesions by amalgamating different datasets as variations in input images subject the model to many more new features to learn and build a model around this concept as it will help both the patients and the doctors to detect. DR detection techniques require a huge amount of dataset. Hence, Datasets from various websites like Messidor, E-ophtha, and KAGGLE can be used. Figure 2 shows the diagrammatic representation of the difference between a normal retina and a DR-affected retina.

2 Literature Survey

In the mid-19th century, the detection of DR was a complex task that involved many processes and was time-consuming. It was difficult to detect the fundus and the accuracy of detecting DR was less. Conventional optical photography and Fundus

Imaging were the common techniques that were used to capture images and identify the lesions.

Film cameras were used in the beginning for color fundus photography, then digital cameras eventually took the role of film cameras. The introduction of the Scanning Laser Ophthalmoscope (SLO) was a result of the Fundus camera's gradual growth which could get better angles of the eye fundus. The development of Ultra-Widefield (UWF) Imaging improved the precision of the process. Later, Optical coherence to tomography (OCT), has made a significant impact on the study of DR. In recent years, the development of artificial intelligence (AI) has been applied to the detection of DR.

Table 1 and Fig. 3 below show the Literature survey and the Evolution of Detecting DR.

3 Proposed Methodology

The present work aims to build an algorithm using image segmentation, feature extraction, and binary classification using CNN [13]. Pre-processing is the initial step in CNN. The images are analyzed and collected from different datasets. Then, the images are augmented to enlarge the features of the retina. Then the images undergo a segmentation process. Segmentation isolates and separates the images that contain retinal features. Contrast-limited Adaptive Histogram Equalization (CLAHE) and Spot Highlighting (SH) are applied to highlight those features by normalizing the intensity and eliminating the background which helps in feature extraction [14]. Then the images are fed to the CNN algorithms for the binary classification of DR.

The input image is first augmented and then goes through the CLAHE and SH process. CNN algorithms train and store the features of this input image for classification. We have selected LeNet, ZFNet, and Googlenet for the CNN algorithms. The pipeline shown below helps us to propose the CNN model, where the system initially takes the retinal images and the CNN algorithm detects the morphological changes in the input image which helps in detecting the DR (Fig. 4).

4 Results and Analysis

We trained the three models on amalgamated datasets and we found out that GoogleNet best suited for our system. The LeNet model is faster but the accuracy varies drastically hence it is eliminated. ZFNet has fewer layers hence feature extraction is less not suitable for our model. Table 2 explains the performance of the three algorithms on the dataset. From Table 2, it is evident that Googlenet gives the best results.

Table 1 Literature table

Paper	dataset	ML/DL method	Objective	Performance
[6]	EyePACS and Messidor	Deep convolution neural network	Binary classification	Eyepacs ACC (%)–75 Messidor AUC (%)–91.2
[7]	DIARETDB E-optha DIARETDB1 and HEI-MED	Multi-depth (CNN), SVM and linear regression, DCNN	Microaneurysms, Hemorrhage, and Exudates detection	AUC (%)–89.32 ACC (%)–32.13 SEN (%)–82.2 F1-score–0.932
[8]	DIARETDB0 and DIARETDB1	SVM for classification	Five-class classification method is based on calculating the areas of several features (hemorrhages, microaneurysms, exudate, and blood vessels)	SEN (%)–82 SP (%)—86 ACC (%)–85.9
[9]	DIARETDB0	K-nearest neighbor (KNN) classifier	DR grade binary classification	SEN (%)–98.68 ACC (%)–96.73
[10]	EyePACS	Deep convolutional neural network (CNN) and PCA because limited availability of the dataset	Detection and stages of diabetic retinopathy	ACC (%)–87.95
[11]	IDRiD	ML-CAD (Machine learning–computer-aided design)	Identifying and diagnosing between healthy and DR grades using fundus images	ACC (%)–94.48
[12]	Messidor IDRiD	Deep convolutional neural network (CNN) and CLAHE	Detection and stages of diabetic retinopathy	Accuracy Normal–97.9% Stage 1–97.65% Stage 2–97.65% Stage 3–98.64%

LeNet is a convolution neural network, made up of a total 7 layers which have 3 convolutional layers, 2 sub-sampling layers, and 2 fully connected layers. It uses SoftMax for the classification. However, since there are fewer layers, some features are not extracted, which decreases the performance of LeNet. ZFNet is another classic convolutional neural network, consisting of 19 deep layers. Among them, 16 are convolutional layers and 3 are fully connected layers. But compared to GoogleNet, its performance is low. GoogleNet is 40 layers deep. The training and validation accuracy is obtained better compared to other models and thus we are using the GoogleNet model for dataset amalgamation and the spot-detection method. The CLAHE process is the process that enhances the contrast of the bright spots like exudates and swollen blood vessels that are observed in the eye fundus images. Spot Highlighting Process

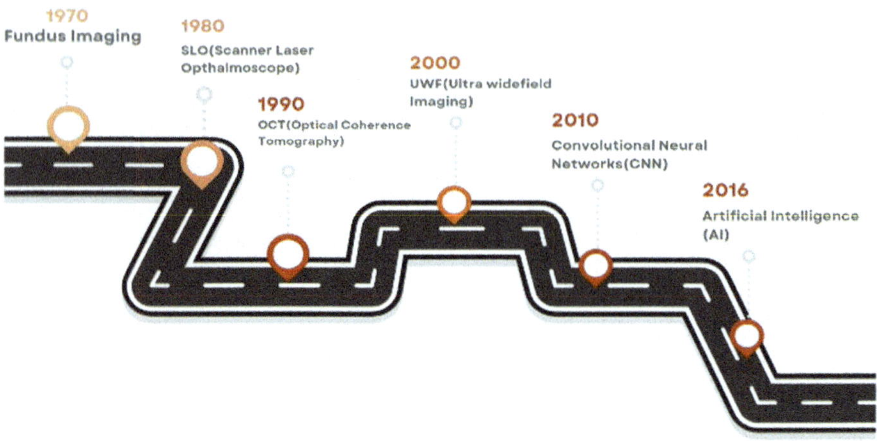

Fig. 3 Evolution of Detecting DR

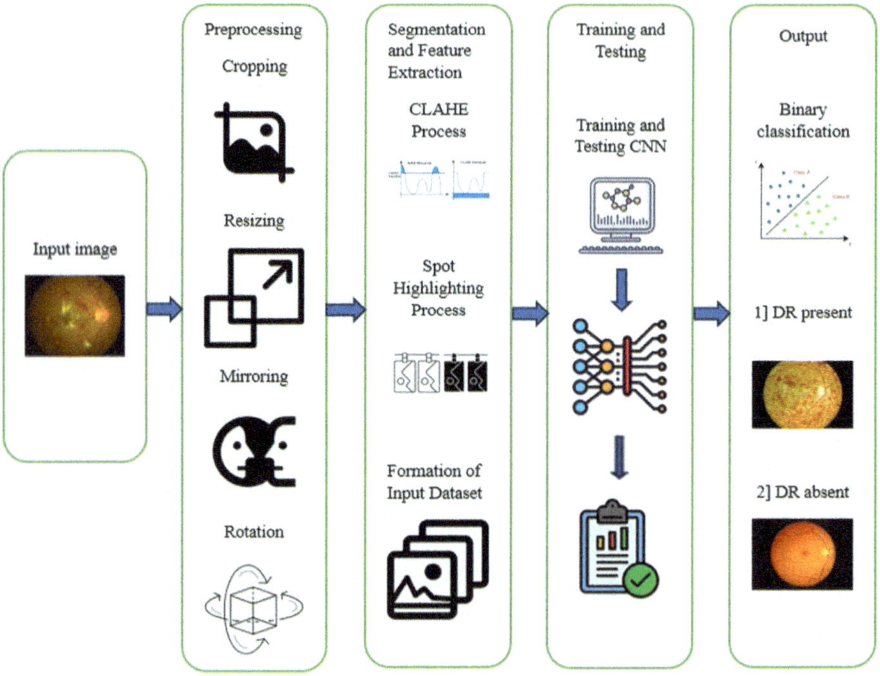

Fig. 4 Proposed pipeline for the binary classification of DR images

Table 2 Comparison table of models

Sl. no	Dataset	ML/DL method	Performance
1	Kaggle	GoogleNet	Test accuracy = 0.95700 Test loss = 0.04300
2	Kaggle	ZFNet	Test accuracy = 0.90899 Test loss = 0.09101
3	Kaggle	LeNet	Test accuracy = 0.84275 Test loss = 0.15724

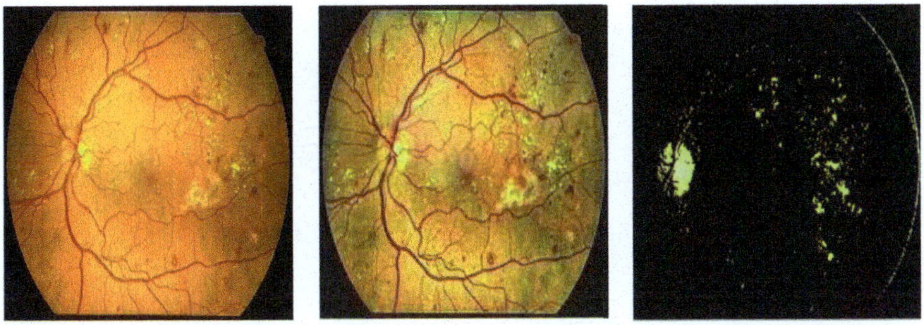

(a) Input image (b) CLAHE image (c) SH image

Fig. 5 Comparison of images

gives out images that only contain features that are highlighted by the CLAHE and nullify the background and noise. Figure 5a–c show us the preprocessing of the image through the CLAHE and SH processes. This is an important process which is also our contribution which increases its accuracy and reduces the false positives.

Figure 6a and b represent the graphs of accuracy and loss of Training and Testing Set which are plotted against epochs. The graphs are plotted for a dataset taken from Kaggle and they show us that the results of the Training set are reliable but the Testing Set results vary drastically. Hence, data amalgamation is necessary for the system.

Our contribution to the model is the spot-highlighting process which comes after the CLAHE process. It captures even the smallest details that are highlighted by the CLAHE process and gives an image. An image helps the algorithm to learn the features faster and clearer. Hence, it increases the performance of the algorithm and gives overall better results. Table 3 explains the performance of the algorithm without using and with using the SH process. Figure 7a and b show us the graphs of accuracy and loss without using spot-highlighting whereas Fig. 8a and b show the graphs of accuracy and loss when we use the spot-highlighting process. From the graphs, it is seen that the spot highlight process gives better results with the data amalgamation.

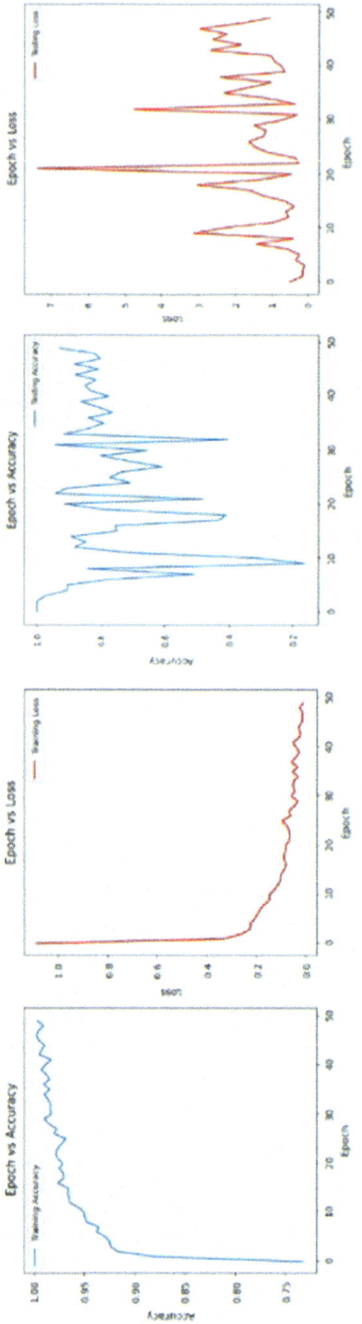

(a) Training Set

(b) Testing Set

Fig. 6 Comparison of training and testing sets

Table 3 Contribution table

Sl. no	Dataset	Method	Performance
1	Kaggle and IDRiD	GoogleNet with CLAHE process	Test accuracy = 0.9974 Test loss = 0.0025
2	Kaggle and IDRiD	GoogleNet with CLAHE and spot-highlighting process	Test accuracy = 0.9987 Test loss = 0.0016

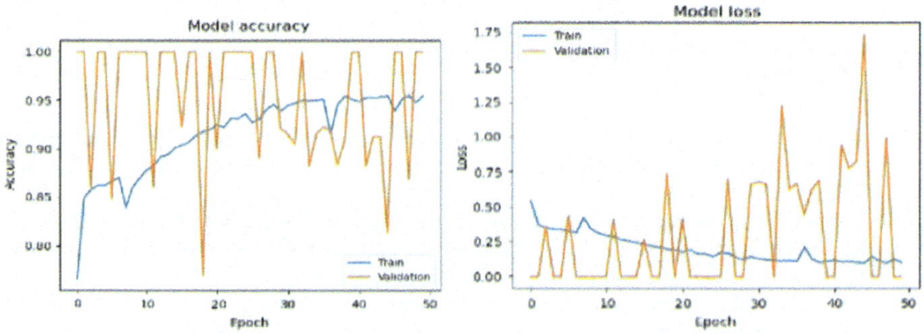

(a) Model accuracy (b) Model loss

Fig. 7 Comparison of accuracy and loss using CLAHE

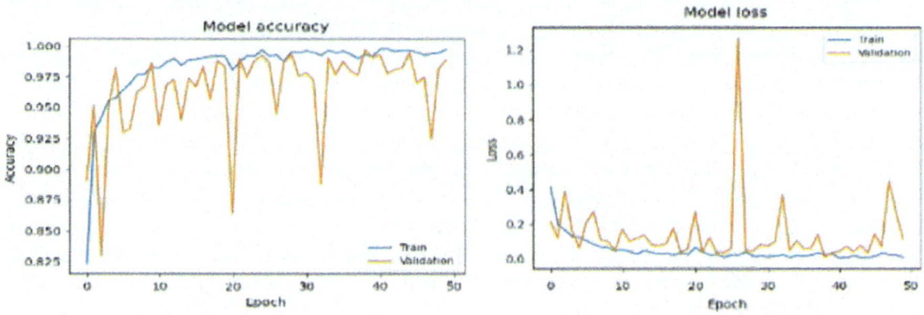

(a) Model accuracy (b) Model loss

Fig. 8 Comparison of accuracy and loss using CLAHE and SH

5 Conclusion

We conclude our present work by stating that Diabetic Retinopathy has been increasing in recent years and computer-aided algorithms are commonly used for their detection. We have used three algorithms namely LeNet, ZFNet, and GoogleNet, and concluded that GoogleNet is best suited for detecting DR. Then Datasets from different databases are amalgamated to avoid overfitting in the model predictions and to increase the performance of the algorithm. Spot Highlighting (SH) process is added along with traditionally used CLAHE processes for the feature extraction. The CLAHE process highlights the features and SH gives a feature-extracted in which the background is nullified. Then GoogleNet is trained to predict the presence of DR accordingly. It is found that the algorithm gives better results when the input images from different datasets are amalgamated and the SH process is used. Its performance also increases and reduces false negatives which is an important factor to be considered in the medical field.

References

1. Sumadithya S, Latheef FN, Sameer M, et al (2023) Detecting diabetic retinopathy using deep learning techniques. In: 2023 3rd international conference on intelligent technologies (CONIT). IEEE, pp 1–6
2. Kanungo YS, Srinivasan B, Choudhary S (2017) Detecting diabetic retinopathy using deep learning. In: 2017 2nd IEEE international conference on recent trends in electronics, information & communication technology (RTEICT). IEEE, pp 801–804
3. Abdelmaksoud E, El-Sappagh S, Barakat S, Abuhmed T, Elmogy M-H (2021) Automatic diabetic retinopathy grading system based on detecting multi-retinal lesions. IEEE Access 9:15939–15960
4. Nazir T, Irtaza A, Rashid J, Nawaz M, Mehmood T (2020) Diabetic retinopathy lesions detection using faster-RCNN from retinal images. In: 2020 first international conference of smart systems and emerging technologies (SMARTTECH). IEEE, pp 38–42
5. Saeed F, Hussain M, Aboalsamh HA (2021) Automatic diabetic retinopathy diagnosis using adaptive fine-tuned convolutional neural network. IEEE Access 9:41344–41359
6. Adal KM, Van Etten PG, Martinez JP, Rouwen KW, Vermeer KA, van Vliet LJ (2017) An automated system for the detection and classification of retinal changes due to red lesions in longitudinal fundus images. IEEE Trans Biomed Eng 65(6):1382–1390
7. Nandhini S, Sowbarnikkaa S, Mageshwari J, Saraswathy C (2023) An automated detection and multi-stage classification of diabetic retinopathy using convolutional neural networks. In: 2023 2nd international conference on vision towards emerging trends in communication and networking technologies (ViTECoN). IEEE, pp 1–5
8. Nagpal D, Panda SN, Malarvel M, Pattanaik PA, Khan MZ (2022) A review of diabetic retinopathy: datasets, approaches, evaluation metrics and future trends. J King Saud Univ Comput Inf Sci 34(9):7138–7152
9. Omar M, Khelifi F, Tahir MA (2016) Detection and classification of retinal fundus images exudates using region based multiscale LBP texture approach. In: 2016 international conference on control, decision and information technologies (CoDIT). IEEE, pp 227–232
10. Palavalasa KK, Sambaturu B (2018) Automatic diabetic retinopathy detection using digital image processing. In: 2018 international conference on communication and signal processing (ICCSP). IEEE, pp 0072–0076
11. Pratt H, Coenen F, Broadbent DM, Harding SP, Zheng Y (2016) Convolutional neural networks for diabetic retinopathy. Procedia Comput Sci 90:200–205
12. Qiao L, Zhu Y, Zhou H (2020) Diabetic retinopathy detection using prognosis of microaneurysm and early diagnosis system for non-proliferative diabetic retinopathy based on deep learning algorithms. IEEE Access 8:104292–104302

13. Deepa V, Sathish Kumar C, Cherian T (2022) Ensemble of multi-stage deep convolutional neural networks for automated grading of diabetic retinopathy using image patches. J King Saud Univ-Comput Inf Sci 34(8):6255–6265
14. Gurcan OF, Beyca ÖF, Dogan O (2021) A comprehensive study of ma-chine learning methods on diabetic retinopathy classification. Int J Comput Intell Syst (1):1132–1141

Music Versus Speech Classification Using 3D CNN

Pramod Hullur, Dhanya Kulkarni, Vinayak Jainapur, Satish Chikkamath, S. R. Nirmala, and Suneeta V. Budihal

Abstract In the Scenario of technological advancements, the accurate classification of audio signals stands as a decisive challenge related to applications like Automatic Speech Recognition (ASR), Music Information Retrieval (MIR) systems, music recognition, speaker identification, and speech enhancement. This paper proposes a novel methodology for the subtle classification of music versus speech, motivated by a three-dimensional Convolutional Neural Network (3D CNN). Unlike traditional methods, the proposed model illustrates the power of using 3D CNN on the 3D representation of Mel-Spectrograms, offering an innovative means to capture intricate temporal and spectral features inherent in audio signals. The proposed model gives good accuracy in the classification between music and speech, demonstrating its efficacy in addressing the challenges associated with audio signal classification.

Keywords 3D spectrogram · Mel spectrogram · 3D CNN · Convolutional neural network · Music · Speech

P. Hullur (✉) · D. Kulkarni · V. Jainapur · S. Chikkamath · S. R. Nirmala · S. V. Budihal
School of Electronics and Communication Engineering, KLE Technological University Hubballi, Hubballi, Karnataka, India
e-mail: pramod2003hullur@gmail.com

V. Jainapur
e-mail: vinay.j777@icloud.com

S. Chikkamath
e-mail: chikkamath@kletech.ac.in

S. R. Nirmala
e-mail: nirmala.s@kletech.ac.in

S. V. Budihal
e-mail: suneeta_vb@kletech.ac.in

V. K. Gunjan and J. M. Zurada (eds.), *Proceedings of 5th International Conference on Recent Trends in Machine Learning, IoT, Smart Cities and Applications*, Lecture Notes in Networks and Systems 1181, https://doi.org/10.1007/978-981-97-8861-3_24

1 Introduction

The audio industry holds significant importance in entertainment, telecommunication, and various other fields. Audio classification is an important requirement for Automatic Speech Recognition (ASR) systems and Music Information Retrieval (MIR) systems [1]. In a classification of music and speech understanding of music can be derived as a diverse form of expressing expressions in a more rhythmic, harmonic, and beautiful manner. Whereas speech is a form of sound with irregular linguistic patterns, pauses, and varying rhythms based on spoken language patterns. There are many classes or types of audio but music and speech form the prominent categories.

Music is the art of arranging sound to create some combination of harmony, melody, and rhythm. Music involves a broad range of frequencies, spanning from low frequencies produced by instruments like double bass (around 40 Hz) to high frequencies produced by instruments like piccolo (around 4,000 Hz) [2]. Different musical instruments and vocal ranges contribute to this wide frequency spectrum. Whereas, speech is a human vocal communication that has pitch modulation, intensity levels, tempo, and other non-representational or paralinguistic aspects of vocalization. Speech signals consist of a range of frequencies, typically between 85 and 255 Hz for male voices, 165 and 255 Hz for female voices, and 255 and 400 Hz for child voices [3]. The fundamental frequency (also known as pitch) determines the perceived pitch of speech. When we talk about pitch the distribution of pitch in music is likely to contain more voiced frequency which is spaced apart by chromatic intervals than the speech. Music is stationary as it is continuous and has non-changing spectrums and speech has changing spectrums due to the punctuations and pauses, unlike in music. The rhythms of music and speech are also very different because speech is very irregular and has linguistic patterns, whereas music has a specific rhythmic structure. The timbre of speech has variations but music has a more diverse palette due to various contributing factors. Many other differentiating factors include amplitude, harmonics and overtones, and temporal patterns. A wide range of data is available on the internet most of which is unlabeled. Therefore, we are required to make a system that can function like the human brain to accurately differentiate between the classes of speech and music.

The main purpose of the classification of music versus speech is to propose a model for better classification to ease human-computer interaction through voice commands [4]. Therefore, a novel approach is necessary to distinguish between the two and achieve improved results in speech versus music models, such as in the music recognition model that can differentiate between music and speech and eliminate speech factors to carry out the further process. And a system that will accurately recognize speech in any language for clear speaker identification, speech enhancement, and emotion recognition models. In this regard, using the 3D CNN [5] could be beneficial as it is expected to have a more precise effect and any further work on music and speech recognition will provide a solid base for designing robust models with more complex features. The idea of a 3D model provides a very new insight into the classification of speech and music as it has a spatial component to it which in turn provides an improved understanding of speech and as well as music.

1.1 Previous Works on Audio Classification

There exist several works and studies that utilize different approaches in the field of audio classification, focusing on different aspects and applications. Some notable previous works are feature-based Power Normalized Cepstral Coefficients (PNCC) as the acoustic feature [6]. The multi-resolution analysis is the most significant statistical method to extract the features from the input audio signal in this study. K-means clustering organizes the feature vectors into k number of groups. The overall accuracy of the pro- posed method K-means using PNCC is 87%. The two prominent approaches based on the time domain features are zero crossing rate (ZCR), and short-term energy (STE) [7]. In the frequency domain features spectral centroid, spectral flux, roll-off of the spectrum, cepstral residual, and their variance were used. And in time-frequency domain approaches spectrogram and evolutionary spectrum were used. In conclusion, more re- search is needed to resolve human problems to give precision and see which is the most suitable method of implementation in audio classification. Further, ZCR and Bayesian classification (ZB) are elementary forms concerning computational form [8]. But, Performance will be affected due to the music segment which contains speech superimposed on it or some rhythmic components. Thus they have proposed a second method that is based on a Multi-Layer Perceptron. In this case, better performance is at the expense of limited growth in computational complexity. Another feature extraction was from the empirical mode decomposition (EMD) and intrinsic mode functions (IMFs) and residual [9]. It was observed that speech follows a lower frequency band and music tends to have a wide frequency range and has higher bands [9]. It can be observed that music is continuous but speech is discontinuous (Table 1).

Table 1 Methods of audio classification

Methods	Year	Accuracy (%)
Frequency and amplitude classification	1950's	90
K-nearest neighbor (KNN)	1950's	71.28
Artificial neural network	1958	91
Hidden Markov model (hmm)	1970's	91.4
Support vector machine (SVM)	1985–1995	98
Mel frequency spectrum	1980	76
Long short term memory (LSTM)	1997	90
Zero crossing rate	2000	93
Convolutional neural network	2010–ongoing	95

The table provides a list of methods used for audio classification over the years. However, data may vary according to various sites and various research papers. Employing the method of the Gaussian mixture model to classify the categories of sounds has been employed over the years broadly and has provided significant results [10]. Using cepstral coefficients speech and non-speech categories have been made the number of segments and segment duration have been recorded. Where speech is characterized by formant structure and music by harmonic structure [10]. In recent years machine learning gained popularity and advancements due to improvements in machine learning CNN which is a complex form of deep learning [11] emerged and had a prominent effect on the audio industry. CNN proved to be a great aid in the audio classification methodologies. It is used in many classification problems. For example, it is used for sentiment analysis, Emotion recognition [12] feature extraction [13], genre classification [14], and prediction [15]. The CNN model is also widely used in audio classification and word sorting. CNN predicts the results through the data given to it in the training stage which is achieving flawless recall of spoken content across all types of media sources [16]. CNN classifies the error into two types which are Scheirer Slaney corpus and Musan corpus. When these errors are analyzed manually they show that vocal singing excerpts without instrumental accompaniment, or singing style close to regular speech, are predicted as speech. MFCC (Mel Frequency Cepstral Coefficients) is the most versatile feature that is used in both speech and music processing [17]. A model Swish-Net was proposed which is a 1D convolutional network that operates on MFCC features. The idea of introducing a 1D model instead of 2D (operates on the spectrogram) is that it is less computationally expensive, and 1D feature maps require less memory during processing [18]. The above are the methods and results obtained due to result of years of research in the audio classification field.

The paper is organized as follows- Section2 describes the proposed methodology. Section 2.1 is about the data set used. Section 2.2 describes the feature extraction process. Section 2.3 describes the model architecture. Section 3 provides the details of result obtained. And Sect. 4 concludes this study.

2 Proposed Methodology

In the domain of music versus speech classification using machine learning and deep learning, several previous works have contributed significantly. However, like any field of research, there are certain drawbacks. And the proposed methodology counters some of these drawbacks.

Music and speech exhibit complex temporal patterns that are not fully captured by traditional feature extraction methods which can limit the performance of models. However, utilizing this methodology of 3D CNN and 3D representation of the spectrogram to capture spatiotemporal features may enhance the representation of temporal dynamics, enabling the model to capture dynamic patterns over time.

Many earlier approaches relied heavily on manual feature engineering that may not fully exploit the richness of the data and can be time-consuming and subjective. But 3D CNN learns hierarchical features automatically from the 3D representation, eliminating the need for extensive manual feature engineering. 3D CNN, trained on a diverse set of 3D mel-spectrograms, can learn to differentiate between signal and noise patterns. The temporal aspect of the 3D representation helps the model distinguish between transient noise and consistent audio features. 3D CNN's ability to capture both spectral and temporal features can contribute to better generalization.

2.1 *Dataset*

Data Collection: The proposed methodology employed two key datasets: the Gtzan Music Genre Collection and LibriSpeech. Each dataset offers unique characteristics and contributes to the generalizability of the model. Gtzan Music Genre Collection has 500 music tracks, representing 10 different music genres: blues, classical, country, disco, hip-hop, jazz, metal, pop, reggae, and rock. Each audio file is 30 sec long. The Gtzan dataset captures the essence of each genre, allowing the model to learn discrete features and patterns of music class. LibriSpeech is a large-scale corpus of approximately 1,000 hours of read English speech, from which a random 500 audio files were chosen, 10–15 sec long each. The LibriSpeech dataset offers diverse speech patterns, thereby facilitating in learning and recognition of varied verbal features. The aim of combining the diverse musical genres found in the Gtzan dataset with the speech data from LibriSpeech is to construct a comprehensive dataset suitable for training a precise model capable of distinguishing between music and speech. This balanced integration removes inherent biases and enhances the model's capacity to effectively handle previously unseen samples in real-world applications.

2.2 *Feature Extraction*

Before initiating feature extraction, the audio data undergoes preprocessing to ensure uniformity through resampling and to reduce computational overhead via downsampling, optimizing it for the chosen feature extraction methods. The feature extraction process is a crucial step in this proposed methodology and was executed using the Python library called Librosa. Commencing with the primary feature extraction step, the computation of a mel-spectrogram was performed, which is a time-frequency representation of an audio signal. The x-axis denotes the time frame, the y-axis denotes the frequency bin, and the third dimension indicates the intensity of spectral energy at each frequency bin over time.

Here is the theoretical method to calculate mel-spectrogram $S(f, t)$ of an audio signal is as follows:

1. The STFT $X(m, \omega)$ of an audio signal $x(n)$ is given by:

$$X(m, \omega) = \sum_{n=0}^{N-1} x(n)\, w(n - mH)\, e^{-j\omega n} \quad (1)$$

where H is the hop size, $w(n)$ is the window function, and N is the window size.

2. Transform STFT to the mel scale using The Mel filterbank:

$$M(f) = 1127 \log\left(1 + \frac{f}{700}\right) \quad (2)$$

3. The mel-spectrogram is then obtained by applying the filterbank:

$$S(f, t) = \sum_{m=1}^{M} |X(m, \omega)|^2 \cdot H(m, f) \quad (3)$$

To further ensure feature stability and accentuate spectral patterns, the mel spectrogram undergoes two critical transformations: logarithmic scaling and normalization. In the final step to construct a representation that seamlessly aligns with the 3D CNN architecture, a novel approach was employed (Fig. 1).

The log-mel spectrogram was quantized to minimize the computational complexity. Then the quantized log-mel spectrogram was precisely segmented into multiple 2D arrays, mirroring the shape of the original spectrogram. This segmentation was achieved by iterating over the unique values present in the spectrogram. For each unique value, a dedicated 2D array was created; this unique value was placed

Fig. 1 3D spectrogram

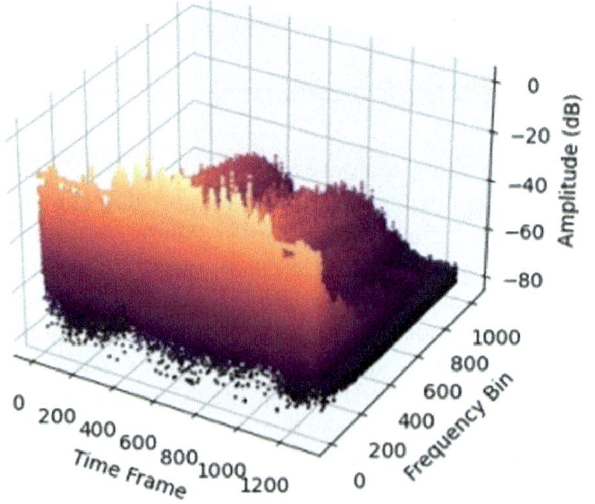

in the 2D array in such a way that its positions in the 2D array had to match the positions in the spectrogram. Then, these 2D arrays were converted into grayscale images and stacked up to form a 3D spectrogram, efficiently serving as input to the 3D CNN architecture. Consequently, the novel design of the feature extraction process yields informative 3D spectrograms that encode the spectral, temporal, and amplitude variations of audio signals.

2.3 Model Architecture

After obtaining the 3D spectrogram in the preprocessing, the model was to be passed to the 3D convolutional neural network for validation and training. The convolutional neural network was built using the Python libraries Tensor-flow and Keras. The CNN model consists of 2 convolution layers (Conv3D) with 32 and 64 filters, respectively, and kernel sizes of (3, 3, 3) using the ReLU activation function. After each convolution layer batch normalization was employed for improved convergence. Dropout layer for regularization with a rate of 0.25 to prevent overfitting. Then the data was flattened for input into fully connected layers, which contain two dense layers with 128 and 64 neurons, utilizing ReLU activation. The final layer has two neurons and a softmax activation function, optimized for binary classification. Lastly, the model is compiled using the Adam optimizer, binary cross-entropy loss, and accuracy as the evaluation metric (Fig. 2).

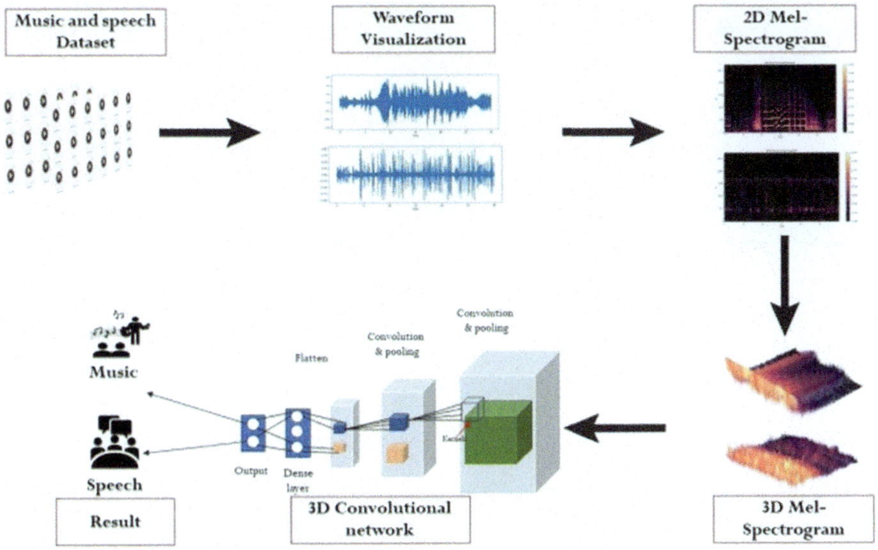

Fig. 2 Pipeline of proposed model

3 Experimentation and Results

In the phase of experimentation and training, the proposed model of a 3D convolutional neural network was trained and validated over a total of 1,000 audio files which contain 500 music and 500 speech as mentioned in the methodology section. The data set is split into 70% train set and 30% validation/test set. The training data achieved an accuracy of 90.5% when passed through the CNN model. These results show the effectiveness of using the approach of quantization and 3D representation of spectrograms, providing a comprehensive representation of temporal and spectral features that played an important role in accurate classification. This model demonstrated good generalization capabilities, performing well on unseen data, suggesting its potential for real-world applications. This success can be attributed to the careful data preprocessing pipeline and the well-designed 3D CNN architecture, highlighting the effectiveness of our approach for music vs. speech classification tasks (Figs. 3 and 4).

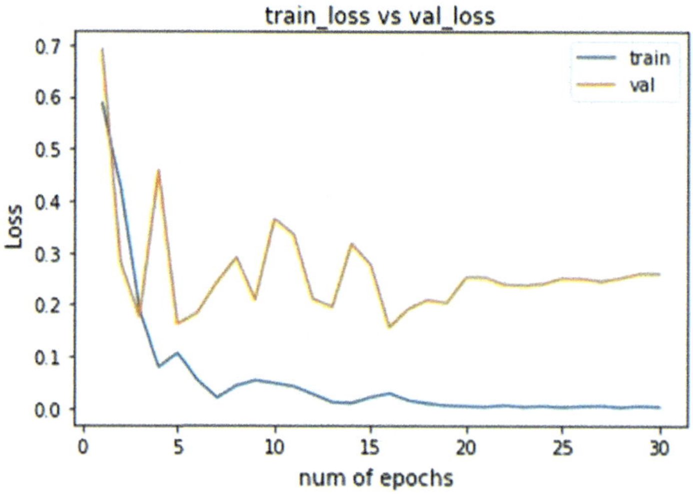

Fig. 3 Loss versus epoch

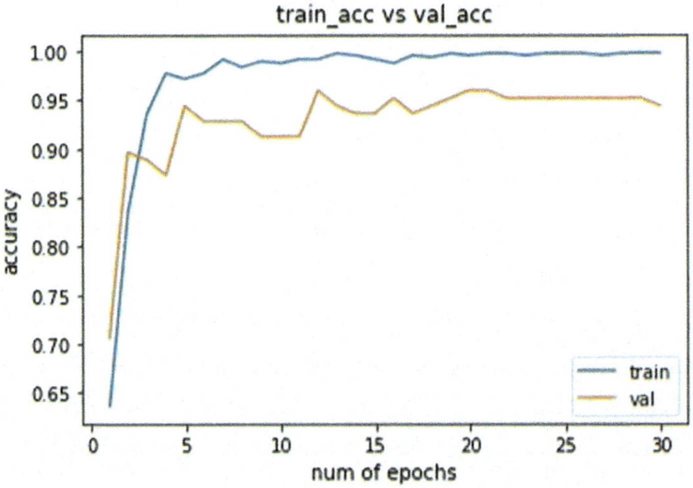

Fig. 4 Accuracy versus epoch

4 Conclusion

In this study, a novel approach for music versus speech classification uses a 3D convolutional neural network (CNN) and 3D representation of a spectrogram. This methodology effectively captured the spectral and temporal features of audio signals through the unique feature extraction process that involves quantization and 3D representation of mel-spectrograms. The model achieved a promising accuracy of 90.5%, demonstrating its ability to distinguish between music and speech. Though the results of the model were good, the model's performance could be further refined and elevated compared to the already existing 2D CNN models if focused on scaling up the dataset, thereby significantly contributing to the field of audio signal processing and machine learning.

References

1. Afouras T, Chung JS, Senior A, Vinyals O, Zisser-man A (2018) Deep audio-visual speech recognition. IEEE Trans Pattern Anal Mach Intell 44(12):8717–8727
2. Treider JM, Kunst JR, Vuoskoski JK (2023) The influence of musical parameters and subjective musical ratings on perceptions of culture. Sci Rep 13(1):20682
3. Hollien H, Hollien PA, de Jong G (1997) Effects of three parameters on speaking fundamental frequency. J Acoust Soc Am 102(5):2984–2992
4. Khudavand AA, Chikkamath S, Nirmala SR, Iyer N (2021) Music/non-music discrimination using convolutional neural networks. In: Soft computing and signal processing: proceedings of 3rd ICSCSP 2020, vol 1. Springer, pp 17–28

5. Rajput M, Chauhan K, Gopal G, Johar AK, Tripathi A, Varma T (2021) Speaker recognition using 3d convolutional neural network and gmm. In: International conference on energy systems, drives and automations. Springer, pp 549–558
6. Thiruvengatanadhan R (2016) Speech/music classification using power normalized cepstral coefficients and k-means. Int J Adv Comput Sci Technol 6(1):13–17
7. Al-Shoshan AI (2006) Speech and music classification and separation: a review. J King Saud Univ-Eng Sci 19(1):95–132
8. Bugatti A, Flammini A, Migliorati P (2002) Audio classification in speech and music: a comparison between a statistical and a neural approach. EURASIP J Adv Signal Process 2002:1–7
9. Pinquier J, Rouas J-L, André-Obrecht R (2003) A fusion study in speech/music classification. In: 2003 IEEE international conference on acoustics, speech, and signal processing, 2003. Proceedings (ICASSP'03), vol 2. IEEE, pp 2–17
10. Ghosal A, Dhara BC, Saha SK (2011) Speech/music classification using empirical mode decomposition. In: 2011 second international conference on emerging applications of information technology. IEEE, pp 49–52
11. Chikkamath S, Nirmala SR (2022) Music detection using deep learning with tensorflow. In: ICDSMLA 2020: proceedings of the 2nd international conference on data science, machine learning and applications. Springer, pp 283–291
12. Roopaei M, Rad P, Jamshidi M (2017) Deep learning control for complex and large scale cloud systems. Intell Autom Soft Comput 23(3):389–391
13. Li TL, Chan AB, Chun AH (2010) Automatic musical pattern feature extraction using convolutional neural network. Genre 10(2010):1x1
14. Nakashika T, Garcia C, Takiguchi T (2012) Local-feature-map integration using convolutional neural networks for music genre classification. In: Thirteenth annual conference of the international speech communication association
15. Sigtia S, Dixon S (2014) Improved music feature learning with deep neural net- works. In: 2014 IEEE international conference on acoustics, speech and signal processing (ICASSP). IEEE, pp 6959–6963
16. Doukhan D, Carrive J (2017) Investigating the use of semi-supervised convolutional neural network models for speech/music classification and segmentation. In: The ninth international conferences on advances in multimedia (MMEDIA 2017)
17. Logan B et al (2000) Mel frequency cepstral coefficients for music modeling. In: Ismir, vol 270. Plymouth, MA, p 11
18. Hussain MS, Haque MA (2018) Swishnet: a fast convolutional neural network for speech, music and noise classification and segmentation. arXiv preprint arXiv:1812.00149

GUI-Based Multi-model Approach for Alzheimer's Detection and Classification

Anirudh Deshpande, Nandan Illur, Roshith Hegde, Satish Chikkamath, Kaushik Mallibhat, and S. R. Nirmala

Abstract Alzheimer's disease (AD) is a challenging medical condition to treat since it requires accurate diagnosis as soon as possible to start treatment. Using a variety of neuroimaging modalities, including magnetic resonance imaging (MRI), positron emission tomography (PET), and electroencephalography (EEG), this study presents a multi-source deep learning architecture designed for AD classification. By collecting a variety of biomarkers and patterns of brain activity, the integration of different modalities provides a thorough knowledge of the course of AD. At the heart of this system is an intuitive Graphical User Interface (GUI) that allows researchers and clinicians to dynamically choose MRI, PET, or EEG data inputs according to specific diagnostic requirements or preferences. This GUI improves flexibility and customisation in the diagnostic workflow by streamlining the data input process.

Keywords Alzheimer's disease · Deep learning · Multi-source · Graphical user interface · Classification

---

A. Deshpande (✉) · N. Illur · R. Hegde · S. Chikkamath · K. Mallibhat · S. R. Nirmala
School of Electronics and Communication Engineering, KLE Technological University, Hubballi, India
e-mail: anirudhdeshpande07@gmail.com

N. Illur
e-mail: nandansillur508@gmail.com

R. Hegde
e-mail: roshithhegde22@gmail.com

S. Chikkamath
e-mail: chikkamath@kletech.ac.in

K. Mallibhat
e-mail: kaushik@kletech.ac.in

S. R. Nirmala
e-mail: nirmala.s@kletech.ac.in

V. K. Gunjan and J. M. Zurada (eds.), *Proceedings of 5th International Conference on Recent Trends in Machine Learning, IoT, Smart Cities and Applications*, Lecture Notes in Networks and Systems 1181, https://doi.org/10.1007/978-981-97-8861-3_25

1 Introduction

Alzheimer's disease is a serious neurological ailment that primarily affects elderly people. It causes cognitive decline, memory loss, and difficulty with daily chores. It is the principal cause of dementia and impairs memory, thinking, and self-sufficient functioning. The illness is caused by toxic protein aggregates building up in the brain, such as amyloid plaques and tau tangles, which impair neuronal transmission and injure brain tissue.

Alzheimer's disease is a significant issue for worldwide public health. There are over 55 million people worldwide living with dementia in 2020. This number will almost double every 20 years, reaching 78 million in 2030 [1]. This number is expected to double to 152 million by 2050, in part due to the aging population.

The following picture presents a graphical depiction of the data patterns of Alzheimer's Disease (AD) in senior citizens in India (Fig. 1) [2].

Research and patient outcomes both depend on the quick and accurate detection of Alzheimer's disease. Our novel strategy puts the needs of the user first and provides a customised, non-invasive path to early diagnosis using machine learning. Users have the option to choose preferred scan modalities (MRI, CT, PET, and EEG), which creates a paradigm in healthcare that is data-driven and patient-centric. Giving people agency and choice over their healthcare experience redefines AI in healthcare.

Our research attempts to categorise and diagnose Alzheimer's disease utilising a variety of data sources, driven by the need to address healthcare resource discrepancies in the diagnosis of the condition. Using datasets from magnetic resonance imaging (MRI), Positron Emission Tomography (PET), and electroencephalography (EEG), we want to train several models for Alzheimer's identification. The four

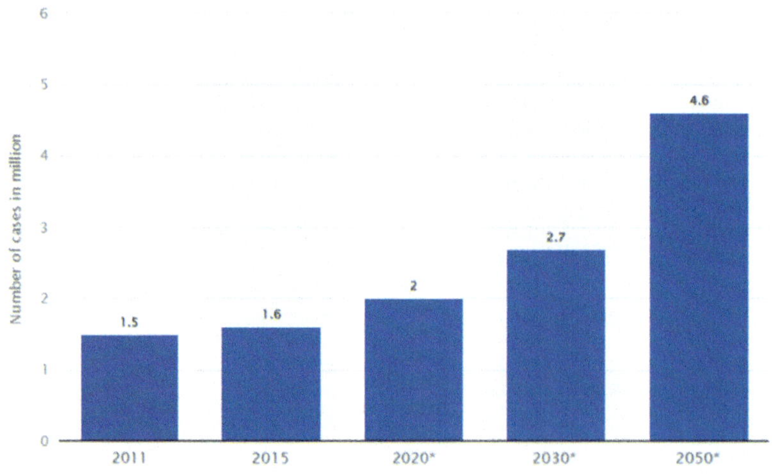

Fig. 1 Alzheimer's statistics in India

different forms of data that are accessible for uploading will be made possible by the integration of these sources into an intuitive Graphical User Interface (GUI).

Creating a versatile machine learning framework for Alzheimer's detection, integrating MRI, CT, PET, and EEG data through a user-friendly GUI, marks a crucial advance in medical diagnostics. Its exceptional flexibility empowers healthcare professionals and researchers to leverage various data types, accommodating differences in availability and preferences. By accepting diverse inputs, it enhances inclusivity and accessibility, tailoring Alzheimer's diagnosis to individual needs. This approach not only offers a comprehensive view but also ensures wider access to early detection tools, advancing our understanding and management of Alzheimer's disease.

2 Literature Review

This section covers the background research necessary for Alzheimer's disease classification.

2.1 Literature Survey

Capsule Network for Dementia Management

The study by Jin et al. [3] addresses early detection and management of Alzheimer's disease-related dementia. Employing a capsule network, the research produces a highly accurate model that surpasses conventional methods. Utilizing data sets from the Open Access Series of Imaging Studies (OASIS), primarily from MRI and PET scans, the model's main limitation is its inability to be trained on CT scan images.

We only used MRI scans. Integrating additional techniques, such as PET scans, genetic testing, cognitive tests, etc., could yield additional information. The web application's clinical viability and field testing were not discussed. In therapeutic settings, user studies are crucial [4]. The ADNI (Alzheimer's Disease and Neurological Institute) Western dataset was the only one used to evaluate the framework. It is crucial to test cross-population generalizability with a variety of datasets. The examination of disease development over time was excluded. They were cross-sectional models.

2.2 Temporal Evolution of Alzheimer's Disease Classification Models

A. Comparative Analysis of MRI, PET, and EEG modalities.

(1) Comparative Evaluation of Metrics for Model Accuracy for Different Algorithms. The comparison of the performance of the different models and their performance accuracy. There are various models, with each model taking a different input source to detect and classify Alzheimer's disease. The following table indicates the various sources and some of the models that were previously employed for Alzheimer's detection (Fig. 2).
(2) Comparative Preprocessing Methods for AD Detection. The comparison of several preprocessing methods used for Alzheimer's Disease (AD) detection using MRI, PET, and EEG data modalities is shown in this image (Fig. 3).
(3) Comparative Feature Extraction Methods in the case of Magnetic resonance imaging, Positron emission tomography, and electroencephalogram. The overview of diverse feature extraction methods utilized in Alzheimer's Disease (AD) detection, spanning MRI, PET, and EEG models (Fig. 4).

3 Proposed Methodology

An overview of the methodology used in this study to use MRI, PET and EEG data to distinguish between brains affected by AD and brains not impacted by AD is given in this section.

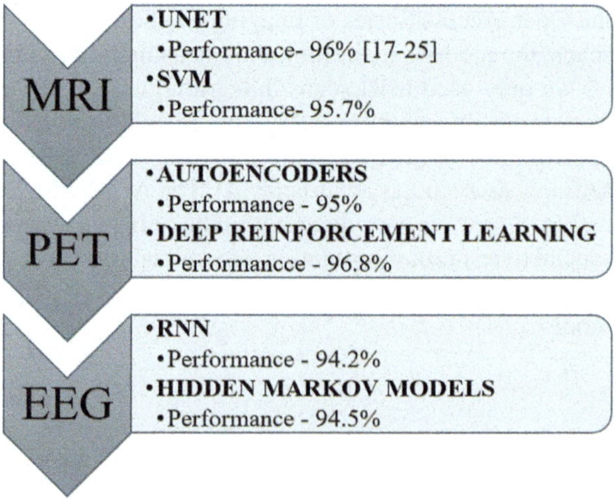

Fig. 2 Models and performance

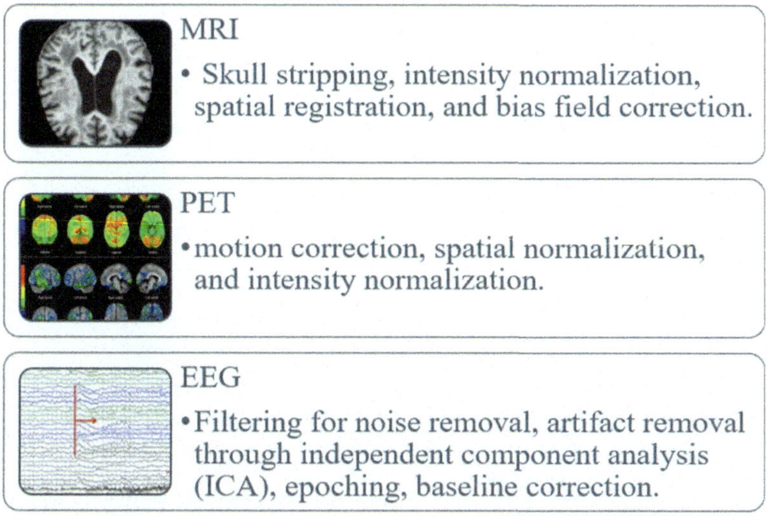

Fig. 3 Various pre-processing steps employed

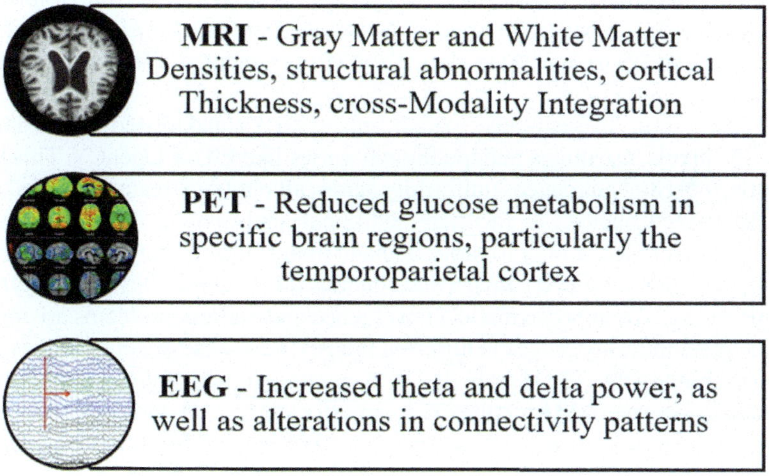

Fig. 4 Features extracted in various models

1. Proposed Framework for AD classification using MRI dataset. The proposed framework work aims to establish a robust framework for Alzheimer's Disease (AD) classification by leveraging Magnetic Resonance Imaging (MRI) data. The study adopts a supervised machine learning approach, employing SVMs as the primary classification model due to their robustness in handling high-dimensional data and their proven track record in medical image analysis. Using characteristics

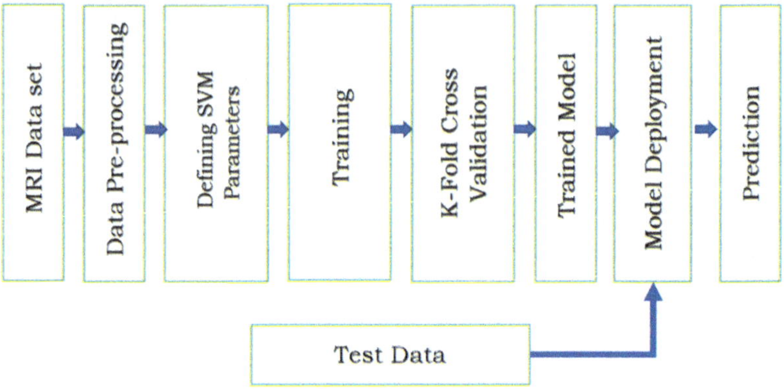

Fig. 5 Proposed Pipeline for data flow

from MRI images, the SVM model aims to discriminate between AD-affected and non-AD brains.

Data Source: The MRI images are obtained from the Kaggle dataset for Alzheimer's classification. Total number of images is 6400. Divided into four classes: Mild Demented: We have 896 images. Moderate Demented: We have 64 images. Non-Demented: We have 3200 images. Very-Mild Demented: We have 2240 images (Fig. 5).

Pre-processing: To resize the loaded images to a standardised dimension, say 256 × 256 pixels, use image manipulation libraries like PIL or OpenCV. This lowers computing cost and guarantees uniformity across all photos. Create numerical arrays from the resized photos. To facilitate additional analysis and model training, the majority of machine learning libraries demand image data in array format. Create an array representation for every image that contains the pixel intensity values (typically a NumPy array). To improve model convergence and address problems arising from different pixel intensity ranges in different images, normalise the pixel values within a standard range (e.g., [0, 1] or [−1, 1]). Assigning suitable labels or identifiers to differentiate between MRI pictures that are affected by AD and those that are not will help you group the arrays that represent MRI images from the two categories (AD stages and non-AD).

Feature extraction: Alzheimer's Disease (AD) detection hinges on intricate details within MRI images, where pixel intensities become critical features for analysis. Converting MRI pictures into resized, normalized high-dimensional feature vectors captures the brain's anatomical subtleties. These vectors, representing sequences of pixel intensity values, highlight differences in brightness and darkness [5]. Leveraging Support Vector Machine (SVM) models, these feature vectors serve as the foundation for discrimination and learning. Examining these representations reveals subtle patterns unique to different stages of Alzheimer's and non-AD brains. These distinctive features enable SVM models [6] to establish an optimal boundary, effectively distinguishing between AD and non-AD brain images.

SVM implementation: The linear Support Vector Machine (SVM) distinguishes Alzheimer's disease (AD) from non-AD brain scans by constructing an ideal hyperplane in a high-dimensional space. The hyperplane equation $w{\cdot}x + b = 0$ uses w as the normal vector and b as the bias term. The SVM maximizes the margin between classes under constraints $y{\cdot}(w{\cdot}x + b) \geq 1$, forming a robust decision boundary. New instance prediction is $w{\cdot}x_{\text{new}} + b$, with positive indicating AD and negative denoting non-AD. Support vectors, critical samples at the decision boundary, are crucial [7]. Leveraging complex pixel intensity patterns in the high-dimensional feature space, the linear SVM effectively identifies AD and non-AD brain images based on this mathematical foundation.

2. Proposed AD Classification Framework Using PET Dataset PET imaging is employed for Alzheimer's disease (AD) diagnosis, utilizing a radioactive tracer to detect brain abnormalities. PET scans measure the accumulation of amyloid protein, a key factor in Alzheimer's identification. Our machine learning model leverages PET datasets from ADNI for AD classification. Preprocessing involves resizing, grayscale conversion, and pixel value normalization. The processed PET images are then fed into CNN networks to extract relevant features, forming a comprehensive framework for AD classification.

The CNN architecture utilizes the Adam Optimizer for weight updates during training, allowing efficient adjustments of learning rates for individual parameters. The input layer features a 2D convolution layer with 32 filters, (3,3) kernel size, and ReLU activation, converting images to grayscale. Maxpooling with a (2,2) pool size downsamples images, reducing spatial dimensions. The model iterates 2D Convolution and Maxpooling, incrementing filters (32, 64, 128, 256, 512) for complex feature capture. Layers are flattened to 1D for Dense layer input. Two Dense layers with 128 units each, ReLU activation, and a 0.5 dropout rate prevent overfitting. The output layer, with units matching classes, uses softmax activation for multi-class classification [8 9. Data is split for training and validation, and the CNN is compiled and trained with Adam optimizer and sparse categorical cross-entropy loss. Evaluation of preprocessed test data yields the test accuracy (Fig. 6).

3. Proposed AD classification framework for EEG Dataset The incorporation of machine learning (ML) principles in this context lays the groundwork for predicting MMSE scores derived from EEG data, essential in cognitive assessment. A strong ensemble learning method capable of capturing intricate, non-linear correlations between EEG variables and MMSE scores is represented by the RandomForestRegressor model selection. It is possible to achieve optimal model convergence during training and minimise the effect of different feature sizes on model performance by utilising Standard-Scaler(), which guarantees constant feature scaling. Through parameter optimisation over several epochs, the iterative training loop strengthens the model's adaptability and increases its resistance to different EEG patterns and generalisation ability [10]. When used as an assessment metric, RMSE allows for accurate prediction accuracy measurement, which adds to a thorough comprehension of the model's functionality.

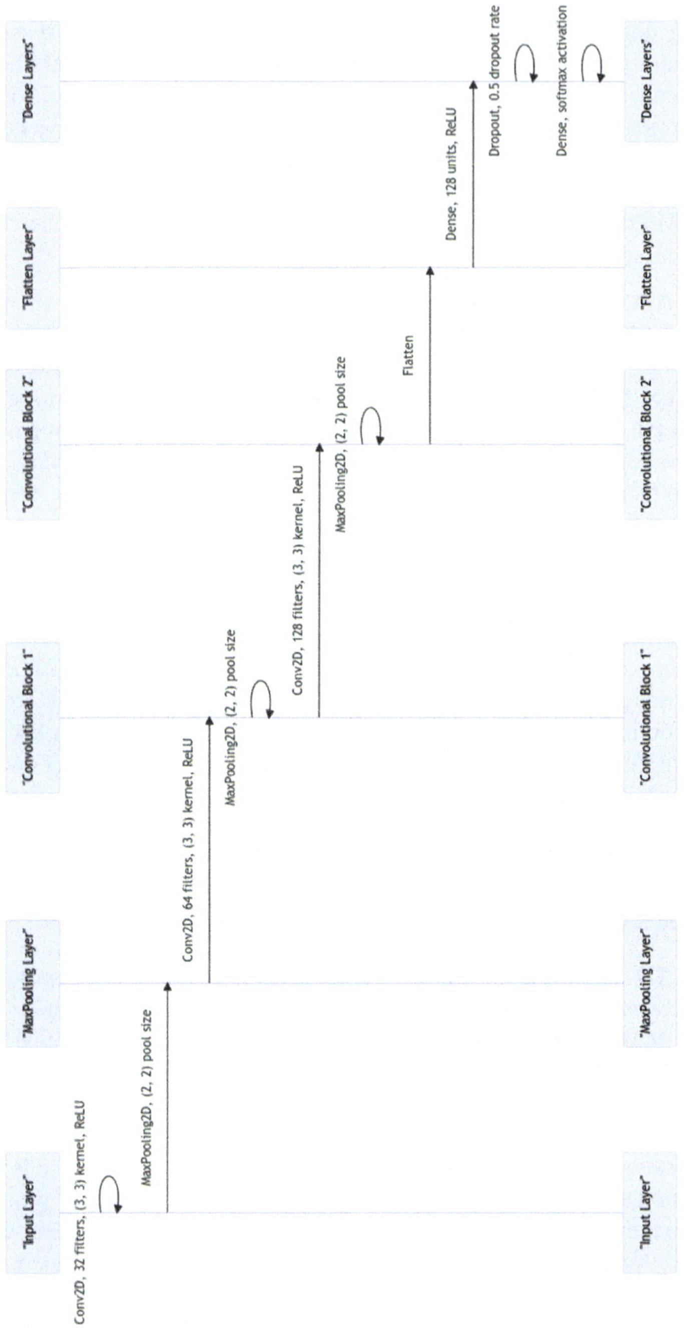

Fig. 6 Model architecture

Additionally, it's noteworthy that MMSE scores are typically obtained through a series of standardized questionnaires, further emphasizing the significance of the model's predictive capacity in cognitive assessment based on EEG data [11].

4 Results

The results for validating the accuracy of classification are discussed below.

4.1 GUI (Graphical User Interface)

The application of Graphical User Interfaces (GUIs) to the classification of Alzheimer's Disease (AD) accelerates medical progress by simplifying the study of large amounts of data, enabling physicians to quickly explore and comprehend complex datasets. Because of its user-friendly interfaces, accessibility promotes early diagnosis and makes it easier for doctors to spot biomarkers and subtle patterns. GUI-based tools make decision-making faster, enabling medical practitioners to start individualised treatment plans and treatments on time, greatly improving the chances of Alzheimer's disease early detection and management (Figs. 7 and 8).

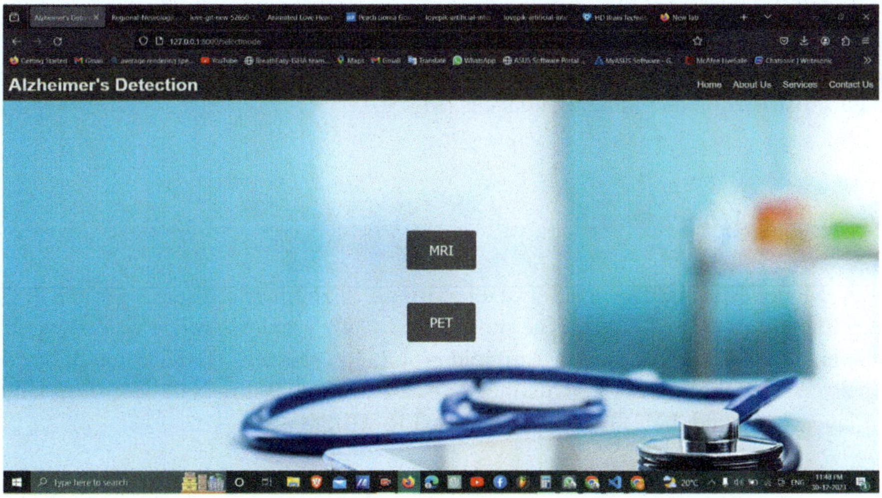

Fig. 7 GUI home page

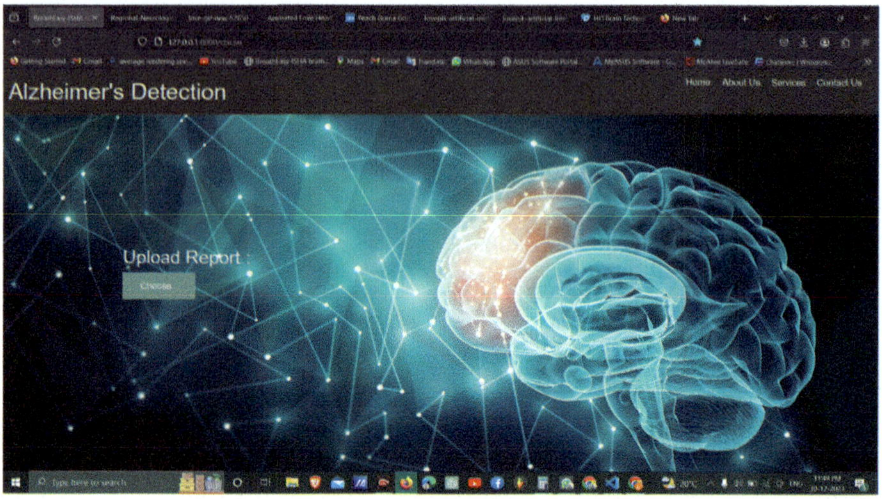

Fig. 8 Data upload choosing page

4.2 MRI Model Results

In Alzheimer's detection from MRI images, the SVM with a linear kernel excelled, achieving 98% detection and 98.67% classification accuracy on the test set. This highlights SVM's prowess in capturing linear separability among Alzheimer's-related MRI features, emphasizing its significance for this task. However, achieving such precision demands careful data preparation and model selection, considering the data's distribution and imaging nuances (Fig. 9).

4.3 PET Model Results

PET Model Results The results of classifying Alzheimer's disease for the PET pipeline are discussed below. The model used CNN architecture for extracting the distinctive features (build-up of Amyloid protein in the brain). The amount of build-up determines the degree of Alzheimer's disease present. Below is the overview table of layers used and their respective outputs (Fig. 10).

4.4 EEG Model Results

EEG Model Results The results of classifying Alzheimer's disease for EEG pipeline is discussed below. The model uses random forest algorithm to determine the respective

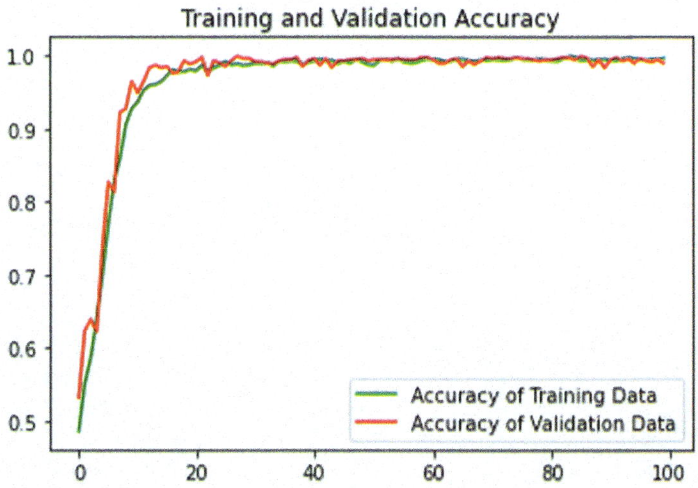

Fig. 9 MRI pipeline results

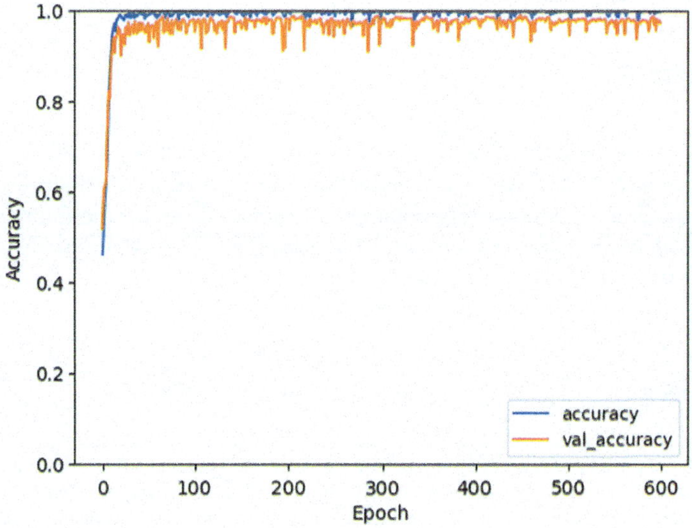

Fig. 10 PET pipeline results

RMSE scores. These RMSE scores are determined by processing the values collected by the electrodes placed on the scalp (Fig. 11).

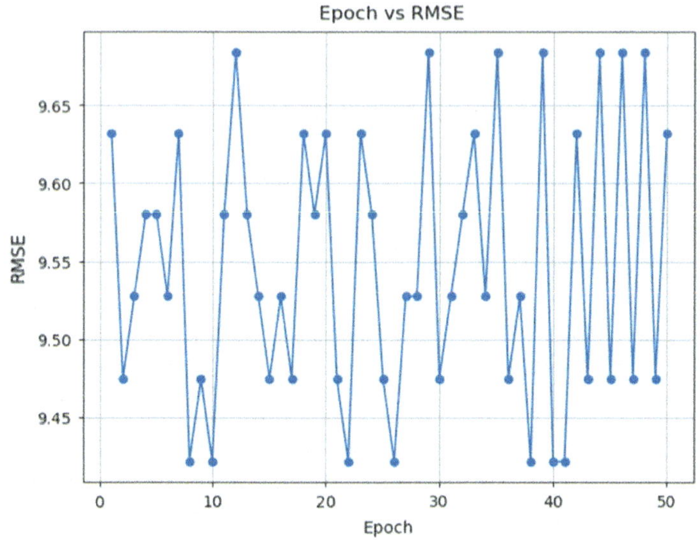

Fig. 11 EEG pipeline results

5 Conclusion

In this paper, we used multi-source data to predict the presence of Alzheimer's and also the classification of Alzheimer's disease. We used separate models for each of the pipelines and the results were obtained as discussed. As for the future scope, we could consider having more sources for the classification of Alzheimer's which will be helpful to researchers worldwide.

References

1. Schachter AS, Davis KL (2022) Alzheimer's disease. In: Dialogues in clinical neuroscience
2. Minhas A (2023) Number of cases of Alzheimer's disease in senior citizens across India from 2011 to 2050. https://www.statista.com/statistics/944170/india-number-of-cases-of-alzheimer-s-disease-in-senior-citizens/statisticContainer
3. World Health Organization (n.d.) Dementia. World Health Organization. https://www.who.int/news-room/fact-sheets/detail/dementia/
4. Sled JG, Zijdenbos AP, Evans AC (1998) A nonparametric method for automatic correction of intensity nonuniformity in MRI data. IEEE Trans Med Imaging 17(1):87–97
5. Maintz JB, Viergever MA (1998) A survey of medical image registration. Med Image Anal 2(1):1–36
6. Manjón JV, Coupé P, Buades A, Fonov V, Collins DL (2010) Non-local means denoising of MR images with the SPM toolbox. Neuroimage 53(4):1430–1445
7. McCollough CH, Kalender W (2006) Technologies for dose reduction in computed tomography. Radiol Clin 44(5):731–747

8. Silva AC, Lawder HJ (2006) Hara AK, Kujak JL (2009) Innovative techniques and advanced strategies in abdominal imaging. Radiol Clin 47(1):1–21
9. Watson CC (2000) New, faster, image-based scatter correction for 3D PET. IEEE Trans Nucl Sci 47(4):1587–1594
10. Rousset OG, Ma Y, Evans AC (1998) Correction for partial volume effects in PET: principle and validation. J Nucl Med 39(5):904–911
11. Rahmim A, Tang J, Zaidi H, Subramaniam RM (2007) Technical aspects of cardiac SPECT imaging. Phys Med Biol 52(14):R299

Malaria Prediction Using Fusion Learning with Enhanced Trust and Interpretability

Anjali Gangone, Swapna Mandu, and G. J. Bharat Kumar

Abstract Malaria, a significant global health threat, is traditionally diagnosed through manual examination of blood smears for parasite-infected cells, a method limited by its reliance on the examiner's expertise. To overcome these limitations, this study employs deep learning for malaria prediction, leveraging models like XCEPTION, AlexNet, DenseNet, and MobileNet, enhanced with transfer learning, to classify blood smears as parasitized or uninfected using Anaconda software. The models were trained on a dataset comprising 80% of the online medical records, with 10% used for validation and the remaining 10% for testing on 224 × 224 pixel images. DenseNet emerged as the top performer with a 97% accuracy rate, followed by MobileNet, AlexNet, and XCEPTION. The study further incorporates the LIME (Local Interpretable Model-agnostic Explanations) technique for interpretability, providing understandable explanations for predictions to bolster trust and transparency in the deep learning model's diagnostic capabilities.

Keywords Deep learning · AlexNet · XCEPTION · MobileNet · DenseNet · Trust and interpretability · LIME

A. Gangone (✉)
Department of Computer Science and Engineering, CMR Institute of Technology, Hyderabad, India
e-mail: anjali.gangone@gmail.com

Swapna M · G. J. Bharat Kumar
Department of Computer Science and Engineering, Sreyas Institute of Engineering and Technology, Hyderabad, Telangana, India
e-mail: drswapnam@meu.edu.et

G. J. Bharat Kumar
e-mail: dr.bharatkumar@meu.edu.et

V. K. Gunjan and J. M. Zurada (eds.), *Proceedings of 5th International Conference on Recent Trends in Machine Learning, IoT, Smart Cities and Applications*, Lecture Notes in Networks and Systems 1181, https://doi.org/10.1007/978-981-97-8861-3_26

1 Introduction

The World Health Organization reported 212 million global malaria cases in 2016, leading to 429,000 deaths, predominantly in Africa, with 90% of cases and 92% of deaths. Despite a decline in global malaria incidence and deaths by 57 and 10% over the past decade, malaria caused 409,000 deaths worldwide in 2019, with the majority of cases and fatalities concentrated in just 13 countries [1, 2]. The advancement of artificial intelligence (AI) and machine learning (ML), including deep learning (DL), has revolutionized medical diagnostics and various cognitive tasks, achieving results comparable to or surpassing human performance. This progress is fueled by increased data availability, technological advancements, and novel algorithms, positioning DL as a leading computational strategy in AI, machine learning, and data science, with significant research investment from companies like Google and Microsoft [3–6].

The concept of Interpretability, crucial for the development of expert machines and human–machine interactions, emerged in the 1990s. It represents a critical ethical criterion in AI, highlighted in 87% of surveyed AI guidelines. Trust in AI systems is essential, especially when human experts might err. Research indicates that increased Interpretability in AI systems boosts user trust, a significant factor in human–computer interaction studies [7, 8]. This study aims to leverage deep learning's potential, enhanced by trust and Interpretability, for accurate malaria prediction.

2 Related Works

Research by authors in [9] explored the link between climate variables and potential malaria outbreaks, employing historical meteorological and malaria case data over 6 years. They tested various algorithms, including KNN, Naive Bayes, and Extreme Gradient Boost, to evaluate their efficacy in predicting malaria outbreaks based on weather patterns. Their findings indicated that certain algorithms, notably, performed well in leveraging climate forecasts for malaria prediction, underscoring the potential of using weather data to prevent malaria-related fatalities.

In another study by Niyitegeka [10], the relationship between malaria transmission in Rwanda and environmental factors like temperature, humidity, and rainfall was examined through Machine Learning and IoT technologies. The research highlighted the effectiveness of Artificial Neural Networks (ANN) and Random Forest algorithms in predicting malaria transmission, with ANN achieving 93.9% and 88.2% accuracy in training and testing phases, respectively, in the Bugesera district.

Furthermore, Stephen et al. [11] determined that Naive Bayes offered the highest accuracy rates for both training (82.65%) and testing (75.6%) phases, making it the most reliable model for predicting malaria incidence from data spanning 2010 to 2020. The study also recognized Support Vector Machine (SVM) as a viable alternative with 77.9% training and 73% testing accuracy, while advising against the use of Linear Regression for such predictions due to its lower efficacy.

2.1 Trust and Interpretability

The healthcare sector is increasingly exploring the potential of deep learning across various domains, including medical imaging applications such as radiography, histology, ophthalmology, dermatology, and parasite detection like malaria. Despite significant progress and innovation in this field, a major challenge remains the opaque nature of deep learning systems. Often referred to as "black boxes," these systems, particularly neural networks, are so complex that understanding their inner workings can be elusive [12]. Consequently, fostering trust and ensuring the Interpretability of deep learning models are essential for their acceptance by healthcare professionals and patients alike. The importance of trust and ethical considerations in AI has been highlighted in a 2019 review of global ethical AI guidelines, underscoring trustworthiness as a key ethical pillar. Similarly, principles like Interpretability and open science are recognized as fundamental to responsible research and innovation, emphasizing the need for accountability in scientific advancements [13]. Trust plays a pivotal role in the adoption of AI-driven solutions, especially in critical areas such as autonomous vehicles and medical diagnostics. For AI technologies to be fully embraced, they must be deemed trustworthy by end users [12]. Interpretability in AI involves creating algorithms that are, on their own or through supplementary tools, sufficiently transparent to be understood by humans. Achieving this level of clarity requires providing stakeholders with comprehensive insights into the AI process. This includes detailed documentation, transparent training and validation methodologies, dataset characteristics, data analysis procedures, code accessibility, and clear explanations of the models and their outcomes [14].

3 Methodology

3.1 Data Collection and Preparation

Collecting and annotating image datasets from the health sector is very difficult. We gathered data from the US department's National Institute of Health and Human Service (NIH) that is already readily available for this investigation. Also, the study uses previously collected blood smear images from https://www.kaggle.com. So, a 1:1 proportion of infectious to uninfected cells was included in the information set with dataset size with total images of 27,560.

For the training as well as classification tests, the images were normalized to the average width and height at 224 * 224 pixels, which is the typical source image size of the majority of pre-trained models for quicker model convergence. The labeling of the data that results from this stage indicates parasitized pictures like one (1) as well as uninfected images with zero (0). The dataset is divided into parasitized and uninfected classes, as seen below in a few representative images. The images of sample parasitized red blood cell (Fig. 1) and unparasitized red blood cell (Fig. 2) are shown.

In this study, we used 80% for training, 10% for validation, and the remaining 10% for testing from a total of 27,560 red blood smear image datasets of parasitized and uninfected classes. Table 1 shows the amount of data in training set, testing set and validation set after the split ratio.

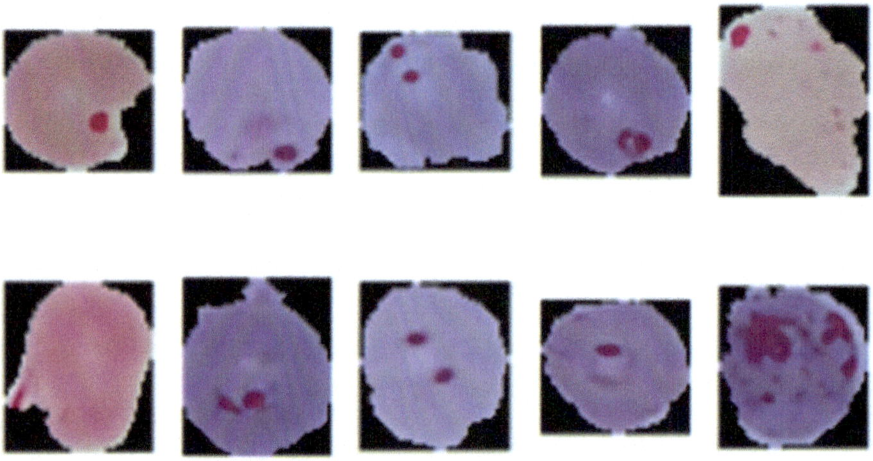

Fig. 1 Parasitized red blood cell sample images

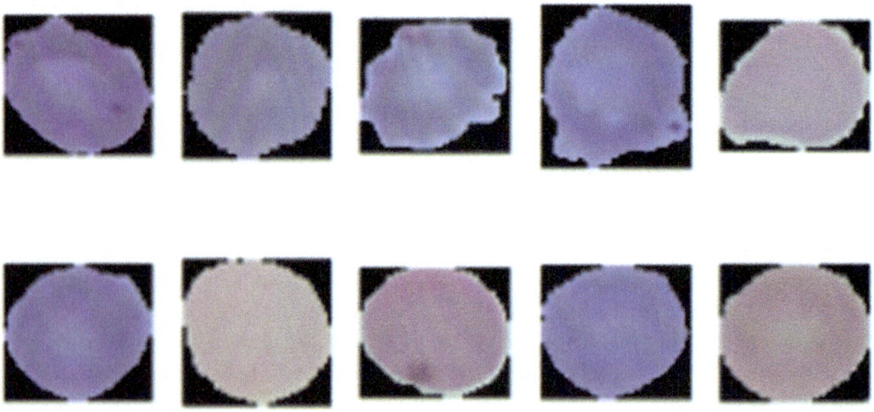

Fig. 2 Uninfected red blood cell sample images

Table 1 Malaria dataset across training, validation, and test data

Class	Parasitized	Uninfected
Training samples	11,024	11,024
Validation samples	1378	1378
Test samples	1378	1378
Total samples of each class	13,780	13,780

3.2 Proposed Model

A total of 27,560 photos of malaria-infected and uninfected red blood cells makes up the input dataset's contents. In the following step, the dataset is divided into training, validation, and test data. After that, the training data will be enhanced and images will be preprocessed. The classification layers are then modified and the layers of the pre-trained models are frozen. Then, the model building was done using transfer learning. Finally, based on the test data the model was trained and tested. The model was also evaluated and it was possible to predict if a person had malaria or not (Fig. 3).

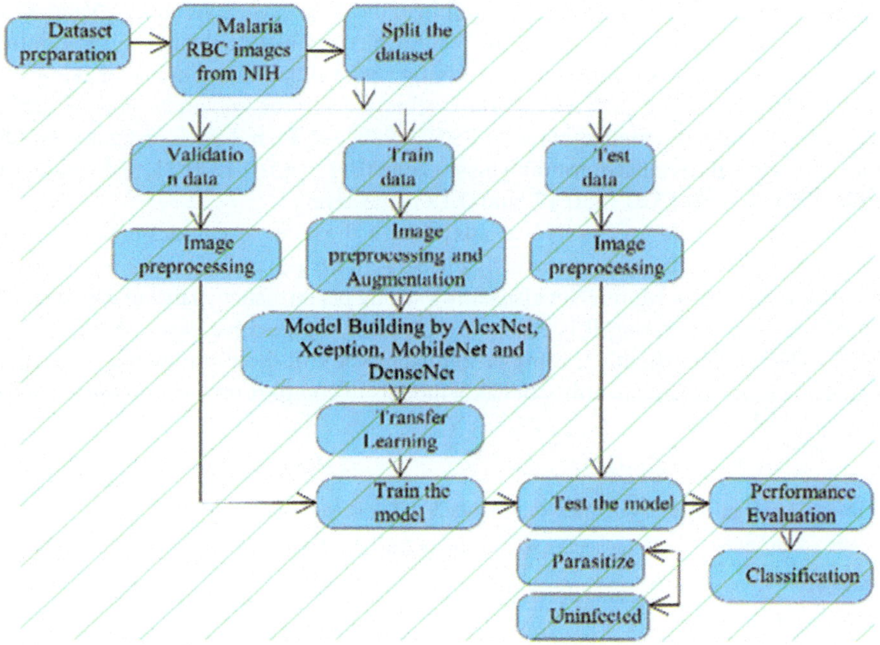

Fig. 3 Proposed malaria prediction model

Algorithms for Ensuring Trust and Interpretability

In this study context of deep learning, Local Interpretable Model-Agnostic Explanations (LIME) were used to approximate the network locally with a simpler model to explain the predictions of a deep neural network. Natural language processing, picture categorization, and medical diagnosis are just a few of the many applications that LIME has been used in. It is a well-known and frequently applied method for deriving explanations for the predictions made by machine learning models, and it can contribute to decision-making that is transparent and based on trust. In general, LIME was utilized as a trust and Interpretability checker for generating explanations.

4 Results and Discussion

All 8 models were implemented using Anaconda Python Distribution using Jupyter Notebook python editor and the hardware environment use was an Intel Core i3 processor Laptop with 8 GB of the Anaconda Jupyter notebook, the model was trained using Python 3.11 with CPU Accelerator throughout 10 iterations. One of the primary metrics for evaluating the performance of the model is accuracy.

The metrics measured during the training of the dataset were Accuracy, Validation Accuracy, Loss, and Validation Loss. These metrics were calculated for training and validation data. In Table 2, the training and validation accuracy of AlexNet, MobileNet, DenseNet, and XCEPTION achieves training accuracy of 95%, 95%, 97%, 96%, and validation accuracy of 95%, 95%, 96%, and 92% respectively. On the other hand, the results of training losses were 0.11, 0.15, 0.12, 0.21, and validation loss results were 0.15, 0.15, 0.17, and 0.18 respectively.

As shown in Table 3, DenseNet stands out with the highest accuracy and precision, indicating its effectiveness in correctly classifying images and minimizing incorrect positive predictions. MobileNet also shows strong performance, particularly in recall and F1-Score, suggesting it is effective at identifying positive cases. AlexNet and XCEPTION, while slightly trailing in some metrics, still show commendable performance, demonstrating the utility of CNN models in complex image recognition tasks like medical diagnosis.

Table 2 Loss and accuracy of AlexNet, MobileNet, DenseNet, and XCEPTION

CNN architecture	Training accuracy	Training loss	Validation accuracy	Validation loss
AlexNet	0.95	0.11	0.95	0.15
MobileNet	0.95	0.15	0.95	0.15
DenseNet	0.97	0.12	0.96	0.17
XCEPTION	0.96	0.21	0.92	0.18

Table 3 Performance comparison of proposed models

Model	Accuracy (%)	Precision (%)	Recall (%)	F1-score (%)
AlexNet	95	91	91	93
MobileNet	95	92	93	94
DenseNet	97	96	94	94
XCEPTION	96	85	91	88

4.1 LIME Explanation for Trust and Interpretability

Lime explanation is a method that helps interpret the decisions made by complex machine learning models, such as CNNs, by highlighting the most influential features or regions contributing to a particular prediction. In this specific study, the lime explanation to gain a deeper understanding of how the CNN models mentioned above make predictions related to malaria is utilized. The models were trained on a considerable dataset, likely consisting of images relevant to malaria diagnosis [15]. The lime explanation results showcased in Figs. 4, 5, 6, and 7 offer valuable insights into the inner workings of the CNN models. By visualizing and analyzing the explanations provided by Lime, researchers were able to identify the specific regions or features in the images that significantly influenced the models' predictions. The use of multiple CNN models, such as AlexNet, MobileNet, DenseNet, and XCEPTION, demonstrates a comprehensive exploration of lime explanation across different architectures. This approach helps establish the robustness and generalizability of the lime technique in interpreting predictions made by diverse CNN models.

Overall, the findings presented in Figs. 4, 5, 6, and 7 and the accompanying analysis provide a compelling understanding of how lime explanation can clarify the predictions of CNN models in the context of malaria diagnosis. This knowledge contributes to the interpretability and trustworthiness of machine learning models, fostering their applications in real-world scenarios.

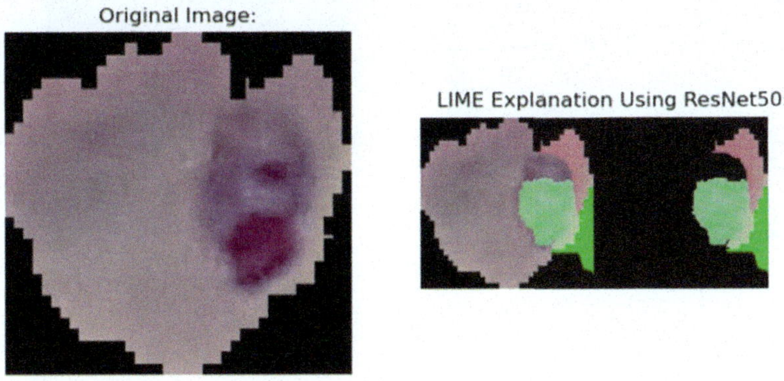

Fig. 4 Lime explanation for AlexNet

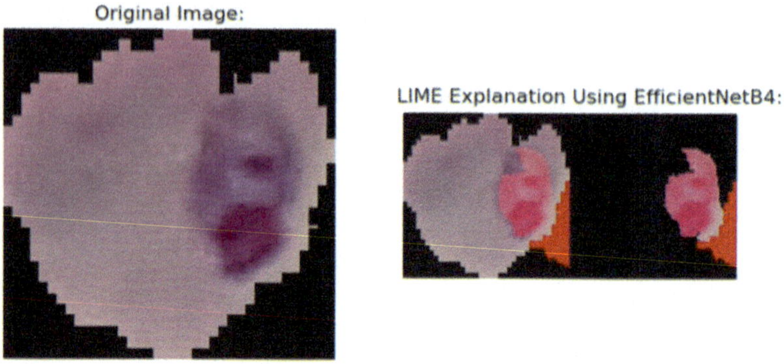

Fig. 5 Lime explanation for MobileNet

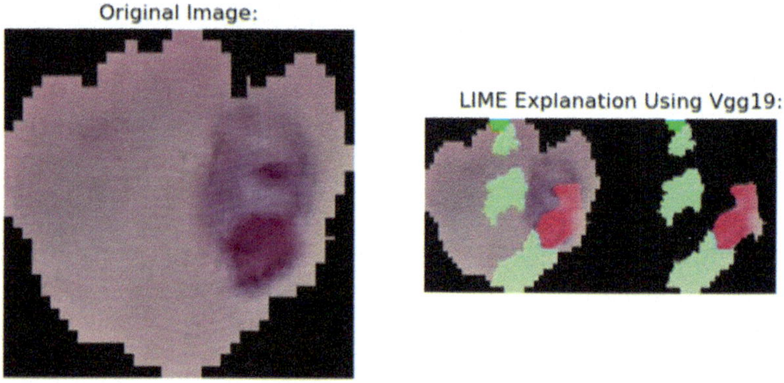

Fig. 6 Lime explanation for DenseNet

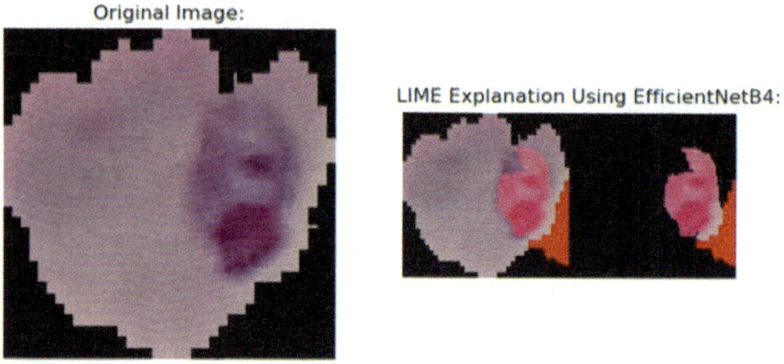

Fig. 7 Lime explanation for XCEPTION

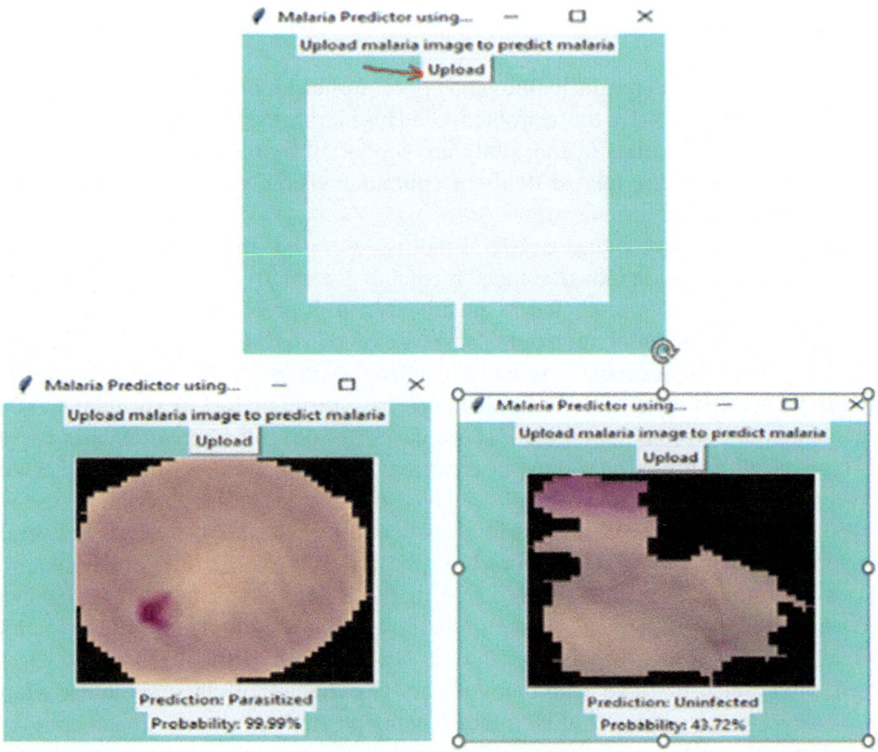

Fig. 8 The GUI of the model for predicting unknown data

4.2 Graphical User Interface (GUI) Representation of the Proposed Work

The classification system can be interacted by being visually intuitive and easily by using GUIs. To facilitate data input, parameter configuration, and the classification process, they offer buttons, menus, and graphical features to users such as academics or healthcare professionals without requiring complicated programming knowledge or command-line interfaces (Fig. 8) [16].

4.3 Comparative Analysis

Table 4 compares the proposed work with other existing work done by other researchers in this area. The comparisons are based on the methods used, Accuracy, precision, Recall, F1-score, Specificity, GUI, and LIME Explanation for trust and Interpretability. It also shows that the proposed model, surpasses all the existing models with highest accuracy with AlexNet and it has LME explanation and the GUI.

This section covers the difficulties and opportunities for image classification in the disease of malaria to solve these problems and highlights the potential and challenges in this field. The findings of this study are supported by the reviewed literature as well as research in the related fields of computer vision, pattern recognition, and image classification.

For the comparison purpose several metrics are used. In the context of malaria classification, precision indicates the percentage of correctly identified malaria cases out of all the samples predicted as positive. A high precision value is crucial in ensuring accurate identification of malaria cases and avoiding unnecessary treatment or alarm. Recall, also known as sensitivity or true positive rate, measures the model's ability to identify all positive samples correctly. In this case, recall indicates the percentage of correctly identified malaria cases out of all the actual malaria cases. A high recall value is important in ensuring that all malaria cases are detected, preventing potential misdiagnosis or delayed treatment. On the other hand, the F1 score in the evaluation of a malaria classification model helps assess the overall performance and takes into account both the ability to correctly identify malaria cases and minimizes misclassifications. A higher F1 score indicates a better balance between precision and recall, indicating a more reliable and effective classification model.

As mentioned in Table 4, in the work of [17], the author classifies malaria into two categories parasitized and Uninfected using an ANN approach. The overall test accuracy achieved was 87.8% which is very far from the results achieved by the

Table 4 Comparative analysis with existing models

Ref	Methodology	Accuracy (%)	Precision (%)	Recall (%)	F1-score	GUI
[17]	ANN	87.8	–	–	–	No
[18]	Deep CNN	93.4	–	–	–	No
[19]	VGG 19 ResNet50 AlexNet	93.8 80.2	93 77	– –	– –	No
[20]	Basic CNN VGG 19 Frozen VGG 19 Fine-tuned	94 92 96	95 93 98	94 90 93	95 91 96	No No No
[21]	ResNet50 EfficientNetB4 InceptionV3 VGG 19	94 93 91 88	93 91 92 85	91 91 91 91	93 91 91 84	No
This work	DenseNet AlexNet MobileNet XCEPTION	97	96	94	94	Yes

proposed work. Similarly, Raj et al. [18], used a deep CNN to classify malaria and obtained 93.47% test accuracy. Even if, the result achieved in this work is better than [17], but still less than the proposed study. Another study by Turuk et al. [19], stated three different deep learning models like VGG19, ResNet50, and AlexNet. As a result, the best accuracy of 93.89% in VGG19 was obtained when compared to ResNet50 and AlexNet. The developed model by Shekar et al. [20] used Basic CNN, VGG19 Frozen CNN, and VGG19 Fine Tuned CNN within 25 epoch processes and achieved 92% accuracy by using the Frozen method. However, the proposed model of this study used Frozen VGG19, within 10 epochs and 94.37% accuracy was obtained by using the Frozen method.

In general, the results stated in Table 4, confirmed that the proposed work outperformed by 9.57% of [17], 3.9% of [17], 3.48% of [17], 5.37% of [20] and 3% of [21] in terms of test accuracy as well as all the three related work do not performed GUI application and LIME explanation.

5 Conclusion and Future Works

This research evaluated a custom Convolutional Neural Network algorithm-based end-to-end deep learning model to improve malaria detection grounded on thin-blood smear images. Therefore, the study used the Convolutional Neural Network algorithm of AlexNet, MobileNet, XCEPTION, and DenseNet with a transfer learning approach to detect parasites or uninfected malaria blood samples' image data set of humans by using Anaconda IDE.

The image data was split into two groups; preprocessing data and validation data. In this procedure, we used 80% for training, 10% for validation and the remaining 10% were used for testing. Therefore, based on this grouping the accuracy value of AlexNet, MobileNet, DenseNet, and XCEPTION was 95%, 95%, 97%, and 96% classified as parasitized and uninfected cell images with a high degree of precision of 91%, 92%, 96%, and 85% respectively. The AlexNet model was a suitable model for malaria disease detection and classification as compared to the MobileNet, DenseNet, and XCEPTION models. Thus, we believe that the results achieved from this work will benefit the development of a new model. In the future we will try to use most state of the models like Large Language Models like Generative Pretrained Model, Google GEMINI and other custom built models.

References

1. Brown BJ et al (2020) Data-driven malaria prevalence prediction in large densely populated urban holoendemic sub-Saharan West Africa. Sci Rep 10(1):15918. https://doi.org/10.1038/s41598-020-72575-6

2. World Health Organization. [Online]. https://www.who.int/publications/i/item/9789240015791. Accessed 03 May 2022
3. Felzmann H, Villaronga EF, Lutz C, Tamò-Larrieux A (2019) Interpretability you can trust: interpretability requirements for artificial intelligence between legal norms and contextual concerns. Big Data Soc 6(1):205395171986054. https://doi.org/10.1177/2053951719860542
4. Scharowsk N (2020) Measuring the effect of human-friendly AI explanations on objective and subjective trust
5. Alzubaidi L et al (2021) Review of deep learning: concepts, CNN architectures, challenges, applications, future directions. J Big Data 8(1):53. https://doi.org/10.1186/s40537-021-00444-8
6. Sarker IH (2021) Deep learning: a comprehensive overview on techniques, taxonomy, applications, and research directions. SN Comput Sci 2(6):420. https://doi.org/10.1007/s42979-021-00815-1
7. Schmidt P, Biessmann F, Teubner T (2020) Interpretability and trust in artificial intelligence systems. J Decis Syst 29(4):260–278. https://doi.org/10.1080/12460125.2020.1819094
8. Kartikeya A (2021) Examining the correlation between trust and Interpretability with explainable artificial intelligence. arXiv. [Online]. http://arxiv.org/abs/2108.04770. Accessed 26 Mar 2023
9. Marcin C, Monika D, Edward H (2015) Parasitic diseases in humans transmitted by vectors. Ann Parasitol 3:137–157. https://doi.org/10.17420/ap6103.01
10. Niyitegeka J, Employing machine learning and internet of things for malaria outbreak prediction in Rwanda
11. Stephen A, Akomolafe PO, Ogundoyin KI (2021) A model for predicting malaria outbreak using machine learning technique
12. Meske C, Bunde E (2020) Interpretability and trust in human-AI-interaction: the role of model-agnostic explanations in computer vision-based decision support. In: Degen H, Reinerman-Jones L (eds) Artificial intelligence in HCI. Lecture notes in computer science, vol 12217. Springer International Publishing, Cham, pp 54–69. https://doi.org/10.1007/978-3-030-50334-5_4
13. Aicardi C, Bitsch L, Dat ta Burton S, Evers K, Farisco M, Mahfoud T, Rose N, Rosemann A, Salles A, Stahl B, Ulnicane I (2020) Trust and interpretability in artificial intelligence
14. Bernal J, Mazo C (2022) Interpretability of artificial intelligence in healthcare: insights from professionals in computing and healthcare worldwide. Appl Sci 12(20):10228. https://doi.org/10.3390/app122010228
15. Anjali G, Abiy A, Swapna G, Bharat KGJ (2023) Malaria prediction using fusion learning with enhanced trust and transparency. In: Intelligent computing, machine learning, artificial intelligence, robotics, sensors. IEEE Publishers. ISSN 2195-3988, ISSN 2195-3996 (electronic), ISBN 978-981-19-2357-9 and ISBN 978-981-19-2358-6
16. Indeed, Editorial Team, What is a GUI (Graphical User Interface)? Definition, elements and benefits. [Online]. https://www.indeed.com/career-advice/career-development/gui-meaning. Accessed 16 Nov 2023
17. Nugroho HA, Akbar SA, Murhandarwati EEH (2015) Feature extraction and classification for detection malaria parasites in thin blood smear. In: 2015 2nd International conference on information technology, computer, and electrical engineering (ICITACEE), Semarang, Indonesia. IEEE, pp 197–201. https://doi.org/10.1109/ICITACEE.2015.7437798
18. Raj M, Sharma R, Sain D (2021) A deep convolutional neural network for detection of malaria parasite in thin blood smear images. In: 10th IEEE international conference on communication systems and network technologies (CSNT). https://doi.org/10.1109/CSNT51715.2021.9509619
19. Turuk M, Sreemathy R, Kadiyala S, Kotecha S, Kulkarni V (2022) CNN based deep learning approach for automatic malaria parasite detection. 49(3)
20. Shekar G, Revathy S (2020) Malaria detection using deep learning
21. Stadlhofer A, Mezhuyev V (2023) Approach to provide interpretability in machine learning models for image classification. Ind Artif Intell 1(1):10. https://doi.org/10.1007/s44244-023-00009-z

Hospital Automation System

Siddhi Darekar, Anushka Dudhane, Riddhi Kanse, Malavika Preeth, and Amruta Mhatre

Abstract Establishing a website for a hospital is imperative. This digital presence facilitates seamless access to critical details about the hospital's services, medical professionals, and facilities, empowering patients, and caregivers with comprehensive and up-to-date information. The incorporation of an online appointment scheduling system not only optimizes the booking process but also contributes to reducing patient wait times. In the realm of healthcare technology, this project also introduces an innovative Seizure Detection System designed for continuous monitoring of tonic–clonic seizures, with a focus on user comfort by placing a piezoelectric sensor discreetly in the palm of the hands. Integrated with ESP technology, the piezoelectric sensor captures the pressure associated with muscle contractions during the tonic–clonic stages of a seizure. A buzzer alert is triggered, providing an immediate notification to caregivers.

Keywords Hospital website · Seizure alarm · Piezoelectric sensor · ESP32 microcontroller

S. Darekar (✉) · A. Dudhane · R. Kanse · M. Preeth · A. Mhatre
Department of Electronics and Telecommunication Engineering, Fr. Conceicao Rodrigues Institute of Technology, Vashi, Navi Mumbai 400703, India
e-mail: ssdarekar26111@gmail.com

A. Dudhane
e-mail: anushkadudhane6@gmail.com

R. Kanse
e-mail: rkanse45@gmail.com

M. Preeth
e-mail: malavikapreeth@gmail.com

A. Mhatre
e-mail: amruta.pabarekar@fcrit.ac.in

V. K. Gunjan and J. M. Zurada (eds.), *Proceedings of 5th International Conference on Recent Trends in Machine Learning, IoT, Smart Cities and Applications*, Lecture Notes in Networks and Systems 1181, https://doi.org/10.1007/978-981-97-8861-3_27

1 Introduction

1.1 Background

Hospitals will shortly be able to replace paper prescriptions with web-based systems that will greatly increase patient management efficiency, streamline physician schedules, and give everyone in the healthcare facility access to information about patients. Modern web technologies provide online solutions for practically any industry. In order to remain competitive in the growing digital market, many industries are shifting their fundamental operations to digital platforms [1].

Seizure detection is a critical area of research and development, especially in the field of healthcare and medical technology. Different types of seizures may require different detection methods. Understanding the characteristics and patterns associated with various types of seizures is necessary, such as tonic–clonic, absence, or focal seizures. Seizures are generally classified into different stages, which can vary depending on the type of seizure. The stages of a seizure may include: Aura (Focal Aware Seizure), Tonic Phase (Tonic Seizure), Clonic Phase (Clonic Seizure) and Postictal Phase [2].

Figure 1 shows that there are 4 stages of seizures. As well as it shows movement of the person going through it.

Detecting tonic–clonic seizures using wearable devices involves monitoring physiological signals and movement patterns associated with seizures. Wearables equipped with sensors, such as accelerometers, pressure sensors and heart rate monitors, can be instrumental in identifying the distinct characteristics of tonic–clonic seizures [4].

The arms, legs, or trunk muscles abruptly stiffen or tense during a tonic seizure. The stiffness typically occurs during sleep and continues for around 20 s. When someone has a tonic seizure while they are standing, they could trip and fall. The person may experience fatigue following the seizure [5].

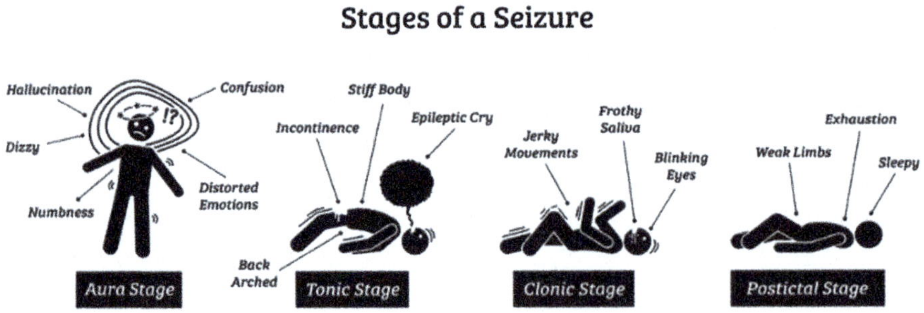

Fig. 1 Stages of seizure [3]

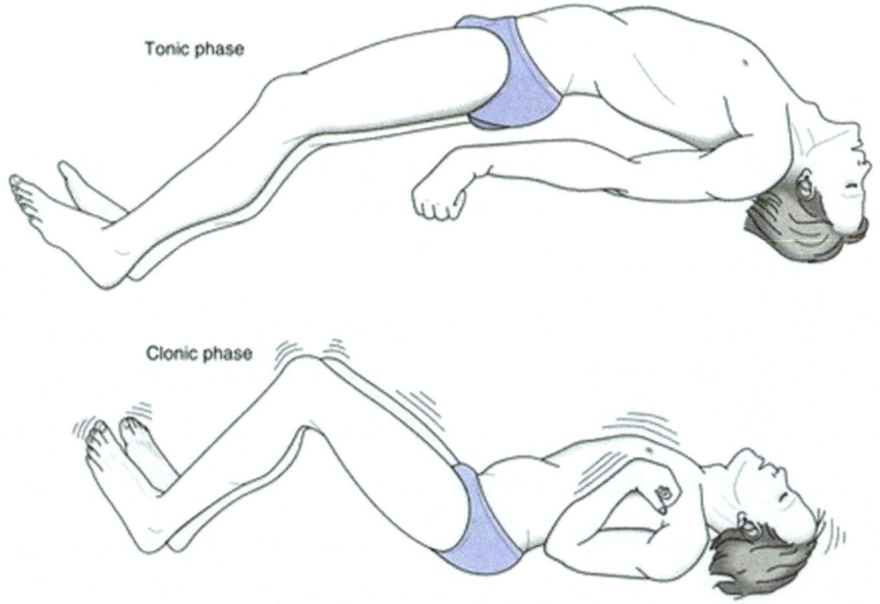

Fig. 2 Tonic and clonic stages of seizure [7]

For focal (partial) seizure, the person may be aware of what's happening. During a generalized seizure, the person may be unconscious. People of all ages, including babies and infants, can experience clonic seizures [6].

Figure 2 mainly focuses on movement of human body during tonic and clonic stages of seizure.

The term "tonic–clonic or convulsive status epilepticus" refers to a prolonged convulsive seizure. Seizures that happen repeatedly may result in death or lasting harm. In order to prevent Asphyxia, quickly turn the person onto one side. As a result, liquids like vomit or saliva cannot enter the lungs through the mouth [8].

1.2 Related Work

A. Rathi, M. Chaudhary, and E. Bisht presented a system that involves patient registration, data entry, and scheduling medical appointments [9]. K. Saimanoj, G. Poojitha, K. Devendra Dixit, and L. Jayannavar use websites in their hospital administration job. They provided the software and the ability to automatically save patient and staff data and assign each patient a unique ID. Using the ID, users can look up a doctor's availability and a patient's data. They developed the Hospital Management System, which requires a username and password to access. Only administrators and receptionists have access to it, and only they are able to add data to the database. It is simple to retrieve the data [10].

Stafstrom CE, Carmant L. proposed an innovative approach for early epilepsy seizure detection [11]. The system they designed utilized Arduino-based hardware devices, including ECG and heart rate sensors to monitor cardiac signals and pulse rates. They compared the acquired data with a predefined dataset of diseased cardiac signals. Through IoT, the information was relayed to an authorized doctor in real-time which allowed them for prompt intervention and comprehensive patient monitoring.

Lin SK, Istiqomaah, Wang LC, Lin CY, Chiueh H proposed a smart headband designed for epileptic seizure detection [12]. The authors created a cloud-connected service app that records patients' health conditions in order to improve patient care and facilitate remote monitoring.

2 Methodology and Procedure

2.1 Proposed Method

Figure 3 is block diagram showing main three components of seizure detection system, such as piezoelectric sensor, esp 32 microcontroller and the buzzer.

In the implemented seizure detection system, a piezoelectric sensor serves as the primary sensor for detecting pressure applied to it. It is set to continuously monitor pressure levels, the sensor triggers a signal if the pressure exceeds a predefined threshold for a specified duration. The ESP32 microcontroller processes the incoming data, detecting seizure-like patterns, and initiates the output upon detection. It activates a buzzer, providing an audible alert for immediate attention. This facilitates timely response and care for individuals experiencing seizures.

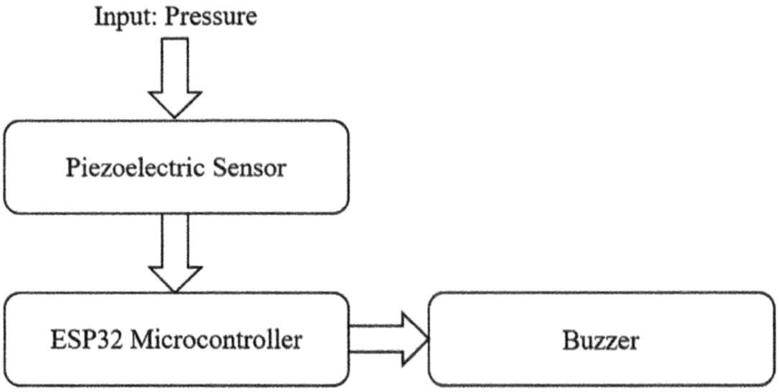

Fig. 3 Block diagram of seizure detection system

2.2 Website Implementation

Index page that is the main webpage, built with responsive HTML, boasts a navigation bar for easy access to various sections. It features key information about the hospital's services, and an expert team. External CSS and JavaScript libraries enhance the site's styling and functionality.

The main home page contains various features of website. Services lead us major services offered. Like, Pulmonary Services Page is focused on showcasing Pulmonary Services, this HTML page provides a visually emphasized title and presents three pulmonary specialists with a user-friendly layout. The page enables easy selection of specialists for appointments or information. Similar for Eye Care Services Page, Neurology Services Page. The next one that is 'Our doctors' show the expert team of doctors.

Book Appointment is a feature through which the appointment is made which is sent to the email connected. Admission in hospital is digitalised by the website. If a new person is there then by selecting new it leads to another page in which all kinds of important patient as well as guardian details are asked, submit button is provided. Then the existing ones have to login to get access to their paper of identification. It also gives facility to view own medical history as well as other reports etc.

PregaDiary Login Page is vital for expectant mothers, the file facilitates access to hospital services. Upon validation of login credentials, users seamlessly navigate to the PregaDiary section, ensuring secure access control. Error handling redirects users to the home page, maintaining security.

PregaDiary Main Page is serving as the central hub for expectant mothers, this user-friendly page allows input of daily health updates. A form captures essential details, which are sent directly to the doctor or receptionist via email. The page also provides an effortless option to navigate back to the home page.

2.3 Seizure Detection

Piezoelectric Sensor: A piezoelectric sensor responds to applied mechanical stress by generating an electrical charge. This type of sensor is adept at converting vibrations into electrical signals. The piezoelectric sensor captures the mechanical vibrations resulting from body movements during a seizure episode. The analog data produced by the sensor is subsequently processed to identify and detect abnormal activity indicative of a seizure.

ESP32 reads data from sensors, processes the information, and executes algorithms to determine if the observed patterns indicate a seizure. Additionally, the ESP32 can be utilized to interface with other components of the system and facilitate communication, such as sending alerts via Wi-Fi or Bluetooth.

Buzzer: The electronic sound-generating device that produces a buzzing or beeping sound when an electrical current is applied serves as an audible alert to notify caregivers, medical professionals, or the individual experiencing the seizure. The activation of the buzzer is typically triggered by predefined conditions, such as the detection of abnormal sensor readings for a specified duration.

Seizure Detection Algorithm: The seizure detection algorithm employed in this system relies on monitoring the analog data obtained from the piezoelectric sensor. The algorithm initiates by continuously reading sensor values from analog Pin 34 on the ESP32 microcontroller. When the sensor reading surpasses a predefined threshold of 1000, a timer is initiated to record the elapsed time. If the conditions persist for over 3 s, the algorithm triggers an alert by activating the buzzer connected to digital Pin 2. This dual-check mechanism helps filter out transient events and ensures that the system responds only to sustained abnormal activity. The threshold value of 1000 serves as a benchmark, and readings beyond this value are indicative of significant mechanical vibrations associated with seizure-like movements. The algorithm's efficiency lies in its ability to provide a timely alert, balancing sensitivity to detect seizures promptly and specificity to avoid false positives. Additionally, the system incorporates a reset mechanism to restart the timer when the sensor reading falls below the predefined threshold, ensuring the algorithm is responsive to real-time changes in sensor data. The seizure detection algorithm, implemented within the ESP32 microcontroller, offers a robust and adaptable solution for the early identification of seizures through the analysis of piezoelectric sensor data.

3 Results

This section displays the outcomes of the software and hardware implementations.

Figure 4 is the home page of website, showing some of the top services. This page makes the web application easy to navigate.

Figure 5 shows the facility of Booking appointment is featured above. It also shows contacts for the website and the other facilities provided.

Figure 6 shows the screenshot of the booking that was received in the provided email.

Figure 7 shows the admission panel which leads to the above page where login is asked.

When special care for pregnancy is needed, the pregadiary section of our website leads pregnant women to this fresh page, where a login is required (Fig. 8).

Figure 9 is the page we get after we login in the pregadiary. Here various questions related to the pregnancy are asked. The questions are general but they are important for an accurate diagnosis.

Figure 10 shows the simulation of seizure detection system done on Proteus.

Figure 11 shows the hardware preparation. The above is the setup of the seizure detection system where piezoelectric sensor, esp 32 and a buzzer is seen.

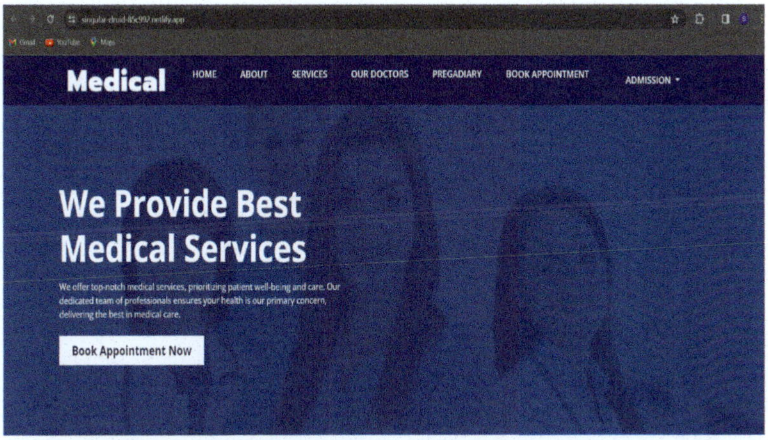

Fig. 4 Main page

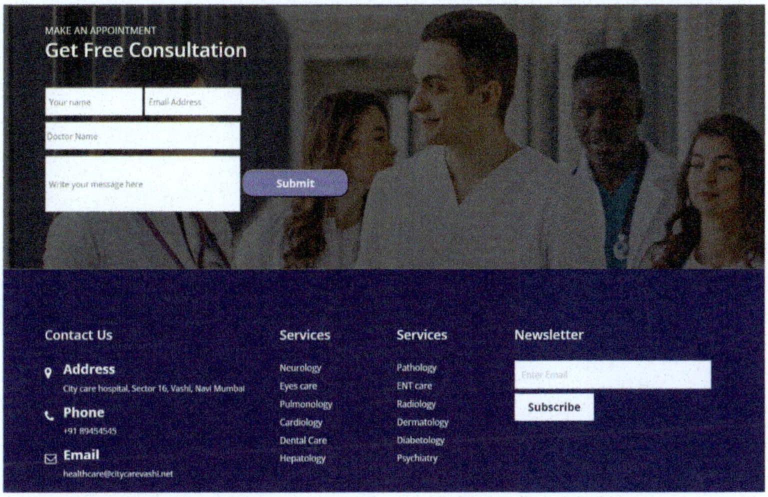

Fig. 5 Booking an appointment

4 Conclusion

The website encompasses a multifaceted platform designed to optimize patient experience and streamline healthcare services. The website features a user-friendly homepage for easy navigation. A categorized list of medical services is available in the "Services" section, allowing visitors to explore specific offerings. Patient reviews contribute to the hospital's credibility, while the "Our Doctors" page offers insights into the medical professionals. The inclusion of a "PregaDiary" feature supports pregnant women in rural areas, enabling them to upload weekly health updates.

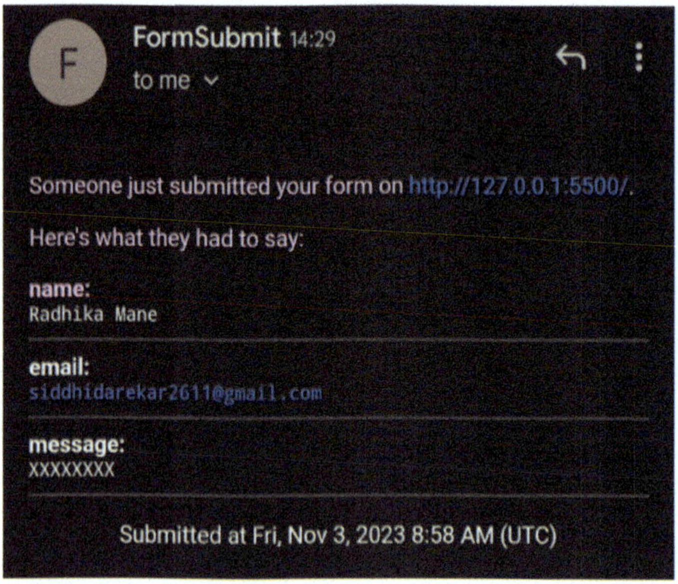

Fig. 6 Email screenshot of the received booking

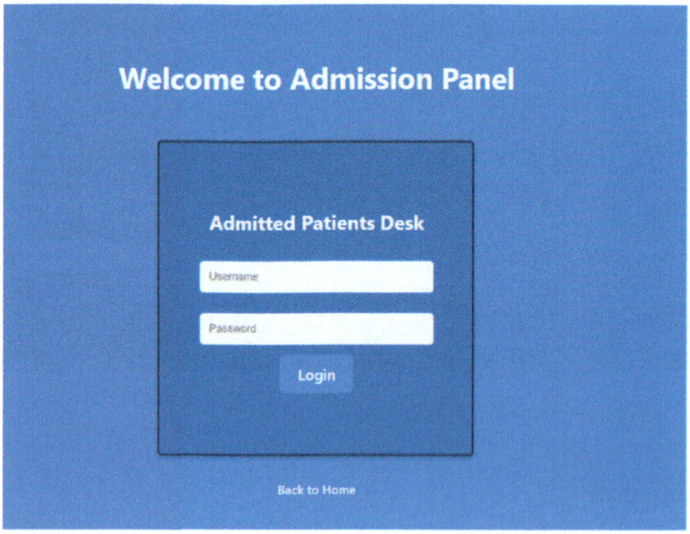

Fig. 7 Admission panel

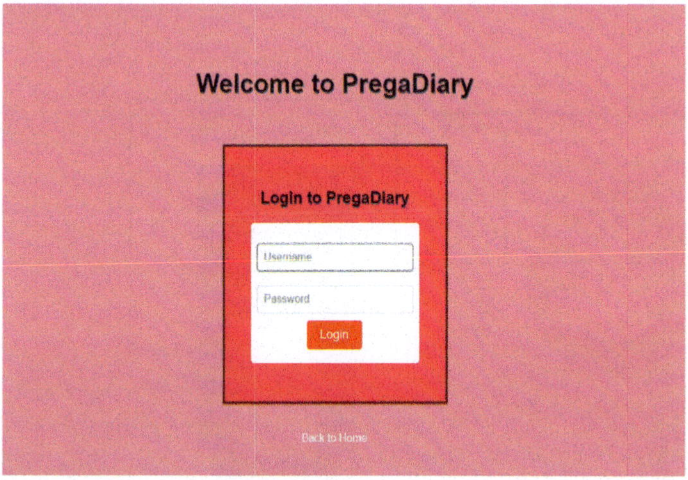

Fig. 8 PregaDiary login page

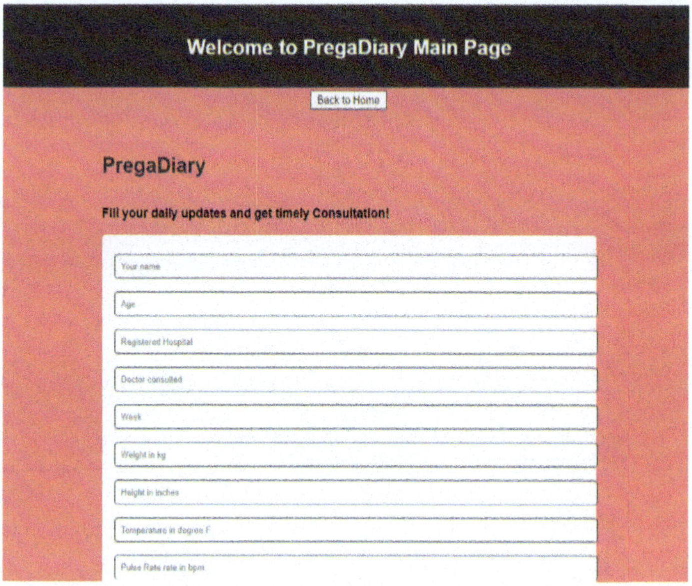

Fig. 9 PregaDiary main page

Additionally, an "Appointment Booking System" simplifies scheduling and sends confirmation emails, enhancing overall efficiency and patient engagement. It also includes paperless admission of patient. Newly admitted patients have to fill up all required details. Admitted patients can upload their reports.

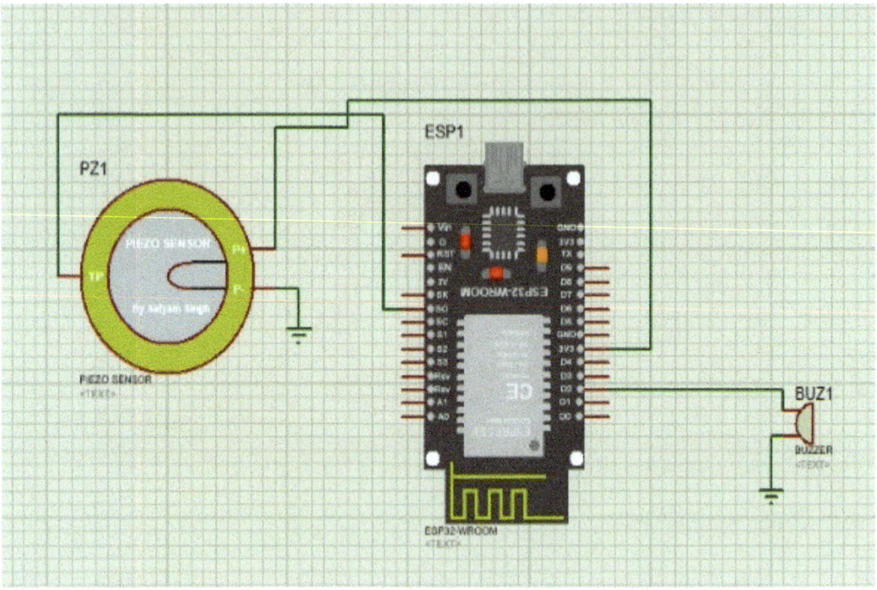

Fig. 10 Simulation of seizure detection system

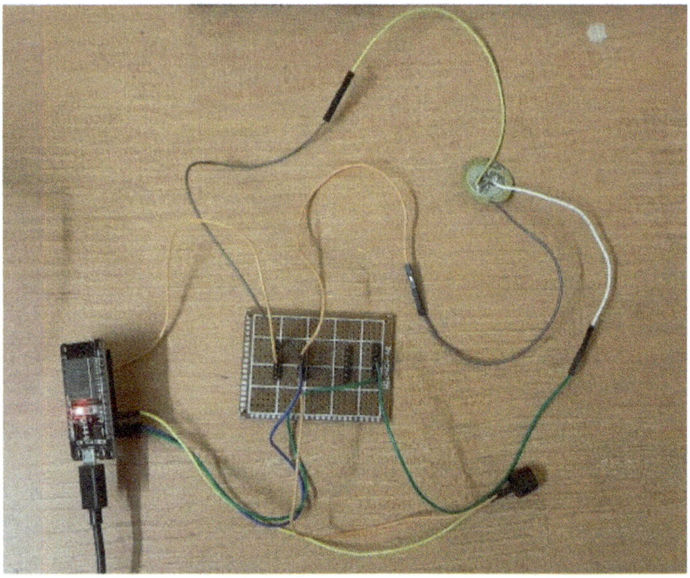

Fig. 11 Hardware setup of seizure detection system

The seizure detection system aims to enhance the monitoring and early detection of tonic–clonic seizures through wearable technology. The system employs a piezoelectric sensor placed on the palm, detecting pressure changes associated with seizures. ESP-based hardware processes the sensor data, triggering a buzzer for immediate alerts. This innovative approach provides real-time awareness and facilitates timely intervention, crucial for individuals with epilepsy.

References

1. https://www.sciencedirect.com/science/article/pii/S1877050919321945. Accessed on 15 Oct 2023
2. https://www.hopkinsmedicine.org/health/conditions-and-diseases/epilepsy/tonic-clonic-grand-mal-seizurestonic-clonicseizures
3. https://www.shutterstock.com/search/clonicimageforstagesofseizure
4. https://epilepsyfoundation.org.au/understanding-epilepsy/seizures/seizure-phases/#:~:text=Seizures%20take%20on%20many%20different,(post%2Dictal)%20stage.Differentphasesofseizures
5. https://www.hopkinsmedicine.org/health/conditions-and-diseases/epilepsy/tonic-and-clonic-seizures#:~:text=A%20tonic%20seizure%20causes%20a,may%20feel%20tired%20or%20confused.Tonicseizures
6. https://www.hopkinsmedicine.org/health/conditions-and-diseases/epilepsy/tonic-and-clonic-seizures
7. https://quizlet.com/559947349/the-causes-assessment-and-care-of-seizures-diorders-flash-cards/imageofTonicandClonicstagesofSeizure
8. https://www.epilepsy.com/what-is-epilepsy/understanding-seizures/how-seriousareseizures#:~:text=Generally%20speaking%2C%20a%20generalized%20tonic,injury%20or%20death%20can%20occur
9. Bisht E, Rathi A, Chaudhary M, Hospital management. Mata Sundri College for women, Delhi University, Department of computer science
10. Saimanoj K, Poojitha G, Dixit KD, Jayannavar L (2020) Hospital management system using web technology. In: Conference: ICAIT-2020, Bengaluru
11. https://pubmed.ncbi.nlm.nih.gov/26033084/
12. Lin S-K, Istiqomah, Wang L-C, Lin C-Y, Chiueh H (2018) An ultra-low power smart headband for real-time epileptic seizure detection. IEEE J Transl Eng Health Med 6:1–10 (2018)

Chatbot Based Event Planning App

Reshma Gopinathan, Archana Haridas, Riya Elizabet Stanly, Vaishnavi Rajan Pol, and Payel Thakur

Abstract Event planning involves the process of conceptualizing, organizing, and executing various types of events, from weddings and corporate conferences to concerts and festivals. It requires careful attention to detail, effective communication skills, and a thorough understanding of logistics and budgeting. The proposed system helps users plan and organize their events with ease. The app uses rule-based menu driven chatbot. The chatbot guides users through the event planning process, offering tips and advice along the way. This system is a cross platform application with user friendly GUI. Users are guided through a series of predefined steps or options, which can be particularly useful in complex processes like event planning. The chatbot can present menu choices for different stages of planning, such as venue selection, catering, and decoration, ensuring that users don't miss any important details. Users can also easily manage their guest list, maintain notes related to the event, plan proper schedule and track budget directly through the app. With the event planning chatbot, users can save time and reduce stress while planning their perfect event.

Keywords Event planning · Chatbot · AI · Neural network

R. Gopinathan (✉) · A. Haridas · R. E. Stanly · V. R. Pol · P. Thakur
Department of Computer Engineering, Pillai College of Engineering, Panvel, Maharashtra 410206, India
e-mail: gopinathan20comp@student.mes.ac.in

A. Haridas
e-mail: archana20comp@student.mes.ac.in

R. E. Stanly
e-mail: eriya20comp@student.mes.ac.in

V. R. Pol
e-mail: pvaishnavi20comp@student.mes.ac.in

P. Thakur
e-mail: payelthakur@mes.ac.in

V. K. Gunjan and J. M. Zurada (eds.), *Proceedings of 5th International Conference on Recent Trends in Machine Learning, IoT, Smart Cities and Applications*, Lecture Notes in Networks and Systems 1181, https://doi.org/10.1007/978-981-97-8861-3_28

1 Introduction

The event planning app offers an efficient and personalized experience for users. The app includes a user-friendly interface with a chatbot that can assist users with various tasks, such as event registration, managing the guestlist and vendor contact, providing recommendations related to themes, gift ideas, dress code, colour combinations etc., graphically creating flow of the event and creating notes related to the event.

The Fundamental Components in our Event Planning App are Budget Tracker, To-Do List, Notes, Event schedule, Vendors list, Guest list, Chatbot for recommendation. Fundamental components in our proposed system starting from the phase of planning an event which comes from recognizing the type of event for example birthday party, house party, business party or so on. The user can give any name to the event, set the date and location of the event. Other components include creating a guestlist, making a to-do list, budget tracking and note finally the event schedule helps the user to create flow of the event and view it graphically.

The event planning app has been enhanced through the integration of artificial intelligence (AI), automating tasks and providing users with increased flexibility during the planning process. A chatbot was seamlessly incorporated into the React Native application using Flask, ensuring smooth communication. The development of a dedicated API facilitated this integration, enabling the chatbot to guide users throughout the planning journey. This strategic amalgamation not only streamlines tasks but also adds an intelligent layer, enhancing overall functionality and user satisfaction.

2 Motivation

A group of college students, driven by the goal to simplify and enhance event planning, is developing an all-in-one app. Rooted in the belief that event planning should be accessible and enjoyable for everyone, the app aims to streamline tasks, foster creativity, and transform the planning process. With a passion for innovation and a focus on practical needs, the team is dedicated to contributing to a seamless and enjoyable event planning experience for users of all backgrounds and occasions.

3 Problem Statement

Organizing events involves numerous steps such as venue selection, catering, decorations, entertainment, and guest list management. Currently, the process requires using multiple apps, resulting in scattered information, communication issues, and potential oversights. To address this challenge, there is a need for a user-friendly event planning app that offers a one-stop solution, eliminating the need for multiple downloads. The app aims to provide personalized recommendations based on user preferences and budget, coupled with seamless communication and collaboration features. The ultimate goal is to make event planning a stress-free and efficient experience for users.

4 Objective

- To create a user- friendly application.
- To create an all-in-one application for event planning.
- To plan an event in a very organized manner. Helps user create detailed schedule
- To make user-specific event recommendations such as themes, decorations, gifts ideas etc.
- To seamlessly track the progress of the event making it easy for the user to be visually updated on the event.

5 Literature Survey

Mehrotra and Lobo [1] has cited in their publication that Event Planning App was Time Saving, Cost-effective and had better communication, networking and planning, but the problem is Complexity of event software, which can be a put-off for clients, especially during data transfers between applications. Technology can also be expensive, and learning to use it properly can be difficult and costly.

Bawankar et al. [2] have cited in their publication that it has a user-friendly GUI and could manage multiple events time saving in finding the location and communication; also it has overcome the problem of delays in events but it does not consider the feedback features which would give better results and solve the existing problem. The notifications regarding the events are not sent to the clients.

Goyal et al. [3] have cited in their publication that it saves time organizing different events and helps to control the problems that usually happen in an event but a new feature to keep in touch with the vendors and facilitators should be inculcated and a vendor directory should be maintained region wise.

Nilesh Verma et al. [4] has cited in their publication that it improves communication and accessibility but there are security risks and technical issues (Table 1).

6 Proposed System

See Fig. 1.

Table 1 Summary of Literature Survey

SN	Paper	Author	Advantages	Disadvantages
1	Technology driving event management industry to the next level (2020)	Mehrotra et al. [1]	Saving time, Cost-effectiveness, Better communication, Better networking, Better planning	Complexity of event software, especially during data transfers, Cost and difficulty in learning technology, Potential unemployment in the events industry due to automation
2	Event management system (2021)	Bawankar et al. [2]	User-friendly GUI for managing multiple events, Time savings in location finding and communication, Overcoming delays in events	– Lack of feedback features – Absence of event notifications to clients
3	Online event management system (2021)	Goyal et al. [3]	Time savings in organizing different events, Problem control in events, Secure online data submission, Real-time results, Easy venue appointment checking, Quick participant registration, Web-based accessibility	Potential need for more features, Innovation and advancements in themes and decorations, Lack of a feature for vendor and facilitator communication
4	An android-based application for event management (2022)	Verma et al. [4]	Improved communication, Time-saving, Increased accessibility, Convenience, Better organization, Increased participation	Technical issues, Dependency on technology, Security risks
5	Online event management system (2022)	Mohana and Anbumani [5]	Overcoming client inconvenience, 24/7 scheduling and booking,—Access for both clients and staff, Reduction in time and cost	Limited benefits for large-scale events
6	A Chatbot application by using natural language processing and Artificial intelligence markup language (2022)	Arya et al. [6]	User-friendliness, Ease of system dialogue, Use of various languages for formal, computer-readable representation	Lack of proper interface, Integration challenges, No use of APIs

(continued)

Table 1 (continued)

SN	Paper	Author	Advantages	Disadvantages
7	Sentiment-based Chatbot using machine learning for recommendation system (2022)	Hsu and Liao [7]	Consideration of user emotions for recommendations, Simple interface	Lack of intelligence
8	Mobile application for college event management (2022)	Pabba et al. [8]	Simplified registration and event attendance, Updated database with server changes	Absence of a feedback option

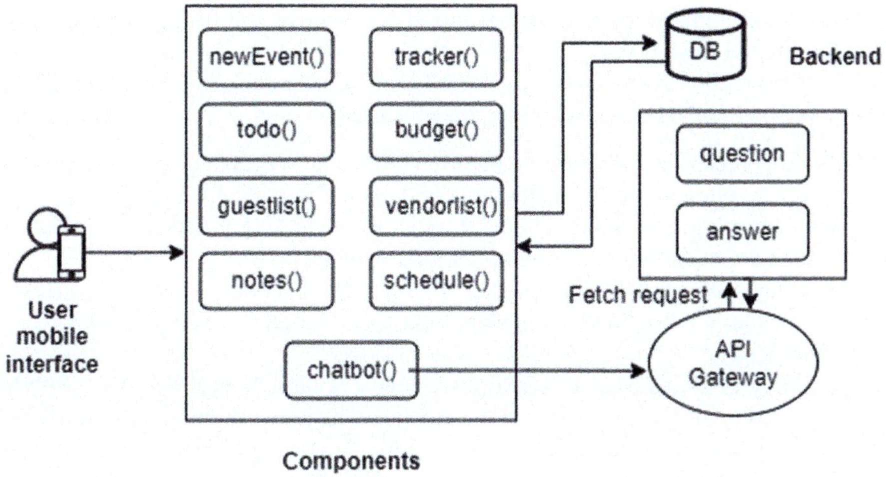

Fig. 1. Proposed system architecture

A. *Introduction*

1. System Overview: The proposed system streamlines event planning by employing natural language processing and artificial intelligence. It understands users' preferences, offering personalized recommendations for venues, catering, decorations, and more. The user-friendly cross-platform chatbot guides individuals through the planning process, providing tips and advice. Artificial intelligence enhances the chatbot's ability to comprehend and respond to user queries naturally, enabling it to analyze data for intelligent decisions and personalized suggestions based on user preferences.

Table 2 Hardware and Software details

Software requirements (minimum)	
Operating system	Android, MAC OS
Programming language	React native, Python
Database	Async storage (local storage)
Hardware requirements (minimum)	
Processor	2 GHz Intel
RAM	4 GB

B. *Details of Hardware and Software*
See Table 2.

C. *Methodology used*
For the design, of the app we have used react native and for the integration of the chatbot flask is used.

The chatbot is implemented using NLP processing concepts and deep learning using pytorch. These are the key concepts used.

1. Tokenization: Tokenization involves breaking user input questions related to event planning into smaller units or "tokens" using the NLTK library.
2. Stemming: Stemming reduces familiar words identified after tokenization to their root form, enhancing text normalization.
3. Bag of Words: Utilizing the Bag of Words (BoW) concept, words are checked against a predefined vocabulary, and their presence is marked with 1 or 0 (Fig. 2).
4. Feed Forward Neural Network: A Feed-Forward Neural Network (FFNN) architecture is employed on the chatbot. The network integrates the Bag of Words output and includes input, hidden, and output layers with Softmax activation for multiclass classification (Fig. 3).
5. NLP Processing Pipeline: A processing pipeline, is designed for text data analysis. The steps include:

   - Tokenization: Breaking input text into individual words or tokens.

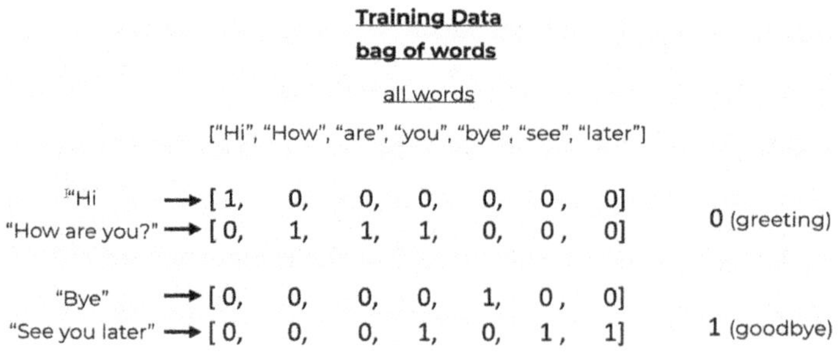

Fig. 2 Bag of words

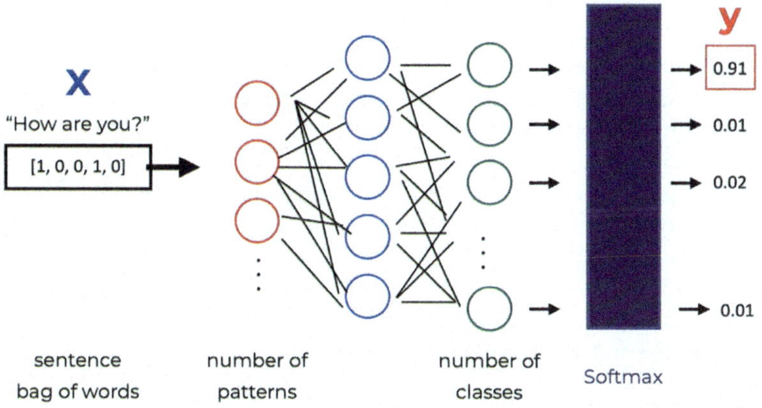

Fig. 3 Feed forward neural network

- Lowercasing and Stemming: Ensuring case-insensitivity and reducing words to their base form.
- Excluding Punctuation: Removing punctuation for a focus on words.
- Bag of Words (BoW): Representing input using a predefined vocabulary, generating a binary value for each word's presence (Fig. 4).

Fig. 4 NLP processing pipeline

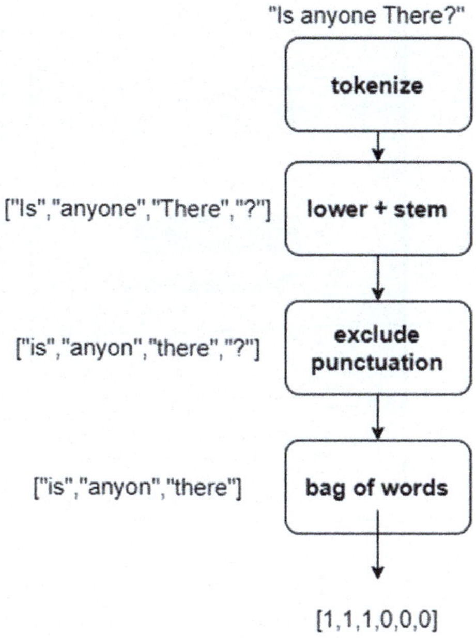

D. *Modules*

1. Creating Events:
   Efficiently organize and plan events with a user-friendly interface for seamless event creation (Fig. 5).
2. To-Do List:
   Stay organized and on track by managing tasks and to-dos in one central location within the app (Fig. 6).
3. Budget Management:
   Keep finances in check with a dedicated budget management module for accurate event cost tracking (Fig. 7).
4. Guestlist and Vendors Contact Management

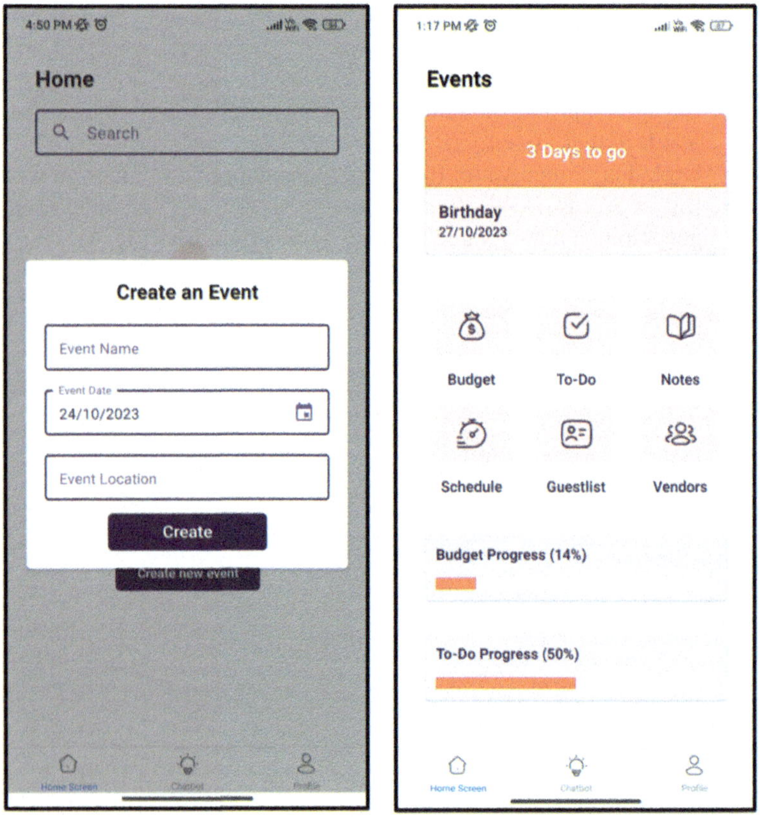

Fig. 5 Creating events—the users can create new events and track their progress through various trackers on the dashboard page of the events

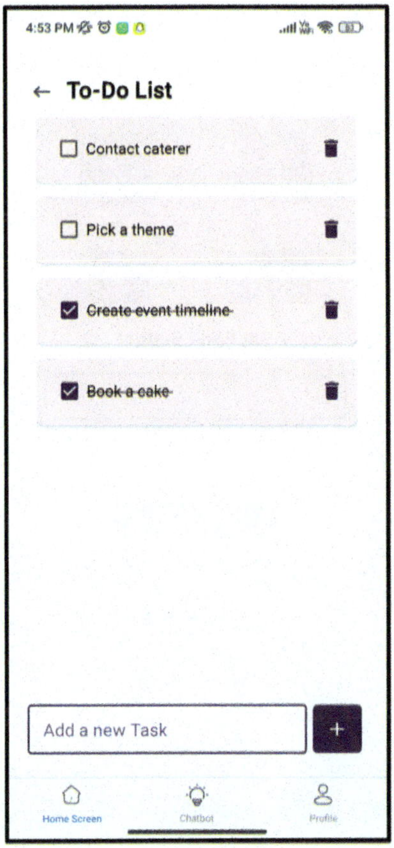

Fig. 6 To-do list page—the user can create and organize to-do lists and check of the tasks they have completed

Have all the contacts be it the guests or the vendors involved in the event in one place (Fig. 8).

5. Notes:
   Jot down and organize important notes and details related to your event effortlessly (Fig. 9).
6. Event Schedule:
   Plan and visualize your event timeline with a comprehensive schedule module for efficient coordination (Fig. 10).
7. Chatbot:
   Receive personalized recommendations and guidance throughout the event planning process with the integrated chatbot feature (Fig. 11).

Fig. 7 Budget management—the user can set any desirable amount for the budget

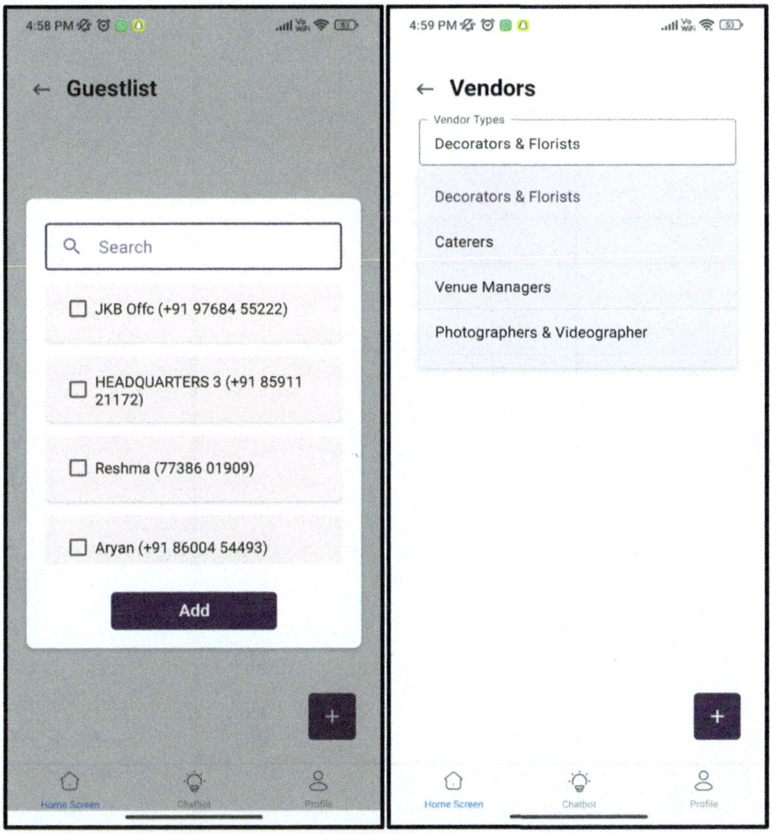

Fig. 8 Guest and vendor contact management—the users can import their contacts on the app and organize them

Fig. 9 Notes—the user can create notes related to the events

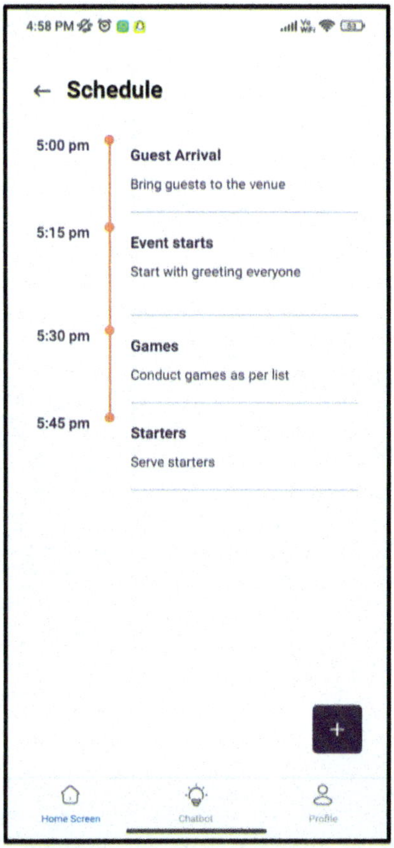

Fig. 10 Event schedule—the detailed time line of the particular event can be created be graphically pictured

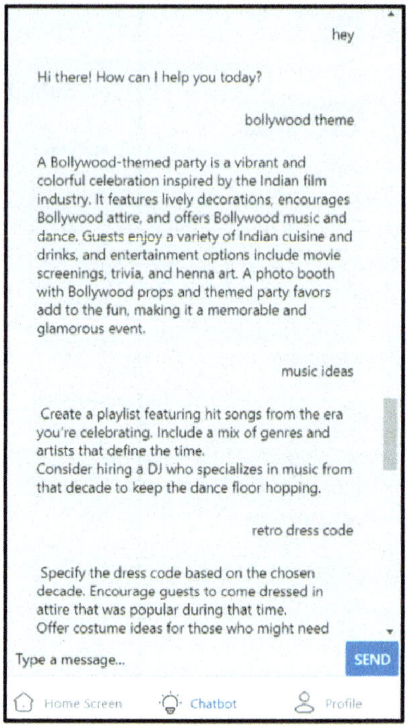

Fig. 11 Chatbot—user can get recommendations on dress idea, gifts and various other topics from the AI generated chatbot

7 Conclusion

In this report, the study of different event management techniques is presented. An event management project typically involves the planning, coordination, and execution of events such as conferences, trade shows, weddings, and corporate events. The project includes various tasks such as identifying the event objectives, selecting a suitable venue, managing budgets, coordinating vendors and suppliers, managing guest lists, creating event schedules, and ensuring smooth execution on the day of the event. The project typically starts with a planning phase where the event objectives are identified and a project plan is created. This includes defining the scope of the event, establishing timelines and deadlines, creating a budget, and selecting a suitable venue. The system uses rule-based menu driven chatbot to make the event planning process more tailored.

Acknowledgements We are grateful to our project guide Mrs Payel Thakur and the Head of the Department for their invaluable support and guidance throughout the completion of our major project in AI ML. Their contributions have been instrumental in the academic success and we are indebted to them for their mentorship and tireless efforts.

References

1. Mehrotra A, Lobo J (2020) Technology driving event management industry to the next level. In: International Conference on Reliability, Infocom Technologies and Optimization (ICRITO), June 4–5, 2020
2. Bawankar NM, Tembhurne AT, Chatap GR, Bandbuche MM, Dhote TS, Surjuse VR (2021) Event Manag Syst 7(2), IJARIIE-ISSN(O)-2395-4396
3. Goyala D, Alib A, Haider MdN (2021) Online event management system. Turk J Comput Math Educ 12(6):5297–5303
4. Verma N, Salvi V, Salvi P, Rathod J, Dalvi S (2022) An android based application for event management. Int J Res Publ Rev 3(2):729–733
5. Mohana S, Anbumani P (2022) Online event management system. Int J Res Publ Rev 3(7):1947–1952
6. Arya V, Khan R, Aggarwal M (2022) A Chatbot application by using natural language processing and artificial intelligence markup language. Int J Soft Comput Eng (IJSCE) 12(3). ISSN: 2231-2307
7. Hsu I-C, Liao A-H (2022) Sentiment-based Chatbot using machine learning for recommendation system. Res Square
8. Pabba P, Viswadha B, Srivani P, Rajashree P, Vinay Kumar S (2022) Mobile application for college event management. Int Res J Eng Technol (IRJET) 09(12). ISSN: 2395-0072

IOsE for Real Time Monitoring of Combustion Flames (RMCF) in Industrial Boilers

K. Sujatha, G. Rohini, G. Durga Devi, R. S. Ponmagal, S. Bhuvaneswari, N. Janaki, B. Latha, N. P. G. Bhavani, and V. Srividhya

Abstract The innovation in the area of Internet of Smart Environment (IOsE) for the quality prediction of combustion process in industrial boilers is highlighted in this article. IOsE is quality reliant. The characteristics of flame in the combustion chamber which is used for identification and control of Carbon Dioxide (CO_2) emissions. Currently, existing methods for flame analysis include non-invasive and high frequency spectral light dependent imaging techniques. Colour of the flame inside the Combustion Chamber (CCh) determines the prediction of Combustion Quality (CQ) and thereby its corresponding Combustion Quality Index (CQI). An online quality detection scheme for smart environment with colour extraction technique from the flame images to identify complete and incomplete combustion conditions is

---

K. Sujatha (✉)
EEE Department, Dr. M.G.R. Educational and Research Institute, Chennai, India
e-mail: sujathak73586@gmail.com

G. Rohini · S. Bhuvaneswari
EEE Department, S.A. Engineering College, Chennai, India
e-mail: dr.rohinig@saec.ac.in

G. Durga Devi
ECE Department, New Prince Shri Bhavani College of Engineering and Technology, Chennai, India

R. S. Ponmagal
CSE Department, SRM Institute of Science and Technology, Kattankolathur, Chennai, India

N. Janaki
Vels Institute of Science, Technology and Advanced Studies, Chennai, India
e-mail: janaki.se@velsuniv.ac.in

B. Latha
Physics Department, Dr. M.G.R. Educational and Research Institute, Chennai, India
e-mail: latha.phy@drmgrdu.ac.in

N. P. G. Bhavani
ECE Department, Saveetha Institute of Medical Sciences, Chennai, India

V. Srividhya
EEE Department, Meenakshi College of Engineering, Chennai, India

V. K. Gunjan and J. M. Zurada (eds.), *Proceedings of 5th International Conference on Recent Trends in Machine Learning, IoT, Smart Cities and Applications*, Lecture Notes in Networks and Systems 1181, https://doi.org/10.1007/978-981-97-8861-3_29

developed. For this, the intensity value of the pixels in Red-Blue-Green (RBG) plane is utilized. The coal is used as the prime fuel to heat water and convert it to steam for operating the industrial boilers. The quality prediction of combustion from flame colour and shape includes Complete Combustion (CC) and Incomplete Combustion (IC) categories. The flame images for CC and IC categories are subjected to preprocessing, attribute extraction and finally the prediction of flame quality is done using Random Forest (RF), Ada Boost (AB), Gradient Boosting (GB), Decision Tree (DT) and Multi Layer Perceptron (MLP). MLP is found to effectively categorize the CQ using the performance metrics like Sensitivity (SEY), Specificity (SPY) and Accuracy (ACY).

Keywords Random Forest (RF) · Ada Boost (AB) · Gradient Boosting (GB) · Decision Tree (DT) · Multi Layer Perceptron (MLP) · Non-invasive and flame quality prediction

1 Introduction

A great significance exists for generation of electrical energy which depends on the quality of coal used for firing the industrial boilers. Burning pulverized coal produces heat energy which heats the preheated water and converts it to steam. The superheated steam drives the turbine coupled to a generator for electrical energy generation. There is a danger of boiler explosion due to unmonitored excessive coal loading into the CC. So, real time monitoring of combustion chambers can be made possible using internet of things to enable smart power plants. Since the flame temperature is 1200 °C during CC, conventional gas analyzers and thermocouples are not suited for effective measurements. Hence, a Burner Flame Detector (BFD) system is integrated with IOsE. IOsE for flame monitoring system displays the status of the flame at the control room so that the inlet fuel supply can be cut down to lower the risk of furnace overloading resulting in boiler explosion. The actual flame shape and the cross section of the burner are studied. The burners are V94.2 Ansaldo for thermal power plants (120 MW) [1–6]. Each combustion chamber has 20 burners out of which 12 are Lignite burners and remaining 8 are oil burners. Two flame detectors in each chamber send command via the monitoring system, which detects the flame out condition. Because of high temperature a drastic effect on the main parts of the detector takes place and hence it is installed out of this region and isolated by an external lens. Cooling and sooth blowing air is used to avoid overheating and clogging of the line of sight. Experts extract some valuable results viewing the flame shape and colour [7–11, 12]. Some of them are hypothetically described below and the various flame types are shown in Fig. 1.

- Appropriate Flame Shape (AFS): Normal burner with uniform fuel flow and no leakage.

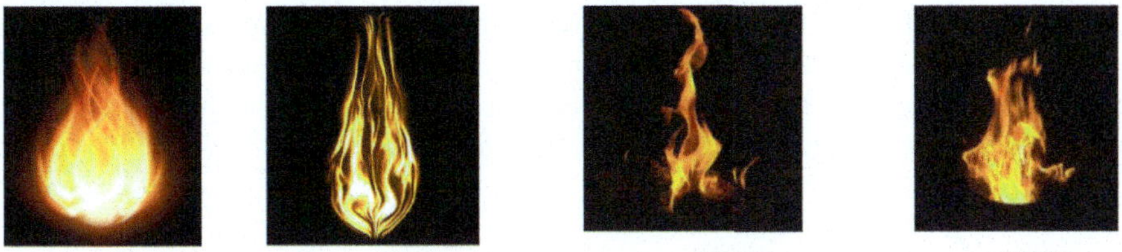

OFS CQI=0.4 | ELF CQI=0.3 | IPELF CQI=0.2 | IPSFL CQI=0.1

CC | IC

Fig. 1 Categorization of flame shapes from a BFD

- Excessively Long Flame (ELF): Cracked burner, fuel nozzle leakage and complete clogging in the return path.
- Irregular Periphery for Excessively Long Flame (IPELF): wear and tear and partial clogging of fuel nozzle.
- Irregular Periphery for Short Flame Length (IPSFL): Partial clogging of lines supplying natural gas.

2 Literature Survey

Industrial Boilers are used to generate steam, where the feed water is superheated to produce steam. The water tube boilers have tubes which carry water and this water will be heated up by the furnace flame, so that it gets converted to steam. The industrial boilers play an important role in power generation plants where the pulverized coal is burnt in the presence of oxygen to facilitate combustion [1–5]. Generally power plant steam is produced and plays an important role in the case of steam power plant. It is a must that an appropriate air to fuel ratio is to be maintained to conform the complete combustion in Industrial boilers that are used for power generation. According to the law of conservation of energy suggest that in the boilers, the potential energy of the water stored along with the chemical and light energy during burning of coal is converted to heat energy which converts water to steam, followed by conversion to mechanical energy and electrical energy. This combustion process takes place inside the CC. When the fuel is burnt inside the combustion chamber, heat energy is produced, and exhaust gases are liberated. The heat energy is used to generate steam due to the combustion process in the boiler furnace within the boiler. The superheated steam heated by the outgoing flue gas and pre-heated water, heated by the same exhaust comes out of the boiler will go to the boiler to generate stream efficiently [6–10]. There two methods: optimizing the performance of the boiler. Cost saving in reducing the operation of the boiler is one of the major goals which tends to optimization. CQ improvement by controlling flame temperature is the most common opportunity to save the energy and cost in a boiler. Another method commonly used is the combustion improvement is injecting correct combination of air and fuel, by optimizing this air fuel mixture increase the boiler performance. Combustion efficiency indicated by the combustion improvement is carried out within the boiler combustion process. The flame color indicates the combustion quality approach where they use statistical indicators to access the patterns [11].

3 Aim for RMCF

The aim of this work is to predict the CQI from the flame colour to predict CC and IC from the flame images. The quality of combustion is inferred from the flame shape and colour at the furnace level itself.

4 Collection of Flame Samples

The images of nearly fifty one (51) images of flame samples were collected from the thermal power plant. The samples of the flame images along with the attributes are displayed in Table 1. The variation in colour is considered to be the major attribute which facilitates the classification of the images using Random Forest (RF), Ada Boost (AB), Gradient Boosting (GB), Decision Tree (DT) Multi Layer Perceptron (MLP), thereby establishing an expert system for prediction of CQI from flame images using AI prediction scheme.

5 Methodology for Combustion Quality Determination of Flame Images Using Colour Image Processing

Contrast based flame image classification can be done with an input flame image which contains many objects. It is possible to recognize these objects inside the sample of flame image with an appropriate identification of every entity along with the exact position in that image by using Random Forest (RF), Ada Boost (AB), Gradient Boosting (GB), Decision Tree (DT) Multi Layer Perceptron (MLP) models. The flowchart and block diagram for the flame shape prediction is illustrated in Figs. 2 and 3 respectively.

Table 1 Attributes extraction

Properties	AFS	ELF	IPELF	IPSFL
Area	289,811	579,623	1,159,245	5,629,801
Centroid	[155.977 117.083]	[311.955 234.166]	[623.9115 468.3331]	[972.5008 1.4485e + 003]
Bounding Box	[0.5000 0.5000 312 467]	[0.5000 0.5000 623 934]	[0.5000 0.5000 1246 934]	[0.5000 0.5000 1944 2896]
Sub array	{[1 × 312 double] [1 × 467 double]}	{[1 × 623 double] [1 × 934 double]}	{[1 × 934 double] [1 × 1246 double]}	{[1 × 2896 double] [1 × 1944 double]}
Major axis length	0.359425 e^{+003}	0.71885 e^{+003}	1.4377e^{+003}	3.3440e^{+003}
Minor axis length	0.269475 e^{+003}	0.53895 e^{+003}	1.0779e^{+003}	2.2447e^{+003}
Eccentricity	0.2439	0.5822	0.6617	0.7412
Orientation	54.3286	7.3289	0.0687	89.9999
Convex hull	[9 × 2 double]	[9 × 2 double]	[9 × 2 double]	[9 × 2 double]
Convex image	[312 × 467 logical double]	[623 × 934 logical double]	[934 × 1246 logical]	[2896 × 1944 logical]
Convex area	290,941	581,882	1,163,764	5,629,824
Image	[312 × 467 logical]	[623 × 934 logical]	[934 × 1246 logical]	[2896 × 1944 logical]
Filled image	[312 × 467 logical]	[623 × 934 logical]	[934 × 1246 logical]	[2896 × 1944 logical]
Filled area	290,928	581,855	1,163,710	5,629,824
Euler number	11	1475	2950	22
Extrema	[8 × 2 double]	[8 × 2 double]	[8 × 2 double]	[8 × 2 double]
Equivalent diameter	0.303725 e^{+003}	0.60745 e^{+003}	1.2149e^{+003}	2.6773e^{+003}
Solidity	0.9883	0.9922	0.9961	1.0000
Extent	0.9883	0.9922	0.9961	1.0000
Pixel Idx list	[289811 × 1 double]	[579623 × 1 double]	[1159245 × 1 double]	[5629801 × 1 double]
Pixel list	[289811 × 2 double]	[579623 × 2 double]	[1159245 × 2 double]	[5629801 × 2 double]
Perimeter	4838	2.2026 e^{+003}	4.4052e^{+003}	9676

6 Results and Discussion

Hyper-parameter optimization and feature reduction is carried out to assess the performance of the model. A set of training data is utilized to construct the model. Note that certain features in the training data set have been modified using feature

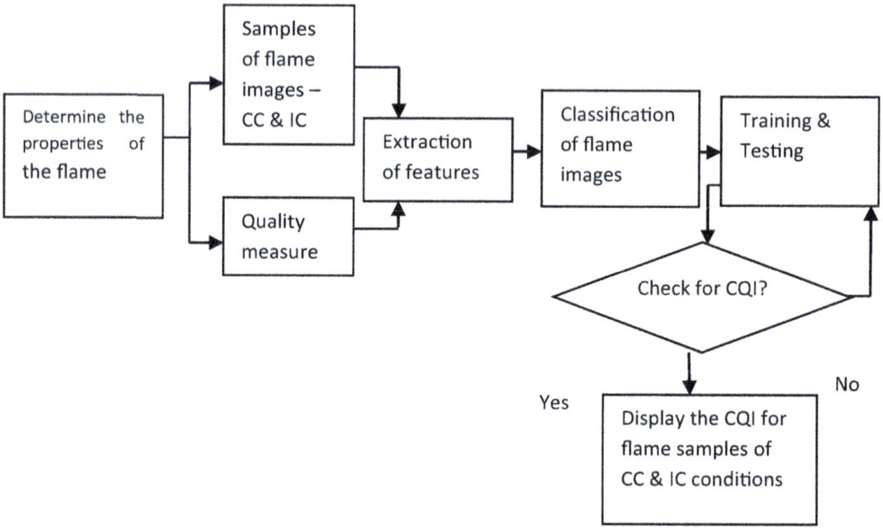

Fig. 2 Flowchart for prediction of CQI from flame shape

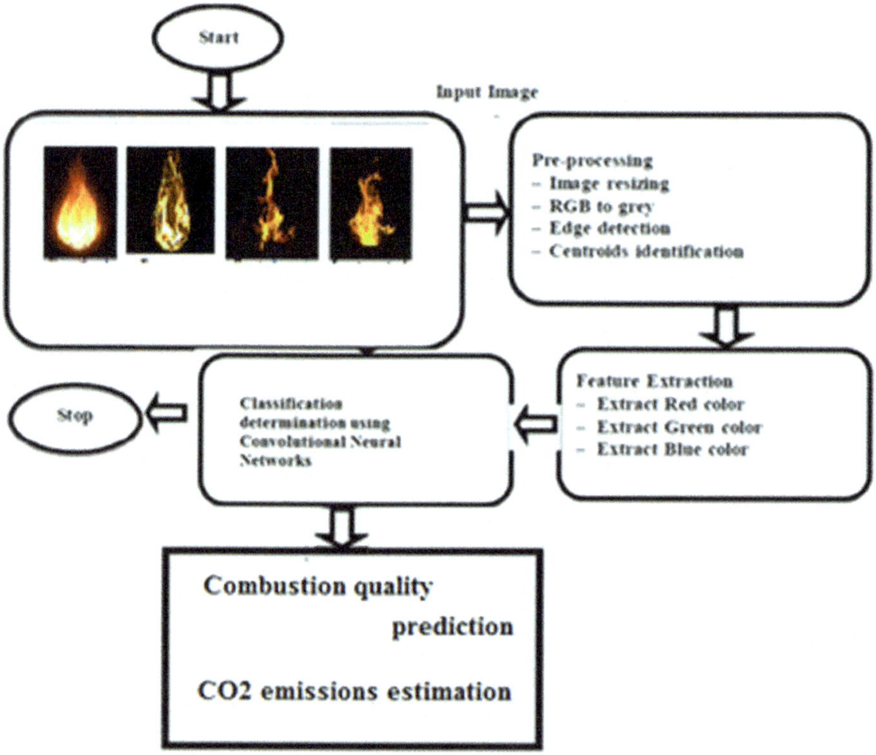

Fig. 3 Block diagram for prediction of CQI

engineering. A methodical, repeatable process that maximizes the utilization of the available training data is necessary for model training. Our issue statement must be created, our data set must.

The flame shape classification and CQI classification and CO_2 emission detection system is trained using a set of input features and target data extracted from the images of the flame samples. According to Table 2, the accuracy, sensitivity, precision and specificity for MLP have the highest rates. Conversely, all the other algorithms have lower values for accuracy, sensitivity and precision, while MLP has the lowest accuracy. Furthermore, RF has the lowest specificity. MLP works the best on our dataset as it has the optimal parameter settings when compared to the other four methods (Fig. 4).

Table 2 Learning models versus real time approach

S. No	Algorithm/method	Accuracy (%)	Sensitivity (%)	Specificity (%)
	Formulae	$A = 1/N\sum(T_p + T_N)/(T_p + T_N + F_p + F_N)$	$SEN = 1/N\sum T_p/(T_p + F_N)$	$SP = 1/N\sum T_N/(T_p + F_N)$
1	MLP	98.72	97.25	99.24
2	BPA	95.23	96.24	96.48
3	RF	91.2	91.6	93.12
4	AB	93.4	90.1	93.46
5	GB	94.36	93.3	91.2
6	DT	92.7	87.2	87.2
7	Flame detector	95.86	96.56	96.62

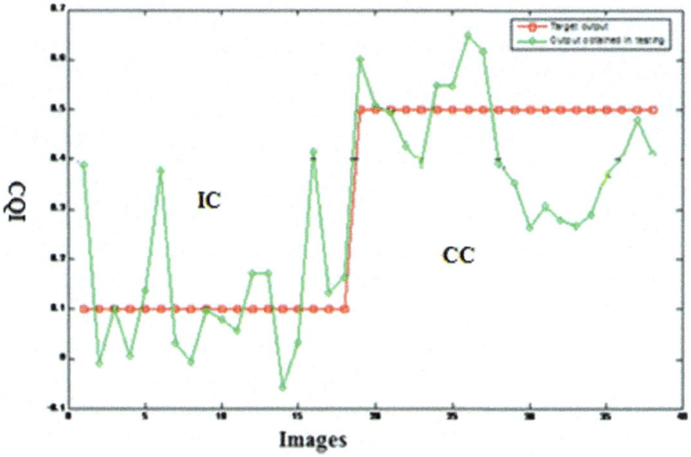

Fig. 4 Output for quality prediction of combustion quality

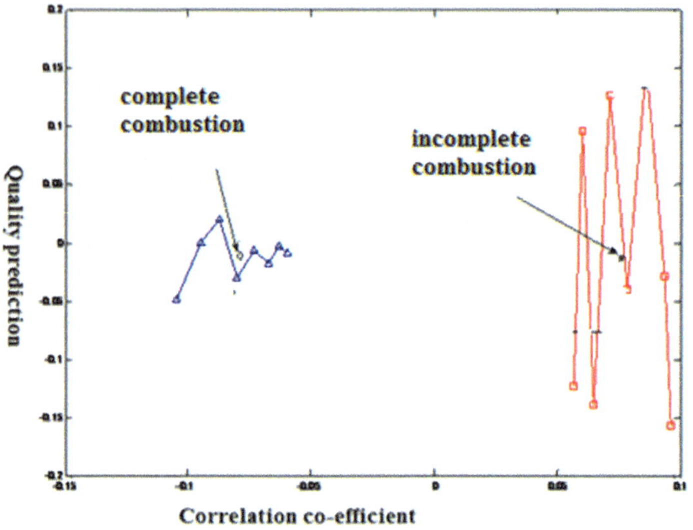

Fig. 5 Determination of correlation coefficient after combustion quality prediction from flame images

While the network is capable of determining three parameters of the flame images corresponding to red, green and blue colours, with hidden nodes and three output nodes. The learning rate of both the classification systems was fixed at one and there is no momentum and bias applied for training the systems. The output of the trained by MLP is very much close to the target value for both systems because of minimum MSE of 0.00011 and higher regression coefficient of about 0.998 can be seen in Fig. 5.

7 Conclusion

The concept of CQI classification system is proposed, because of its simplicity in implementation. As a result, the proposed technique offers a thorough and effective means of identifying systemic disorders based on flame colour and shape using MLP. The program prioritizes user authentication and data protection, uses Custom Vision for precise flame irregularity recognition, and embraces continuous learning to guarantee its efficacy and dependability. This technique has the potential to have a huge influence in power stations by offering onset identification of CQI and possibly prevention of systemic irregularities in flame shape through flame image analysis.

References

1. Sujatha K, Pappa N (2011) Renand, Combustion monitoring of a water tube boiler using a discriminant radial basis network. ISA Trans 50(1):101–110
2. Toyofuku N, Haff RP (2012) In: Sun DW (ed) Computer vision technology in food and beverage industry. Woodhead Publishing, pp 181–205
3. Joo M, Wu S, Lu J, Toh HL (2002) Face recognition with Radial Basis Function (RBF) neural networks. IEEE Trans Neural Netw 13:697–910
4. Fisher RA (1936) The use of multiple measurements in taxonomic problems. Ann Eugen 7:178–188
5. Foley DH (1972) Considerations of sample and feature size. IEEE Trans Inform Theory 18:626–681
6. Han P, Zhang X, Zhen C, Wang B (2006) Boiler flame image classification based on hidden Markov model. In: Proceedings of the IEEE international symposium on industrial electronics, vol 9–13. IEEE Xplore Press, pp 9–12
7. Hong ZQ, Yang YJ (1991) Optimal discriminant plane for a small number of samples and design method of classifier on the plane. Pattern Recognit 24:317–324
8. Jiang F, Liu S, Liang S, Li Z, Wang X et al (2009) Visual flame monitoring system based on two-color method. J Therm Sci 18:284–284
9. Lippmann RP (1987) An introduction to computing with neural nets. IEEE Trans Acoust Speech Signal Process Mag 35:4–22
10. Lu G, Gilabert G, Yan Y (2005) Vision based monitoring and characterisation of combustion flames. J Phys 2005(15):194–200
11. Krokida MK et al (2001) Color changes during deep fat frying. J Food Eng 48(3):219–225

Quantum Cryptanalysis: Analyzing Shor's Algorithm and Its Impact on RSA Security

Vaishnavi Khandu Purkar, Roshani Sunil Gavande, Shailaja Satishrao Kulkarni, and H. S. Thakar

Abstract The security of long-standing cryptographic protocols such as the RSA algorithm is under threat from the emergence of quantum computing, which is transforming the landscape of cryptography. The interaction between RSA and quantum-based threats—specifically, Shor's Algorithm—is examined in this review. The paper emphasizes how RSA's security depends on the challenge of factoring big integers after providing a thorough analysis of the algorithm's creation, encryption, and decryption steps. Large integer factoring is made possible by quantum computing, which breaks this security paradigm by using concepts like superposition. This process is accelerated by Shor's Algorithm, a quantum instrument that poses a danger to RSA encryption. To counter this threat, the paper emphasizes how urgent it is and supports post-quantum cryptography alternatives. The influence of Shor's Algorithm on RSA at different key lengths is examined along with vulnerability assessments.

Keywords Quantum computing · RSA algorithm · Shor's algorithm · Cryptography · Post-quantum cryptography · Encryption · Quantum threats · Computational complexity

V. K. Purkar (✉) · R. S. Gavande · S. S. Kulkarni · H. S. Thakar
Savitribai Phule Pune University, Pune, India
e-mail: vaishnavipurkar7@gmail.com

R. S. Gavande
e-mail: roshanigavande202@gmail.com

S. S. Kulkarni
e-mail: kulkarnishailaja87@gmail.com

H. S. Thakar
e-mail: hsthakar@pict.edu

V. K. Gunjan and J. M. Zurada (eds.), *Proceedings of 5th International Conference on Recent Trends in Machine Learning, IoT, Smart Cities and Applications*, Lecture Notes in Networks and Systems 1181, https://doi.org/10.1007/978-981-97-8861-3_30

1 Introduction

The advent of quantum computing has ushered in a new era of computational capabilities that challenge the foundations of classical cryptography. In this rapidly evolving technological landscape, the security of data transmission and digital communication is of paramount importance. This survey paper, titled "Quantum Cryptanalysis: Analyzing Shor's Algorithm and its Impact on RSA Security," delves into the intricate intersection of quantum computing and cryptography, focusing on the vulnerabilities introduced by Shor's algorithm to the widely adopted RSA encryption. Data communication plays a pivotal role in our daily lives, and as digital communication becomes increasingly prevalent, safeguarding information from unauthorized access and interception becomes a critical concern. Classical cryptographic methods, particularly the RSA algorithm, have long been relied upon for secure data transmission. However, the advent of Shor's algorithm, a groundbreaking development in quantum cryptography, poses a potential threat to the security assumptions underpinning RSA.

This survey aims to comprehensively analyze the vulnerability of the RSA encryption algorithm to Shor's quantum algorithm, shedding light on the theoretical aspects and practical implications of this intersection. The RSA algorithm, a cornerstone of secure data transmission for decades, relies on the computational infeasibility of factoring large integers. Shor's algorithm, leveraging the power of quantum computing, challenges this assumption by efficiently factoring large integers in polynomial time. The paper begins by providing a foundational understanding of classical cryptographic principles, including the RSA encryption algorithm. It then explores the principles of quantum computing, quantum gates, and the fundamental concepts behind Shor's algorithm. An in-depth analysis of Shor's algorithm is conducted, breaking down its steps and elucidating how it efficiently factors large integers—a task considered computationally infeasible for classical computers.

The survey extends its focus to the implications of Shor's algorithm on RSA security. The vulnerabilities introduced by the quantum algorithm are analyzed in terms of computational complexity, highlighting the potential risks posed to the widely deployed RSA encryption method. Additionally, the paper explores the burgeoning field of post-quantum cryptography, investigating alternative cryptographic techniques designed to withstand attacks from quantum computers. Through a combination of theoretical analysis, computational complexity assessments, and, where applicable, practical simulations, this survey aims to contribute to the growing body of knowledge in the field of quantum-resistant cryptography. By addressing the vulnerabilities of classical cryptographic systems in the face of advancing quantum computing capabilities, this paper offers insights into the urgent need for post-quantum cryptographic solutions and provides recommendations for securing digital infrastructure in the quantum era.

2 Principal and Working of RSA

The key components of the RSA algorithm for encryption and decryption include the generation of two large prime numbers, the calculation of the public key and private key, and the use of modular arithmetic to encrypt and decrypt plaintext and ciphertext. The encryption process involves raising the plaintext to the power of the encryption key and taking the result modulo the public key, while the decryption process involves raising the ciphertext to the power of the decryption key and taking the result modulo the public key. The security of the RSA algorithm is based on the difficulty of prime factorization, which makes it computationally infeasible to determine the private key from the public key. The security of the encryption and decryption algorithm in RSA depends on the confidentiality of the keys. If the private key is kept confidential and not leaked, it ensures the security of the encryption and decryption process. The confidentiality of the private key is crucial because anyone with access to the private key can decrypt the ciphertext and potentially compromise the confidentiality of the information. Therefore, maintaining the confidentiality of the private key is essential for the security of the RSA encryption and decryption algorithm. If a key in an encryption system is leaked, it means that anyone can decrypt the ciphertext and access the confidential information. This compromises the security of the encryption system and renders the encryption algorithm ineffective. The potential implications of a leaked key include unauthorized access to sensitive information, loss of confidentiality, and the possibility of tampering, forgery, or counterfeiting of the encrypted data. Therefore, the implications of a leaked key in an encryption system is severe and can have detrimental effects on information security [1] (Fig. 1).

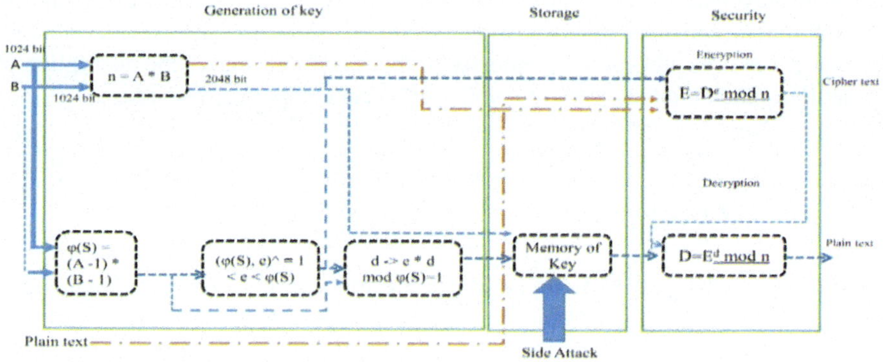

Fig. 1 RSA algorithm

3 Classical Decryption Complexity of the RSA Algorithm

The classical approach to decrypting the RSA algorithm is to factor the modulus N into its prime factors p and q, which can then compute the private key. The time and space complexity of this approach depends on the size of the modulus N and the specific factoring algorithm used. The most efficient factoring algorithm currently known is the General Number Field Sieve (GNFS), which has a time complexity of O(exp(sqrt(log(N) * log(log(N))))), where N is the size of the modulus. This means that the time required to factor an RSA modulus grows exponentially with the size of the modulus. For example, factoring a 1024-bit RSA modulus using GNFS would require approximately 2^{90} operations, which is currently beyond the capabilities of even the most powerful supercomputers. The space complexity of factoring algorithms can also be significant, requiring large amounts of memory for the computation of the factorization. For example, factoring a 1024-bit RSA modulus using GNFS would require several terabytes of memory. In practice, the security of RSA relies on the fact that factoring large integers is currently believed to be a computationally infeasible problem for sufficiently large moduli. As a result, the size of the RSA modulus is typically chosen to be large enough to resist current factoring attacks. For example, a 2048-bit RSA modulus is currently considered to be secure against all known factoring attacks.

It's important to note that there are other attacks on RSA, such as timing attacks and side-channel attacks, which do not rely on factoring the modulus. These attacks can have lower time and space complexity than factoring attacks, but they require access to specific implementation details of the RSA algorithm. As a result, these attacks are typically more difficult to carry out in practice. The decryption of RSA involves factoring the modulus N into its prime factors p and q, which are then used to compute the private key. The classical approach to RSA decryption involves factoring the modulus N, and there are several methods for factoring large integers given below. However, the time and space complexity of these methods makes the decryption of large RSA module infeasible with current classical computing resources.

3.1 *Trial Division*

Trial division is the most straightforward method for factoring an integer N. It involves dividing N by each integer from 2 to the square root of N to check for divisibility. While this method is simple, it becomes impractical for large integers due to its time complexity of O(sqrt(N)).

Example: Factoring a 20-bit RSA modulus using trial division would require checking divisibility by all primes up to approximately 2^{10}, which is computationally feasible. However, for larger moduli, the number of divisions becomes prohibitively large.

3.2 *General Number Field Sieve (GNFS)*

The GNFS is the most efficient known algorithm for factoring large integers. It has a time complexity of O(exp(sqrt(log(N) * log(log(N))))), making it the best method factoring large RSA moduli. However, even with the most powerful classical computing resources, the GNFS is currently infeasible for factoring RSA moduli of practical sizes.

Example: Factoring a 1024-bit RSA modulus using GNFS would require approximately 2^{90} operations, which is currently beyond the capabilities of classical computing.

3.3 *Infeasibility of Classical RSA Decryption*

The infeasibility of classical RSA decryption is primarily due to the exponential time complexity of factoring algorithms such as GNFS. For example, factoring a 2048-bit RSA modulus using GNFS would require an estimated 2^{110} operations, which is currently beyond the capabilities of classical computing resources. Additionally, the space complexity of these algorithms often requires large amounts of memory, further limiting their practicality.

As a result, the security of RSA relies on the infeasibility of factoring large integers with current classical computing resources. The size of the RSA modulus is chosen to be large enough to resist all known factoring attacks. For example, a 2048-bit RSA modulus is currently considered secure against all classical factoring attacks.

While there are several methods for factoring large integers, the time and space complexity of these methods makes the classical decryption of large RSA module infeasible with current computing resources. As a result, the security of RSA relies on the computational infeasibility of factoring large integers with classical computing [2].

4 Principle and Working of Shor's Algorithm

Shor's algorithm is a quantum algorithm for integer factorization, which has the potential to efficiently factorize large numbers, a task that is classically very challenging. The algorithm consists of several steps and a simple explanation is provided below with the mathematics involved.

Step 1 Choose the Number to Factorize

Let's choose a large composite number N to factorize. For this example, let's say, N = 15.

Step 2 Choose a Random Number

Choose a random number "a" that is less than N. For this example, let's choose a = 7.

Step 3 Calculate the Greatest Common Divisor (GCD)

Calculate the greatest common divisor of "a" and N using a classical algorithm such as Euclid's algorithm. If the GCD is not 1, then it is a non-trivial factor of N, and the factorization is complete. If the GCD is 1, proceed to the next step.

Step 4 Quantum Fourier Transform

Use quantum operations to perform the Quantum Fourier Transform (QFT) on a quantum register. The QFT is a key step in Shor's algorithm and involves complex quantum operations.

Step 5 Measure the Quantum Register

Measure the quantum register to obtain a value "r". This value "r" is related to the period of the function $f(x) = a^x \bmod N$.

Step 6 Calculate Potential Factors

If "r" is odd or if $a^{(r/2)} \equiv -1 \pmod N$, then go back to Step 3 and choose a different "a". Otherwise, calculate the potential factors of N using the formulae.

$$\text{Factor 1} : \gcd(a^{r/2} + 1, N)$$
$$\text{Factor 2} : \gcd(a^{r/2} - 1, N)$$

The significance of the probability "p" in Shor's algorithm for factorization lies in its relationship to the success rate of the algorithm. In the context of Shor's algorithm, "p" represents the probability of observing a particular state |c, xk (mod n) > with certain conditions, and it is used to estimate the success probability of the algorithm.

The success probability of Shor's algorithm is crucial because it determines the likelihood of obtaining the correct factors of the composite number N. A higher success probability indicates a greater likelihood of successfully factorizing N using the algorithm (Fig. 2).

Therefore, the significance of the probability "p" is that it directly influences the effectiveness and efficiency of Shor's algorithm for factorization. Understanding and analyzing the probability "p" is essential for evaluating the performance and practical applicability of the algorithm in real-world factorization tasks.

There are φ(r) possible values of "c" and "r" possible values of "xk (mod n)" in Shor's algorithm for factorization. Here, φ(r) represents Euler's totient function, which is the number of positive integers less than or equal to "r" that are relatively prime to "r". In the context of Shor's algorithm, "r" is the order of the randomly chosen number "a" modulo "N". Therefore, the total number of possible states |c, xk (mod n) > that enable us to compute "r" is rφ(r). This information is used in the analysis of the algorithm's complexity and success probability.

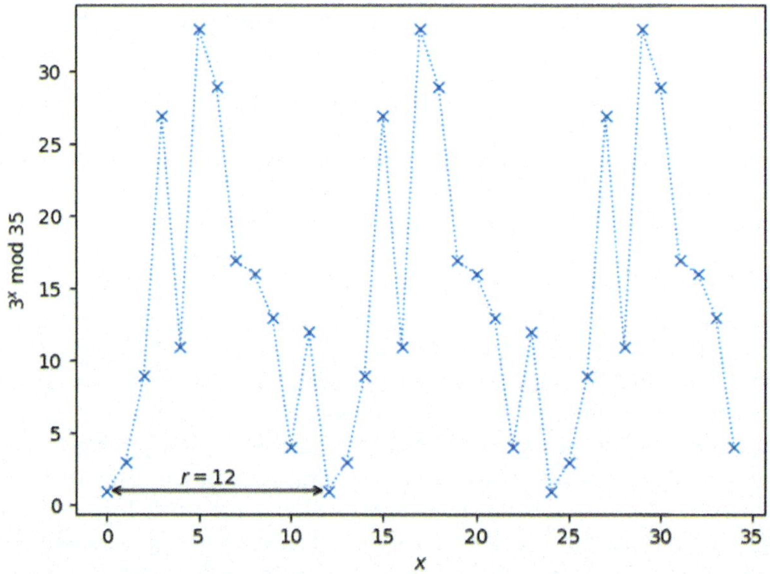

Fig. 2 Period function in Shor's algorithm

The mathematics involved in Shor's algorithm includes modular exponentiation, quantum Fourier transform, and number theory concepts such as greatest common divisor and period finding [3] 4 (Fig. 3).

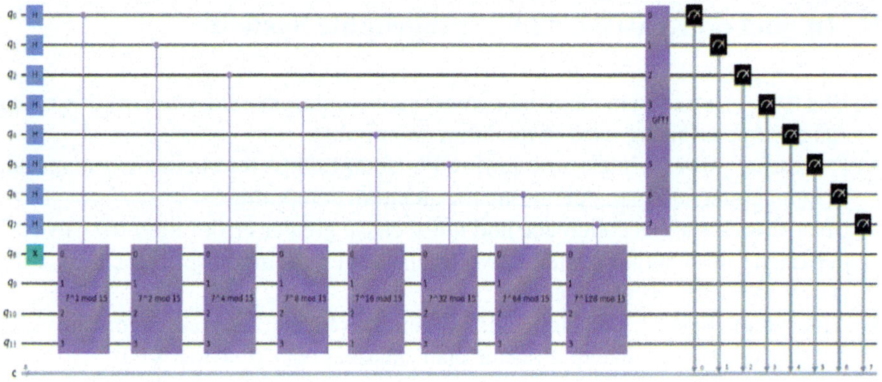

Fig. 3 Shor's algorithm

5 Application of Shor's Algorithm in Decrypting RSA

Shor's algorithm is a quantum algorithm that efficiently factors large numbers, which has significant implications for decrypting RSA encryption. The RSA encryption scheme relies on the difficulty of factoring the product of two large prime numbers. The security of RSA encryption assumes that factoring large numbers is a computationally difficult problem for classical computers.

Shor's algorithm exploits the quantum properties of superposition and entanglement to factorize large numbers exponentially faster than the best-known classical algorithms.

The application of Shor's algorithm in decrypting RSA involves using a quantum computer to efficiently factorize the large number that forms the basis of the RSA encryption key. By finding the prime factors of the large number, an attacker could potentially decrypt the RSA-encrypted message. It's important to note that while Shor's algorithm demonstrates the potential of quantum computing to break RSA encryption, practical implementation of Shor's algorithm for large numbers is currently limited by the scale and stability of quantum computers. However, as quantum computing technology advances, the potential threat to RSA encryption posed by Shor's algorithm becomes increasingly significant.

In summary, Shor's algorithm's application in decrypting RSA demonstrates the power of quantum computing to efficiently solve complex mathematical problems, potentially undermining the security of RSA encryption and other cryptographic systems based on the difficulty of factoring large numbers.

6 Impact on Real-World Cryptographic Systems

- Bits (0 or 1) are used in classical computing to represent data, whereas qubits that can superposition are used in quantum computing.
- Processing power is measured in FLOPS (fetaFLOPS) for classical systems and in qubit count and gate operations for quantum computing.
- Error Correction: Decoherence and noise cause greater error rates (10^{-3} to 10^{-5} per qubit) in quantum systems, while classical systems have lower error rates (10^{-15} to 10^{-18} mistakes per bit).
- Algorithmic Faster Than Classical: For some problems, quantum algorithms such as Shor's exhibit exponential faster than classical ones ($O((\log N)^3)$).
- Computational Properties: Interference, superposition, and entanglement are used by quantum computing to increase computational capacity.
- Cryptographic Implications: The development of quantum-resistant algorithms is necessary because quantum cryptanalysis poses a danger to the security of existing cryptographic techniques.
- Quantum-Resistant encryption is Necessary: In order to protect against quantum attacks, research is concentrated on post-quantum encryption.

- Data Security Issues: Sensitive information sent and maintained by existing cryptographic methods may be subject to long-term security issues in light of quantum cryptanalysis.
- Transition Period: To maintain data security while conventional and quantum technologies coexist, the switch to cryptographic algorithms that are resistant to quantum fluctuations needs to be carefully planned.

7 Summary of Research

See Table 1.

8 Conclusions

In conclusion, the RSA algorithm, a cornerstone of modern encryption, is encountering a significant challenge from the rise of quantum computing and Shor's Algorithm. While RSA's security has long relied on the difficulty of factoring large numbers, quantum computers introduce a novel paradigm that can efficiently factor large integers, posing a threat to RSA's security. The potential of Shor's Algorithm on quantum computers is of paramount concern, as it significantly reduces the time required to break RSA encryption. The asymmetry between quantum and classical computing capabilities is evident, with quantum computers offering exponential speedup and inherent parallelism in factorization tasks.

This dynamic underscores the urgency in developing and implementing post-quantum cryptographic solutions to secure digital communication in the quantum era. As the field of cryptography evolves, addressing the quantum threat is critical to ensuring the continued confidentiality and integrity of data in our interconnected world.

Table 1 Summary of research papers in related domain

Author	Implementation and modifications	Performance and results
Shor [2]	• Exploration of quantum Turing machines and potential speedup over classical computation	• Algorithm for factoring and its reduction
Cao and Liu [5]	• Complexity analysis and probability calculations in Shor's algorithm	• Raises concerns about the soundness of the argument for the success probability estimation in Shor's algorithm
Swathi and Rudra [6]	• Design and implementation of reversible logic gates using quantum gates with IBM QE	• Introduction of quantum gates, employing CNOT, Swap, and Pauli gates in the IBM Quantum Experience
Hamdi et al. [7]	• This paper introduces quantum computing, focusing on Shor's algorithm, and compares it with classical approaches to prime factorization	• In both standalone and distributed scenarios, classical computers, even with the most efficient algorithms, are unable to factor 1000-decimal-digit integers
Humble et al. [3]	• Genetic algorithm-based heuristic methods for test set generation in reversible circuits	• Typical range of bit error probability is 10^{-3} to 10^{-2} per qubit per operation
Barila [8]	• Quantum algorithms outpace classical counterparts in solving specific problems	• Generalization and application of Shor's factoring and Grover's search algorithms
Mailloux et al. [9]	• Exploration of the influence of quantum computing on modern cryptography	• Recent quantum computing (QC) developments pose a significant threat to modern cryptography
Patro and Piedrafita [10]	• Google's quantum supremacy: Sycamore processor, 53 qubits, rectangular lattice, 86 couplers	• Paper overview: "Quantum Algorithms: from Quantum Supremacy to Shor Factorization"
Wang et al. [11]	• Introduction of basic concepts of cryptography and principles of RSA algorithm • Use of C language to simulate the RSA algorithm	• The experimental results showed that the encryption and decryption of plaintext can be realized using the RSA algorithm
Bhatia and Ramkumar [1]	• Comparative analysis of various quantum factoring algorithms • Highlights the potential of Shor's algorithm for breaking RSA encryption	• The paper highlights the potential of Shor's algorithm for breaking RSA encryption • It discusses limitations and alternative techniques for quantum factoring
Wang et al. [4]	• Discussion of existing methods for attacking RSA	• Based on QFT-1 and phase estimation
Soni and Rasool [12]	• Use of MATLAB simulator for implementing Shor's Algorithm for quantum numbers	• RSA algorithm for cipher design and its reliance on the principle of integer factorization

References

1. Bhatia V, Ramkumar KR (2020) An efficient quantum computing technique for cracking RSA using Shor's Algorithm. In: 2020 IEEE 5th International conference on computing communication and automation (ICCCA), Greater Noida, India, pp 89–94. https://doi.org/10.1109/ICCCA49541.2020.9250806
2. Shor PW (n.d.) Algorithms for quantum computation: discrete logarithms and factoring. In: Proceedings 35th annual symposium on foundations of computer science. https://doi.org/10.1109/sfcs.1994.365700
3. Humble TS, Thapliyal H, Muñoz-Coreas E, Mohiyaddin FA, Bennink RS (2019) Quantum computing circuits and devices. IEEE Des Test 36(3):69–94. https://doi.org/10.1109/MDAT.2019.2907130
4. Wang Y, Zhang H, Wang H (2018) Quantum polynomial-time fixed-point attack for RSA. China Commun 15(2):25–32. https://doi.org/10.1109/cc.2018.8300269
5. Cao Z, Liu L (2009) On the complexity of Shor's algorithm for factorization. In: 2009 Second international symposium on information science and engineering Shanghai, China, pp 164–168. https://doi.org/10.1109/ISISE.2009.86
6. Swathi M, Rudra B (2022) A novel approach for asymmetric quantum error correction with syndrome measurement, vol 575025. National Institute of Technology Karnataka, Surathkal, India. https://doi.org/10.1109/ACCESS.2022.3170039
7. Hamdi SM, Zuhori ST, Mahmud F, Pal B (2014) A Compare between Shor's quantum factoring algorithm and General Number Field Sieve. In: 2014 International conference on electrical engineering and information & communication technology. https://doi.org/10.1109/iceeict.2014.6919115
8. Barila A (2014) From classical computing to quantum computing. In: 2014 International conference on development and application systems (DAS). https://doi.org/10.1109/daas.2014.6842455
9. Mailloux LO, Lewis CD II, Riggs C, Grimaila MR (2016) Post-quantum cryptography: what advancements in quantum computing mean for IT professionals. IT Prof 18(5):42–47. https://doi.org/10.1109/mitp.2016.77
10. Patro S, Piedrafita A (2020) An overview of quantum algorithms: from quantum supremacy to Shor factorization. In: 2020 IEEE international symposium on circuits and systems (ISCAS). https://doi.org/10.1109/iscas45731.2020.9180793
11. Wang Z, Wu S, Gu Z (2019) A RSA algorithm simulation method using C language. In: 2019 3rd International conference on electronic information technology and computer engineering (EITCE). https://doi.org/10.1109/eitce47263.2019.9094826
12. Soni KK, Rasool A (2018) Cryptographic attack possibilities over RSA algorithm through classical and quantum computation. In: 2018 International conference on smart systems and inventive technology (ICSSIT)

Remote Fish Nourishing and Water Quality Tracking Using a Mobile Application

Singamaneni Rajesh, Deepika Jaya Sri Bezawada, and Venkata Mamatha Kasireddy

Abstract Generally, in many households, offices, and private spaces it is common to have an aquarium. However, a major problem occurs when dealing with monitoring and feeding of fish in the absence of an aquarium caretaker. There are many automatic fish feeder devices that can be used by setting up it with specific instructions. These devices are not operated remotely. There is no specific mobile application for both water quality checking and feature of fish feeding. The research presents an innovative approach to the development of an automatic fish feeder system integrated with a mobile application. Taking data from temperature and turbidity sensors, the system uses advanced machine learning techniques, specifically Long Short-Term Memory (LSTM) networks, to analyze environmental conditions and adapt feeding schedules accordingly. This system allows users to schedule feeding sessions, adjust portion sizes, and ensure the well-being of their aquatic pets. Thus, this project brings up a modernistic approach by connecting fish feeders with IoT devices and mobile applications.

Keywords Remote Feeder · Temperature sensor · Turbidity sensor · LSTM · Schedule feeding sessions · Water quality monitoring

---

S. Rajesh (✉) · D. J. S. Bezawada · V. M. Kasireddy
Velagapudi Ramakrishna Siddhartha Engineering College, Vijayawada, Andhra Pradesh, India
e-mail: srajesh@vrsiddhartha.ac.in

D. J. S. Bezawada
e-mail: 218w1a5408@vrsec.ac.in

V. M. Kasireddy
e-mail: 218w1a5424@vrsec.ac.in

V. K. Gunjan and J. M. Zurada (eds.), *Proceedings of 5th International Conference on Recent Trends in Machine Learning, IoT, Smart Cities and Applications*, Lecture Notes in Networks and Systems 1181, https://doi.org/10.1007/978-981-97-8861-3_31

1 Introduction

Various fields like industries and farming have been transformed by technology. It has significantly reduced the need for human intervention. In India 961.7 thousand people possess aquariums, but if they are distant, they are unable to fees, leading to a decrease in their usage. The utilization of this advanced technology and alterations in the fish feeding system's structure resulted in the creation of remote fish feeders. This development was very useful to the people who are rearing fish as a hobby and for the people who are involved in aqua farming. This feeder helps people when they actually want to feed their fish but can't do it because of their absence. Apart from feeding, the environment around the fish should also be monitored. Temperature fluctuations and variations in turbidity, indicative of water quality, can directly impact the fish activity, metabolism, and feeding habits. Therefore the integration of modern technologies, such as temperature and turbidity sensors, along with machine learning algorithms like Long Short-Term Memory (LSTM) networks, helps in the modern fish-feeding approach metabolism. Also proper feeding schedule with fixed proportions of feed is very crucial as living organisms need a suitable amount of food at the right time to sustain. Various species of fish have distinct dietary requirements determined by their biomass. This was done according to the number of servo motor rotations. The present paper suggested and designed a microcontroller for the device that monitors water quality with the help of temperature and turbidity sensors, maintaining and ensuring the welfare of the fish while also monitoring the environmental quality within the aquarium.

2 Literature Survey

These days, the number of fish keepers is constantly growing. This makes the situation hard for the caretaker of the fish to continuously monitor the aquarium. Authors in [1] explained that the well-being of fish depends on the proper maintenance of the aquarium [2] in the context of water and light. The system can be accessed through mobile phones. The history of data and status of the aquarium are shown in [3]. LDR sensors are installed here to measure turbidity levels. The previous system needed to manually monitor and regulate the fish feeding, lighting, heating, oxygenation, and related aspects. Generally fish consume a small quantity of food and also it is fed two or three times a day, excluding some rare fish which have low chemicals and are fed only a single time a day. Authors in [4] explained that low temperature indicates a low digestive. The authors in [5] devised an automated fish feeder that relies on microcontroller technology. The food is dispensed in time intervals to reduce human labor. In addition to this, a mobile application can be an extension to make it easy. Maintaining the water quality of the aquarium is the most defining feature. Reference [6] includes several factors such as temperature, salinity, ammonia, chlorine, hardness, pH, and dissolved oxygen. Many filtration methods are discussed. Reference [7] suggests using the LSTM model to find the predictions in the water quality parameters which are taken from the sensor readings. This has a very large scope and the model can be trained on different water samples. Sensors are deployed and included in the [8] that measure parameters like dissolved oxygen, temperature, pH, etc. A simplified device can be created using only required parameters like turbidity and temperature. Reference [9] takes sensors as their input and operates fish feeding using an ESP8266 module for communication [10]. All these can be included in a mobile application to make it easy. ATMEGA 328 module also combined to the control system. Reference [11] is equipped with IoT components to operate the aquarium regarding water quality. Light and dissolved oxygen are monitored for water quality track. An automatic feeder can be added to make it more useful. Using deep-learning techniques, based on the movement of water [12] it is decided on whether to feed or stop. In addition to this computer vision, adding water monitoring and automatic fish feeding will make a complete smart aquarium.

3 Methodology

The recommended solution operates an automatic fish feeder and monitors the water quality parameters in an aquarium. The proposed solution includes both hardware implementation, software implementation, and device design. The entire hardware covers sensors for reading the data about temperature and turbidity. The design was divided into phases. The model training on the sensor data is also included in the software implementation.

3.1 Hardware Implementation

ESP32. ESP32 is a dual-core Xtensa 32-bit LX6 microprocessor and integrated Wi-Fi and Bluetooth makes it applicable for a broad spectrum of Internet of Things (IoT) applications[10]. As it is designed to be more power-efficient GPIO handling and precise pulse-width modulation (PWM).

Temperature Sensor. The degree of warmth influences the rate of synthetic and organic reactions. It impacts the concentration of dissolved oxygen in the water, the photosynthetic processes of aquatic plants, and the metabolic rate of fish. To convert sensor readings.

$$Temperature(^{\circ}C) = Digital\ Value \times (0.0625/2^{12}) \tag{1}$$

Turbidity Sensor. An elevation in the levels of total suspended solids and total dissolved solids in water leads to a corresponding increase in turbidity. If a is the regression-estimated coefficient and b is the regression-estimated coefficient, approximately equal to 1. The mg/L measurement is converted into NTU.

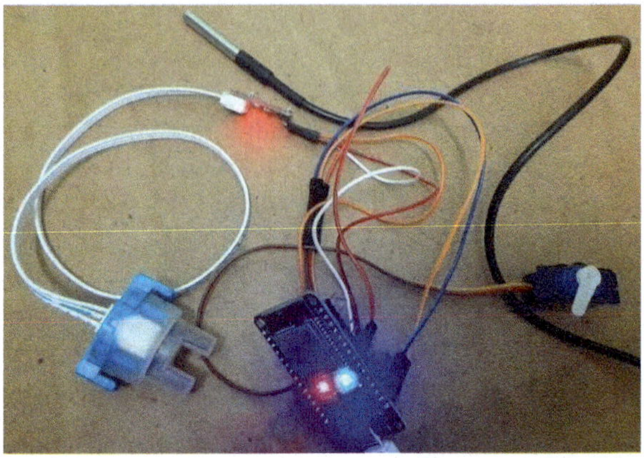

Fig. 1 Connections of sensors and motor with ESP32

$$NTU = a * TSSb \tag{2}$$

Servo Motor. A servo motor is an electric motor that can be controlled with a high degree of precision. In the context of automatic feeders, the servo motor is programmed to dispense food. The amount of food that should be dispensed per rotation should be fixed and finally, the total amount of food dispensed should be calculated.

$$F = R * D \tag{3}$$

The below Fig. 1 shows the overall connections of two sensors and the servo motor to the micro-control. VCC, ground, and output pins of the turbidity sensor are connected to the 5 V, ground, and vp pins of ESP32 respectively once this is done turbidity values will be displayed on the serial monitor. Similar connections are made between the temperature sensor and ESP32. Finally, the D5, 5 V, and ground pins of the ESP32 are connected to the 3 provided pins of the servo motor. These entire connections will help us to monitor the temperature and turbidity values along with the rotation of the servo motor which helps in dispensing food.

3.2 Software Implementation

The entire software implementation part will focus on mainly the Long Short-Term Memory (LSTM) algorithm as well as the developed app.

LSTM Algorithm. The main motive of this LSTM network is to Overcome the drawbacks of the traditional recurrent Neural networks (RNNs). In traditional RNN the network's weights don't change accordingly thus the entire network will fail to identify the underlying patterns in it. But that's not the case in LSTM. The main components of the LSTM network are a memory cell, an input gate, a forget gate, and an output gate. The memory cell helps in maintaining and storing the information for a long time thus updating the network weight correctly at the time of backpropagation. The information flow in the network is achieved using the gates. Thus selectively updating and utilizing the information based on the given input sequence. The LSTM network is trained on the dataset which contains the temperature and turbidity parameters of the water. And the final class label is, is water quality good or normal or bad? After the complete training, validation, and testing of this LSTM model [7], it is deployed into the app. The continuous values from both the sensors are monitored and given to this model so that it can predict the water quality.

The output from the model will decide whether the fish has to be fed or not because the water quality is an essential feature while feeding the fish. Also, the graphs are drawn between the temperature/turbidity values on the Y-axis and Time (in hours) on the X-axis. The graphs are obtained according to the comparison of both predicted and actual values. The below Figs. 2 and 3 depict the images of graphs.

Communication Protocol. In order to monitor the values from the sensors and display them in the Flutter app there must be a bridge between the ESP32 microcontroller and the App. And it is achieved through the communication protocols. Different protocols are used for the communication. However, this paper makes use of the Hypertext Transfer Protocol (HTTP). Specifically HTTP_GET requests. The asynchronous web server is set up and when the mobile app accesses the "/ toggleServo" endpoint with the help of an HTTP GET request, the server invokes a function that helps toggle the state of a servo motor on ESP32, changing its position based on its current state. On the ESP32 when it receives the same HTTP GET request it responds with a message about the servo motor position. The responses convey the outcome of the servo state change, facilitating straightforward and effective communication between the Flutter app and the ESP32 microcontroller.

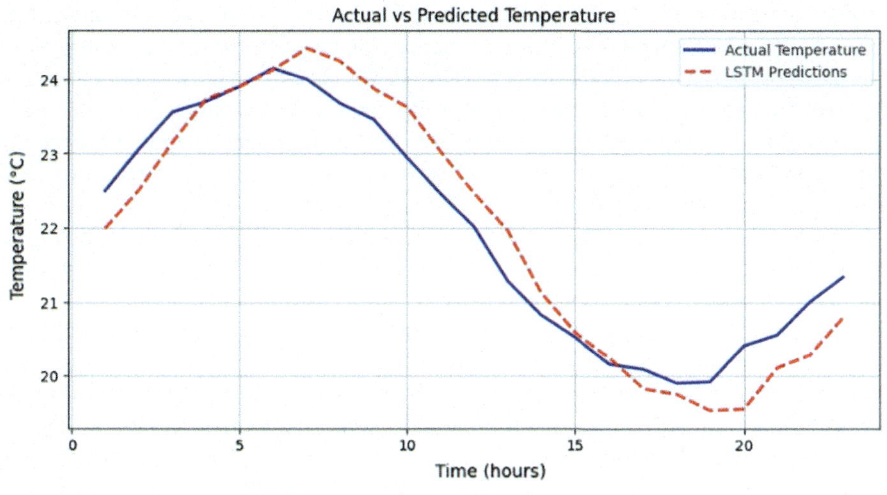

Fig. 2 Turbidity predicted and actual values

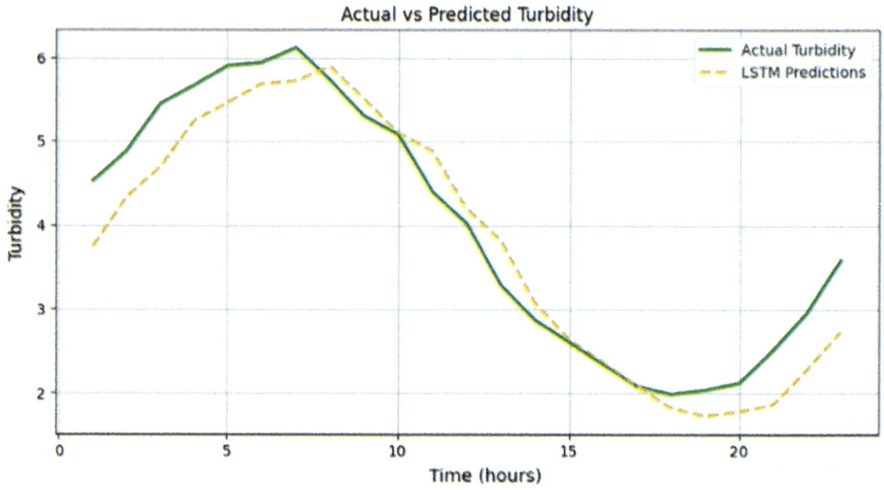

Fig. 3 Temperature predicted and actual values

Protocol: HTTP(Hyper Text transfer protocol).

Endpoint: "/toggleServo".

Method: HTTP_GET.

Data Format: Plain text("Servo state:ON" or "Servo state:OFF").

The App opens up with the option called Login and registers pages. By providing their credentials /they may either log in or if they are new users they have to register first.

By importing all the added dependencies into the main file and using these keys a connection is established. When the user tries to log into the app the details are authenticated using the signInWithEmailAndPassword method. The amount of food that is going to be dispensed also plays a crucial role. In order to meet that requirement this paper proposed a mathematical calculation. We have to set the amount of food to be dispensed per rotation of a servo motor. By using below formula we can achieve it. The below Fig. 4 shows the image of how the total amount of food dispensed is calculated. The values may differ according to the fish type and the amount of food dispensing per rotation.

Another motive of this paper was to allow users to set a fixed time at which food should be dispensed into the aquarium. This can be achieved using the set feed time button. The Flutter code is designed for scheduling feed times in an automatic fish feeder. When the user interacts with the SetFeedTime button, a time picker dialog is presented, enabling them to select a specific time. The chosen hour and minute are then captured for further processing. When selecting a time, the program invokes the schedule function with the recorded hour and minute. This function determines the DateTime for the selected time on the current day and calculates the time difference between the present moment and the scheduled time. Following this calculation,

Fig. 4 Rotations verses food dispensed

Servo Motor Rotations (R)	Amount of food dispensed per rotation(in Grams) (D)	Total amount of food dispensed (in Grams) F = R * D
1	5	5
2	5	10
3	5	15
4	5	20

the program asynchronously waits until the scheduled time arrives. Once the scheduled time is reached, an action is triggered, potentially involving the control of a servo motor, indicated by the toggleServo. This introduces an asynchronous delay, which helps in suspending execution until the scheduled time is reached. Thus the integration of temporal calculations, asynchronous delays, and servo motor control mechanisms collectively facilitates the orchestration of automated and scheduled time feeding of fish through a mobile app.

3.3 *Workflow*

In this streamlined workflow, power flows into the feeder, energizing the ESP32 module at its core. The ESP32, serving as the brain of the operation, efficiently processes data from the inward-connected temperature and turbidity sensors. The data is analyzed using the LSTM algorithm and given the respective graphs.

These sensors work in tandem to ensure the optimal environmental conditions for feeding. Simultaneously, the ESP32 communicates with the outward-connected servo motor, precisely controlling the feeder's operation. This synchronized dance of data and instructions ensures a seamless and controlled release of feed. Crucially, the ESP32 acts as the bridge, connecting the entire system to a user-friendly app. Through this connection, users can remotely monitor and manage the feeder [3], making adjustments on the go for a personalized and responsive feeding experience. The below Fig. 5 shows the entire workflow diagram of this paper.

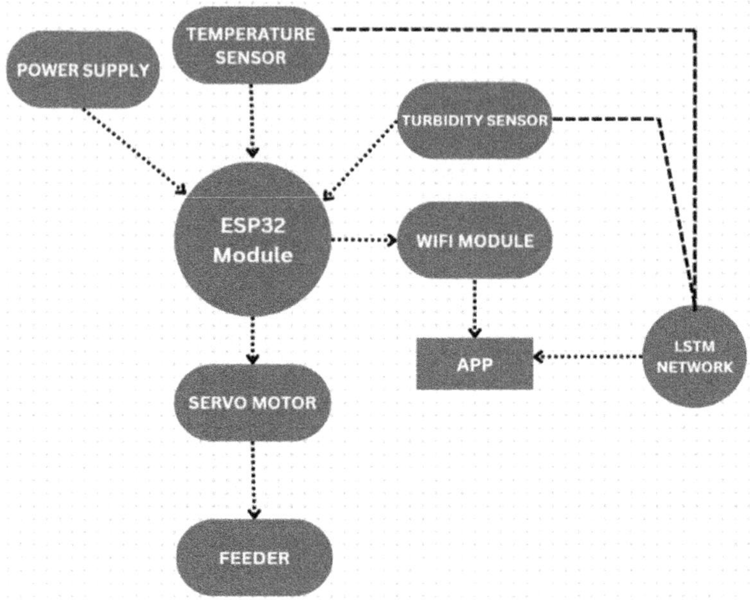

Fig. 5 Work flow

4 Result

The results of the automatic fish feeder along with controlling it through the app were shown. An aquarium monitoring and control system live utilizing IoT technology has been developed. Graphs are shown by comparing both actual and predicted values. Users can watch those graphs and whether to feed the fish or not. The schedule time along with the quantity setting option is given. The sensors which were placed in the aquarium are continuously monitoring the water quality and record the readings. These readings are then stored in a database and the code which was written helps to reflect these values in the designed app. The feeder model (Fig. 6) is prepared by connecting all the sensors to the aquarium and the feeder. The food is dispensed according to the quantity set in the scheduled or required time through the trigger given to the servo motor in the code.

Fig. 6 Feeder model

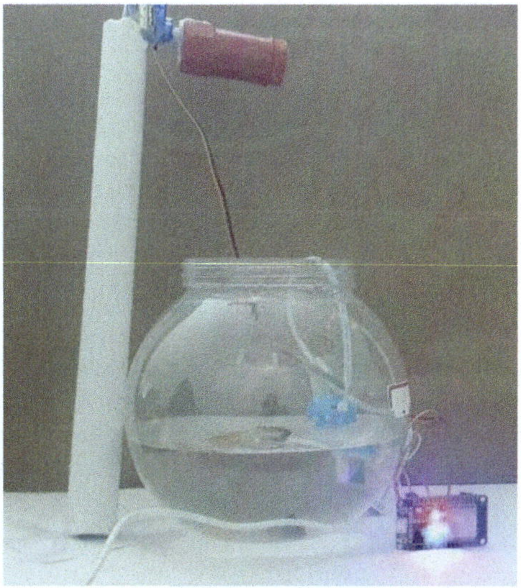

5 Conclusion

The paper presents an automatic feeder where we can feed our fish at any time according to the user's requirements without their actual intervention. The LSTM network model is trained on a large amount of temperature and turbidity data. After the data from the actual sensors are collected and given to the network, it will validate the data and predict the water condition. Thus the continuous monitoring of water quality is achieved. The graphical view depicts the entire process in a very understandable manner so keep in mind this paper also displays all the sensor data on to the app. Through the app, we can control the feeder and feed the fish. Also, a feeding scheduler was designed such that food is dispensed according to the time the user has set. Thus it overcomes the problem of the absence of remote operation in the existing devices.

6 Future Work

• Add deep learning algorithm to analyze the fish behavior to know whether feeding is needed or not. This can reduce over-feeding and insufficient feeding.

• Improving more features in the app like choosing the different fish in the aquarium to enable separate operations for each kind.

• Replacing the power supply with a solar system makes the proposed system more cost-efficient. This can be more eco-friendly.

• More water quality parameters like pH, dissolved oxygen, and also some work can be done including water motor and light.

References

1. Aakash S (2020) Smart aquarium management system. Adv Parallel Comput 37:523–527. https://doi.org/10.3233/APC200196
2. Ma Y, Ding W (2014) Design and implementation of aquarium remote automation monitoring and control system. In: Li D, Chen Y (eds) Computer and computing technologies in agriculture VII. CCTA 2013. IFIP advances in information and communication technology, vol 419. Springer, Berlin, Heidelberg. https://doi.org/10.1007/978-3-642-54344-9_12
3. Devi R, Rodhiyah M, Arif P (2021) IoT prototype development of automatic fish feeder and water replacement. Int J Eng Technol Nat Sci 2:51–55. https://doi.org/10.46923/ijets.v2i2.71
4. Mahendra Y, Anurag S (2021) Food and feeding of ornamental fishes. May edition 17–18
5. Liz O, Achonu A, Ibrahim J (2011) Design and construction of a microcontroller-based automatic fish feeding device. Neighbor Algorithm Learn Cllfication 1255–1260. https://doi.org/10.1109/ICCS45141.2019.9065747
6. Subhendu D (2014) Aquarium water quality management. https://doi.org/10.13140/2.1.2747.6164
7. Ping L, Wang J, Sangaiah AK, Xie Y, Yin X (2019) Analysis and prediction of water quality using LSTM deep neural networks in IoT environment. Sustainability 11(7):2058. https://doi.org/10.3390/su11072058
8. Lin Y-B, Tseng H-C (2019) FishTalk: an IoT-based mini aquarium system. IEEE Access 7:35457–35469. https://doi.org/10.1109/ACCESS.2019.2905017
9. Hardyanto R, Ciptadi W, Prahenusa, Andik A (2019) Smart aquarium based on internet of things. J Bus Inf Syst 1:48–53 (e-ISSN: 2685-2543). https://doi.org/10.36067/jbis.v1i1.12
10. Marek B, Petr F, Pavel S (2019) Using the ESP32 microcontroller for data processing. 1–6. https://doi.org/10.1109/CarpathianCC.2019.8765944
11. Amrita et al (2021) IoT based smart water quality management in aquarium. Biot Res Today 3:000–000
12. Atoum Y, Srivastava S, Liu X (2015) Automatic feeding control for dense aquaculture fish tanks. IEEE Signal Process Lett 22(8):1089–1093. https://doi.org/10.1109/LSP.2014.2385794

Evaluating Women's Fertility Rate Using Machine Learning and Statistical Tools

Amrita Jha and Harpreet Kaur

Abstract Fertility, the likelihood of live births, is crucial for human existence and long-term development. In India, migration, mortality, and fertility impact population dynamics. Understanding fertility rates, trends, and differences is essential for effective political management and long-term development. The study examines fertility rates in India between 1950 and 2020 using Python programming. It explores the impact of migration, mortality, and fertility on population dynamics, predicting a potential population decline. Low birth rates can strain governmental finances and hinder economic progress, leading to increased GDP per capita.

Keywords Fertility rate · Specific age rates of fertility · Rate of overall fertility · Rate of gross reproduction · Linear regression · Correlation · Trends

1 Introduction

The term "fertility" describes a collection of females' real reproductive performance. Everyone has always had serious concerns about women's fertility. Fertility is the likelihood of live births. 35% of infertile couples have variables affecting both the male and female sexes [1]. It has an impact on every facet of human existence. Analysis of fertility rates, trends, and differences is crucial for effective political management and the long-term development of all societal segments, not just in the context of the widening gap between the supply and demand for resources (economic growth) and population growth (population growth). Migration, mortality, and fertility all have a big impact on population dynamics.

The project is centered on utilizing the Python programming language to compare the reduction in fertility rates in the recent past and the present. The data on fertility from 1950 to 2020 have been examined. The information is drawn from https://ourworldindata.org/grapher and covers the years 1950 to 2020. The NFHS-5 (National

A. Jha (✉) · H. Kaur
Department of Mathematics, Chandigarh University, Gharuan, India
e-mail: amritajha598@gmail.com

V. K. Gunjan and J. M. Zurada (eds.), *Proceedings of 5th International Conference on Recent Trends in Machine Learning, IoT, Smart Cities and Applications*, Lecture Notes in Networks and Systems 1181, https://doi.org/10.1007/978-981-97-8861-3_32

Family Health Survey), whose results serve as the project's foundation, is a key component. To conduct our analysis for this report, we focused on the population's (fertility rate). The total fertility rate (TFR) in India has dropped by thirty-eight percent over the past 28 years as women are becoming more educated and having fewer children. The TFR for the quintile with the lowest wealth is children as opposed to 1.6 children. Female sterilization continues to be the most common method of contraception, with the median age at first birth for women aged 25–49 who have no formal education or fewer than five years being less than 19.9 years. In the last 28 years, there has been a 56% reduction in infant death rates [2].

2 Research into a Population Trend

The study of fertility and mortality trends indicates upward or downward trends in a community, country, or area, requiring a structured approach to analyze changes in size, distribution, and characteristics:

- The research aims to understand the methodology, and current knowledge on a demographic trend, identify gaps, and guide future studies through a comprehensive literature review.
- Collect data from various sources like government statistics, surveys, and censuses, and analyze it using statistical analysis and data visualization to identify trends and patterns.
- The analysis provides insights into demographic trends, revealing notable alterations, trends, or contributing elements.
- The study examines population trends influenced by various factors like societal changes, laws, and economic conditions. It investigates potential reactions from individuals, organizations, and policymakers, and suggests policy recommendations based on the findings. It also explores factors affecting the public.
- Research must adhere to ethical standards, safeguard privacy, and implement ongoing monitoring and updating systems to ensure legitimacy, accuracy, and up-to-date population trends.

3 Fertilisation Rate

One can quantify the number of oocytes fertilised by sperm cells using the fertilisation rate. A complete failure of fertilisation occurs when there is no oocyte fertilisation in one cycle, or a zero fertilisation rate. The calculation involves dividing the percentage of fertilized eggs by the total number of eggs treated. Conception rates are impacted by sperm and egg quality, in vitro fertilization, lifestyle choices, and health issues. Studies on fertility pinpoint the variables that affect conception and provide strategies for addressing them. The likelihood of a safe pregnancy is not, however, correlated with fertility rate.

4 Rate of Total Fertility

An estimated population's RATE OF TOTAL FERTILITY is the mean number of offspring a woman has throughout her lifetime if.

- They were to live their entire lives experiencing the precise present specific-age fertility rates.
- and Their life span was to begin at birth and end at the end of their ability to reproduce [2].

  It is calculated by adding together the age-specific single- year rates at a certain moment.

  The rate of total fertility does not take into account the fertility of any actual group of women because doing so would require them to wait until they had finished having children. It is also not predicated on calculating the total number of children they have in reality during their lifespan.

  Demographic measure predicts a woman's expected childbirth, influencing population growth and stability. Factors include economic conditions, social norms, healthcare, family planning, government policies, and cultural values.

  Policymakers, demographers, and demographic trends depend on the rate of total fertility because they offer information on potential population growth or decline and how it may affect the labor force, healthcare, and education.

5 Rate of Replacement

In order to maintain population stability over time, a woman should typically have the number of children known as the replacement level fertility rate [3]. In industrialized nations, it is typically 2.1 children per woman. Factors like family planning access, cultural norms, economic conditions, and government policiescontribute to population growth. Understanding TFR is crucial for policymakers, but population dynamics are influenced by immigration, emigration, and age structure (Fig. 1).

6 Fertility Levels

Fertility level is the quantity of live births within a community, measured by the general fertility rate [19]. It is closely related to birth rate and is commonly measured using the rate of total fertility.

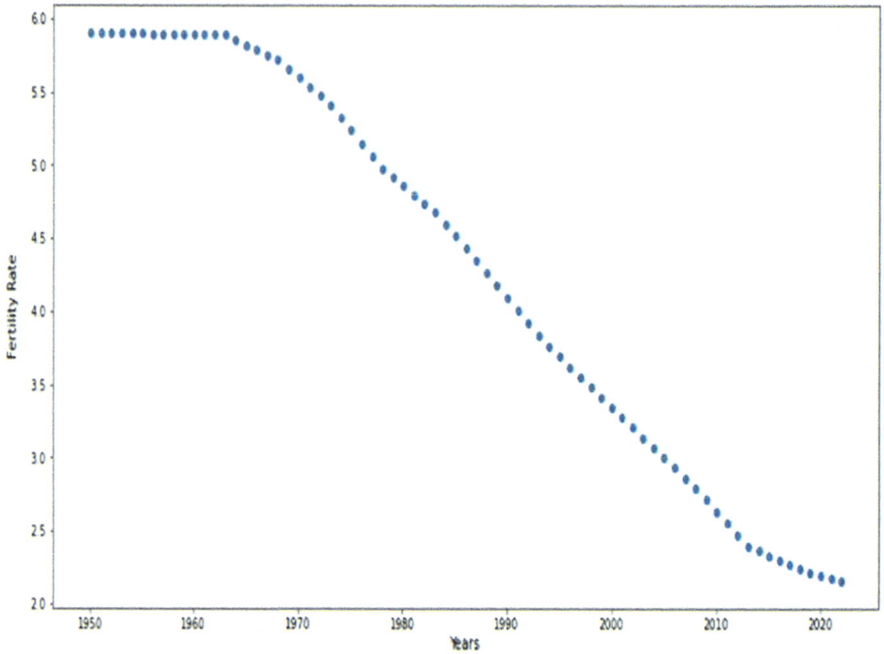

Fig. 1 Rate of replacement

Fertility levels, influenced by factors like the rate of total fertility, specific age fertility rate, etc. reflect population growth. India's fertility rate has dropped below the replacement ratio, with 1.6 children per woman in Delhi. Factors like religion, caste, and education impact fertility differently (Fig. 2).

7 Why is the Fertility Rate Declining?

Fertility decline is influenced by various factors.

7.1 Higher Level of Education Among Females

Lower fertility rates are linked to women with higher education levels, affecting fertility rates across various cultures. Education empowers women, reduces gender wage gaps, improves child health, and promotes gender equality. Educated women participate in decision-making and community leadership, leading to social and cultural change. Nonetheless, fertility varies between women who have received

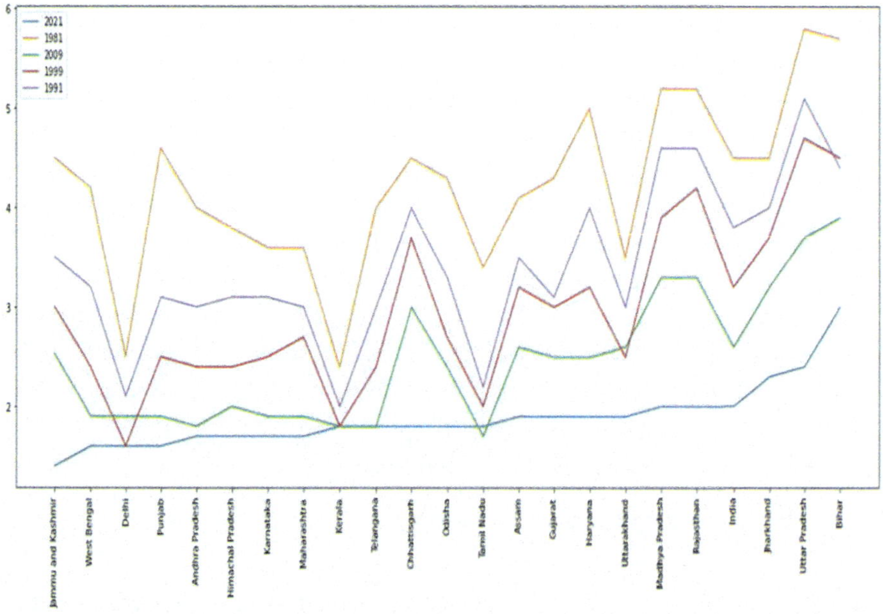

Fig. 2 Fertility level

more and less education in almost every nation, while it is unclear exactly what mechanism causes the rate of total fertility to decrease with increasing education [4].

7.2 Greater Availability of Family Planning Techniques and a High Prevalence of Contraception

Family planning, despite cultural and familial barriers, has significantly reduced reproduction rates worldwide. It improves maternal health, reduces unwanted pregnancies, powers individuals to make informed reproductive decisions, and enhances educational opportunities. Smaller families can invest in their children's education, contributing to economic empowerment. Family planning also aids in population control, environmental sustainability, gender equality, and health-care resource distribution. Addressing cultural, social, and economic barriers is crucial (Fig. 3).

7.3 Financial Self-Reliance

Higher education and job aspirations can contribute to fertility decline in high-income and low-income countries. Financial self-reliance, which allows informed

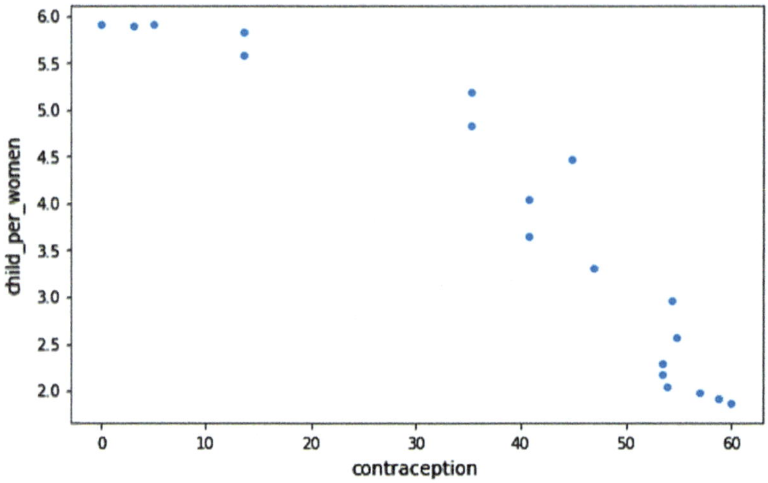

Fig. 3 Family planning techniques and contraception

decisions about childbirth, impacts family planning, reproductive choices, and overall well-being. It contributes to economic stability, reducing unintended pregnancies, improving maternal and child health, and influencing population growth. Governments can promote financial self-reliance in family planning by providing education, employment, affordable healthcare, and family planning services (Fig. 4).

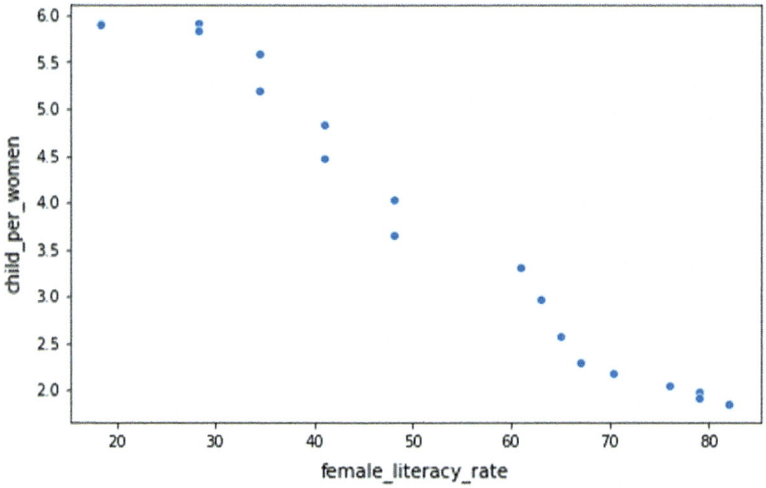

Fig. 4 Financial self-reliance

7.4 *The Decreasing Neonatal Mortality Rate*

The infant mortality rate in NFHS-5 has decreased from 25 to 47 per 1,000 live births, largely due to improved maternal and neonatal healthcare interventions.

- **Increased Access to Healthcare**: Access to healthcare services, especially during pregnancy and childbirth, significantly reduces neonatal mortality by addressing potential health issues through prenatal care, skilled birth attendance, and postnatal care.
- **Better Health for Mothers**: Maternal health, including, passing proper nutrition and antenatal care, can lead to healthier pregnancies and lower neonatal mortality rates, as it is closely linked to newborn well-being.
- **Progress in Health Technology**: Medical advancements and neonatal care, including NICUs, have significantly improved the survival chances of premature or critically ill newborns.
- **Inoculations**: The widespread availability of vaccines for pregnant women and infants can protect newborns from various diseases and decrease neonatal mortality.
- **Professional Birth Attendance**: A skilled birth attendant, like a midwife or healthcare professional, can significantly reduce neonatal mortality by ensuring safe deliveries and prompt responses to complications.
- **Dietary Interventions**: Pregnant women and newborns who receive adequate nutrition can prevent low birth weight and associated complications, leading to reduced neonatal mortality.
- **Health Promotion Campaigns**: Pregnant women and newborns who receive adequate nutrition can prevent low birth weight and associated complications, leading to reduced neonatal mortality.
- **Plans for Community Health**: Community-based health programs and outreach efforts can enhance maternal and neonatal health outcomes by providing education and healthcare services to underserved populations.
- **Enhanced Facilities**: Investing in healthcare infrastructure, such as clean water, sanitation, and electricity, can lead to safer births and lower neonatal mortality.
- **Policies of the Government**: Government policies, investments in maternal and child healthcare, monitoring, and data collection are crucial in reducing neonatal mortality rates.

Society's commitment to improving healthcare, addressing health disparities, and promoting public health has led to a decrease in neonatal mortality rates (Fig. 5).

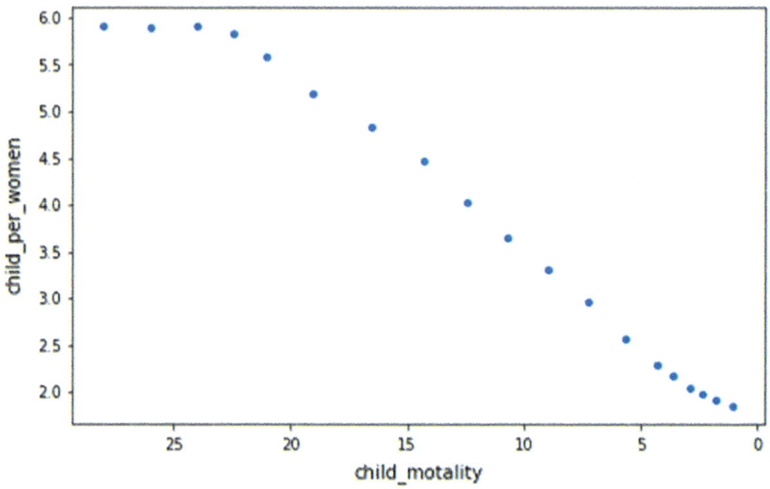

Fig. 5 Decreasing infant mortality rate

8 Variables

Variables affecting fertility rates are crucial for understanding population growth dynamics, aiding researchers, policy-makers, and demographers in studying and analyzing fertility trends. Here are some key variables related to the fertility rate:

- **Rate of Overall Fertility**
  This is the main variable that affects fertility rates; it shows how many children generally a woman should have based on current particular age birth rates.
- **Specific age Rates of Fertility**
  These are fertility rates for specific age groups, calculated in five-year intervals, providing detailed information about childbearing timing.
- **Sub-Replacement Fertility**
  A general reproduction rate below the replacement level may have long-term demographic impacts, such as population decline.
- **Birth Periods**
  Fertility rates are greatly influenced by birth intervals, where shorter intervals lead to higher fertility rates and longer intervals lower fertility rates.
- **State of Marriage**
  Women's marital status, including the percentage of married and single women, as well as cultural norms and family formation regulations, have an impact on fertility rates.
- **Prevalence Rate of Contraceptives**
  The proportion of reproductive-age women who use contraception is a critical determinant of family planning's impact on reproduction rates.

- **Level of Education**
  The educational attainment of both men and women is significantly linked to fertility rates, with higher levels of education, especially for women, resulting in lower fertility rates.
- **Economic Elements**
  Economic conditions, employment opportunities, and income levels significantly impact fertility decisions, often leading to delayed childbearing and smaller family sizes due to economic stability and security.
- **Healthcare Availability**
  By improving mother and child health outcomes, access to healthcare services, such as prenatal care and maternal and child health services, can dramatically increase fertility rates.
- **Customs and Social Standards**
  Cultural and social factors, including gender roles, religious beliefs, and family values, significantly influence fertility behaviors and decisions.
- **State Policy**
  Policies related to family planning, maternity leave, and childcare can influence fertility rates, with government incentives and programs potentially encouraging or encouraging childbearing.
- **Development**
  The transition from rural to urban areas can significantly impact fertility rates due to variations in living conditions, healthcare access, and cultural norms.
- **Fertility Issues and Reproductive Wellness**
  Infertility and reproductive health, including access to fertility treatments, are factors that can impact fertility rates.

  Demographic research is crucial for analyzing variables affecting fertility rates, assessing trends, making predictions, and developing policies to address fertility-related issues.

9 Correlation Between Variables

See Fig. 6.

9.1 Boxcox Transformation

A statistical technique called the Box-Cox transformation is used to reduce variance and improve the normality of the data. Regression modeling and data analysis benefit greatly from its use, especially when handling data that defies the assumptions of continuous variance and normalcy. The transformation has the names of its 1964 introducers, statisticians George E. P. Box and Sir David Roxbee Cox.

It entails applying a regular form to irregular independent variables.

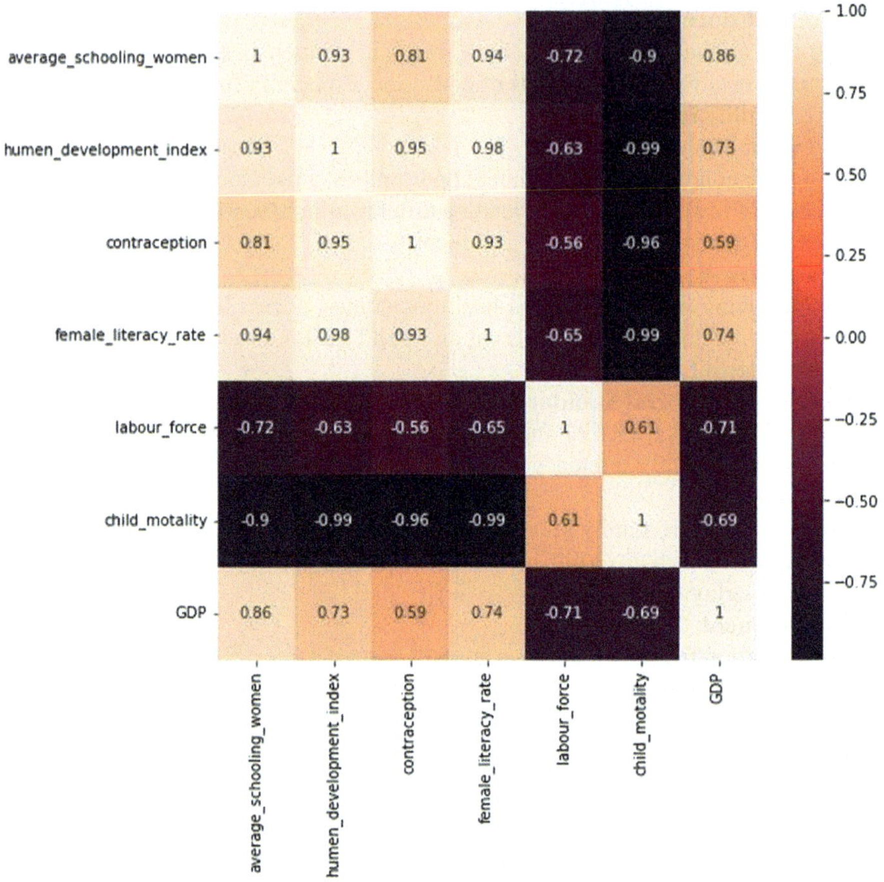

Fig. 6 Correlatoion

Applying a box-cox suggests that you can run more tests if your data is not typical [5]. A fundamental premise of many statistical techniques is normality.

Standardization is an additional scaling technique that centers the values around the mean value of zero with a unit standard deviation.

The formula for Boxcox Transformation (Fig. 7).

where

"y" represents the original data.

Fig. 7 Box-cox transformation

$$\begin{cases} y = \dfrac{x^{\lambda} - 1}{\lambda} \text{ where } \lambda \neq 0 \\ y = \ln x \text{ where } \lambda = 0 \end{cases}$$

"λ" (lambda) denotes the transformation parameter. Any real value can be used for this argument. When λ is zero, the natural logarithm is represented (log transformation).

The Box-Cox transformation maximizes normality and homoscedasticity in data, useful in regression analysis, where normality and constant variance assumptions are often made [6–9].

Steps to Perform a Box-Cox Transformation:

1. Choose a scale of λ Values: To begin, choose a range of values that λ will take into account. Statistical software or manual labor can be used for this.
2. Use the Transformation: Take your data and apply the Box-Cox transformation for every value of λ in the given range.
3. Look Over Normality: Determine the normality of each modified dataset using statistical tests or graphical tools like histograms and normal probability plots. Finding the transformation that produces the most regularly distributed data is the aim.
4. Analyze the homogeneity of variance: Use statistical tests or visual aids like histograms or normal probability plots to evaluate each modified dataset's normality. The transformation that yields the most regularly distributed data is what is sought after.
5. Decide on the Ideal λ: Select the value of λ that maximizes homoscedasticity and normalcy. This is regarded as the ideal Box-Cox transformation value.
6. Applying the chosen λ value

   The Box-Cox transformation is useful for addressing data distribution and variance issues, but it may not be suitable for all datasets. The parameter should be chosen based on practical interpretation, not solely based on statistical significance (Figs. 8, 9, 10, 11, 12, 13, 14).

9.2 Model of Linear Regression

A regression model characterizes variables by fitting a line to the data line, aiming to find a linear function between X and Y variables, assuming Y can be computed using input variables [10].

To forecast results, comprehend correlations, and measure the impact of independent variables, the link between a dependent variable and independent factors is modeled using a linear regression model [11].

Formula of Multiple Linear Regression:- yi = BO + ß1xi1 + ß2xi2 + ... + Boxi + ei Where:

yi is the representative of the dependent or predicted variable.

Xi represents independent variables.

p refers to the coefficient for every independent variable that represents the model's residual term.

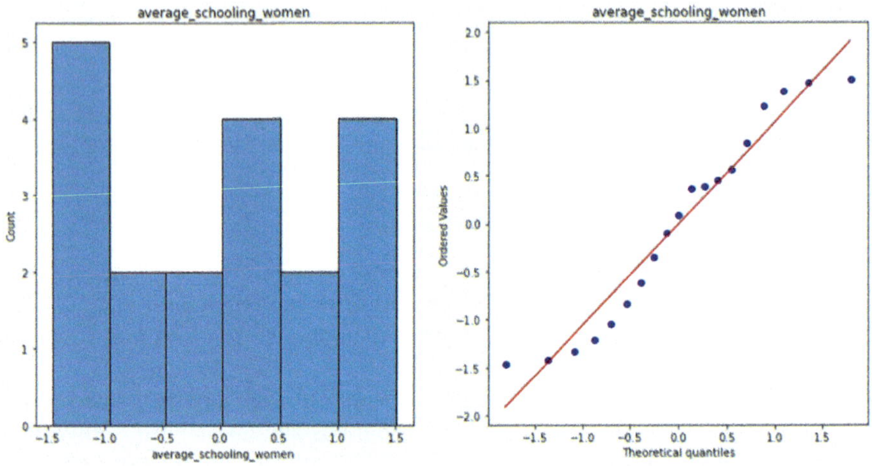

Fig. 8 Graph for average schooling for women using boxcox method

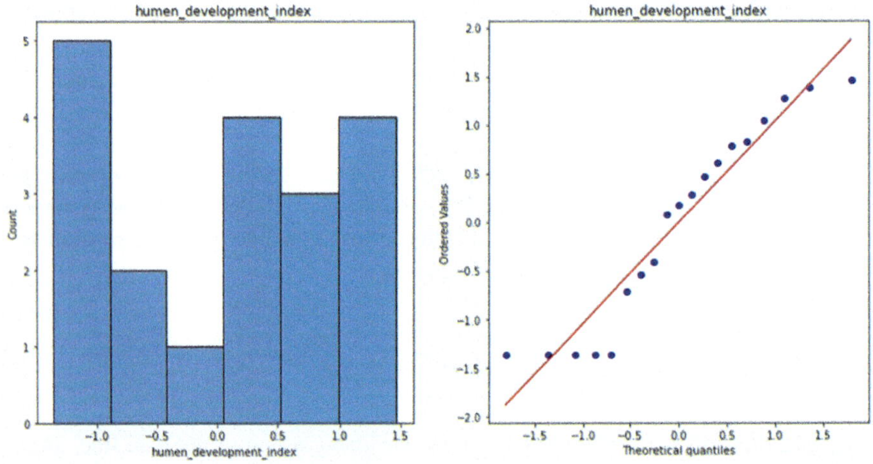

Fig. 9 Graph for human development index using boxcox method

Assumptions

- The independent and dependent variables have a linear connection with one another.
- The independent variables exhibit minimal or no strong correlation, indicating little or no multicollinearity.
- The variance of the residuals remains constant due to the concept of'Hetroscedacity'.
- The observation lacks bias, as demonstrated by the concept of autocorrelation.

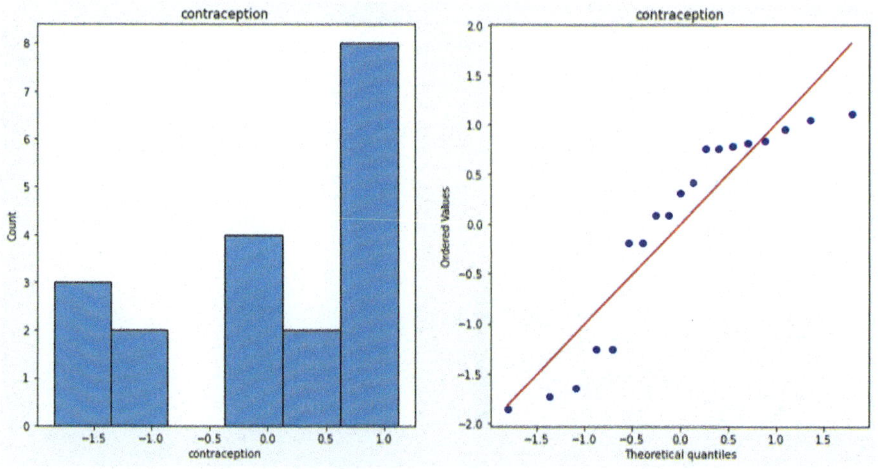

Fig. 10 Graph for contraception using boxcox method

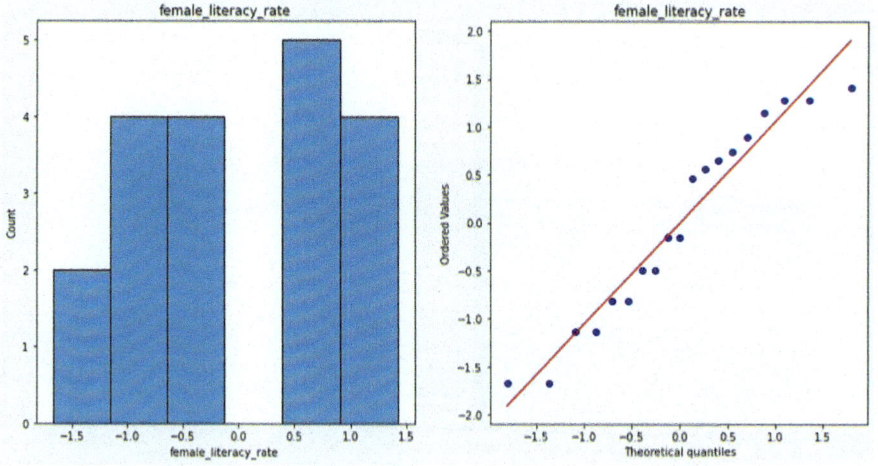

Fig. 11 Graph for female fertility rate using boxcox method

- Multiple-variable normality

  Linear regression models can be structured in various ways.:

Simple Linear Regression

In this, the value of dependent variable can be found by a single independent variable.

Multiple Linear Regression

In this instance, several independent factors are used to predict the dependent variable.

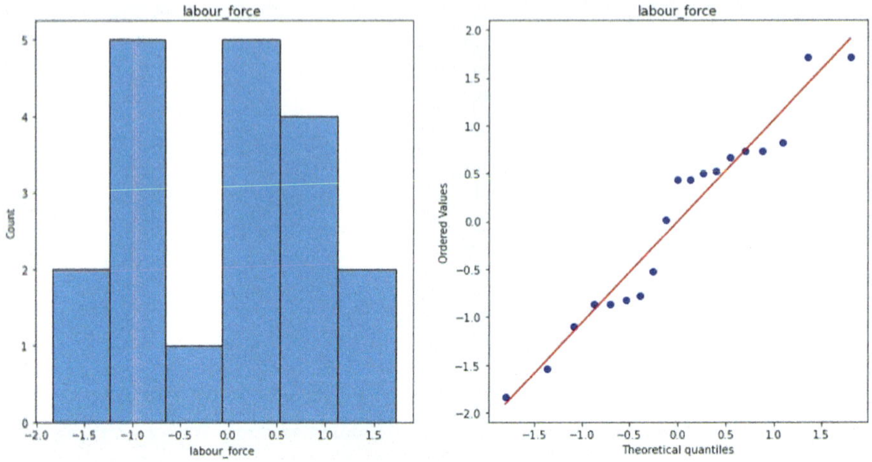

Fig. 12 Graph for labour force using boxcox method

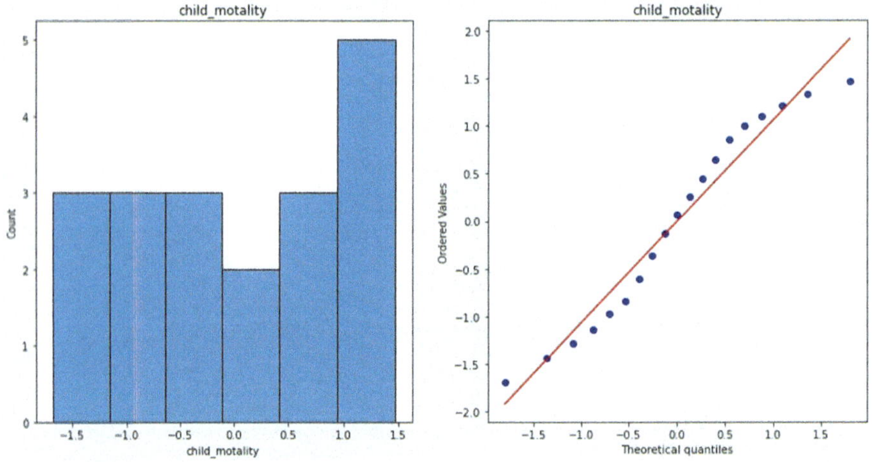

Fig. 13 Graph for child motality using boxcox method

The model is represented as Y = 0 + 1X1 + 2X2 + … + pXp +,
where the symbol p stands for the number of independent variables.

Polynomial Regression

To capture nonlinear relationships such as 2X2 and 3X3, the quadratic regression approach incorporates polynomial terms from independent variables, extending linear regression.

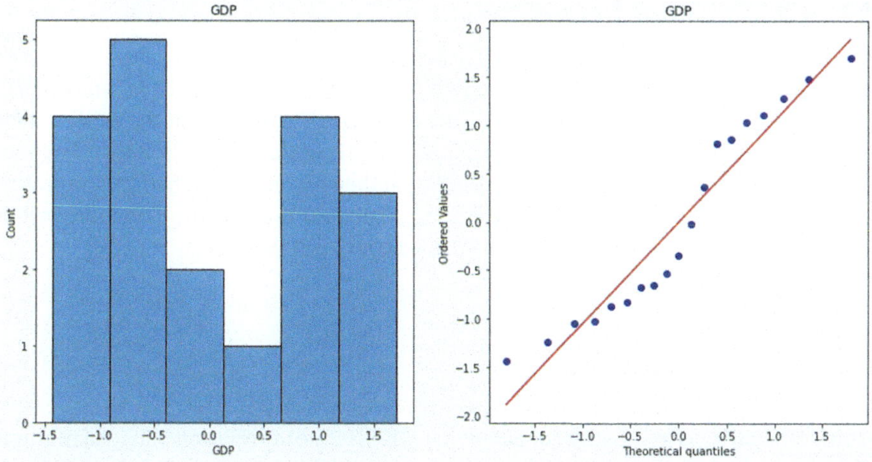

Fig. 14 Graph for GDP using boxcox method

Logistic Regression

Logistic regression is a classification model used for binary classification problems, predicting the probability of a binary outcome based on independent variables.

Linear regression models are widely used in fields like economics, social sciences, and machine learning to analyze and predict relationships between variables, but may not be suitable for complex situations.

9.3 OLS (Ordinary Least Square)

One popular technique for calculating the coefficients of linear regression equations including independent and dependent variables is ordinary least squares regression (OLS).OLS is a technique that minimizes residuals by estimating model coefficients by minimizing squared differences between observed and predicted values. It is popular due to its ease of use and good results [12].

The model of OLS is $Y = XB + E$, where.

The column matrix of error terms, denoted by E, is a crucial component of statistical analysis.

and matrix B represents the actual, but unseen, matrix of regression parameters. In cases where the independent variable exhibits unitary fluctuation, parameter B indicates the variation in the dependent variable.

OLS is a crucial tool in statistics, econometrics, and data analysis, used for understanding and modeling variable relationships, making predictions, and estimating impact. It assumes linear relationships and normal distribution.

9.4 Coefficient of Determination

A model's ability to fit the data is shown by its coefficient of determination, or R squared. It is a part of the variation of the dependent variable that the independent variable can explain and is obtained by squaring the correlation coefficient (r) between the observed and anticipated values. Higher numbers, which go from 0 to 1, represent a better fit [13]. A model with an R squared of 0 always predicts the data mean, while a model with less accurate predictions has a negative R squared.

9.5 Multicollinearity

Multicollinearity is a statistical concept that refers to the correlation between multiple independent variables in a model. A pair of variables is considered completely collinear if their correlation coefficient is less than or equal to 1.0. When independent variables are multicollinear, statistical conclusions will be less reliable [14].

This can cause issues with analysis and lower the accuracy of coefficient predictions. The book emphasizes several crucial multicollinearity concepts.

9.5.1 Multicollinearity's Causes

Multiple sources can lead to multicollinearity, such as:

Data collection issues, model similarities, correlated derived variables, and overlapping factors can affect data accuracy, while the dependent variable is influenced by related factors.

9.5.2 The Multicollinearity Effects

Multicollinearity can lead to various issues.

Coefficient instability and multicollinearity can lead to unreliable estimates, loss of statistical significance, challenging interpretation, and decreased model performance due to individual effects on dependent variables.

9.5.3 Finding Multiple Collinearities

There are various methods for locating multicollinearity, including:

The correlation matrix analyzes correlation coefficients, while the variance inflation factor assesses variance inflated by multicollinearity, and the condition number measures model degree.

9.5.4 Handling Multiple Collinearity

There are various methods to tackle multicollinearity:

Restructuring a model, transforming data, using regularization techniques like ridge and Lasso regression, or using principal component analysis (PCA) can help reduce multicollinearity by removing highly correlated variables.

9.5.5 Analysis of the Findings

Focus on the model's overall significance and relationship direction, not individual coefficients, and consider the practical significance of variables in the research question.

9.5.6 Avoidance

Multicollinearity can be effectively addressed during data collection and variable selection stages, with careful selection of variables preventing or mitigating issues.

Restructuring a model, transforming data, using regularization techniques like ridge and Lasso regression, or using principal component analysis (PCA) can help reduce multicollinearity by removing highly correlated variables.

9.6 Serial Correlation

Serial correlation or Autocorrelation is the degree to which two subsequent time intervals' worth of data are correlated with the same factors. It calculates the relationship between the original value and the lagged value of a variable in a time series [15]. Autocorrelation is another name for the statistical concept of serial correlation.

10 Test and Methods to Remove Multicollinearity

10.1 Regression of Lasso

One method of regularization is a regression of lasso. For a more accurate forecast, regression procedures are not advised. This model capitalizes on shrinking. The process by which data values go closer to the mean is called shrinkage. The lasso approach promotes sparse, simple models [16].

10.2 Test for Autocorrelation

Autocorrelation measures the correlation between two successive time intervals of the same variables, comparing the lagged version of a variable to its original version in a time series.

11 Dimensionality Reduction

11.1 Analyzing the Principal Components

Principal component analysis, or PCA, is used to reduce the dimension of the datasets which are larger in size [17]. It does this by taking a large collection of variables and breaking them down into smaller ones that still has most of the information [18]. Accuracy decreases with fewer variables.

Explained Variation: [0.89511874 0.07802755]

Coefficiennts of first PCA : [0.39549653 0.39361475
0.37413335 0.39431014 -0.29299972 -0.39702389
0.38693264]

Coefficiennts of second PCA : [-0.11264624 -0.16251639
-0.30452502 -0.15482802 -0.91034282
0.06252904
0.10750864]

To forecast the fertility rate's future value, new coefficients have been developed. The updated values are listed below.

[1.69009574 1.37699006 1.13067799 0.82490642].

12 Future Implications

We can see from the forecasted values that the new values are below the replacement rate, which demonstrates how these factors will impact the fertility rate shortly. Similar to affluent nations like Japan, Germany, and Russia, India's low fertility rates will eventually cause a population drop. If fertility falls below replacement level, the population ages and shrinks, which can impede economic growth and create pressure on public finances.

As the fertility rate falls, the GDP per capita rises, which is indirectly proportional to the fertility rate. However, GDP and inflation are inversely related; rising GDP causes rising inflation (Fig. 15).

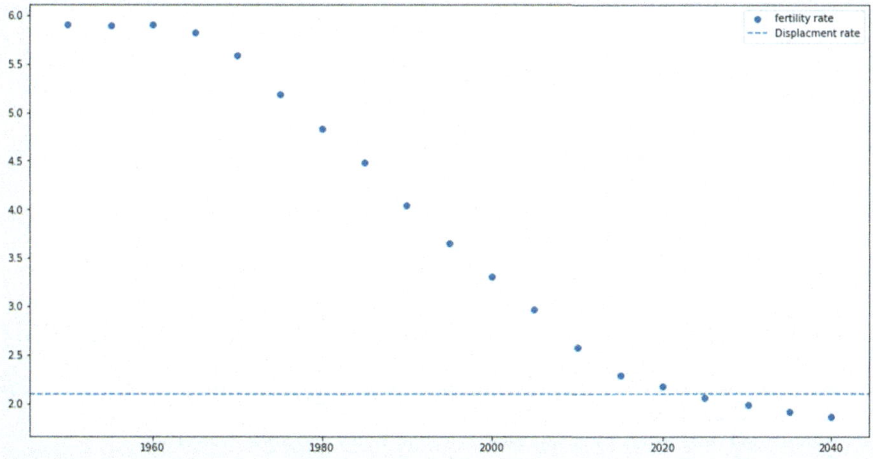

Fig. 15 Principal component analysis

13 Conclusion

The reproductive capacity of females is referred to as fertility, and it is essential for long-term growth and efficient political administration. Using Python programming, the study examines fertility rates between 1950 and 2020. The National Family Health Survey, or NFHS-5, serves as the basis. The impact of migration, mortality, and fertility on India's population dynamics is covered in this research. There may be a population decline since the predicted numbers are less than the replacement rate. Low birth rates can put pressure on governmental finances and inhibit economic progress. GDP per capita increases in direct proportion to declining fertility rates.

Acknowledgements The author expresses gratitude to Mrs. Harpreet Kaur and her family for their invaluable guidance and support during the research paper's completion. Their knowledge and input significantly influenced the investigation's course, and their emotional fortitude provided the necessary emotional fortitude.

References

1. Infertility—Reproductive health—CDC. www.cdc.gov. 2022-03-03. Retrieved 2023–03–02
2. Skirbekk V, Blekesaune M (2014) Personality traits increasingly important for male fertility: evidence from Norway. Eur J Pers 28(6):521–529. https://doi.org/10.1002/per.1936. ISSN: 0890-2070. S2CID 143134400.
3. www.investopedia.com/terms/m/multicollinearity.asp

4. Smarr MM, Sapra KJ, Gemmill A, Kahn LG, Wise LA, Lynch CD et al (2017) Is human fecundity changing? A discussion of research and data gaps precluding us from having an answer. Hum Reprod 32(3):499–504. https://doi.org/10.1093/humrep/dew361.PMC5850610. PMID28137753
5. www.mygreatlearning.com/blog/understanding-of-lasso-regression/
6. Schultz PT (1976) Fertility determinants: a theory, evidence, and an application to policy evaluation. Popul Dev Rev 2(2):293. https://doi.org/10.2307/1972043. JSTOR 1972043
7. NFHS-5 (National Health Survey)
8. Definitions and notes—the world Facebook. www.cia.gov. Archived from the original on 2022-04-18. Retrieved 2023-02-22
9. Lauren Arcuri. August 20, 2019.volusonclub.net empowered women's- health
10. Bongaarts J (2017) Trends in the age at reproductive transitions in the developing world: the role of education. Popul Stud (NY). 71:139–154. https://doi.org/10.1080/00324728.2017.129 1986
11. Box GEP, Cox DR (1964) An analysis of transformations. J R Stat Soc, Series B, 26, 211–252. Available online here
12. Agresti A (1990) Categorical data analysis. John Wiley and Sons, New York
13. Klein G (2013) The cartoon introduction to statistics. Hill Wamg
14. Levine D (2014) Even you can learn statistics and analytics: an easy to understand guide to statistics and analytics 3rd edition. Pearson FT Press
15. machinelearningmastery.com/linear-regression-for-machine-learning/
16. www.investopedia.com/terms/m/mlr.asp
17. December 5, 2023, by Ajitesh Kumar
18. https://corporatefinanceinstitute.com/
19. By TIM SMITH Updated March 19, 2023
20. en.wikipedia.org/wiki/Principalcomponentanalysis

Voltage Stability in Wind-Diesel Adaptive Reactive Control

S. Aravindaraj, P. Sivakumar, A. Aswini, and A. Selvi

Abstract A Wind power technology has revolutionized the energy landscape, emerging as a vital player in the global transition towards sustainable energy solutions. The variability of wind resources and the fluctuating load dynamics in these setups necessitate a sophisticated approach to bolster system stability. To address this, we introduce a pioneering methodology based on UAC strategies, executed through MATLAB-based simulations. This paper is driven by two core objectives: firstly, to improve wind-diesel's dynamic stability microgrids by orchestrating RTC of reactive power, and secondly, to mitigate the impacts of VWP generation. Leveraging advanced control algorithms, we optimize GES (Generator Excitation Systems), enabling meticulous regulation of reactive power flow. Through a comprehensive analysis of WTOs (Wind Turbine Outputs) and LDRs (Load Requirements), our approach adeptly adapts to evolving conditions, ensuring steady VPs (Voltage Profiles) while curtailing power losses. Simulation outcomes affirm the efficacy of our UAC approach, highlighting its capability to maintain desired VP and RP (Reactive Power) levels, even amidst the most challenging VWCs (Variations in Wind Conditions). In order to keep up the stability of a system bus voltage, a second adaptive the voltage compensation the controllers for the static var compensator (SVC) using a double sliding optimal strategy is suggested. This is due to the fluctuation in voltage caused by variable wind energy and reactive load changes of isolated wind-diesel systems.

Keywords Component · Formatting · Style · Styling · Insert

S. Aravindaraj (✉) · P. Sivakumar · A. Aswini · A. Selvi
Rajalakshmi Engineering College, Chennai, India
e-mail: 230911001@rajalakshmi.edu.in

A. Aswini
e-mail: aswini.a@rajalakshmi.edu.in

V. K. Gunjan and J. M. Zurada (eds.), *Proceedings of 5th International Conference on Recent Trends in Machine Learning, IoT, Smart Cities and Applications*, Lecture Notes in Networks and Systems 1181, https://doi.org/10.1007/978-981-97-8861-3_33

1 Introduction

This integration of renewable energy sources, such as wind power, into isolated power systems has attracted considerable interest because of its capability to lower greenhouse gas emissions and increase energy sustainability. Wind-diesel isolated power systems, combining wind turbines and diesel generators, offer a promising solution for providing reliable electricity in remote and off-grid areas. However, the dynamic and intermittent nature of wind power generation, coupled with varying load demands, poses challenges for maintaining stable voltage profiles and ensuring optimal reactive power control. Voltage stability is a critical aspect of power system operation, especially in isolated environments where grid support is limited. The interaction between wind power variability, diesel generators, and varying loads can lead to voltage fluctuations, which, if not effectively managed, can result in system instability and potential blackouts. Traditional methods of reactive power control may not be sufficient to address the uncertainties introduced by wind power fluctuations and load variations. An increasing number of people are interested in exploring advanced control strategies that incorporate uncertainty management and excitation control techniques. The concept of uncertainty-controlled excitation involves adapting the excitation system of generators in response to the uncertain and dynamic nature of wind power generation and load changes. By doing so, it becomes possible to enhance the system's ability to regulate reactive power, maintain voltage stability, and ensure reliable operation under varying conditions. This journal paper aims to delve into the realm within wind-diesel isolated electrical systems using reactive electrical regulation a focus on uncertainty-controlled excitation. The key objectives of this study are twofold: [1] to develop a comprehensive mathematical model that captures the intricate interactions among wind power, diesel generators, and loads, while accounting for uncertainties; and [2] to propose and analyze a novel control strategy that optimizes the excitation systems of generators to enhance reactive power regulation and voltage stability. The value of this research rests in its capacity to offer useful perceptions and remedies for improving the performance and reliability of wind-diesel isolated power systems [3]. By addressing the challenges of uncertainty and dynamic power generation, the proposed approach offers a more robust and adaptable means of reactive power control. The findings of this research have significance for both academics and professionals working in the fields of stability of power systems and renewable energy integration. We will examine the available literature in more detail in the next parts of this paper related to voltage stability, reactive power control, and uncertainty modeling in power systems. Following this, we will present the system description, mathematical model formulation, uncertainty-controlled excitation strategy, stability analysis, and simulation results [4]. Through a thorough exploration of these aspects, we aim to provide valuable insights into the potential of uncertainty-controlled excitation for enhancing reactivity of the balance around the moment of common connection is the objective of this research between the isolated power system network that is powered by wind and diesel [5]. The point of common coupling includes the wind-driven induction generator. According to the availability of the wind pattern (a function of uncertainty), the real power generated by the IG changes instantly. As a result, the reactive power need changes instantly. Voltage fluctuation (sag and swell) results from this and develops at the point of common connection. The current excitation method does not take into consideration the controller's need for unpredictable reactive power [6]. Therefore, the goal of our suggested project was to reduce diesel consumption when a synchronous generator is connected while still achieving the requisite power extraction from IG by incorporating an uncertainty control feature into the controller [7] (Fig. 1).

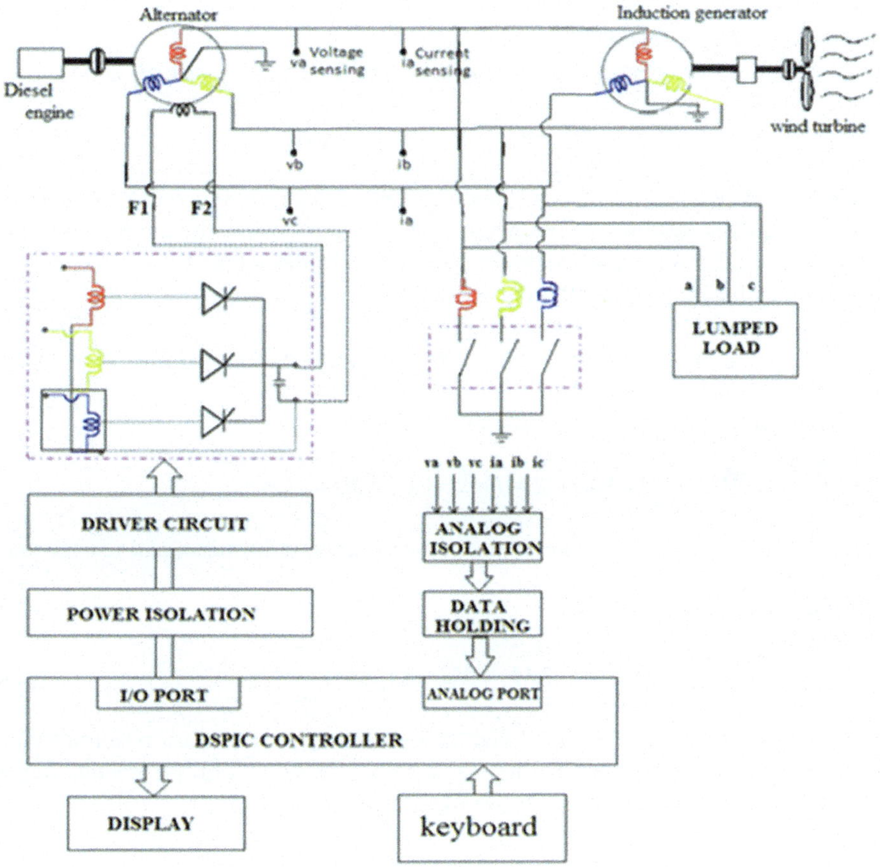

Fig. 1 System block diagram for proposed system

2 Mathematical Modeling of the System

For computational modeling, a wind-diesel system is taken into account in the block diagram, with the SG taken into account with an type-I stimulation connected to the DG set, thereby acting grids, the systems also feature a stimulation that can supply the necessary reactive power. Data about wind-diesel systems are provided. Minor fluctuations in reactive power are primarily reliant on the voltage, while small changes in actual power primarily depend on frequencies. The transient of the excitem is significantly faster in decaying and has no impact on the Load Frequency Control (LFC) dynamic because the stimulation time constants are substantially less than prime-mover time constants.

A system's reactor powers balancing equations in a steady-state situation is

$$QL + QSG + QEXC = QIG \tag{1}$$

where, QL is reactive-power-load demand; QSVC is reactive power generated by EXC and QIG is reactive- power needed by generator. Allow the hybrid system to undergo as change in a reactive power-load for magnitudes [8].

This may be mathematically represented by

$$QSG + \Delta QSVC - QL - QIG = d/dt(\Delta EM) + DV\,V \tag{2}$$

A IG receives its electromagnetic-energy from

$$EM = \tfrac{1}{2} LMIM2 = 1/2LM(V/XM)2 \tag{3}$$

where, IM, LM, and XM stand for the current drawn, inductance, and reactance of the IG.

It may also be expressed by

$$EM = V2/(4\pi fXM) \tag{4}$$

where, the voltage at the terminal and the electromagnetic- radiation stored by the IG, respectively, are denoted by the letters Vo and EoM. All of the reactive power load connected to the source see a rise in voltage [9].

2.1 Synchronous Generator Modeling

The synchronous machine can be used for either driving or generating. The mechanical system define is implemented by the condensed synchronous machine [10].

$$\Delta w(t) = \int (Tm - Te)dt/(2H) - Kd\,\Delta w(t) \tag{5}$$

$$w(t) = \Delta w(t) + wo \tag{6}$$

where, Δw is the inertia constant, indicates speed variation, and Kd is the damping factor, and Te stands for mechanical and electromagnetic torque.

2.2 Excitating Systems

The subsequent function of transfer between exciter voltage (Vfd) and regulator voltage output (Ef) represents the exciters by

$$Vfd\backslash/\backslash Ef\backslash = \backslash 1/\backslash(Ke\backslash + \backslash sTe) \tag{7}$$

where, K_e—exciters gains, T_e—exciters time constants.

2.3 Induction Generator

The total Y IG admission to the IG is provided. GIG and BIG are this admission's respective real and imagined components. Keep in mind that the coefficient of slip and the generator conductance GIG are both negative for generator operation [11].

$$YIG\backslash = \backslash 1/\backslash jXm\backslash\backslash\backslash\backslash ast\backslash 1/\left[\backslash(Rs\backslash + \backslash Rr/Sg\backslash)\backslash + \backslash j\backslash(Xs\backslash + \backslash Xr)\right] \tag{8}$$

$$YIG = \backslash GIG - j2BIG \tag{9}$$

$$GIG\backslash = \frac{Rt\backslash}{R2t\backslash} + X2l \tag{10}$$

$$Rt = Rs + Rr/Sg \tag{11}$$

where, Vb and Rt are the two primary factors in the reactive power expressions.

$$SIG = Vb2Y * IG = Vb2GIG + jVb2BIG = PIG + jQ \tag{12}$$

$$QIG\backslash = \backslash BIG\backslash V2b\backslash\backslash = \backslash R2t\backslash + \backslash(X2l\backslash Xm\backslash Xl\backslash/\backslash Xm(R2t + X2l))\backslash\backslash ast\backslash V2b \tag{13}$$

where, kr the torque-slip curves at the speed specified is used for evaluation and Pw represents the incremental change of the wind turbine when using steady-state power from the input [12, 13].

2.4 *Load of a System*

A 3-phase, 3- wire system with switch to disconnect any phase is used to link resistive inductive loads that are connected in both balance and unbalance delta configurations [14]. The 3-phase inductive-resistive load can be represented by the following equations.

$$Pilab = \{vab - \ilab(Rs + \Rl)\}\/\(Ls + \Ll) \tag{14}$$

$$Pilbc = \{vbc - \ilbc(Rs + \Rl)\}\/\(Ls + \Ll) \tag{15}$$

$$Pilca = \{vca - \ilca(Rs + \Rl)\}\/\(Ls + \Ll) \tag{16}$$

where, ilab, ilbc and ilca are the 3-phases' load current, Rs and L1 are suppling resistances and inductance parameters and Vab, Vbc, and Vca are supply line voltages [15]. Rl and L are load resistances and inductance parameters.

The three-phase loads line currents (i_{la}, i_{lb}, i_{lc}) are given by

$$ila = ilab - ilca \tag{17}$$

$$ilc = ilca - ilbc \tag{18}$$

3 Results and Discussion

We present the results of our study on reactive power regulation in wind-diesel isolated power systems with uncertainty-controlled excitation. Through simulations, we analyzed the proposed control strategy's performance and discussed its implications of voltages stability and regulations of reactive-energy in remote settings. This is an example for a wind diesel energy system model considering wind turbines, diesel generators, and varying loads. The uncertainty-controlled excitation strategy was evaluated against diverse wind power and load conditions. Results demonstrated its effectiveness in mitigating voltage fluctuations caused by wind power variability and load changes, thereby improving voltage stability margins and maintaining acceptable voltage levels. Additionally, the strategy consistently achieved precise and responsive reactive power compensation, enhancing power factor control and reducing imbalances. It exhibited robustness to uncertainties, adapting excitation levels dynamically based on real-time wind power data and load demands. Comparative analysis with conventional methods highlighted the strategy's superiority in voltage stability enhancement and reactive power regulation. Although practical implementation may pose challenges, advancements in technology suggest feasibility with careful design and integration. Overall, our study underscores the potential of uncertainty-controlled excitation to enhance reactive power regulation in wind-diesel isolated power systems, contributing to resilient power generation in remote regions. Further research could explore optimization techniques and real-world validation (Figs. 2 and 3).

Implementation of Three-Phase Half Bridge Rectifier in Hardware.

Structural diagram of implementation made by.

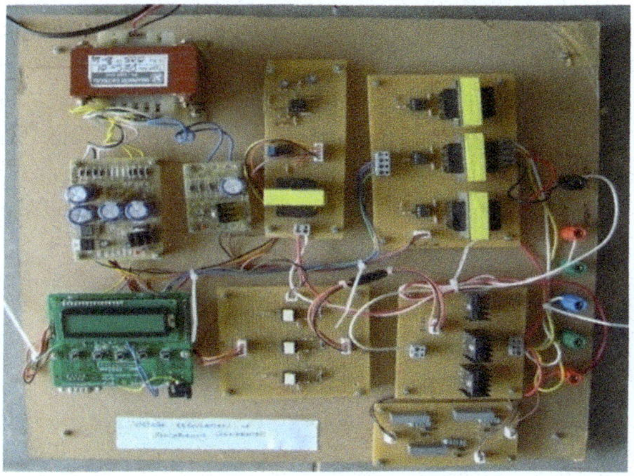

Fig. 2 Three phase half bridge rectifier

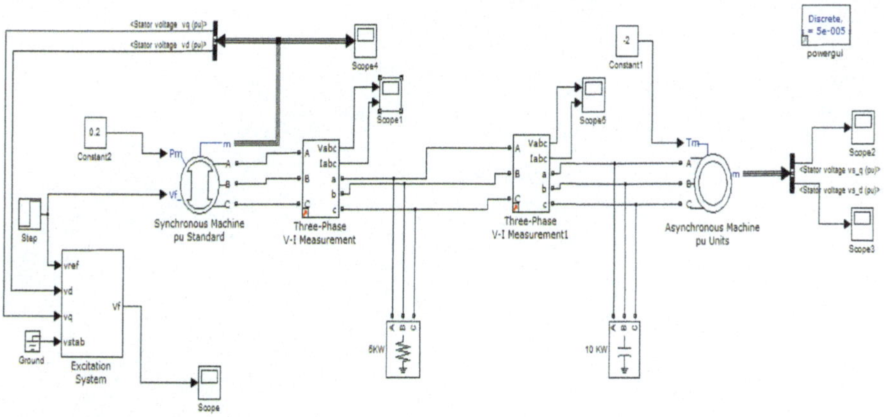

Fig. 3 Simulation circuit diagram

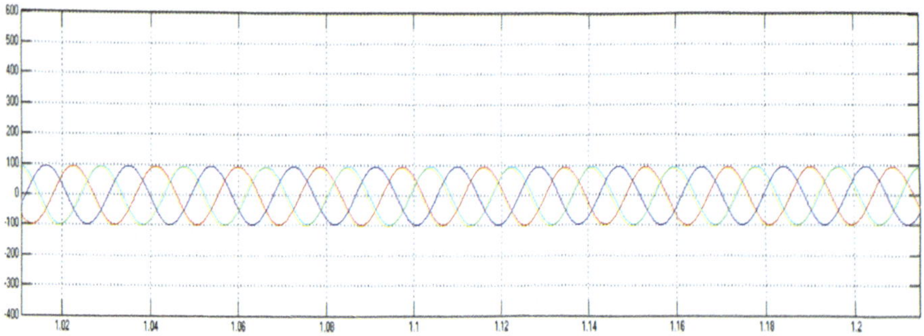

Fig. 4 Simulation output of wave form

Output of current and voltage waveform in simulation (Fig. 4).

The synchronous generator is a crucial component in power systems, and its operation is governed by the balance between mechanical and electrical power. When the generator is operating at synchronous speed, it is producing real power and reactive power, which are essential for the stability and reliability of the power system. When the load on the generator increases, the speed of the generator decreases, and the amount of real power and reactive power produced also decreases. To maintain the generator at synchronous speed, it must be driven by more torque, which is provided by the prime mover (Tables 1 and 2).

Hardware implementation refers to the process of designing and building physical electronic circuits or systems that perform specific functions (Fig. 5).

The magnitude and phase angle of each phase voltage or current. This information helps to determine the degree of imbalance in the system. Difference between two simulations output is provided (Fig. 6).

Table 1 Rating of 3-Phase synchronous generator

Parameters	Max. rating
Voltage	415 V
Rated current	7 A
Power	5 KVA
Frequency	50 Hz
RPM	1500

Table 2 Excitation system rating s for synchronous generator

Parameters	Max. rating
Voltage	220 V
Current	1.55A

Fig. 5 Real time of an hardware implementation

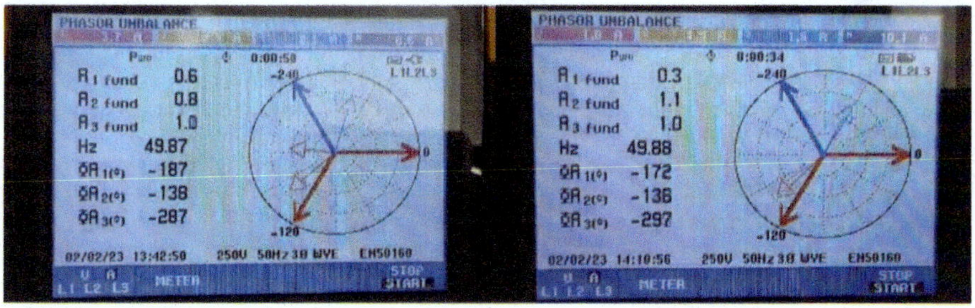

Fig. 6 Simulation output of different values

4 Conclusion

A modified digital excitation system technique for hybrid autonomous system is presented. The induction generators have been taken into consideration for the production of electrical energy, and the system's changeable reactive power needs are met using the digital stimulation technique. Induction generator and synchronous of the alternator synchronization describes the load sharing concept which in turn reduces the diesel consumption of the diesel engine. Therefore, the proposed project is highly reliable and economically efficient.

References

1. Pabitra KG (2021) Application of posicast controller with SVC in wind-diesel hybrid power system model. SN Appl Sci 3(4). https://doi.org/10.1007/s42452-021-04454-2
2. Yiwen X (2021) The frequency-voltage stability control for isolated wind-diesel hybrid power system. Electr Power Syst Res 192(1):106984. https://doi.org/10.1016/j.epsr.2020.106984

3. Mousavi Y (2020) Enhanced fractional chaotic whale optimization algorithm for parameter identification of isolated wind-diesel power systems. IEEE Access 8:140862–140875. https://doi.org/10.1109/ACCESS.2020.3012686
4. Yang M (2019) The adaptive sliding mode reactive power control strategy for wind–diesel power system based on sliding mode observer. IEEE Trans Sustain Energy PP(99):1–1. https://doi.org/10.1109/TSTE.2019.2952142
5. Abhik B (2019) Seeker optimized SVC-PID controller for reactive power control of an isolated hybrid power system. Energy Syst 10(4). https://doi.org/10.1007/s12667-018-0301-0
6. Sebastian R (2019) Review on wind diesel systems dynamic simulation. In: IECON 2019–45th annual conference
7. Felix D (2019) Control of hybrid wind–diesel standalone microgrid for water treatment system application. IEEE Trans Ind Appl PP(99):1–1. https://doi.org/10.1109/TIA.2019.2938727
8. Kuo MT (2019) Applying reactive power compensators to large wind farms to improve the stability of isolated power systems. In: 2019 IEEE/IAS 55th industrial and commercial power systems technical conference (I&CPS), pp 1–8. https://doi.org/10.1109/ICPS.2019.8733340
9. Haixin W (2019) Analysis and suppression for frequency oscillation in a wind-diesel system. IEEE Access PP(99):1–1. https://doi.org/10.1109/ACCESS.2019.2894011
10. Sharma P (2018) Automatic reactive power compensation of an isolated wind-diesel hybrid grid. In: IEEE international conference on industrial technology (ICIT), pp 1171–1176
11. Ma C (2017) The SVC additional adaptive voltage controller of isolated wind-diesel power system based on double sliding-mode optimal strategy. Trans Sustain Energy PP(99):1–1. https://doi.org/10.1109/TSTE.2017.2713700
12. Sebastian R (2017) Simulation of a wind diesel power system with flywheel energy storage. In: 2017 IEEE 26th international symposium on industrial electronics (ISIE), pp 2115–2120. https://doi.org/10.1109/ISIE.2017.8001584
13. Aidoo IK (2016) Optimal controllers designs for automatic reactive power control in an isolated wind-diesel hybrid power system. Int J Electr Power Energy Syst 81:387–404. https://doi.org/10.1016/j.ijepes.2016.02.022
14. Mi Y (2016) Reactive power control of an isolated wind-diesel hybrid power system based on SVC by using sliding mode control. In: 35th Chinese control conference (CCC), pp 9882–9887
15. Sebastian R (2011) Simulation of an isolated wind diesel system with battery energy storage. Electr Power Syst Res 81(2):677–686. https://doi.org/10.1016/j.epsr.2010.10.033

A Comprehensive Approach for Early Detection of Leaf Infection

L. Priya, A. Kiruthika, and P. Kumar

Abstract The agriculture industry is fundamental to the foundation of a country and is essential to the promotion of economic prosperity. It is crucial to ensure that Monitoring the health and detecting leaf infections in plants is crucial for maintaining sustainable agriculture. The actions taken to identify leaf infection represent a significant challenge in the agricultural industry. Manually recognizing plant diseases by analysing leaf images can be a time-consuming undertaking. In response to this challenge, continuous advancements in technology, particularly in image detection, have become instrumental. These advancements enable the accurate and prompt recognition of leaf infections and pests, facilitating the development of early treatment strategies. The paper objective is to precisely recognize diseases in leaf images. The essential steps in this procedure involve pre-processing, training, and recognition. Artificial Intelligence emerges as a ground breaking innovation, empowering the training of machines with image datasets to autonomously identify diseases in crops. Using this proposed methodology this system aims to improve the speed, accuracy, and overall effectiveness in identifying and managing crop infections across diverse agricultural settings through the utilization of machine learning.

Keywords Agriculture · Leaf infection · Yield · Image detection · Machine learning

L. Priya (✉) · A. Kiruthika
Department of Information Technology, Rajalakshmi Engineering College, Chennai, India
e-mail: Priya.l@rajalakshmi.edu.in

A. Kiruthika
e-mail: 221011005@rajalakshmi.edu.in

P. Kumar
Department of Computer Science and Engineering, Rajalakshmi Engineering College, Chennai, India
e-mail: kumar@rajalakshmi.edu.in

V. K. Gunjan and J. M. Zurada (eds.), *Proceedings of 5th International Conference on Recent Trends in Machine Learning, IoT, Smart Cities and Applications*, Lecture Notes in Networks and Systems 1181, https://doi.org/10.1007/978-981-97-8861-3_34

1 Introduction

Agriculture is a broad and essential sector that involves the cultivation of crops, livestock farming, and the production of food, fiber, and other goods used to sustain and enhance human life. Agriculture encompasses a wide range of practices, technologies, and methods that vary across regions and are influenced by factors such as climate, soil types, and cultural practices. Leaf infections, stemming from diverse pathogens, adversely impact plant health. Contributing factors include environmental conditions (humidity, moisture, and temperature fluctuations) poor air circulation, and overcrowded plantings. Vectors like insects or contaminated tools can aid disease spread, and weakened plant defences due to stress or nutrient deficiencies heighten susceptibility. Typically, in agriculture, the use of detection and classification models is a common strategy employed to diagnose crop diseases. Grasping these factors is vital for implementing preventive measures such as sanitation practices, employing disease-resistant plants, and timely treatments to reduce the impact on crop yields and overall plant vitality. Leaf infection detection is a crucial aspect of plant disease management that involves the identification and monitoring of infections on plant leaves. Plants are susceptible to various diseases caused by bacteria, fungi, viruses, and other pathogens. Early detection of leaf infections is essential for effective disease control and preventing the spread of diseases in agricultural crops, ornamental plants, and natural ecosystems. Conventional approaches to identifying leaf infections frequently depend on visual examination by experts, a process that can be time-consuming and may not be feasible for extensive agricultural operations [1]. Technological progress, especially in computer vision and image processing, has opened avenues for more streamlined and automated methods for detecting leaf infections. The initial manifestation of plant diseases typically appears on the leaves (Fig. 1).

Fig. 1 Leaf detection process

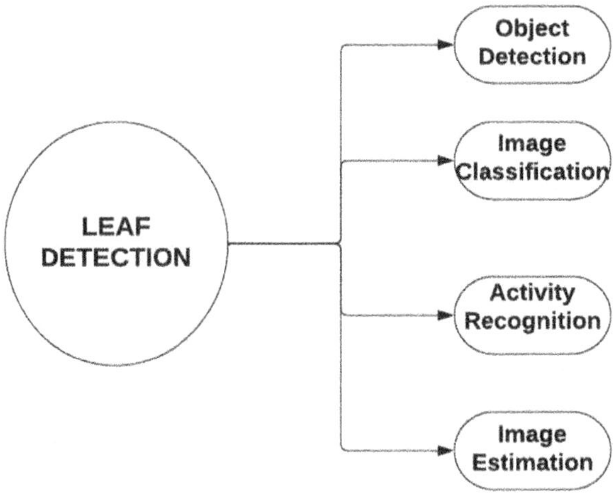

The drive behind detecting leaf infections is rooted in the necessity to safeguard and improve agricultural methods. The early recognition of leaf infections is vital for halting disease spread, minimizing crop harm, and optimizing overall crop well-being. Numerous methods in image processing and pattern recognition have been created to identify diseases occurring on various parts of plants, such as leaves, stems, and lesions etc [2]. Through the utilization of advanced technologies such as computer vision and machine learning, farmers can implement decisions guided by data, ensuring prompt interventions and resource-efficient management. This not only enhances crop yields and financial returns but also adheres to sustainable agriculture principles, contributing to global food security and promoting the integration of modern tools into traditional farming practices. Ultimately, the motivation for leaf infection detection arises from the desire to establish resilient, efficient, and environmentally conscious agricultural systems.

In traditional farming methods the recognition of crop infections heavily depends on manual observation, resulting in inefficiencies and the risk of incorrect diagnoses. The lack of an automated and precise detection system for crop infections presents considerable challenges for farmers. This manual method not only demands considerable time but also necessitates a high level of expertise, causing delayed responses and suboptimal resource utilization. To address these issues, there is acritical need for a Detection system that utilizes advanced technology.

In modern agriculture, farmers commonly grow less-utilized crops, and the weather is unpredictable, often resulting in crop damage due to frequent rainfall. Climatic changes in temperature, rainfall, and other factors can harm crops, False seasonal crops are grown, Diseases that affect crops and how they spread to other crops, Agricultural disease causes agricultural waste (underutilization), Crop yield decrease [3]. To prevent this, it is necessary to have sound recommendation on selecting the appropriate fertilizer, precise identification of bacterial infections in crops, and the ability to differentiate between multiple interconnected diseases in images by incorporating bounding boxes. This study incorporated yolov5, the fifth iteration of the yolo model series, into the Object Detection model. Utilizing it for object detection can identify and mark all infected areas on a leaf with bounding boxes. Similarly, if a leaf is deemed healthy, the identical bounding box will be applied to signify its healthiness.

The proposed model's main goal is to aid farmers in preventing crop diseases in their early stages by utilizing deep learning techniques, including object detection and computer vision [4]. The model identifies the presence of diseases in plant leaves, pinpointing the affected areas using bounding boxes. This enables farmers to promptly apply the required treatments to the identified regions or remove the affected part of the plant, thereby preventing the spread of the disease to the entire plant. Developing a user-friendly Crop Infection Detection system enables quick and accurate identification of infections in crops. The primary goal is to streamline the detection process, empowering farmers with an accessible tool that enhances efficiency, minimizes manual labour, and contributes to the timely and targeted management of crop infections.

2 Related Work

Researchers have devised multiple image processing and pattern recognition techniques to identify diseases on plant leaves, stems, and lesions. Detecting and identifying diseases as soon as they appear is crucial to implementing timely measures and preventing losses. Therefore, there is a need for the development of a swift, precise, and cost-effective system. Researchers have explored diverse methods to accurately detect and identify diseases in plants.

Kulkarni et al. [4]. Addressing the issue of plant disease detection involves the utilization of digital image processing techniques and backpropagation neural networks. The authors elucidated various methods for identifying plant diseases by utilizing leaf images. They applied Otsu's thresholding, followed by boundary detection and spot detection algorithms, to effectively segment the infected portion of the leaf.

Kamlapurkar et al. [5]. This paper identifies Disease identification currently relies on manual observation and pathogen detection. The project's objective is to accurately identify and classify diseases from leaf images, utilizing the CCM method for feature extraction, and employing BPNN for the identification process.

Manjunatha et al. [5]. The image segmentation occurs through the implementation of the K-Means cluster algorithm, followed by classification using an Artificial Neural Network (ANN) block. This system is primarily utilized for the detection of diseases in leaves and skins.

Maryam et al. [6] and shetty et al. [7]. In this paper, the suggestion is to employ Image Processing, K-Means, GLCM, SVM, and Plant Disease Detection in MATLAB for disease detection. The proposed system will be a standalone desktop application that users can download and access on their computers.

Satwik et al. [8] and liu et al. [9]. By employing the YOLO (You Only Look Once) Object Detection technique, we can identify all the areas of infection on a leaf. The captured image is converted into pixel data and processed using.

Convolutional Neural Network (CNN) architecture, fitting into the structure of an Artificial Neural Network. This system can process up to 45 frames per second. YOLO also possesses the capability to comprehend generalized object representation. The architecture will be fit into Artificial Neural Network

Mehul et al. [10]. The research establishes a foundation for indexing digital video using the YOLOv4 Darknet Neural Network for detection. The threshold values are determined through experimental observations across various videos employing the Binary Search Pattern Algorithm (BSPA).

Dr. Sridhathan et al. [11] and Halder et al. [12]. This paper suggests the detection of infections using K-means clustering and GLCM techniques. Segmented image texture analyses are employed for the classification of infections. The system demonstrated the capability to identify and classify infections with an accuracy of 98.27%.

Andrew et al. [3] and Mohanty et al. [13]. In this study, pre-trained models based on convolutional neural networks (CNN) were employed for effective identification of plant diseases. The emphasis was placed on fine-tuning the hyperparameters of well-known pre-trained models, including DenseNet-121, ResNet-50, VGG-16, and Inception V4.

3 Implementation

A Convolutional Neural Network commonly known as CNN, is a type of artificial neural network designed specifically to process structured grid data, particularly images. Predominantly employed in computer vision tasks, image recognition, and pattern detection, CNNs demonstrate proficiency in activities such as image classification, object detection, and facial recognition. Notably, CNNs require minimal data pre-processing when compared to alternative labeling techniques. The algorithm is inspired by the neural structure of the human brain, particularly the neurons in the Visual Cortex situated in the occipital lobe, responsible for processing visual information (Fig. 2).

The key components in the leaf infection detection system using CNN and yolo. It begins by establishing a clearly defined data pipeline responsible for gathering and preprocessing a diverse dataset that includes both healthy leaves and various infections. This prepared dataset is then input into the CNN, acting as a powerful feature extractor to identify intricate patterns related to leaf infections. Concurrently, the yolo model is incorporated to enable real-time object detection, swiftly recognizing and categorizing objects within the images. A fusion mechanism combines the features extracted by the CNN with the dynamic results obtained from yolo, thereby enhancing the overall accuracy of the system. To achieve a subtle balance between

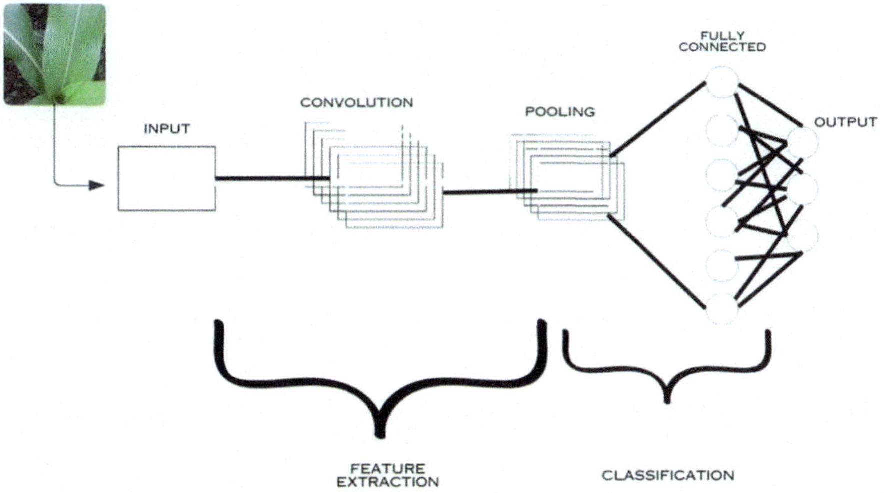

Fig. 2 Working of convolution neural network

precision and recall, confidence threshold optimization is applied. The real-time processing module, furnished with an interactive interface, ensures user engagement by accepting input from images during the system's operation. This comprehensive Fig. 3 aims to deliver an effective and user-friendly tool for the early detection and intervention of leaf infections, contributing to enhanced crop health and sustainable agricultural practices.

The system integrates pre-processed data through feature extraction in each phase to yield the final result. It starts by preprocessing the input image for refined analysis. Feature extraction is then applied to discern essential characteristics from the pre-processed data, and the trained model is saved. Transitioning to the testing phase, the system evaluates performance and accuracy using dedicated test data. Figure 4 Once validated it seamlessly moves to the prediction stage, making informed predictions, then presented with clear outcomes, showcasing the achieved results.

The Block diagram involves translating the conceptual design into a functional system. This newly created system might be entirely ground breaking, replacing an existing manual or automated system, or it could involve substantial modifications to an existing system Fig. 5. The implementation is carried out using Roboflow and a dataset.

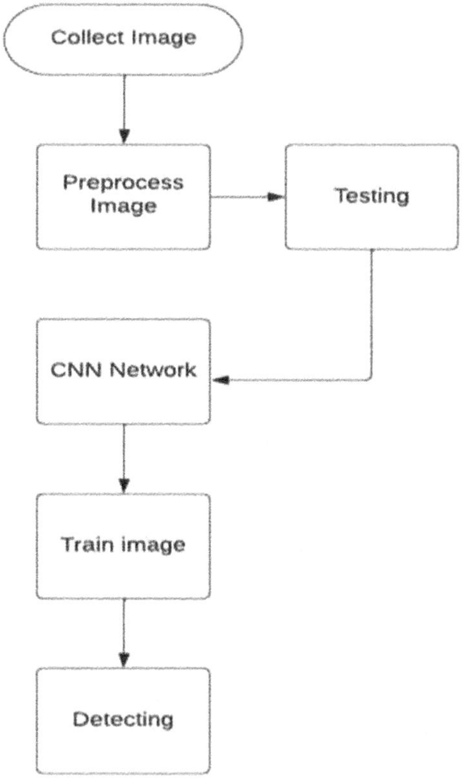

Fig. 3 Use case diagram

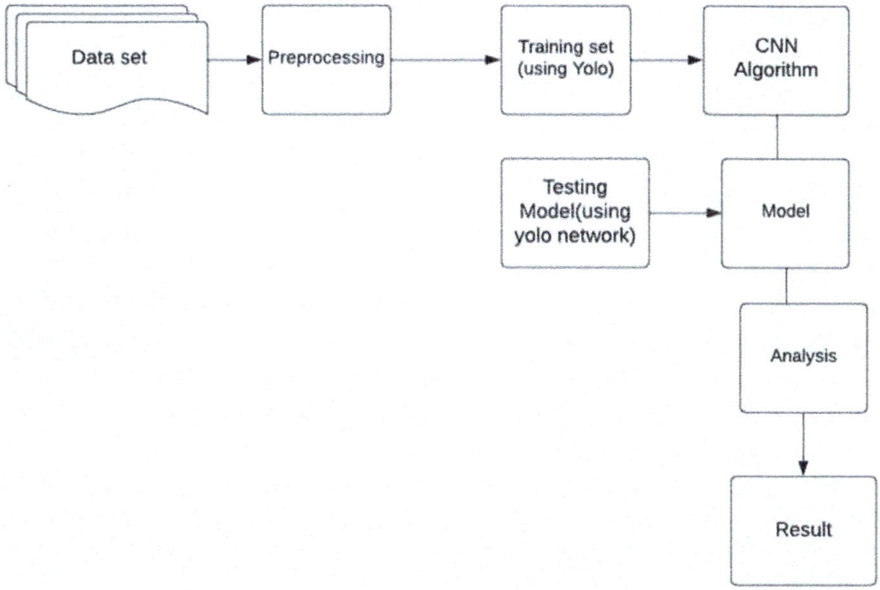

Fig. 4 Flow chart

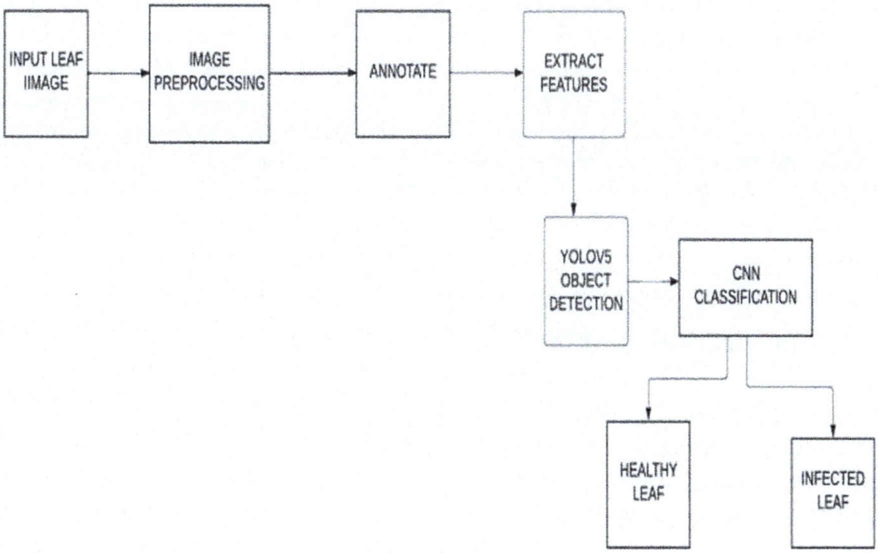

Fig. 5 Block diagram

Fig. 6 Schematic diagram

After completing the feature extraction step, convolution and pooling layers are employed, and the images are subjected to testing. After the testing phase, predictions are generated, and the outcome of these predictions is presented as the final output.

The schematic diagram for leaf infection detection comprises several key components. The process initiates with the Input Module, where leaf images are received. The Preprocessing Module takes center stage, undertaking essential tasks such as image resizing and normalization to ensure the uniformity and optimal condition of the images. The Feature Extraction Module comes into play, diligently extracting essential features from the identified regions. Subsequently the Feature Extraction Module extracts essential features, while Convolution and Pooling Layers enhance representation. The Testing Module assesses the model's accuracy using test images, and the Prediction Module generates conclusive results, determining the health status of leaves. Figure 6, like this the Output Module displays these results visually providing a comprehensive overview of the leaf infection detection process (Fig. 7).

4 Experimental Results

Displays the training graphs depicting the plotted data during the model training, specifically illustrating the relationship between Loss and Epochs Fig. 8. The proposed model successfully detected numerous healthy and infected spots, yielding impressive results during training and on the test data. Further improvements are anticipated by augmenting the quantity of images per category and the augmentation of training epochs, specifically to enhance the confidence of bounding box predictions. Once the model generates output inclusive of bounding box information, continue refining its performance and precision in identifying healthy and infected areas. This integrated approach not only showcases the model's effectiveness in detection but also opens avenues for proactive agricultural management through targeted fertilizer recommendations (Figs. 9, 10, 11 and 12).

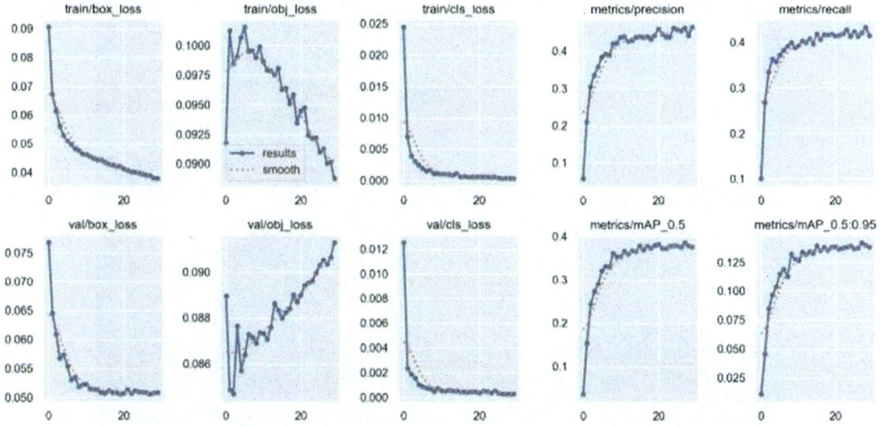

Fig. 7 Training plot

Fig. 8 Model prediction- healthy maize

Fig. 9 Model prediction-infected maize

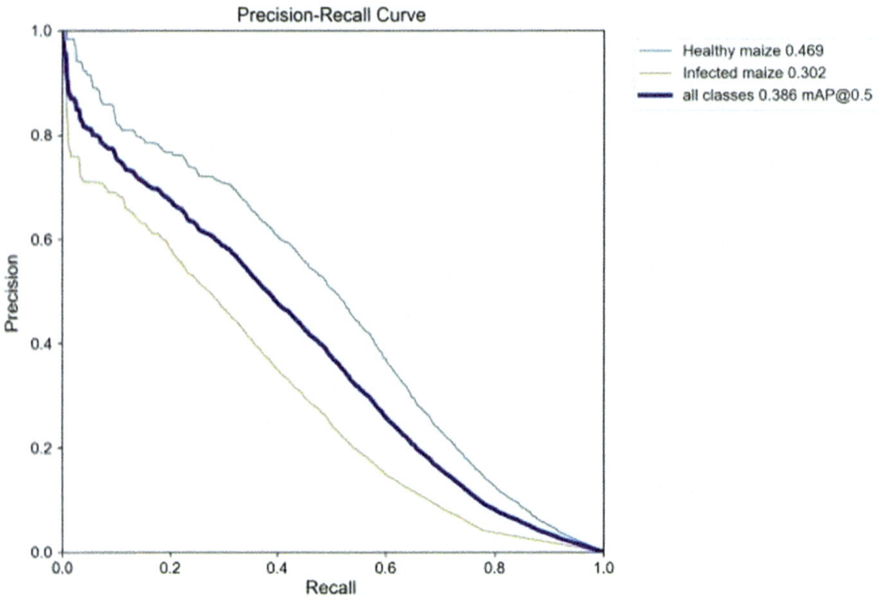

Fig. 10 Precision-recall cure

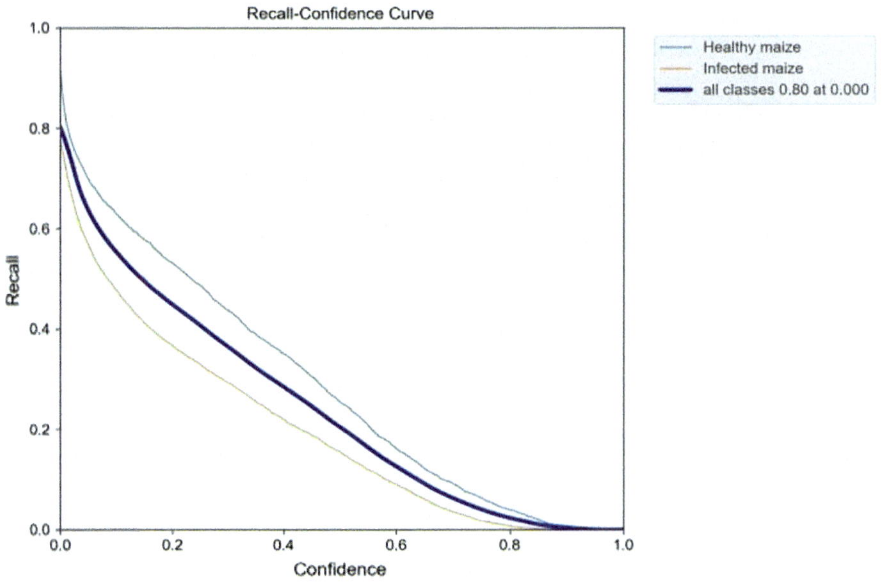

Fig. 11 Recall-confidence curve

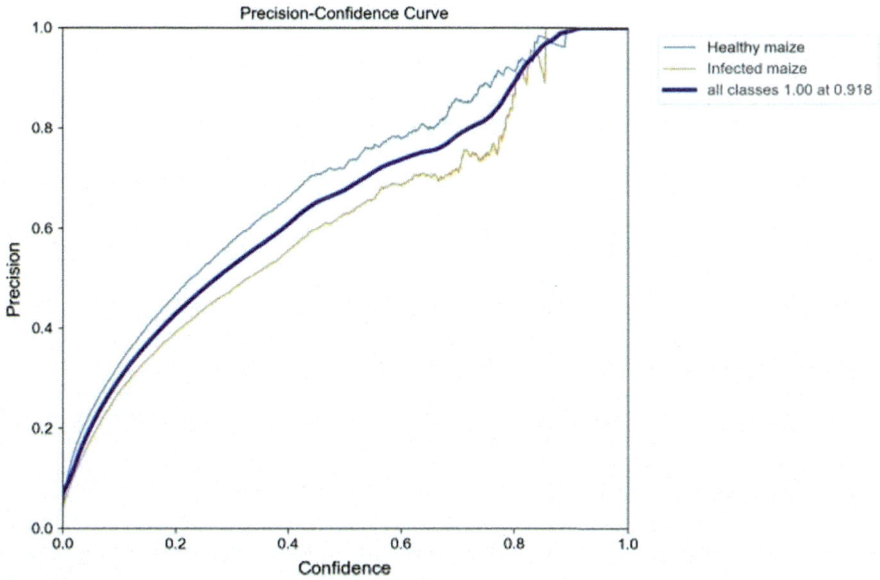

Fig. 12 Precision-confidence curve

5 Conclusion

In conclusion, the utilization of yolov5 for leaf infection detection presents a robust and efficient solution. The suggested system can identify diseases in their early stages, detecting them as soon as they appear on the leaf. This capability enables the prevention of losses and reduces the reliance on experts to a certain extent. The system is designed to assist individuals with limited knowledge about diseases, aligning with the objective of extracting features corresponding to the specific disease. The model's real-time object detection capabilities, coupled with its accuracy, contribute to the system's effectiveness in identifying and classifying infections in plant leaves. For future work we can develop the webpage. This technology holds great promise in agricultural practices, aiding in early detection and intervention, ultimately contributing to improved crop health and yield.

References

1. Sinha A, Shekhawat RS (2020) Review of image processing approaches for detecting plant diseases 4:229–242. https://doi.org/10.1049/iet-ipr.2018.6210, https://doi.org/10.1049/iet-ipr.2018.6210
2. Garcia J, Barbedo A (2013) Digital image processing techniques for detecting, quantifying and classifying plant diseases 51(1):837–841. https://doi.org/10.1186/2193-1801-2-660, http://www.springerplus.com/content/2/1/660

3. Andrew J, Eunice J, Popescu DE, Chowdary MK, Hemanth J (2022) Deep learning- based leaf disease detection in crops using images for agricultural applications 12(12).https://doi.org/10.3390/agronomy12102395
4. Kulkarni P, Karwande A, Kolhe T, Kamble S, Joshi A, Wyawahare M (2018) Plant disease detection using image processing and machine learning, pp 1590–1595. https://doi.org/10.48550/arXiv.2106.10698
5. Kamlapurkar SR (2016) Detection of plant leaf disease using image processing approach. Int J Sci Res Publ 4(1):41–49. ISSN 2250–3153. https://doi.org/10.1016/j.inpa.2016.10.005
6. Takti M, El Azhari MEH (2022) Plant disease detection using image processing 3(1):272–278. https://doi.org/10.1109/ICCUBEA.2015.153, published in IEEE https://ieeexplore.ieee.org/stamp/stamp.jsp?tp=&arnumber=71.55951&isnumber=7155781
7. Shetty AV, Ashika, Amshitha R, Deeksha, Gulappagol L (2022) Leaf disease detection using image processing 3(1):272–278. https://doi.org/10.1109/ICCUBEA.2015.153. https://doi.org/10.1016/j.gltp.2022.03.010
8. Kodandaram SR, Honnappa K (2021) Crop infection detection using YOLO 12(06):1–5. https://doi.org/10.13140/RG.2.2.27776.97281 https://ieeexplore.ieee.org/stamp/stamp.jsp?tp=&arnumber=9579783&isnumber=9579470
9. Liu J, Wang X (2021) Plant diseases and pests detection based on deep learning: a review. https://doi.org/10.1186/s13007-021-00722-9
10. Maharishi M, Morwal S, Muzzafar AW, Bhatia S, Dhadeech P, Rahmani MKI (2021) Video index point detection and extraction framework using custom yoloV4 darknet object detection model. IEEE Access 9:143378–143391. https://doi.org/10.1109/ACCESS.2021.3118048 https://ieeexplore.ieee.org/stamp/stamp.jsp?tp=&arnumber=9559985&isnumber=9312710
11. Sridhathan C, Kumar MS (2018) Plant infection detection using image processing. IJMER‖ISSN: 2249–6645 8(7):768–771. https://ieeexplore.ieee.org/stamp/stamp.jsp?tp=&arnumber=7155951&isnumber=7155781
12. Halder M, Sarkar A, Bahar H (2021) Plant disease detection by image processing, pp 1590–1595. https://doi.org/10.1109/ICCUBEA.2015.153. https://doi.org/10.48550/arXiv.2106.10698
13. Mohanty SP, Hughes DP, Salathé M (2016) Using deep learning for image-based plant disease detection, vol 7. https://doi.org/10.3389/fpls.2016.01419

A Customized Mobile Application for Exploring Chennai City: An Ultimate Sightseeing Companion

L. Priya, M. Sandhiya, and K. Poornimathi

Abstract "TourChennai" is a customized Tourism Suggestion Android application designed to provide a seamless and user-friendly experience for exploring the city of Chennai. Developed using Android Studio, Java, and XML, the app offers comprehensive information about tour attractions, events, hotels, and other points of interest in Chennai. Its intelligent features provide real-time suggestions and personalized recommendations to enhance the user's journey. Rigorous testing with multiple users confirmed the app's robustness and user-friendliness, making it an ultimate sightseeing companion for travelers in Chennai.

Keywords Application · Events · Android studio · Database · Tourism

1 Introduction

This paper presents a comprehensive review of technology-driven tour guide services, acknowledging the rising demand for immersive experiences in the rapidly growing tourism industry. It explores recent developments in technology's role in enhancing tour guide services, discussing their benefits, challenges, and potential to revolutionize the sector. The paper also analyzes the implications of technology on the role of tour guides and the overall tourist experience.

In this context, the paper introduces the "TourChennai" application, a free open-source platform designed to cater to visitors exploring Chennai city based on their

L. Priya (✉) · M. Sandhiya
Department of Information Technology, Rajalakshmi Engineering College, Chennai, India
e-mail: priya.l@rajalakshmi.edu.in

M. Sandhiya
e-mail: sandhiya.m@rajalakshmi.edu.in

K. Poornimathi
Department of Computer Science and Business Systems, Rajalakshmi Engineering College, Chennai, India
e-mail: poornimathi.k@rajalakshmi.edu.in

V. K. Gunjan and J. M. Zurada (eds.), *Proceedings of 5th International Conference on Recent Trends in Machine Learning, IoT, Smart Cities and Applications*, Lecture Notes in Networks and Systems 1181, https://doi.org/10.1007/978-981-97-8861-3_35

preferred criteria, such as budget, duration of stay, and weather conditions. The app aims to provide an interactive and informative guide, showcasing historical facts, reviews, and ratings from fellow travelers [1]. Developed for Android devices using Android Studio and Java, the Tour Chennai app will offer a user-friendly interface, enabling tourists to access relevant information about popular tour destinations, local cuisine, cultural events, and accommodation options in Chennai [2].

The proposed application's key features include detailed destination information, insights into local cuisine and dining establishments, updates on cultural events and festivals, and accommodation recommendations with reviews and ratings [3]. By leveraging various libraries and frameworks, the Tour Chennai app seeks to provide a seamless and enjoyable experience for tourists, enhancing their stay in this vibrant city [4].

Moreover, the paper acknowledges the versatility of Android application development, supporting multiple programming languages such as Kotlin, Java, C#, and C++ [5]. The TourChennai app will be compiled using Android SDK and built with Java programming language [6]. As an essential aspect of mobile application development, the paper emphasizes the need for robust security measures [7].

To conclude, the paper highlights the significance of the Tour Chennai app in bridging the gap between technology and tourism, offering tourists an interactive and personalized experience during their visit to Chennai [8]. The proposed application's architecture, supported by Java and SQL, ensures the efficient delivery of tourism information to mobile users [9]. The paper's content is structured with a literature review, system design, results, and discussion, aiming to enrich the understanding of technology's impact on tour guide services and contribute valuable insights for developers, tour operators, and researchers in the field.

2 Related Work

There have been numerous works related to tourist application development. In [10], the authors built electronic tourist guide systems using two different technologies to showcase the power of Android. A comparison between the two systems was made. These applications were customized to user preferences, allowing access through a simple interface or automatic retrieval of information based on GPS positioning.

In [11], the authors developed a Smart Traveling application with several features to assist tourists visiting Indonesia. The application recognizes tourist attractions, displays nearby events, provides information on the nearest police stations and hospitals for emergencies, and saves the history of recognized objects. The tourist attraction recognition is achieved through image recognition using the landmark detection feature from Google Cloud Vision API, aiding tourists in accessing relevant information about visited attractions.

In [2], the authors designed a tourist application for finding details about tourist places that users wish to visit. The application is developed using Java as the front end and SQL as the back end. The Tamil Nadu Tourism E-Guide proposes a mobile tourist guide system architecture for Android mobile phones, conveniently providing tourism information to mobile users.

In [12], the app significantly enhances the tourism experience in Tamil Nadu, making it a valuable tool for both domestic and international tourists.

The main objective of [13] is to develop a state-of-the-art mobile travel guide application with numerous functionalities, integrating features from various travel apps into a single all-in-one app. This application will provide information about cities, landmarks, monuments, restaurants, picnic spots, and current weather information. It also includes additional features such as currency converter, magnetic compass, world clock, holiday details, and shopping facilities.

In [14], the Android application serves as a personalized travel guide, allowing users to arrange tours based on their budget and estimated travel duration. The front-end of the application is designed using the React Native framework, and the back-end is connected through the Apollo Client library and an express server. Various security functions, including Bcrypt, JSON Web Token, and cookie-parser, have been implemented to ensure the application's security.

In [15], the objective is to develop and implement a comprehensive mobile application that enhances the tourism experience in Bangladesh by providing valuable information and tools for tourists.

The Tour Chennai application is specifically designed for the Chennai Metropolitan city. It offers a curated list of popular attractions, events, and tours in Chennai. The home screen allows users to browse attractions based on categories such as historical sites, cultural experiences, food tours, and various filters like location, ratings, and price. The application provides detailed information about attractions, including images, reviews, visitor ratings, and navigation with maps.

3 Methods

The "TourChennai" application utilizes a range of data structures to optimize the user experience. Geographical data is stored using geospatial databases and GIS, while user profiles and preferences are managed through user-friendly structures like JSON or XML. Location data and real-time updates are efficiently organized using arrays, linked lists, and real-time data streaming structures. User-generated reviews and ratings are stored either in arrays or linked lists, while navigation and route planning rely on graph structures. Data caching employs hash tables or in-memory stores for improved performance, and search and query structures like binary trees or hash maps expedite place searches. Data security is maintained through encryption algorithms, and user-generated content is managed using stacks and queues, ensuring a secure and efficient exploration experience of Chennai City.

3.1 Architecture

Figure 1 depicts the use-case diagram for the TourChennai application, which includes functionalities like login/registration, location of Chennai city, item search, reviews, rankings, and logout page.

Figure 2 illustrates the system block diagram for the TourChennai application, outlining the components responsible for handling detailed application information, reviews, and rankings.

User Registration:

To register, users enter their details into the UI, including a unique name, email, password, city, state, and country. The server-side application validates the input and stores the user's data in the database. Upon successful registration, the server-side application sends a response to the UI indicating success and redirects the user to the home page.

User Authorization:

For existing users, the application requires login credentials, including their email and password. The server-side application verifies the credentials against the database. If valid, the server sends a response indicating successful login and redirects the user to the home page. If the credentials are invalid, the application indicates login failure.

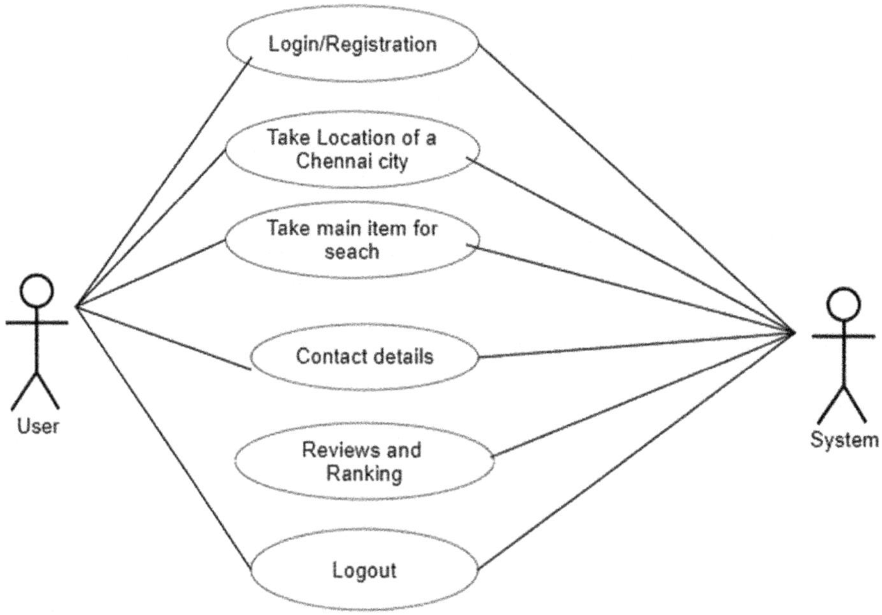

Fig. 1 Use case diagram

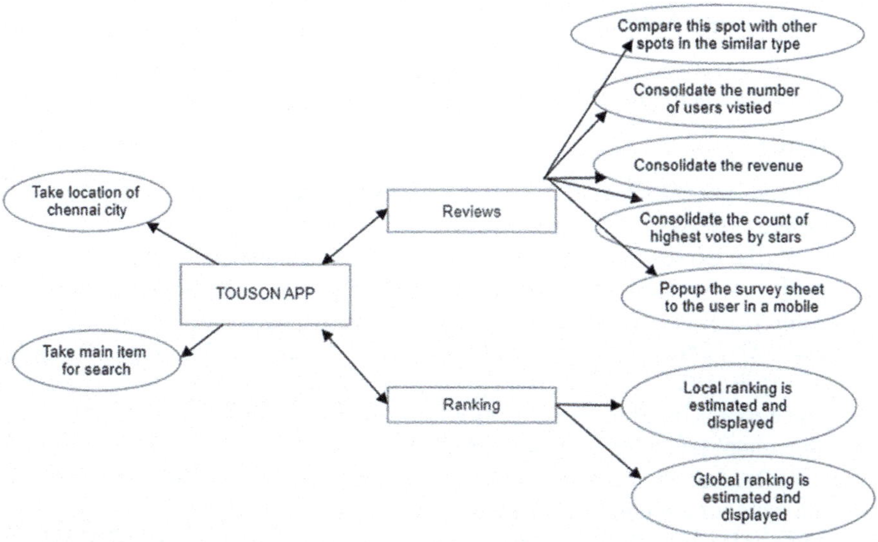

Fig. 2 System block diagram

Categories of Tourist Attractions:

TourChennai provides categories of various tourist attractions, such as hotels, places, restaurants, shopping, and events, to classify and organize attractions based on their characteristics and appeal.

Browsing Tourist Attractions:

Users can browse through the list of tourist attractions provided by TourChennai and select the attractions they are interested in visiting.

Viewing Attraction Information:

TourChennai displays comprehensive information about selected attractions, including their names, descriptions, locations, contact details, and ratings.

Getting Directions:

The application offers users directions to the attractions, assisting them in navigating to their desired destination. Navigation features are provided using Google Maps.

Visiting Attractions:

Users follow the location information provided by TourChennai to visit the selected attractions.

Providing Feedback:

After visiting an attraction, users have the option to provide feedback on their experience, such as leaving a rating.

3.2 *Database Schema*

Database Schema for TourChennai Application with SQLite Connectivity (Table 1).

Description

1. Name: Stores the name of the user as a VARCHAR data type, allowing up to 50 characters to accommodate most names.
2. Email_ID: Serves as the primary key of the table, uniquely identifying each user based on their email-id. It is stored as a VARCHAR data type with a maximum length of 100 characters to handle longer email addresses.
3. Password: Stores the user's password as a VARCHAR data type. To ensure security, it is essential to encrypt passwords before storing them in the database.
4. City: Records the city where the user lives, stored as a VARCHAR data type with a maximum length of 50 characters.
5. State: Records the state where the user lives, stored as a VARCHAR data type with a maximum length of 50 characters.
6. Country: Records the country where the user lives, stored as a VARCHAR data type with a maximum length of 50 characters.

This User table is crucial for the authentication process when users log in to the TourChennai Application. It will allow the application to verify user credentials and provide personalized experiences based on their location information.

4 Result and Discussion

The TourChennai application, developed using Android Studio, aims to offer an interactive and informative experience for tourists exploring Chennai's popular destinations, local culture, and cuisine [16]. This application can lead to various positive outcomes.

Table 1 User table schema

Column name	Data type	Description
Name	VARCHAR(50)	The name of the user
Email_ID	VARCHAR(100)	The email-id of the user (Primary Key)
Password	VARCHAR(50)	The password created by the user
City	VARCHAR(50)	The city where the user lives
State	VARCHAR(50)	The state where the user lives
Country	VARCHAR(50)	The country where the user lives

Enhanced Tourist Experience: By providing accurate and up-to-date information about tourist attractions, cultural events, and local cuisine, the TourChennai application enhances the overall tourist experience. Tourists can plan their itineraries effectively, ensuring a memorable and enjoyable visit to Chennai.

Increased Tourism: The Tour Chennai application plays a vital role in promoting Chennai's tourist destinations [17]. By showcasing the city's attractions and facilitating easy access to relevant information, the application can attract more tourists to Chennai. This, in turn, can contribute to increased revenue for the tourism industry and local businesses.

Improved Engagement: The application's user-friendly interface and rich content can keep tourists engaged throughout their journey. Interactive features and personalized recommendations can further enhance user engagement, making the visit to Chennai more enjoyable.

Valuable Data Collection: The TourChennai application can collect valuable data on tourist preferences and behaviors. This data can be analyzed to gain insights into visitor patterns, allowing tourism stakeholders to make informed decisions on marketing, promotions, and resource allocation.

The TourChennai application has the potential to bring numerous benefits to Chennai's tourism industry, businesses, and tourists alike. Its provision of accurate information, seamless user experience, and promotion of local attractions can contribute to a thriving tourism ecosystem in the city. User testing results have been presented Fig. 3.

To evaluate the performance of a Tour Chennai application showing the effective places and monuments in Chennai the following metrics were considered:

User engagement: This metric measures the level of user engagement with the application. It can be measured by the number of active users, the frequency of application usage, and the duration of each session. It can be calculated by various metrics, but one commonly used formula is:

$$\text{User Engagement} = (\text{Number of Active Users/Total Users}) * 100 \quad (1)$$

Fig. 3 User testing result cloud

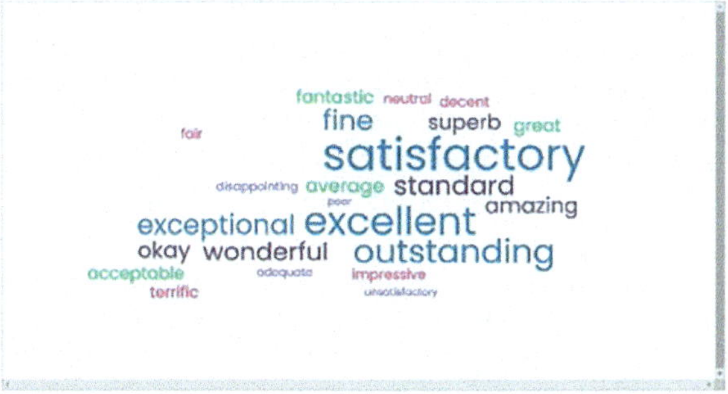

Tour Chennai application with a total of 1,000 registered users. Out of these, 500 users are actively using the application.

$$\text{User Engagement} = (500/1{,}000) * 100$$

$$\text{User Engagement} = 0.5 * 100$$

$$\text{User Engagement} = 50$$

User feedback: User feedback can be collected through application store reviews, surveys, and other feedback channels. This metric can provide insights into user satisfaction with the application and identify areas for improvement. It has been calculated by using the formula:

$$\text{Average Rating} = \frac{\text{Sum of all ratings}}{\text{Total number of ratings}} \tag{2}$$

Rating Scale
1 star: Poor
2 stars: Below Average
3 stars: Average
4 stars: Above Average
5 stars: Excellent

Sample User Rating:

$$\text{Sum of All Ratings} = 4 + 3 + 5 + 4 + 2 + 5 + 3 + 4 + 5 + 1 = 36$$

$$\text{Total Number of Ratings} = 10$$

$$\text{Average Rating} = 36/10 = 3.6$$

Application store ratings: Application store ratings are a key metric for evaluating the success of the application. High application store ratings indicate that users are satisfied with the application's features, user interface, and overall experience. It can be calculated by one commonly used formula is:

$$\begin{aligned}&\text{Average Rating} = (5 * \text{Number of 5-Star Ratings} + 4 * \text{Number of 4-Star Ratings}\\&+3 * \text{Number of 3-Star Ratings} + 2 * \text{Number of 2-Star Ratings} + 1 * \text{Number of 1-Star Ratings})\\&/\text{Total Number of Ratings}\end{aligned}$$

$$\text{Average Rating} = (5 * 3 + 4 * 3 + 3 * 2 + 2 * 1 + 1 * 1)\,/\,10$$

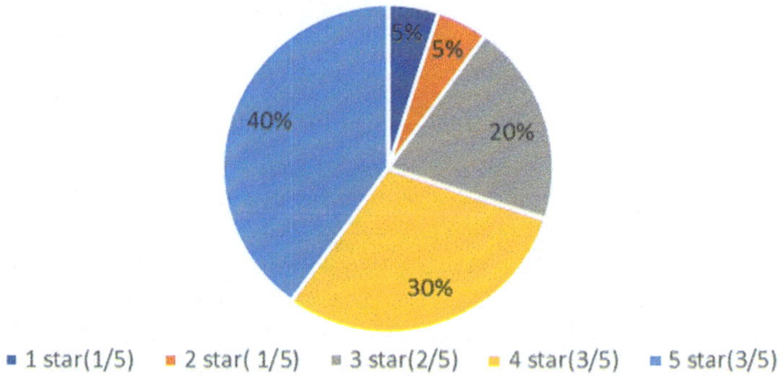

Fig. 4 Sample survey report of tour Chennai

$$\text{Average Rating} = (15 + 12 + 6 + 2 + 1)\,/\,10$$

$$\text{Average Rating} = 36/10$$

$$\text{Average Rating} = 3.6$$

Retention rate: This metric measures the percentage of users who return to the application after their initial usage. A high retention rate indicates that users find the application useful and engaging. The formula for calculating retention rate is as follows:

$$\text{Retention Rate} = \frac{(\text{Number of Users at the End of a Period} - \text{Number of New Users During the Period})}{\text{Number of Users at the Beginning of the Period} \times 100} \tag{3}$$

Sample survey report of a tour in Chennai, represented in Fig. 4, using the aforementioned formula.

4.1 Main Features

(a) App Icon: The app icon is designed to be easily recognizable and memorable so that users can quickly find and access the TourChennai application. It typically consists of a simple and distinctive design, with a limited color palette and clear typography.

(b) App Loading page: The loading page serves as a visual indicator to the user that the application is launching and provides an opportunity for the application to showcase its branding and design.

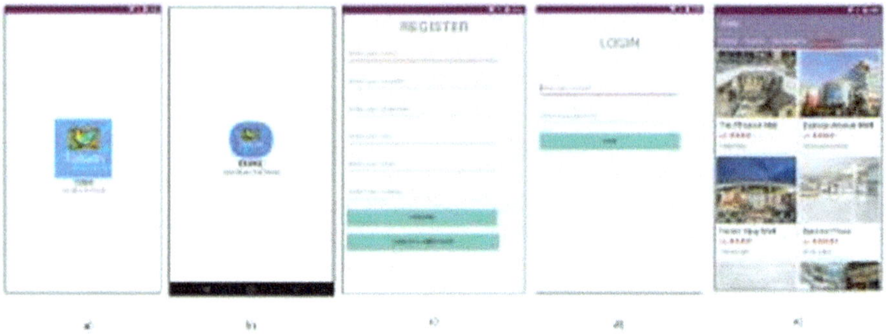

Fig. 5 User screens in tour Chennai

(c) Register: The register page is used to collect relevant information about users such as name, email, password, city, state, and country which will be used to create an account and authenticate their access to the application.
(d) Login: In the Login page user can log in using their previously created account credentials by using email and password.
(e) Shopping Mall Page: On this page, users can view various hotels to be visited and provide detailed information about that hotel.

User screens in the Chennai tour app, as shown in Fig. 5, include the App Icon, App Loading, Login, Register, and Shopping Mall.

5 Conclusion and Future Work

Developing a tourist guide app for Chennai offers an excellent opportunity to showcase its attractions, culture, and history, enhancing the visitor experience. To ensure its success, comprehensive audience research is vital, focusing on user-friendly interfaces, easy navigation, and up-to-date information. Regular updates, push notifications, and user engagement strategies should be employed for continuous enhancement, offering benefits in terms of user engagement, revenue, and brand recognition. Potential future improvements include implementing augmented reality for immersive experiences, personalization based on user preferences, voice-enabled assistants for convenience, multilingual support, gamification for engagement, social media integration for sharing experiences, and sustainability features to promote responsible tourism, further elevating Chennai's tourism industry.

References

1. Liu L, Jing Y (2012) Android city tour guide system based on web service, In: 2012 2nd international conference on consumer electronics, communications and networks (CECNet). IEEE, pp 3118–3121. https://doi.org/10.1109/CECNet.2012.6201621
2. Vasoya M, Vaghasiya T, Patel A, Kansagra Z (2019) City Guide Survey. Int J Sci Res Comput Sci Eng Inf Technol. https://doi.org/10.32628/CSEIT1952291
3. Wali M, Akbar R, Iqbal T, Al-Bahri FP (2019) Development of an android-based tourism guide (a case study: Sabang City, Indonesia). Int J Sci Technol Res 8(11):887–893
4. Raja P, Raghu A (2020) Sustainable tourism a case study–of Chennai and Singapore. Int J Manag (IJM) 11(12). https://doi.org/10.34218/IJM.11.12.2020.281
5. Palomba F, Di Nucci D, Panichella A, Zaidman A, De Lucia A (2017) Lightweight detection of android-specific code smells: the doctor project. In: 2017 IEEE 24th international conference on software analysis, evolution and reengineering (SANER), pp 487–491. https://doi.org/10.1109/saner.2017.7884659
6. Hou S, Lu T, Du Y, Guo J (2017) Static detection of android malware based on improved random forest algorithm. In: 2017 IEEE international conference on intelligence and security informatics (ISI). https://doi.org/10.1109/isi.2017.8004913
7. Jinendra DR, Bhagyashri JR, Pranav GY, Seema VU, Parag AN (2012) Smart travel guide: application for android mobile. Int J Electron, Commun Soft Comput Sci Eng (IJECSCSE) Amravati 2:115–120
8. Smyth N (2017) Android studio 2.3 development essentials android, 8 edn. PayloadMedia Inc. Neil Smyth Payload Media Inc—Computers, 800 pp
9. Kovalev VS (2017) Development of a flexible interface for the audio player application for the Android platform. Doctoral dissertation, South Ural State University; Jithin PK, Vishnuram M, Prasath P, Thirukrishna JT (2018) Tourism guide for Tamilnadu. Android Appl 4(11):112–116
10. Irmanti D, Hidayat MR, Amalina NV, Suryani D (2017) Mobile smart travelling application for Indonesia tourism. Procedia Comput Sci 116:556–563. https://doi.org/10.1016/j.procs.2017.10.059
11. Nurhasan U, Pradibta H, Suryadi SB, Alfinda A (2019) Guide application for the introduction of Malang city tourism potential using hybrid technology. In J Phys Conf Ser 1402(6):066051. https://doi.org/10.1088/1742-6596/1402/6/066051
12. Jithin PK, Vishnuram M, Prasath P, Thirukrishna JT (2018) Tourism guide for Tamilnadu (android application). Int J Innov Res Sci Technol 4(11):112–116. ISSN: 2349–6010
13. Florendo J, Concepcion V, Feliciano CG (2021) Isabela travel app: an android based tour guide application for the province of Isabela. Link J Comput Technol 2(1)
14. Irbid J (2013) Building mobile tourist guide applications using different development mobile platforms. Int J Adv Sci Technol 54:13–22
15. Pavel MA, Rana M, Al Roman A, Hassan Y, Khan R (2021) Android application for tourism planning in Bangladesh, In: 2021 IEEE 19th student conference on research and development (SCOReD). IEEE, pp 157–162. https://doi.org/10.1109/SCOReD53546.2021.9652756
16. Shinde KA, Olsen DH (eds) (2020) Religious tourism and the environment. Cabi. https://doi.org/10.1080/14616688.2020.1866057
17. Arun Krishna KV, Sabarish S (2021) Journey companion, an android travel and tourism application. Ind J Softw Eng Proj Manag (IJSEPM) 1(1) ISSN: 2582–8339

Multilingual Voice-Assisted for Traffic Sign Detection and Classification in Adverse Weather Conditions

K. Krishnendhu, Neha Elizabeth Mathew, R. Gunavathi, and Amala Johnson

Abstract In a world where millions of people are wounded in auto accidents each year due to negligence, a lack of understanding of traffic laws, and bad weather, there is an urgent need for greater road safety. This is particularly the case in India, where a disproportionately high number of traffic accidents lead to numerous fatalities. Ignoring traffic signs raises these risks and endangers not only vehicles but also passengers and pedestrians. This project addresses the significant issue of traffic sign recognition in bad weather and offers voice-based instruction in many languages to increase road safety. Using a mix of state-of-the-art technologies, including YOLOv8 for real-time sign detection and the Google Translate API, which supports NLP tasks, this research offers a full solution. The model's remarkable precision and efficacy underscore its capacity to revolutionize traffic safety and furnish a more secure and expedient driving encounter. With the world moving towards more autonomous mobility, this study is laying the groundwork for safer and more effective driving in the future.

Keywords Road safety · Traffic laws · Traffic signs · Voice-based guidance · Multiple languages · Inclement weather · Google translation API · YOLOv8

1 Introduction

Vehicle accidents are a common and horrifying occurrence that cause injuries to millions of people worldwide. Numerous factors commonly contribute to these collisions, the most common ones being carelessness, a disregard for traffic signs, ignorance of the law, and carelessness in inclement weather. These crashes have detrimental effects on society as a whole in addition to the drivers who are involved. The seriousness of traffic accidents in India serves as a stark reminder of the urgent need

K. Krishnendhu · N. E. Mathew (✉) · R. Gunavathi · A. Johnson
Christ (Deemed to be) University, Bengaluru, India
e-mail: nehamathew@msds.christuniversity.in

V. K. Gunjan and J. M. Zurada (eds.), *Proceedings of 5th International Conference on Recent Trends in Machine Learning, IoT, Smart Cities and Applications*, Lecture Notes in Networks and Systems 1181, https://doi.org/10.1007/978-981-97-8861-3_36

for tighter enforcement of safety laws. Ignoring traffic signs and signals is a significant issue that raises the risk of accidents. In addition to the individuals who choose to disregard these signs, it also endangers other vehicles, passengers, and pedestrians who share the road. In addition to being essential for maintaining orderly traffic flow, traffic signs and signals are also essential for lowering the frequency and severity of collisions on the road [1]. Auto-pilot modes for cars have been introduced as a result of numerous technical advancements. There are now automobiles that operate on their own. The market for self-driving cars has expanded rapidly. These amenities, however, are limited to a few expensive vehicles that are out of reach for most people [2]. The current study aims to create a system that would make driving easier. Studies have discovered that surveying the number of traffic accidents in India is concerning. According to reports, there are roughly 53 accidents on the roadways per hour. Moreover, these accidents result in about 16 fatalities every hour [3].

The present study desires to address the issues concerning traffic sign board recognition in adverse weather conditions and provide the driver with voice-based assistance in the language of their choice. The main objective of this study is to solve these problems. An explanation of the paper's structure is provided below:

2 Literature Review

People regularly disregard traffic signs in the rush of today's fast-paced society, which increases the risk of breaking the law and being involved in accidents. Given the importance of this problem, a great deal of study has been done to reduce the likelihood of accidents [4]. To properly classify traffic signs and swiftly inform drivers, researchers have implemented many Convolutional Neural Network (CNN) architectures and investigated various classification techniques. Our system's design aims to provide further benefits, such as early warnings for drivers, while also improving the recognition process. We aim to improve traffic safety by refining the traffic sign recognition system, even in the face of challenges brought on by modern, hectic lives [3, 5].

According to research by Jeong, YOLOv5x was trained to recognise traffic conditions using smartphone photographs from the IEEE BigData Cup Challenge 2020 dataset. Apart from CSPNet and Spatial Pyramid pooling, YOLOv5x provides speed and accuracy with features like multi-GPU training, complex activation functions, data augmentation, and a readable manual [6]. The model performed better when datasets from several nations were combined, yielding F1 values of 0.568 and 0.571 for separate datasets. Test-time Augmentation (TTA) and training with the India dataset were two experiments that resulted in higher F1 scores. TTA did not, however, considerably improve performance, and although F1 scores were acceptable, they might have yet to satisfy some standards for practical use [7–9].

Jose et al. presented a camera-based system for recognizing traffic signs that integrates deep learning with the Viola-Jones Framework [10]. They used Indian road signs and the German Traffic Sign Recognition Benchmark (GTSRB) to evaluate the system, and they were able to obtain an accuracy of 90% for the Indian dataset and 71.56% for the GTSRB [10]. In order to lessen distractions, the system efficiently informed drivers of recognized signs using voice prompts. The system's capacity to recognize road signs outside these classes may need to be improved because its efficacy is restricted to a predetermined selection of classes within both datasets [10]. Enhancing traffic sign recognition speed and accuracy in high-resolution photos was the goal of Avramovic et al.'s paper, "Neural-Network-Based Traffic Sign Detection and Recognition in High-Definition Images" [11]. Building upon the YOLO architecture, they used effective preprocessing and parallelized CNNs on GPUs. They implemented moveable Regions of Interest (ROIs). They used larger ROIs across different architectures, improving accuracy by at least 0.4% using the DFG dataset, which includes a wide range of traffic sign pictures. Detection accuracy was improved by fine-tuning on a larger dataset with movable ROIs. Though it processed more slowly than TinyYOLO, which was quicker but less accurate than YOLOv3 and YOLO, YOLOv4 with movable ROIs performed more accurately than other architectures.

This work provides ideas for enhancing high-definition traffic sign identification [11].

A YOLOv3-based approach for enhancing traffic sign recognition was presented by Gisela Pereyra Doval et al. in their paper "Traffic Sign Detection and 3D Localization via Deep Convolutional Neural Networks and Stereo Vision" [12]. To overcome the shortcomings in the available datasets, they established the LSITSD dataset. The study demonstrated the algorithm's efficacy in precisely calculating traffic signs' 3D placements and detecting them, especially at lower resolutions. The evaluation techniques included ground truth distances determined by LiDAR for 3D localization, as well as average precision (AP) and F1 score for detection. Notably, the altered design demonstrated the trade-off between accuracy and speed when resizing high-resolution photographs, as it performed better on lower-resolution images but at lower Frames Per Second (FPS). This work advances algorithms for spatial localization and traffic sign recognition [12].

3 Methodology

3.1 Dataset

The Ceit-TSR Dataset was used in the suggested system, along with photos that the authors took and obtained from Google. The incorporation of different traffic indicators for model training is shown in Fig. 1. The training and test datasets together contain 2242 distinct photos that have been divided into eight classes including 3406 annotations. A breakdown of the image count by class is shown in Fig. 2. Crucially, there is no visual ambiguity because every single one is meticulously focused on a different traffic sign, guaranteeing individuality. Each of the included classes has its own folder containing the training dataset, along with a CSV file that contains information about the picture location, class, and other properties like width and height.

Fig.1 Preview of the dataset

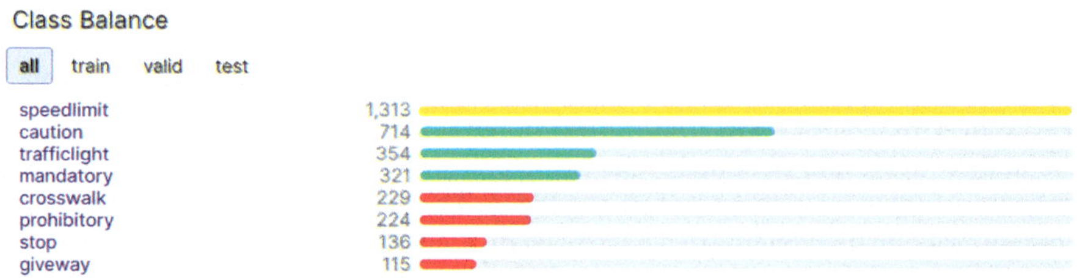

Fig.2 Class balance

3.2 Data Preparation

The dataset used in this study was originally made up completely of unlabeled data. Labeling the data was a crucial step in preparing it for model training. Roboflow, a computer vision development environment, and Labell-mg, two specialized tools, were used to help with labeling. By giving each data point a relevant label, these technologies were essential to the thorough annotation of the dataset. The labeled data was carefully divided into sets for training, testing, and validation after the labeling procedure. With a precise ratio in mind, 70% of the data were partitioned for training, 20% for testing, and 10% for validation. The dataset was systematically divided to enable the development and thorough evaluation of machine learning models for optimal performance.

3.3 Model

The model architecture used in this study is an example of a carefully planned framework that skillfully combines multiple elements to meet the goals of the investigation [2]. It combines the Marian neural machine translation framework with YOLOv8, an innovative image detection algorithm, to ensure a reliable and effective system for multilingual voice-guided directions and traffic sign recognition.

Image Detection with YOLOv8:YOLOv8, an evolution of the YOLO series, is in the model's center and is well-known for its remarkable real-time object detection capabilities. This allows it to be used for image detection [5]. A key component of the model's first step is YOLOv8, which quickly and correctly recognizes traffic signs in various weather scenarios. Its skills are crucial for guaranteeing timely and accurate detection, which is a crucial part of improving road safety [12].

Class Output: The algorithm classifies the detected traffic signs into eight classes after detecting the images. In this step, key traits and features are extracted, which enables the model to categorize each indication accurately [13]. The class output makes sure that traffic signs are represented in an organized manner, which makes voice advice and translation possible later on [14].

Translation with Marian: The model's ability to offer multilingual translations for the recognized traffic signs is largely due to the Marian neural machine translation framework. Marian's proficiency with natural language processing makes it possible to translate material accurately and contextually [15, 16]. Providing directions in the driver's preferred language greatly improves the user experience and helps to make navigation safer and easier to use [17, 18].

Voice Output with gTTS: The last part of the model is in charge of producing voice-guided instructions, which adds usability and usefulness to the system. This is accomplished by integrating the Google Text to Speech (gTTS) library, which uses the Text to Speech API from Google [17]. Customers may easily choose the audio language they want to hear, guaranteeing that the translated text is clear and understandable [19].

YOLOv8's Val mode provides a precision-driven evaluation approach throughout the validation stage. It offers precise measures, such as mAP50, and mAP50-95, to evaluate the model's performance in all aspects [12]. By maintaining training settings and enabling flexibility in validation datasets and image sizes, the Val mode streamlines the validation procedure. This adaptability is especially useful for adjusting the model's performance to match certain domains and real-world applications [10].

3.4 Proposed Solution

The precision and efficiency of the readied network are shown in Fig. 3. With its carefully crafted model, the suggested solution has continuously proven more accurate than other models. This accomplishment highlights the model's remarkable capability in multilingual voice-guided directions and traffic sign recognition, which adds significant value to navigation systems and road safety [15, 16, 20].

3.5 Proposed Methodology FlowChart

4 Implementation

A full-stack web application with a trained prediction model, constructed with Node.js and Flask, is presented in the study paper. This system divides incoming photos into one of eight classes using a convolutional neural network (CNN). The system notifies the user if a traffic sign is missing from an image. It determines the highest probability class by examining the model and predict; output array. To improve functionality and user experience, the system also includes speech output via gTTS and machine translation using the Marian model. Following multiple testing iterations, the results showed that the model continued to classify images into one of the eight groups, even in cases where they did not fit into any of the specified categories. The anticipated values produced by the and "model. Predict" function are noticeably low in these circumstances, however.

This technique makes it easier for the user to identify traffic signs that are nearby. It uses Google Text-to-Speech (gTTS) and the Google Translation API to deliver speech alerts, assisting users in adhering to traffic sign boards even in adverse weather conditions. In addition, the system offers assistance in the language the user requires [15, 16, 20].

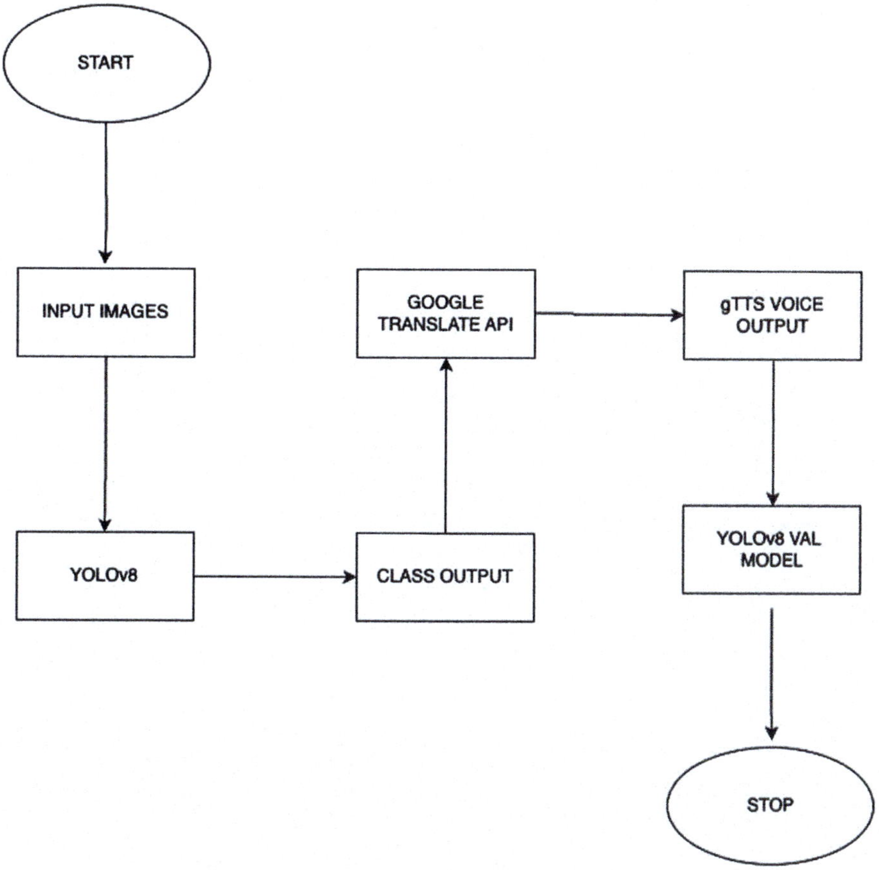

Fig. 3 Flow diagram

5 Results and Analysis

Using the pre-trained YOLOv8x model, we trained the model in our implementation on eight classes of photos throughout 100 training epochs using 640 pixels as the image input size. After pretraining on the COCO dataset, the YOLOv8x algorithm processed data at a speed of 479.1 ms and attained an accuracy of 55.9%. After 100 training epochs of Auto-Batch Size and GPU acceleration, the model and performance was evaluated again. The results showed a Mean Average Precision (mAP) of 89.1%, Precision of 87.1%, and Recall of 84.9%.The graphs in Figs. 4, 5, 6 and 7 respectively represent the Models Accuracy measured on Validation Set, Box Loss, Object Loss and CLass Loss. The generated categories were then translated into the user's preferred language using the Google Translate API. The gTTS model was then employed to provide voice output. These outcomes prove our model's efficacy and Precision and highlight its potential for practical uses.

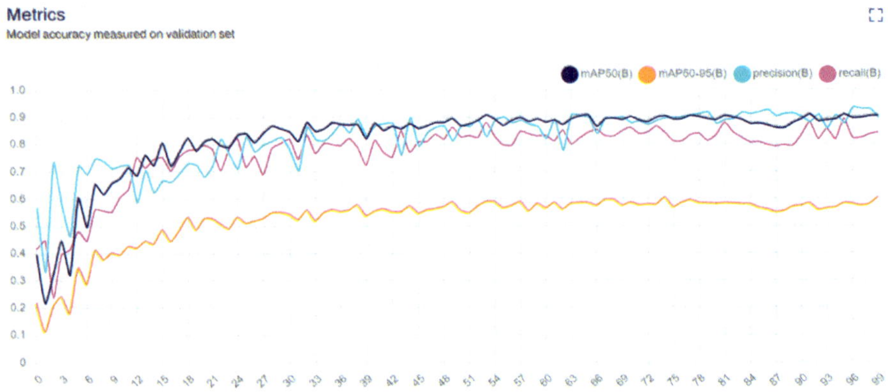

Fig. 4 Models accuracy measured on validation set

Fig. 5 Box loss

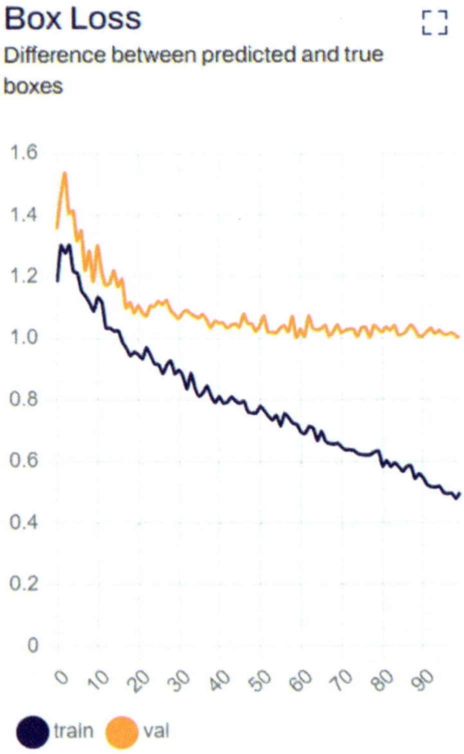

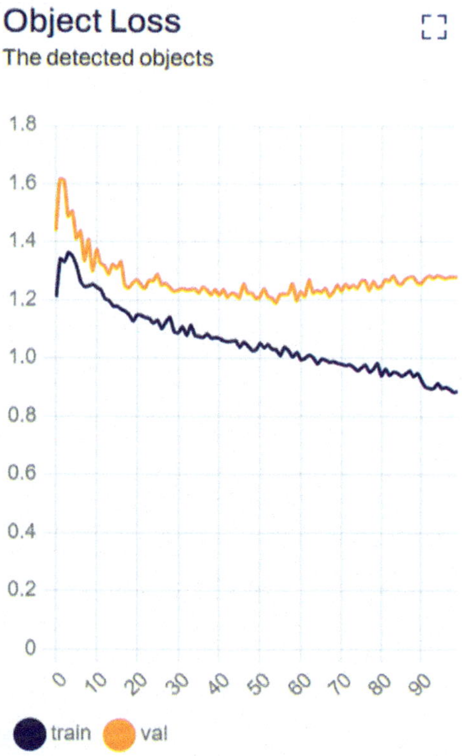

Fig. 6 Object loss

6 Further Scope

The study establishes a solid framework for enhancing traffic safety with cutting-edge technology, with a particular emphasis on voice-based guiding and traffic sign recognition. There are, nevertheless, several directions for further study and possible gaps that need to be filled. First off, even if the study provides a thorough answer, actual testing in a variety of environmental settings is required to confirm the system's viability. The system's usefulness would be increased by including self-driving vehicles and extending the coverage of traffic signs to include temporary or non-standard signs. Further enhancing the system's adaptability and usability might be user customization choices, strong user verification techniques, and cloud-based updates for sign recognition and translations. Future research must address data privacy and ethical issues and work with traffic authorities to integrate this system with the current traffic management infrastructure. The system can develop into a comprehensive, internationally applicable solution for improving traffic sign recognition and road safety by filling in these gaps, ultimately making driving safer and more effective.

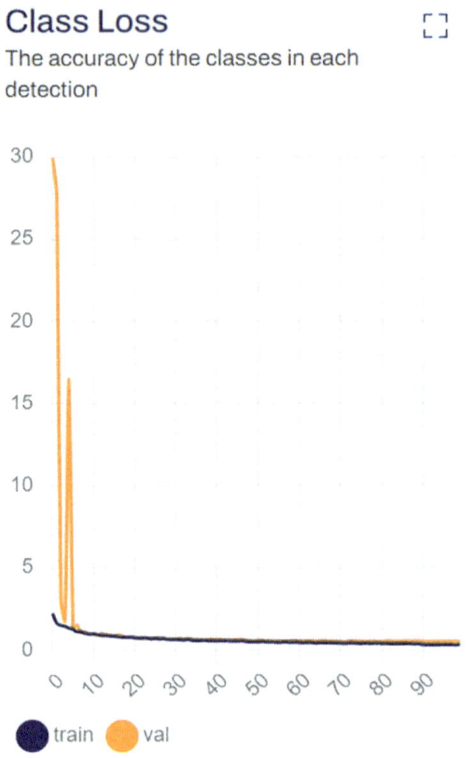

Fig. 7 Class loss

7 Conclusion

In conclusion, this research study has shed light on traffic safety, emphasizing the value of voice-activated navigation devices and effective traffic sign identification. People all throughout the world are impacted by the devastating and frequent problem of road accidents. Negligence, a failure to follow traffic laws, a disdain for traffic signals, and bad weather are often the causes of these incidents. The shockingly high frequency of traffic accidents in India and the urgent need for stricter safety regulations are two topics covered in detail in this study. The approach proposed in this study aims to increase road safety by offering a complete way to lessen the dangers related to disobeying traffic signs. A comprehensive examination of the literature is explored in order to accomplish this, with a focus on the advancements achieved in Convolutional Neural Network (CNN) designs for the detection of traffic signs and their potential to alert drivers. The YOLOv8 image detection algorithm for real-time traffic sign identification and the Marian neural machine translation framework for multilingual voice-based guidance are integrated into the system's design. The model was reevaluated after 100 training epochs with GPU acceleration and Auto-Batch Size. The findings revealed an 89.1% Mean Average Precision (mAP), 87.1% Precision, and 84.9% Recall. A detailed evaluation of the system's performance demonstrates its accuracy and efficacy. However, there are still intriguing directions to go in the future, such as testing in real-world settings, interacting with autonomous vehicles, and addressing moral and privacy concerns. By addressing these shortcomings, the proposed system can grow into a complete traffic safety solution that would ultimately result in safer and more efficient route navigation.

References

1. Wang C (2018) Research and application of traffic sign detection and recognition based on deep learning. IEEE. https://doi.org/10.1109/icris.2018.00047
2. Dhawan K, Srinivasa PR, Nadesh RK (2023) Identification of traffic signs for advanced driving assistance systems in smart cities using deep learning. Multimed Tools Appl 82(17):26465–26480. https://doi.org/10.1007/s11042-023-14823-1.
3. Krish S, Radha S, Jay S, Krushna S (2021) Traffic sign board recognition and voice alert system using convolutional neural network. In: IEEE conference publication. IEEE Xplore. https://ieeexplore.ieee.org/stamp/stamp.jsp?tp=&arnumber=9456302
4. Rahman MK (2012) Speed limit sign board detection and extraction of digits in different weather conditions. https://dspace.bracu.ac.bd/xmlui/handle/10361/2380
5. Yadav S (2019) Indian traffic signboard recognition and driver alert system using machine learning. Yadav Int J Appl Sci Smart Technol. https://e-journal.usd.ac.id/index.php/IJASST/article/view/1843/1490
6. Gupta N, Bisantz AM, Singh T (2002) The effects of adverse condition warning system characteristics on driver performance: an investigation of alarm signal type and threshold level. Behav Inf Technol 21(4):235–248. https://doi.org/10.1080/0144929021000013473
7. Abdulrahman K, Mehmet OE (2021) Recognition of traffic signs with artificial neural networks: a novel dataset and algorithm, In: IEEE conference publication. IEEE Xplore. https://ieeexplore.ieee.org/document/9415238
8. Zhao Z-Q, Zheng P, Xu S-T, Wu X (2019) Object detection with deep learning: a review. IEEE Trans Neural Netw Learn Syst 30(11):3212–3232. https://doi.org/10.1109/tnnls.2018.2876865
9. Jeong D (2020) Road damage detection using YOLO with smartphone images, In: 2020 IEEE international conference on big data (Big Data). https://doi.org/10.1109/bigdata50022.2020.9377847
10. Jose AT, Thodupunoori H, Nair BB (2019) A novel traffic sign recognition system combining Viola–Jones framework and deep learning. In: Advances in intelligent systems and computing, pp 507–517. https://doi.org/10.1007/978-981-13-3600-3_48
11. Avramović A, Sluga D, Tabernik D, Skočaj D, Stojnić V, Ilc N (2020) Neural-network-based traffic sign detection and recognition in high-definition images using region focusing and parallelization. IEEE Access 8:189855–189868. https://doi.org/10.1109/access.2020.3031191
12. Doval GN, Al-Kaff A, Beltrán JR, Fernández FG, López GF (2019) Traffic sign detection and 3D localization via deep convolutional neural networks and stereo vision. Int Conf Intell Transp Syst. https://doi.org/10.1109/itsc.2019.8916958
13. Wang G, Chen Y, An P, Hu H, Hu J, Huang T (2023) UAV-YOLOV8: a small-object-detection model based on improved YOLOV8 for UAV aerial photography scenarios. Sensors 23(16):7190. https://doi.org/10.3390/s23167190
14. Cao J, Zhang J, Jin X (2021) A traffic-sign detection algorithm based on improved sparse R-CNN. IEEE Access 9:122774–122788. https://doi.org/10.1109/access.2021.3109606
15. MarianMT. https://huggingface.co/docs/transformers/model_doc/marian
16. Sign in to Roboflow. https://app.roboflow.com/traffic-res/traffice-signs-detection/health
17. Chen L et al (2021) Deep neural network based vehicle and pedestrian detection for autonomous driving: a survey. IEEE Trans Intell Transp Syst 22(6):3234–3246. https://doi.org/10.1109/tits.2020.2993926

18. Natarajan SP, Annamraju AK, Baradkar CS (2018) Traffic sign recognition using weighted multi-convolutional neural network. IET Intel Transp Syst 12(10):1396–1405. https://doi.org/10.1049/iet-its.2018.5171
19. Liu C, Li S, Chang F, Wang Y (2019) Machine vision based traffic sign detection methods: review, analyses and perspectives. IEEE Access 7:86578–86596. https://doi.org/10.1109/access.2019.2924947
20. https://hub.ultralytics.com/models/iA3yRzF7wS2QRBKY0T7a

A Survey of Ethics in Aadhaar, Cybersecurity, and Healthcare Data

Priya Vijay, Poojitha Ramesh, and Sai Keshav I

Abstract In the face of raising enterprises over data breaches and sequestration contraventions, this study intricately examines ethical confines within the Aadhaar scheme, cybersecurity, and digital healthcare. Checking the impact of the Aadhaar frame on individual sequestration, the exploration critically analyses both its graces and limitations. Addressing emergent pitfalls in the cybersecurity sphere, the study proposes strategies for visionary data breach forestalment. Shovelling into digital healthcare, ethical nuances girding health data, encompassing collection, storehouse, and innovative technologies like telemedicine, are explored. A relative analysis of global nonsupervisory fabrics enhances our understanding of data protection. This check offers perceptivity to guide policymakers and interpreters, emphasizing the imperative for a comprehensive approach to responsibly manage information in our decreasingly digitized society.

Keywords Data privacy · Aadhaar · Cybersecurity · Digital healthcare · Ethical considerations · Regulatory frameworks

1 Introduction

In a period marked by unknown digital integration, enterprises girding data sequestration, cybersecurity, and the ethical counteraccusations of evolving technologies have come consummate. This preface provides a comprehensive overview of the current

P. Vijay
Department of Information Technology, Rajalakshmi Engineering College, Chennai, India
e-mail: priya.vijay@rajalakshmi.edu.in

P. Ramesh (✉) · S. Keshav I
Department of Information Technology-Data Science, Rajalakshmi Engineering College, Chennai, India
e-mail: 231011009@rajalakshmi.edu.in

S. Keshav I
e-mail: 231011012@rajalakshmi.edu.in

V. K. Gunjan and J. M. Zurada (eds.), *Proceedings of 5th International Conference on Recent Trends in Machine Learning, IoT, Smart Cities and Applications*, Lecture Notes in Networks and Systems 1181, https://doi.org/10.1007/978-981-97-8861-3_37

state of exploration in these cutting disciplines specifically, the Aadhaar scheme, cybersecurity practices, and the geography of digital healthcare.

The rapid-fire digitization of particular information, instanced by enterprise similar as the Aadhaar scheme, has led to heightened apprehensions about individual sequestration and the ethical ramifications of ubiquitous data collection. Coincidently, the raising frequency and complication of cyber pitfalls emphasize the imperative for robust cybersecurity measures to guard sensitive information. Meanwhile, the integration of digital technologies in healthcare introduces ethical complications related to the collection, storehouse, and application of health data.

This study aims to contribute to the ongoing converse by intricately examining the ethical confines within the environment of Aadhaar, cybersecurity, and digital healthcare. Through a critical analysis of the Aadhaar frame, we aim to delineate both its graces and limitations, slipping light on its impact on individual sequestration. Addressing emergent pitfalls in the cybersecurity sphere, our exploration proposes visionary strategies for data breach forestalment. Shovelling into the ethical nuances of digital healthcare, encompassing the realms of data collection, storehouse, and innovative technologies like telemedicine, we seek to unravel the ethical complications girding health data.

The motivation is to discover the threats related to the privacy of a person traveling places and how safe can the information and data be kept. In the current era, as each and every action is done digitally, the data likewise is also stored in a digital format, i.e.: cloud storage, etc. The goal is to identify and understand the measures that can be taken in order to protect the data of each individual and henceforth implement those measures effectively.

2 Impact of Aadhar Scheme

S. no	Paper/author	Methodology used	Findings	Issues/ drawbacks	Inference
[1]	India's "Aadhaar" Biometric ID: Structure, Security, and Vulnerabilities. – Pratyush Ranjan Tiwari, et al.	– Thorough analysis of Aadhar structure with the help of public sources and direct communications with the developers – Identified limitations on security and privacy, researched potential risks, and examined the Central Identities Data Repository (CIDR)	– Bringing to light the infrastructure's weaknesses, potential risks and misuse of data by external unauthorized parties – Need for enhanced security, due to 17 security breaches and 10 privacy breaches	– Criticized for neglect in complying with security requirements, probable data manipulation, and lack of openness in data access	– Though Aadhar possesses a strong security, there is still lack of transparency making it necessary to adhere to strict security measures
[2]	Evaluating Aadhar Scheme with Special Reference to The Right to Privacy.—Dr Khaleeq Ahmad, et al.	– Utilized legal research, policy assessment, and critical inquiry to examine Aadhaar's influence on privacy rights in India	– Highlighted the privacy difficulties, data leaks and possible exploitation of personal data – Government services are inaccessible due to technical or logistic issues	– Criticized Aadhaar for privacy concerns, data leakage, and lack of transparency in UIDAI operations	– Data security flaws, privacy problems, and lack of transparency in the Aadhar governance were pointed out. The research thus demanded for strict rules and a secured infrastructure

(continued)

(continued)

S. no	Paper/author	Methodology used	Findings	Issues/ drawbacks	Inference
[3]	Becoming data: biometric IDs and the individual in 'Digital India.—Vijayanka Nair	– Employed an ethnographic technique involving fieldwork, participant observation, and docu-mentation analysis	– Aadhar has the ability to build a digital society but concerns are raised about privacy and data leaks	– Capture and storage of biometric data for Aadhar has raised concerns related to privacy and surveillance	– Though Aadhar supports diversity, threats were imposed upon the privacy, data security, and possible misuse of data

(continued)

(continued)

S. no	Paper/author	Methodology used	Findings	Issues/ drawbacks	Inference
[4]	Seeing Like an Infrastructure: Low-resolution Citizens and the Aadhaar Identification Project.—Ranjit Singh, et al.	– Utilized ethnographic fieldwork over eighteen months to study the implementation of Aadhaar, assessing tactics, mechanisms, and barriers in enrolment, seeding, and authentication procedures	– In all cases Aadhar acts as an un-biased system but there are exceptional cases where Aadhar is unbiased in citizenship – Aadhar requires biometric details like thumb print which make it a challenge for people with physical disabilities to make use of the biometric system, citizens without a permanent resident will also face difficulties in acquiring an Aadhar	– Aadhaar's implementation is inconsistent, leading to exclusion and discriminatory treatment	– The goal of Aadhar is to establish a citizenship but obstacles arise when it comes to implementing the practice, maintain record for data and establish access to every corner of the society. In order to establish a fair system every citizen must have equal rights

(continued)

(continued)

S. no	Paper/author	Methodology used	Findings	Issues/ drawbacks	Inference
[5]	Protecting Mental Health Data Privacy in India: The Case of Data Linkage With Aadhaar.—Ameya Bondre, et al.	– Offered guidelines based on GDPR, stressing ethical standards in AI systems used for mental health research and clinical treatment	– People diagnosed with mental health conditions face a serious threat related to privacy of their personal information – With the involvement of AI, issues and threats that arise due to privacy issues will only complicate matters especially in a country like India where there is a lack in data and information – Opting for an Aadhar is optional, but implementing strong legal frameworks for data privacy is a must and giving a clear view of AI biases is necessary while using mental health data	– Addressed concerns about Aadhaar's centralized data structure, potential AI biases, and the absence of particular privacy rights in Indian data protection laws	– The AI models must be trained in such a way that they maintain fair and accurate data throughout the system and is not and will not be biased in any way

The research infers many issues and threats which imposes a potential risk on privacy of data. Privacy issues, lack of transparency, risks associated to data security and poor data quality are the prominent issues under the Aadhar scheme. In order to tackle the issue associated to transparency it is necessary to tighten the security force, impose strict regulations, bring in policy measures and laws and safeguarding the policies simultaneously, and likewise the policies must be stated clearly without any misleads which will lead to lack of uncertainty. Correspondingly respecting the data associated to mental health is necessary to enhance the trust upon the digital services associated with data privacy. On top of that is the AI system, developers must ensure the development of an unbiased and trustable AI algorithm, securing and ensuring genuine data quality across diverse systems.

3 Impact on Digital Healthcare

S. no	Paper/author	Methodology used	Findings	Issues/drawbacks	Inference
[6]	Privacy Issues in Wearable Technology: An Intrinsic Review. Vidhi Kapoor, et al.	– A review was done on wearables which are used in tracking certain health factors like heart rate, pulse rate, etc. to highlight the associated privacy issues – Reviewed and analysed relevant material for specific problems – Emphasized testing for privacy in remote patient monitoring equipment – Provided future directions, highlighting the importance for continuing study	– Identified privacy concerns: data authentication, unprotected access, time/location-based privacy, third-party restrictions, and IoT app accessibility – Healthcare data sensitivity amplifies unauthorized access threats – Emphasized continual research for tackling privacy issues	– Ambiguity in health data authentication threatens privacy breaches and illegal access – Unsecured data access exposes personal health information to unwanted access – Time/location tracking weakens user privacy, causing issues -Inadequate third-party policies create problems concerning personal health data sharing – Insecure IoT app accessibility can lead to privacy breaches	– Balancing privacy and data utility is a difficulty in wearable devices – Existing authentication techniques may have faults, threatening breaches – Urged stringent privacy protections and continuous research for wearable device privacy

(continued)

(continued)

S. no	Paper/author	Methodology used	Findings	Issues/ drawbacks	Inference
[7]	Privacy and Security Concerns with Electronic Health Records-Shreds of Evidence from India.—Vinaytosh Mishra	– Utilized secondary literature study for privacy vulnerabilities in Indian personal health details and Electronic Health Records (EHR) – Conducted narrative literature review to identify these concerns – Employed Best–Worst Method (BWM), a multicriteria decision-making approach devised by Jafar Rezaei of Delft University in 2015 – Panel of six healthcare specialists and two IT professors scored concerns; price and government policies regarded most relevant	– Narrative review revealed six main EHR privacy concerns in India: cost, government policy top-ranked – HIPAA adherence increases Indian healthcare sector's worldwide competitiveness – Patients' perspectives on data privacy crucial owing to increased empowerment – Implementing security measures in resource-limited India presents ethical difficulties	– Resource limits and government backing key in EHR privacy and security rollout in India – Top concerns: price and government regulation in EHR privacy and security rankings – HIPAA can be incorporated in the Indian Healthcare to compete globally but it is denied due to its high complexity – Patient empowerment underscores the importance of their perspective on privacy challenges	– In order to maintain Electronic Health Record in India there is a need for resources, government assistance and monetary funds – HIPAA can be approved for its effectiveness but denies for its complexity -Patient empowerment highlights the importance of considering patients' perspectives on privacy issues – The study provides a valuable EHR privacy framework for policymakers in India and other developing nations

(continued)

(continued)

S. no	Paper/author	Methodology used	Findings	Issues/ drawbacks	Inference
[8]	A Comprehensive Survey on Data Utility and Privacy: Taking Indian Healthcare System as a Potential Case Study.—Prathamesh Churi, et al.	– Utilized literature review and analysed research studies, including the NITI Aayog report – Identified factors contributing to privacy issues: complacency, culture, politics, budget, and population – Explored data mining techniques (SVM, decision tree, KNN) and AI applications in medical fields	– Privacy issues stem from complacency, cultural factors, politics, budget constraints, and India's large population – Concerns include doctor-patient relationships, consent management, and lack of security policies – High data volume raises breach risks, necessitating immediate privacy protection – Telemedicine industry faces data breaches, phishing attacks, and unauthorized access due to employee busyness and lack of security training	– Unstructured data storage raises concerns about privacy and data breaches – Cultural factors affect healthcare disclosure, leading to data discrepancies – Challenges in implementing privacy models due to India's large population and high costs – AI-based healthcare apps raise privacy concerns regarding sensitive data use for training and individual identification	– Balancing privacy and data utility is a challenge in healthcare digitization, especially in culturally diverse nations like India – Effective privacy models and stringent regulations are essential to protect sensitive healthcare data – A clear view on the risks and threats imposed by data breaches and unauthorized access in the health sector must be stated to the society. Via training and awareness drives

(continued)

(continued)

S. no	Paper/author	Methodology used	Findings	Issues/ drawbacks	Inference
[9]	Digitization of healthcare sector: A study on privacy and security concerns.—Metty Paula, et al.	– Privacy and Security issues still prevail in the healthcare sector, as it can be affected by a serious threat due to technology – Before implementing any digital strategies it is necessary to consider the citizens of the environment – Explored methodologies like predictive analytics, big data algorithms and evidence-based medicine	– In the fast paced era, the healthcare sector is expected to surge as high as 38 Billion by 2025 – Operational Intelligence and patient care is improved with the use of AI/ ML and IOT – Common challenges in the health care sector include organizing and storing patient data which can lead to data leaks and privacy issues – Cybersecurity threats include ransomware attacks, software vulnerabilities and phishing	– Cybersecurity challenges: endpoint leakage, authentication deficiencies, excessive user permissions – Integrating digital technology: prioritizing customer privacy, ensuring secure communication – Global expansion challenges: rising complexity costs – Solutions: advanced authentication methods, secure communication protocols	– Medical knowledge searches and related researches, patient observation, clinical support, real-time monitoring, better decision-making can be enhanced by digitalizing the healthcare sector – IOT and AI play a major role in the digitalization of the healthcare and privacy, security of the patient data must also be ensured to prevent any issues and risks

Healthcare is digitalized with the help of AI and IOT not only for storing patient data, monitoring patients, maintaining electronic health records, conducting medical knowledge searches and researches, real-time monitoring, making decisions and also keeping track of a normal person's health factors like pulse rate, heartbeat, etc. in the form of wearables. But in the digital era, there is a need to maintain privacy of the patient the data, follow strict privacy protections, following a continuous research for privacy in wearable devices. Electronic Health Records can be implemented by following HIPAA but India lacks the necessary government assistance, resources and monetary funds, moreover the infrastructure of HIPAA make it a disapproval for the system. As issues related to data security and privacy are common in the digital era, the society must be aware of such happenings in the digital system by organizing training and awareness camps, campaigns and programs.

4 Cybersecurity and Data Breach Prevention

S. no	Paper/author	Methodology used	Findings	Issues/drawbacks	Inference
[10]	Developing a Framework to Measure Cyber Resilience Behaviour of Indian Bank Employees.—Tanvi Godbole, et al.	– Quantitative approach, 8 identified factors as independent variables, cyber resilient behaviour as dependent variable, focused on regulated Indian banking sector	– Addressed gaps in employee-focused cyber resilience, emphasized unique challenges in Indian banking, and proposed culture fostering cyber resilience	– Lack of framework limitations detail	– Employee behaviour crucial, focus on both IT and behaviour vital for cyber resilience, culture influences cyber resilience
[11]	Secured and Privacy-Based IDS for Healthcare Systems on E-Medical Data Using Machine Learning Approach.—Ibrahim Osamah, et al.	– Hidden Markov Model (HMM) for attack classification, machine learning-based IDS, evaluation through precision-recall curves, focus on DDoS detection	– High precision and recall rates, robust against diverse attacks, emphasis on security, privacy, and awareness	– Dependent on training dataset quality, algorithm choice critical, adaptability and scalability challenges, need for continuous monitoring	– Machine learning enhances IDS accuracy, robust training crucial, holistic approach covering various attacks necessary, privacy awareness key
[12]	Confrontation-Wi-Fi Risks and Data Breach.—Ravi Kamurthi, et al.	– Experiments on Wi-Fi signals and interference, introduction of Wi-Fi channel setting, disaster recovery plan suggestion, and new password manager with PWNAGE check methodology	– Wi-Fi channel 7 avoids intrusions, emphasizes disaster recovery planning, minimization of customer data collection, new password manager introduced	– Lack of detailed methodology, missing information on drawbacks and challenges	– Awareness vital for data breach prevention, Wi-Fi channel optimization aids security, disaster recovery and data minimization essential, new tools enhance data safety

(continued)

(continued)

S. no	Paper/author	Methodology used	Findings	Issues/drawbacks	Inference
[13]	Cybercrime and Cybersecurity in India: Causes, Consequences and Implications for the Future.—Nir Kshetri	– Multidisciplinary approach: economics, criminology, institutional studies, international relations – Analytical framework: formal and informal institutions, prosperity factors, international relations dynamics	– Rising cybercrime, challenges in prosecution and reporting – Successful initiatives by NASSCOM – Lack of awareness among cyber-offenders, resource limitations, unsupportive law enforcement attitudes, poor cybersecurity orientation	– Low prosecution and reporting rates – Resource constraints and caseload pressures – Informal and formal institutions contributing to cybercrime escalation	– Unique cybercrime risks in developing economies – Need for awareness among cyber-offenders – Challenges in prosecution, reporting, and cybersecurity orientation
[14]	Indian Government Interventions in Micro, Small and Medium Enterprise Development: An Enterprise Lifecycle Perspective.—Kalai Socrates, et al.	– Analysis from an enterprise life cycle perspective – Comparison with international policies and programs	– Focus on creation and growth stages, lacking interventions for revival and exit stages	– Poor enterprise development tactics – Enterprises often face difficulties and challenges which are unable to be fixed by the existing system – Competition from third parties, unorganized working infrastructure, lack of funds and poor handling of working capital	– Need for interventions in revival and exit stages – Supportive ecosystem to encourage new entrepreneurs – Improve existing enterprises' quality through government initiatives

(continued)

(continued)

S. no	Paper/author	Methodology used	Findings	Issues/drawbacks	Inference
[15]	Machine Learning for Intelligent Data Analysis and Automation in Cybersecurity: Current and Future Prospects.—Iqbal H. Sarker	– Utilizing machine learning algorithms for automating cybersecurity and for data analysis – Classification methods to build prediction and classification models in cybersecurity – Ensemble learning-random forest techniques to detect anomalies, denial of service attacks and cyber-attacks	– Training machine learning algorithms with real world cyber related use cases to identify non-redundant association rules – Contribute to the growth in knowledge based systems – Clustering method to solve security related security issues like anomaly detection and cyber-attack detection	– Security solutions are not sufficient to put forth contemporary security issues due to cyber-attacks and threats – Training machine learning algorithms to treat real world cyber-attacks – Need of large quantities of data	– Machine learning algorithms can strengthen cybersecurity through data analysis, automation and protection – Traditional security solutions are enough to put forth security related issues due to cyber-threats ad attacks – Developing and training machine learning algorithms with real world cyber data to develop solutions for the target problem

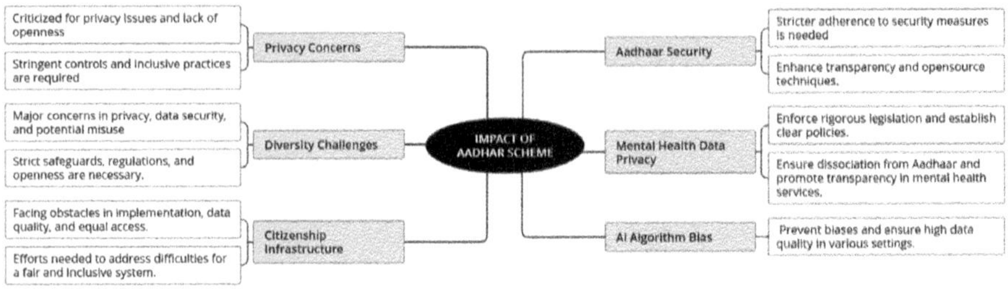

Fig. 1 Impact of Aadhar scheme

The employees and the technology architecture must work hand in hand in order to create a well-functioning secured environment and system as maintaining security is crucial for everyone involved within and outside the organization. Cyber-attacks must be taken seriously and strict regulations must followed in order to prevent attacks by following protocols. To avoid cyber-attacks machine learning tops the list as it helps identify and sort out the issues that arise with cyber-attacks. But ML alone will not help in preventing cyber-attacks; the systems must also follow other measures to prevent cyber-attacks from happening by prioritizing the privacy of data in any case. The company and employees alone working for the security of data is not enough but also effort from the public also plays a major role. The public must be made aware of risks and how to tackle them. Taking small measures like making Wi-Fi safer, planning for disasters, and using less data will make a big impact in preventing digital attacks. Cybercrimes which are high in number in certain places need immediate and strict action to put a stop to these intruders. Upcoming entrepreneurs require a supportive ecosystem and ongoing enterprises can be improved with the help of government support.

5 Conclusion

In this fast-paced digital era it has been identified that Aadhar, digital healthcare and cybersecurity give rise to a vast field of new discoveries and barriers. In Aadhar every detail and information is digitalized and likewise in healthcare every patient's personal data has also been digitalized in the form of monitoring and storing them. The main concern for both Aadhar and the digital healthcare is that storing the data and how effectively that data is used must go hand in hand and there should also be strict regulations which are to be followed while securing and storing the data. The employees and the security system must undergo frequent checks and sessions to know how to handle and safeguard the data. The more advanced the technology is, the more well trained employees and a well secured system is needed. Each discipline helps in understanding the happenings and changes in the digital era by supporting ongoing researches, investigations and analyses (Figs. 1, 2 and 3).

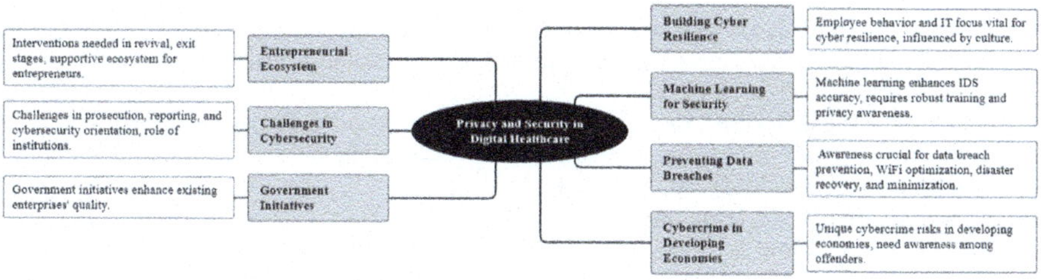

Fig. 2 Privacy and security in digital healthcare

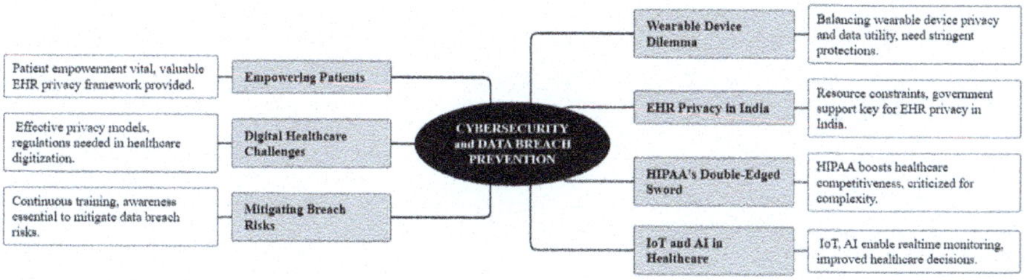

Fig. 3 Cybersecurity and data breach prevention

References

1. Tiwari PR et al (2022) India's 'Aadhaar' biometric ID: structure, security, and vulnerabilities. Financ Cryptogr Data Secur 672–693. https://doi.org/10.1007/978-3-031-18283-9_34
2. Prajapati S, Khan MH, Qureshi N, Ahmad DK Evaluating Aadhar scheme with special reference to the right to privacy. J Surv Fish Sci
3. Nair V (2021) Becoming data: biometric ids and the individual in 'digital India.' J Roy Anthropol Inst 27(S1):26–42. https://doi.org/10.1111/1467-9655.13478
4. Singh R, Jackson S (2021) Seeing like an infrastructure: low-resolution citizens and the Aadhaar identification project. Proceed ACM Hum Comput Interact 5(CSCW2):1–26. https://doi.org/10.1145/3476056
5. Bondre A, Pathare S, Naslund JA (2021) Protecting mental health data privacy in India: the case of data linkage with Aadhaar. Glob Health: Sci Pract 9(3):467–480. https://doi.org/10.9745/ghsp-d-20-00346
6. Kapoor V, Singh R, Reddy R, Churi P (2020) Privacy issues in wearable technology: an intrinsic review. SSRN Electron J. https://doi.org/10.2139/ssrn.3566918.MicrosoftWord-ResearchPrivacy.docx
7. Mishra V Privacy and security concerns with electronic health records-shreds of ..., ResearchGate
8. Churi P, Pawar A, Moreno-Guerrero A-J (2021) A comprehensive survey on Data Utility and privacy: taking Indian healthcare system as a potential case study. Inventions 6(3):45. https://doi.org/10.3390/inventions6030045
9. Paul M, Maglaras L, Ferrag MA, Almomani I (2023) Digitization of healthcare sector: a study on privacy and security concerns. ICT Express 9(4):571–588. https://doi.org/10.1016/j.icte.2023.02.007
10. Godbole T, Gochhait S, Ghosh D (2021) Developing a framework to measure cyber resilience behaviour of Indian bank employees. ICT Intell Appl 299–309. https://doi.org/10.1007/978-981-16-4177-0_31

11. Sengan S et al (2022) Secured and privacy-based IDS for healthcare systems on E-medical data using machine learning approach. Int J Reliab Qual E-Healthc 11(3):1–11. https://doi.org/10.4018/ijrqeh.289175
12. Kamurthi RT, Chopra SR, Sharma R (2021) Confrontation-Wi-Fi risks and data breach. In: 2021 International conference on emerging smart computing and informatics (ESCI). https://doi.org/10.1109/esci50559.2021.9397047
13. Kshetri N (2016) Cybercrime and cybersecurity in India: causes, consequences and implications for the future. Crime Law Soc Chang 66(3):313–338. https://doi.org/10.1007/s10611-016-9629-3
14. Socrates K, Gopalakrishna BV (2022) Indian government interventions in micro, small and medium enterprise development: an enterprise life cycle perspective. Colombo Bus J 11(1):132–164. https://doi.org/10.4038/cbj.v11i1.60
15. Sarker IH (2022) Machine learning for intelligent data analysis and automation in cybersecurity: current and future prospects. Annals Data Sci. https://doi.org/10.1007/s40745-022-00444-2

Forecasting Breast Cancer with Integrated Pre-trained CNN and Machine Learning Framework from CT Images

Jagendra Singh, Nazeer Shaik, Dinesh Prasad Sahu, Mohit Tiwari, Mohammad Shabbir Alam, and Kamal Upreti

Abstract This article investigates machine learning techniques' effectiveness at using computed tomography (CT) images to forecast breast cancer, hoping to expedite early identification and plan treatment. Drawing on many different machine learning models, such as CNN, SVM, VGG16, RNN and RF, we did extensive work to measure their performance distinguishing between malignant and benign breast tissue regions. The dataset includes 2,430 CT pictures, with 70% for training and 30% for testing. It has been carefully selected and prepared in order to guarantee robustness and consistency. The precision, and in-sensitivity measure the accuracy, sensitivity, specificity is used as analytic indicators to measure the models' ability to predict regions of breast cancer accurately. Our findings show that the proposed CNN model achieved an accuracy of 98.75%, superior performance. Other machine learning models are also highlighted in this study, demonstrating how breast cancer can be predicted using various methods. This research will determine the forms and technologies suitable for breast cancer forecasting. Medical imaging and clinical decision-making can move forward because of this research, offering a glimpse into

J. Singh (✉) · D. P. Sahu
School of Computer Science Engineering and Technology, Bennett University, Greater Noida, India
e-mail: jagendrasngh@gmail.com

N. Shaik
Department of Computer Science and Engineering, Srinivasa Ramanujan Institute of Technology—Autonomous, Anantapur, India

M. Tiwari
Department of Computer Science and Engineering, Bharati Vidyapeeth's College of Engineering, Delhi, India

M. S. Alam
Department of Computer Science, College of Computer Science and Information Technology, Jazan University, Jazan, Saudi Arabia
e-mail: shabbir@jazanu.edu.sa

K. Upreti
Department of Computer Science, CHRIST (Deemed to Be University), Ghaziabad, India

V. K. Gunjan and J. M. Zurada (eds.), *Proceedings of 5th International Conference on Recent Trends in Machine Learning, IoT, Smart Cities and Applications*, Lecture Notes in Networks and Systems 1181, https://doi.org/10.1007/978-981-97-8861-3_38

how integrated machine-learning systems can bring greater precision to diagnosis and prognosis. By careful experimentation and analysis, we hope to prepare people for early intervention and personalized treatment methods. This will make for improved patient outcomes in fighting breast cancer.

Keywords Breast cancer · Forecasting · Computed tomography · Machine learning · Convolutional neural networks

1 Introduction

Breast cancer presents a major global health problem. Early detection is useful for improving patient prognosis, and survival rates. Computed tomography (CT) imaging is an effective modality for diagnosing and staging breast cancer as it can provide detailed anatomical information. However, breast cancer can be accurately diagnosed from CT images and poses a serious obstacle due to the numerous and complex tumour characteristics [1, 2]. Manual interpretation and subjective assessment are both characterized by inconsistency and less than ideal outcomes.

To address these challenges, interest in machine learning techniques has been growing. They are intended to make automated breast cancer diagnosis from CT images more sensitive. By learning how to identify subtle indications of breast cancer from large numbers of labeled CT images, these models are expected to have some effect in aiding early detection and treatment planning in addition to Key image zones for prediction. And when trained together in large datasets of labeled CT images algorithms can recognize very slight patterns which hundreds of linear equations would all be necessary to solve [3, 4].

Research aims to look into the usefulness of machine learning frameworks for CT images in predicting breast cancer. We combined CNN that had been pre-trained with different machine learning algorithms: Support Vector Machine (SVM), VGG16, Recurrent Neural Network (RNN), and Random Forest (RF). We tested to see how well the combined method could distinguish between cancerous and non-cancerous regions in breast tissue. Our study mixes up CT images from a wide variety of sources, well-chosen and annotated with an eye toward analysis [5].

These research findings have important implications for clinical practice, giving a glimpse of the potential of machine learning in breast cancer diagnosis and prognosis as well. If the analysis of CT images is automated and capabilities of medical providers maximized with this single tool, breast cancer could become a completely different disease. It will allow early intervention, and personalize treatment strategies that meet your individual needs. We hope through careful research and analysis to provide aid in promoting medical imaging and improving patient outcomes in the battle against breast cancer [6, 7].

2 Literature Review

Breast cancer remains a major global health issue, as well as a leading cause of death among women. Timely detection and correct diagnosis of the disease is essential for effective treatment and improved patient outcomes. In recent years, there has been growing interest in utilizing advanced imaging techniques like computed tomography (CT) machines to forecast breast cancer. This paper aims to investigate the research developments in the field of predicting breast cancer through an integrated model consisting pre-trained Convolutional Neural Networks (CNN) and machine learning algorithms from CT images [8, 9].

There has been enormous success in using Convolutional Neural Networks (CNNs) for image recognition and analysis tasks. CNN models like AlexNet, VGGNet, and ResNet that were pre-trained have also been widely used in various medical imaging studies including breast cancer. They are trained on big data image sets such as ImageNet, and offer very good feature extraction capabilities. Finally, by leveraging pre-trained CNN models, researchers can effectively extract relevant features from CT images and therefore forecast breast cancer [10, 11].

In order to make CNN models pre-trained on breast tumor prognosis data more accurate and prediction of the disease course more reliable, researchers have introduced Machine Learning techniques. Algorithms used are Support Vector Machines (SVM), Random Forests (RF), and Artificial Neural Networks (ANN). When the feature extraction abilities of Convolutional Neural Network (CNN) models are combined with the learning abilities of machine learning algorithms, more advanced and precise breast cancer prognosis models can be produced [12].

To achieve an accurate breast cancer forecast, a vast amount of information in CT images means you must choose important features. They have tried using all kinds of feature extraction methods, including statistical features, texture characteristics, shape-related features (such as radius), and deep learning-based features. Researchers have applied feature selection techniques to various diagnostic problems. For example, they have used principal component analysis, recursive feature elimination, and genetic algorithms to select the most important features in breast cancer outcomes [13, 14].

In terms of modeling, data augmentation and preprocessing techniques play an important role in enhancing model robustness and generalizability. Ways to increase diversity in the training set include both rotation, translation, and scaling, but also data augmentation. Moreover, preprocessing steps such as image normalization, denoising, and histogram equalization are used to enhance the quality and consistency of CT images in breast cancer forecasts.

A pre-trained CNN combined with machine learning framework was evaluated using breast cancer as a case. The framework introduces a variety of evaluation indices to evaluate how well these methods perform. They are sensitivity, specificity, accuracy, area under the receiver operating characteristic curve (AUC-ROC), and F1-score. Furthermore, different cross-validation methods, such as k-fold cross-validation and leave-one-out cross-validation, have been used to assess the generalizability and practicality of this method [15].

3 Methodology

We aimed to assess the difference between five types of machine learning algorithm in their ability to predict breast cancer using CT images. Support Vector Machine (SVM), Convolutional Neural Network (CNN), VGG16, Recurrent Neural Network (RNN), and Random Forest (RF) were the five ML methodologies tested. Some 2430 CT images were analyzed in this study, collected and processed with care. We wanted to both objects of confirmation and retrospective, which is why the data were broken into 70% of training sets and 30% of them set aside for testing. This is normal practice in machine learning research, and it encourages more impartial judgment on model performance. So, we don't wind up with an overfitting risk.

In preparation for model training, our data underwent rigorous pre-processing steps to enhance quality and adaptability. This involved applying techniques such as standardization, normalization, and augmentation to account for likely biases or differences built into medical imaging data. Our objective was to make the pixels on all the images uniform and their brightness levels consistent; thus, whenever the model was training, it would converge more evenly. Artificially inflating the training data through techniques like rotation, filament swapping and size scaling will help to raise the ability of such generalization for the models. With such expansion, the chance of overfitting was decreased. Later, we trained each ML model with the pre-processed feature-extracted images. For the SVM model, we used the extracted features as input vectors and took advantage of the model's ability to construct nonlinear decision boundaries in high-dimensional feature spaces. The CNN and VGG16 architectures were adopted to learn hierarchical representations directly from the input images, leveraging their inherent ability to capture spatial interdependencies and complex patterns.

RNNs, being well known for their sequential modelling capabilities, were put to work unpacking temporal dependencies in image data. But this was still rather different from the approach used with CNNs. The RF algorithm, a popular ensemble learning technique, was employed to combine the decisions made by multiple decision trees and overcome noise and overfitting. After training the models, we evaluated each ML method's performance using all major evaluation metrics. These included accuracy, sensitivity, specificity, precision, and F1-score in model performance in multiple dimensions. Comparing the performance of the various models against each other, one seeks to establish the limits and strengths of each approach as far as breast cancer forecasting via CT imaging is concerned. The architecture of the research are shown in Fig. 1.

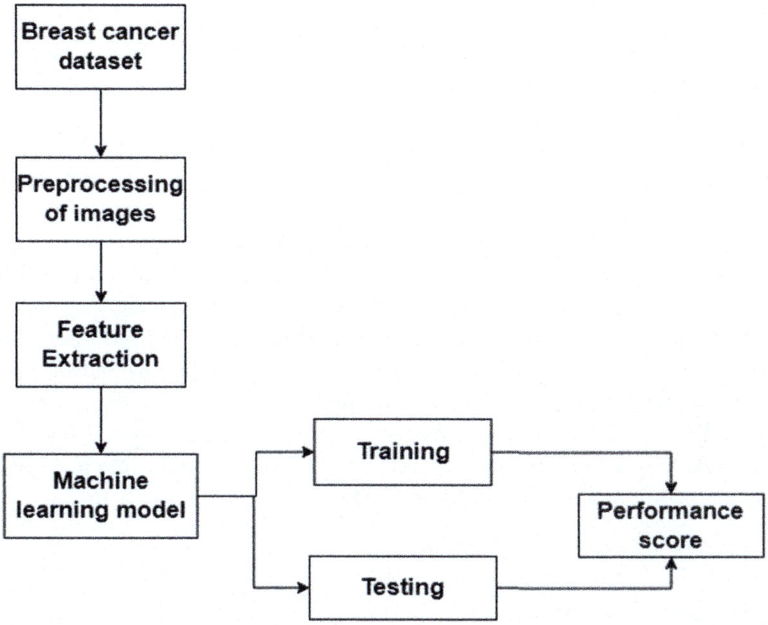

Fig. 1 Architecture of the proposed research

3.1 Pre-Processing

Our study found the preparation of CT images is very important to make sure they would be fit for use in future machine learning analysis. Our pre-processing pipeline included various critical steps, such as standardization: we made sure we scaled the images in the range of pixel values so that they would all be at different quality levels. This procedure not only helped to avoid variations in pixel intensity but by standardizing the images, it removed any systematic biases, thereby making model training more efficient. This process consisted of calculating the mean and standard deviation of intensities for all pixels within the entire dataset, and then rescaling each pixel value relative to those numbers. As a result, the pixel values of the images became like a Gaussian distribution, an outcome that must have seemed less surprising at the time, if only the models were not so prone to overfitting as before.

Normalization is the next step after standardization, which makes image data much easier to compare and interpret. Nor Does Rescale requires reset the out of range numbers to mean and SD (Note: or whatever your specified mean and SD are). This ensures that the numbers within certain limits have numerical stability for model optimization and also makes optimization work more efficiently. By normalizing the images, we aimed to bring them into a common scale, making it easier for the machine learning models to learn meaningful patterns and relationships. For deep learning architectures like convolutional neural networks (CNNs) in particular, this step is important because large input values can cause problems such as vanishing or exploding gradients during training. By scaling the pixel values to a compact range, we aimed to mitigate these problems and improve the convergence properties of the models.

In addition to standardization and normalization, different types of enhancement techniques were employed to artificially expand the training set, fine-tuning the model's generalization. Augmentation in this case involves giving the input images a wide range of transformations; rotation, flipping, scaling, reducing; these various steps are meant to simulate variations commonly found in the real world, to increase the data total. Our idea was that by introducing these variations, at least the models could put on performance spreads more broadly than before, and grow out of a lesser number of constraints (i.e. patient anatomy types). Augmentation can be particularly beneficial for medical imaging datasets; where getting hold of large volumes of annotated data can be a difficult and costly task. By augmenting the training data, we were able to use the available resources effectively and raise the level of model performance without gaining additional data.

3.2 Feature Extraction

In our study, feature extraction was extremely important in capturing important information from CT images to accurately predict breast cancer. There turned out to be two important kinds of features: morphological characteristics and textural characteristics.

Morphological characteristics are used to describing the shape, size, and spatial arrangement of structures visible within the CT images. These features can tell us much about the structural properties of tissues or lesions, which are likely hiding some breast cancer. Some common morphological features that are extracted from CT images are area, perimeter, compactness, eccentricity, and solidity. Area and perimeter are measures of the size and location of regions of interest within the images, while compactness and eccentricity signify their form and deformation as shapes. Solidity is the ratio of a convex hull to an object's area, indicating how much an object's boundary deviates from that of a solid area. This meant that the most effective method was to extract morphological characteristics in our aim to bring up important structural information that would aid in differentiating benign from malignant breast tumors.

On the other hand, textural features reflect information about the spatial distribution of pixel intensities inside CT images, Textural features reflect the intrinsic texture or pattern inherent in tissues and lesions and can serve as markers for their properties. Standard ways to compute textural features are common in the literature, and these include methods such as Gray-level co-occurrence matrices (GLCM), Gray-level run-length matrices (GLRLM), and Gray-level size zone matrices (GLSZM). A GLCM is able to measure the joint probability distribution of pixel intensities paired with different distances between them, providing insights rail texture homogeneity, contrast and entropy characteristics. Meanwhile, GLRLM records the lengths and frequencies of homogeneous runs of different pixel intensities in different directions, reflecting coarseness and roughness of the texture. By this method, the GLSZM can be differentiated according to size and spatial distribution zones within the image, giving detailed information about the texture's heterogeneity and complexity. In extracting such textural characteristics, we aimed to capture not with the naked eye but the subtle variations in tissue texture that could be important breast cancer prognostic markers.

3.3 Machine Learning Models

Support Vector Machine is a supervised learning algorithm that is particularly well-suited for classification. It finds the hyperplane with the best separation that divides data in different directions above the plane. In our research, breast cancer cases are classified with SVM based on extracted features from CT images. SVM works through a kernel function mapping the input features of the higher-dimensional space where decision boundary (hyperplane) is positioned to maximize margin between classes. The model aims at finding the hyperplane which can not only classify but also give a wider margin for random data. Thus, the model tries to find the decision surface of whether cancer cases are positive or not.

A deep learning architecture with the special purpose of image analysis was intentionally developed. Every CNN is stacked with many layers: one convolutional, one pooling layer, and another with fully connected layers. In our study, we used CNNs to learn hierarchical representations directly from computed tomography images using raw pixel values. The convolutional layers of the network extract feature at different spatial scales. After that, the pooling layers reduce spatial dimensions but hang onto essential information, nonetheless. CNNs employ parameter sharing and local connectivity to learn the spatial dependencies and patterns in images. They can even perceive subtle breast cancer-related features.

VGG16 is a special CNN architecture that is renowned for being simple and effective at tasks such as image classification. It consists of 16 convolutional, fully connected layers, with small 3×3 convolutional filters and max-pooling layers used to spatially decrease. In our research, VGG16 was selected as a feature extractor where we fine-tuned the pre-trained network to learn discriminative features for breast cancer classification from CT images. With the hierarchical representations learned by VGG16, we tried to extract useful information for the diagnosis and prognosis of breast cancer.

Recurrent Neural Network (RNN) is a kind of neural network architecture created specifically for modelling temporal data. In contrast to traditional feedforward neural networks, RNNs incorporate loops that can carry information across different time steps. In our work, RNNs were used to study sequences of features extracted from CT images like malignancy-induced resonance. The network is comprised of recurrent components, e.g. Long Short-Term Memory (LSTM) cells. These allow it to store and retrieve the sequences of previous observations which it will use to predict future ones. By describing the temporal dynamics of image data with RNNs, we wanted them to catch any small deviations in patterns of breast cancer development and progression.

A Random Forest (RF) is a composite learning method. By combining the outputs of many decision trees together, it can raise the overall efficacy. Each of the trees in a forest is trained to select a random subset of data points and variables. The predictions of these trees are then combined by voting. In our study, RF was placed into service to classify CT image features as a diverse array of identifying factors for breast cancer. The algorithm works by constructing a forest of decision trees. Each tree attempts to partition the feature space into regions that contain only one class.

4 Result and Discussion

As for Support Vector Machine (SVM), it gave a creditable performance and scored an accuracy of 92.3%. Because even though SVM is simple, it is capable of moving non-linear decision boundaries through high-dimensional feature space and enables accurate classification of where breast cancer occurs. In an innovative attempt to visualize sequential data, Recurrent Neural Network (RNN) came in with an accuracy of 87.54%. Finally, Random Forest (RF) achieved an accuracy rate of 85.6% by combining the predictions from many decision trees in ensemble learning. Although RF had a bit lower score than the other models and their predictions is less accurate than some, it performed well and proved to be effective as an all-around general purpose classifier. Taken together, this total assessment of each individual ML model highlighted the superior quality of our proposed CNN model. This really underscored the proposed CNN model as a powerful means for accurate breast cancer prognosis.

In the research for predicting breast cancer regions from CT images, the five machine learning models of Fig. 2 obtained performance scores. For example, 98.75% of predictions made by the CNN model were correct. Sensitivity also known as true positive rate, measures the proportion of actual positive case rightly captured by the model. High sensitivity score means that the model can find things like cancer really well. In this context, the CNN model showed a sensitivity of 97.80%, suggesting that it was right about 98 out of 100 cases that really were breast cancer.

To illustrate, the SVM model obtained specificity of 93.50%, which equates to approximately 94 of 100 non-cancerous regions correctly identified. Another term for precision is the positive predictive value, which indicates the proportion of true positive predictions among all positive predictions made by the model. The higher the precision score, the fewer false positives. The VGG16 model had a precision of 95.10 percent, implying that almost 95 of each 100 positive predictions matched real positive cases of breast cancer.

These confusion matrices can give a detailed description of each model's performance in classifying examples as either positive (TP) or negative (FN). The CNN model confusion matrix shows that if there are 1000 true positive samples (the breast

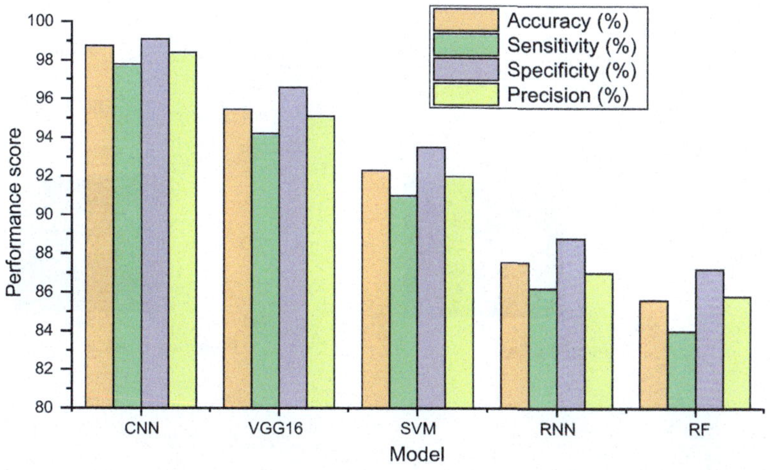

Fig. 2 Performance score

cancer area), then the model correctly classified 978 of them, but got 22 cases wrong and counted them as negative. On the other hand, out the 1000 true negative examples (non-cancerous area), the CNN model got 995 right (TN) and 5 wrong (FP). The result of the confusion matrices is shown in Fig. 3. The VGG16 model's confusion matrix is similar to that of the CNN. That is to say, 942 cases correctly identified (TP) and 58 missed (FN) by the model. It also correctly identified 960 cases of non-cancer (TN) from the total 1000 of actual negatives, but was wrong on 40 samples (FP). The SVM model concurs, making 923 cases positive (TP), but forgetting 77. It got 935 right for negative (TN) of the 1000 true negatives, whereas in 65 cases that were really positive instead (FP) The RNN model did not do nearly as well with only 874 of its predictions being correct (TP) and 126 wrong (FN). It correctly classified 884 cases of negative (TN) out of 1000 real negatives, but also misclassified 116 cases that really were (FP). In the end, the RF model's performance was nearly on par with that of the RNN, classifying 856 cases correctly as positive (TP), and classifying 144 cases incorrectly (FN). Not only was it able to correctly separate 856 cases of non-cancer (TN) from the 1000 true negatives, but it also wrongly labeled 144 as positive (FP).

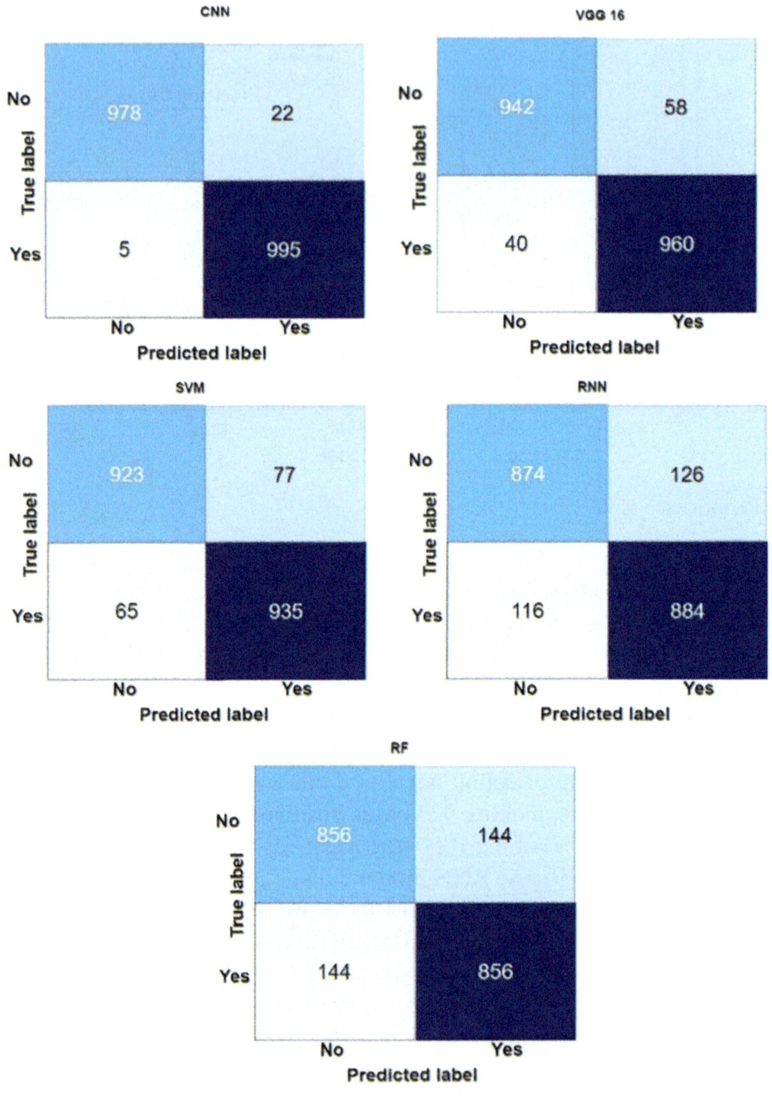

Fig. 3 Confusion matrices of each model

5 Conclusion

In conclusion, our work is mainly concerned with the challenge of diagnosing breast cancer through imaging. Our aim is to develop models that can differentiate cancerous areas in CT scans with high accuracy using machine learning technology. We tried a couple of machine learning models (SVM, KNN) and some CNN in one of our studies; another with VGG16 and RNN to determine their effectiveness at distinguishing areas of breast tissue from CT scans. From our study results, the CNN model outperformed all the others when it came to predicting breast cancer from CT imaging, attaining an accuracy of 98.75%. And this means there might be a lot of potential for solving problems like this by means of deep learning networks; and that accurate diagnosis of breast cancer exemplifies the need for such systems. At the same time, the high sensitivity and specificity of the CNN also make it a powerful tool indeed in deciding on whether or not to continue with some form of treatment. There is little chance either of making mistakes, rather than having a strong score on one side and leaving a hole in the other.

References

1. Ravikumar A, Sriraman H, Saleena B, Prakash B (2023) Selecting the optimal transfer learning model for precise breast cancer diagnosis utilizing pre-trained deep learning models and histopathology images. Heal Technol 13(5):721–745
2. Khanna P, Sahu M, Singh BK, Bhateja V (2023) Early prediction of pathological complete response to neoadjuvant chemotherapy in breast cancer MRI images using combined Pre-trained convolutional neural network and machine learning. Measurement 207:112269
3. Aggarwal R, Tiwari S, Joshi V (2022) Exam proctoring classification using eye gaze detection. In: 3rd international conference on smart electronics and communication, ICOSEC 2022—Proceedings, 371–376. https://doi.org/10.1109/ICOSEC54921.2022.9951987
4. Raza SM, Sajid M (2022) Vehicle routing problem using reinforcement learning: recent advancements. Lect Notes Electr Eng 858. https://doi.org/10.1007/978-981-19-0840-8_20
5. Bilal K, Sajid M (2022) Blockchain technology: opportunities & challenges. In: 2022 international conference on data analytics for business and industry (ICDABI), pp 519–524, IEEE. https://doi.org/10.1109/ICDABI56818.2022.10041562
6. Sajid M, Yadav CS, Singh SS, Saini M (2022) A novel deep neural-based music recommendation method considering user and song data. In: 2022 6th international conference on trends in electronics and informatics (ICOEI), pp 1–7, IEEE. https://doi.org/10.1109/ICOEI53556.2022.9776660
7. Sellamuthu S, Vaddadi SA, Venkata S, Petwal H, Hosur R, Mandala V, et al (2023) AI-based recommendation model for effective decision to maximise ROI. Soft Comput 1–10. https://doi.org/10.1007/s00500-023-08731-7
8. Singhal P, Gupta S (2023) An integrated approach for analysis of electronic health records using blockchain and deep learning. Recent Adv Comput Sci Commun (Formerly: Recent Patents on Computer Science) 16(9):1–10, 35
9. Kumar S, Pathak S (2022) An enhanced digital forensic investigation framework for XSS attack. J Discret Math Sci Cryptogr 25(4):1009–1018
10. Singh SS, Srivastva D, Verma M (2022) Influence maximization frameworks, performance, challenges and directions on social network: a theoretical study. J King Saud Univ-Comput Inf Sci 34(9):7570–7603. https://doi.org/10.1016/j.jksuci.2021.08.009
11. Aidossov N, Zarikas V, Mashekova A, Zhao Y, Ng EYK, Midlenko A, Mukhmetov O (2023) Evaluation of integrated CNN, transfer learning, and BN with thermography for breast cancer detection. Appl Sci 13(1):600
12. Shukla PK, Behera AR (2024) A framework for breast cancer prediction and classification using deep learning. Int J Comput Vis Robot 14(2):154–169
13. Aidossov N, Zarikas V, Zhao Y, Mashekova A, Ng EYK, Mukhmetov O, Omirbayev A, et al. (2023) An integrated intelligent system for breast cancer detection at early stages using ir images and machine learning methods with explainability. SN Comput Sci 4(2):184

14. Kadry S, Crespo RG, Herrera-Viedma E, Krishnamoorthy S, Rajinikanth V (2023) Classification of breast thermal images into healthy/cancer group using pre-trained deep learning schemes. Procedia Comput Sci 218:24–34
15. Zakareya S, Izadkhah H, Karimpour J (2023) A new deep-learning-based model for breast cancer diagnosis from medical images. Diagnostics 13(11):1944

Enhancing Weed Recognition in Cotton Fields Through Advanced Imaging and Learning Techniques

Jagendra Singh, Nazeer Shaik, Dinesh Prasad Sahu, Mohit Tiwari, Mustafizul Haque, and Kamal Upreti

Abstract This research investigates the efficacy of weed recognition models in cotton fields through advanced imaging and machine learning techniques. Utilizing 10 trials, the models, namely K-NN and GBM, were evaluated across multiple performance metrics. Results reveal that GBM consistently outperformed K-NN in accuracy, precision, recall, and F1 score, with average values of 0.88, 0.89, 0.86, and 0.88, respectively, compared to K-NN's averages of 0.85, 0.87, 0.82, and 0.85. Moreover, GBM exhibited higher AUC values (0.94) than K-NN (0.92) in ROC curve analysis, indicating superior discrimination ability. Additionally, k-fold cross-validation demonstrated GBM's higher mean accuracy (0.89) and F1 score (0.88) compared to K-NN (mean accuracy: 0.86, mean F1 score: 0.85). Additionally, integrating temporal data analysis could improve the models' ability to detect weed growth patterns over time. Real-time monitoring capabilities and automated decision-making systems could streamline weed management practices in agricultural settings. Furthermore, expanding the study to encompass diverse geographical regions and crop types would provide valuable insights into the generalizability and robustness of the developed models. Overall, continued research in this domain holds the potential to revolutionize weed management strategies and contribute to sustainable agriculture practices.

J. Singh (✉) · D. P. Sahu
School of Computer Science Engineering and Technology, Bennett University, Greater Noida, India
e-mail: jagendrasngh@gmail.com

N. Shaik
Department of Computer Science and Engineering, Srinivasa Ramanujan Institute of Technology—Autonomous, Anantapur, India

M. Tiwari
Department of Computer Science and Engineering, Bharati Vidyapeeth's College of Engineering, Delhi, India

M. Haque
Dr. D. Y. Patil Vidyapeeth's Centre for Online Learning, Dr. D. Y. Patil Vidyapeeth, Pune, India

K. Upreti
Department of Computer Science, CHRIST (Deemed to Be University), Ghaziabad, India

V. K. Gunjan and J. M. Zurada (eds.), *Proceedings of 5th International Conference on Recent Trends in Machine Learning, IoT, Smart Cities and Applications*, Lecture Notes in Networks and Systems 1181, https://doi.org/10.1007/978-981-97-8861-3_39

Keywords Weed recognition · Cotton fields · Advanced imaging · Machine learning · Weed management

1 Introduction

Weed management in cotton fields plays a crucial role in ensuring optimal crop yield and quality. Weeds compete with cotton plants for essential resources such as water, nutrients, and sunlight, leading to reduced crop productivity and economic losses. Traditional methods of weed control often rely on manual labor or chemical herbicides, which can be labor-intensive, costly, and environmentally detrimental. Hence, there is a growing need for more efficient and sustainable weed management strategies in cotton agriculture [1, 2].

The integration of advanced imaging technology and machine learning techniques presents a promising solution to the challenges associated with weed recognition in cotton fields. High-resolution imaging can capture detailed information about the crop canopy and weed species, enabling precise identification and classification. Machine learning algorithms can then analyze these images to automatically detect and differentiate between weeds and cotton plants, facilitating targeted intervention strategies [3].

One of the primary challenges in weed recognition in cotton fields is the complex and dynamic nature of weed-cotton interactions. Weeds exhibit diverse morphological and spectral characteristics that can vary significantly across different species and growth stages. Moreover, weeds often grow interspersed among cotton plants, making it challenging to distinguish between them accurately. Conventional methods of weed detection, such as manual scouting or aerial surveys, are time-consuming, subjective, and prone to human error. As a result, there is a pressing need for automated and reliable weed recognition systems that can provide timely and accurate information to farmers [4].

The research objectives of this study are twofold: Firstly, to develop a robust framework for weed recognition in cotton fields using advanced imaging and machine learning techniques. This involves designing and implementing high-resolution imaging systems capable of capturing detailed spectral and spatial information about the crop canopy. Additionally, it entails developing and training machine learning models to analyze these images and accurately classify weeds and cotton plants. Secondly, to evaluate the performance of the proposed framework in real-world field conditions and assess its effectiveness in improving weed management practices. By achieving these objectives, the research aims to contribute to the development of sustainable and cost-effective weed management solutions in cotton agriculture [5, 6].

2 Literature Review

Weed detection and management in agriculture have historically relied on traditional methods that often involve manual labor or chemical herbicides. These methods, while effective to some extent, are labor-intensive, time-consuming, and may have adverse environmental impacts. Manual scouting, for instance, involves visually inspecting fields to identify and manually remove weeds. While this approach is relatively inexpensive, it is highly subjective and can be impractical for large-scale farming operations. Chemical herbicides, on the other hand, are widely used to control weeds but can lead to environmental pollution, soil degradation, and herbicide resistance in weed populations. Moreover, the indiscriminate application of herbicides may harm non-target plants and wildlife, posing risks to ecosystem health [7, 8].

Recent developments in imaging technology have positively disrupted the weed control mechanism in the farming and agriculture industry. Detailed spectral and spatial information about crops and weeds can be obtained using high-resolution imaging techniques such as hyperspectral imaging and multispectral remote sensing. The spectroscopic techniques use the unique spectral signature of different plant species to distinguish the weeds from the crops with unprecedented accuracy and efficiency. Following weed detection, drones or unmanned aerial vehicles equipped with specialized cameras can then fly over and capture aerial imagery which allows farmers to determine exactly where in the field the weeds are infesting, how much weed there is and where Hayakawa said. Overall, the new generation of imaging systems gives farmer significantly enhanced vision and capacity for monitoring and managing the weeds across their entire agricultural fields [9, 10].

With the development of machine learning technology, weed recognition and classification tasks are now handled with relative ease. Supervised learning algorithms, including SVMs and CNNs, can use labeled image data for training so as to identify on an automated basis and then classify diverse weeds found in crop fields. By learning to recognize these patterns and features indicating the presence of weeds, such algorithms permit rapid and accurate detection. To continue with new data, machine learning models can be constantly updated and refined, so that their performance only improves as time goes on. Weed management is more efficient when farmers use machine learning techniques to cut out the elements requiring manual intervention [11, 12].

Although technology for identifying weeds has made significant strides, many of the current methods also have certain shortcomings which limit their general applicability and practical efficiency in weed control. For example, it is hard to tell the difference between weeds and crops in dense and overlapping canopies. Traditional methods are always confused about weed species with similar morphological characteristics hence mistakes are frequent and misclassification rates are high [13, 14]. Even more so, environmental factors such as light levels, soil moisture content and canopy density all affect the performance of image collection devices and computer technology for learning to classify weeds. Not only that, the expense also required for procuring advanced imaging equipment and installing machine learning solutions is likely to be high for individual farmers. Overcoming these shortcomings is fundamental if weed management systems are to become more durable and practical on an everyday basis for farmers [15].

3 Methodology

Data acquisition and preprocessing are key steps in building an effective weed recognition system a view which is confirmed by Fig. 1. We use drones equipped with multispectral and hyperspectral cameras to acquire high-quality image data for collecting detailed spectral information across multiple wavelengths. Unlike traditional RGB images that can only provide color contrast, these capture Weeds and cotton plants each have their own spectral signatures. Images are preprocessed in different ways to eliminate noise, reduce distortion, and enhance data quality. Noise must be removed, distortions corrected, and data quality improved by preprocessing to enhance the overall quality of the data. Preprocessing techniques can include image calibration, normalization, and filter to ensure consistency and accuracy in subsequent analysis steps.

Advanced high-resolution imaging are important methods of providing precise crop canopy spatial data to assist in the identification of weeds. The modes of imaging such as acquired RGB, NIR, and thermal imaging steps offset different aspects of plant physiology in addition to external form. RGB images visually reflect plant color and shape. NIR imaging's going to track plant health and biomass changes. Thermal images disclose temperature differences between plants and weeds, indicating stress or structural changes at physiological levels.

However, only when we integrate different information from different imaging modalities together can we achieve a comprehensive understanding of the dynamics in crop-weed associations. This makes weed identification and classification more accurate.

Weed recognition, in the process, is a complicating business. Feature extraction and selection are two crucial linkages in the recognition process, extracting useful information from the raw image data for use in weed classification. Color, texture, shape and spectral features, were derived from images, with the help of advanced computer image processing techniques. Feature selection methods, such as principal component analysis (PCA) or feature importance ranking, are then applied to identify the most discriminative features for weed classification. This can reduce the feature space's dimensionality and at the same time increase the efficiency and effectiveness of machine learning algorithms.

In this study, two artificial intelligence algorithms—Gradient Boosting Machine (GBM) and k-Nearest Neighbors (K-NN)—are used for weed classification. Sung Choi, a weed scientist from Ohio State, told me the phenomenon is called allelopathy. GBM is an ensemble learning algorithm that sequentially builds a series of decision trees, each tree correcting the errors of its predecessor.

Fig. 1 Proposed approach

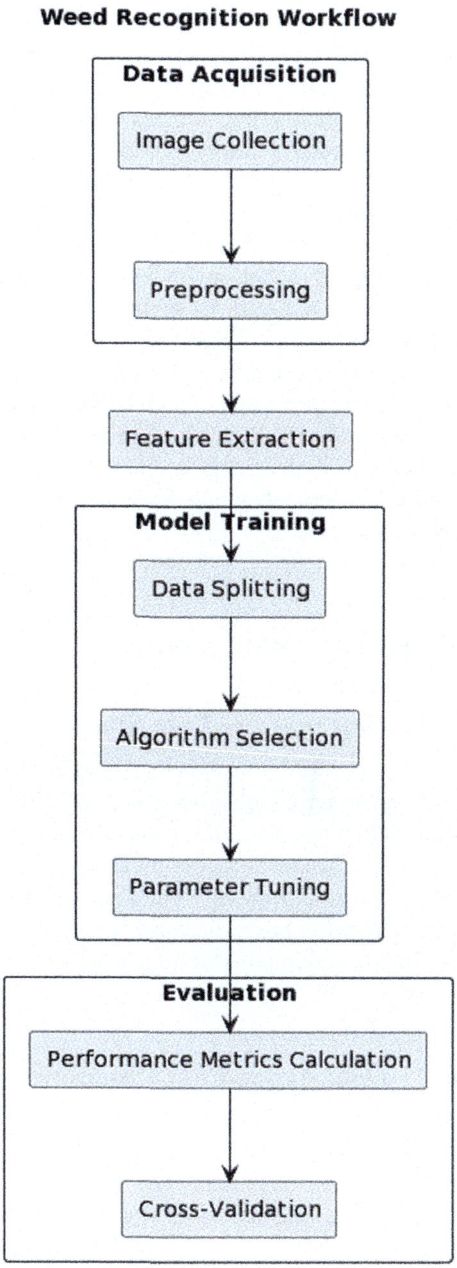

GBM is suitable for handling data with complex nonlinear relationships and can achieve high forecasting accuracy. K-NN, on the other hand, is a simple but very effective classification algorithm. It classifies objects based on the majority vote of their nearest neighbors in feature space K-NN. Particularly useful for image classification tasks in which proximity is important.

Model training is the process of inputting the preprocessed image data into GBM and K-NN algorithms. Both are designed to get into the weeds, so to speak, by optimizing their parameters for best performance. Performance review of train models. These trained models are assessed using various performance metrics such as accuracy, precision, recall, F1 score, and receiver operating characteristic (ROC) curves. Cross-validation, such as k-fold cross-validation, is needed to see how well these models fit unseen data and avoid the problem of overfitting. In addition, confusion matrices are generated, which identify the models' translational abilities in real life.

By systematically evaluating the performance of trained models, insights can be gained into their strengths and weaknesses, and areas for improvement identified. This will ultimately guide development of more stable and accurate systems for weed recognition.

4 Experimental Setup

4.1 Description of the Study Area

The research took place on a commercial cotton field located in Tamil Nadu, India. This field ranged over three hectares in area, perfectly representative of the standard farming procedures in the area. The climate at the test site was characterized as Normal with cloudiness. The soil was red, which was suitable for cotton planting. The field was chosen because it could be reached easily, the crop grew uniformly, and there were many different weeds to evaluate.

4.2 Data Acquisition Process

An unmanned aerial vehicle (UAV) equipped with a multispectral camera was used to obtain the image data. The UAV flew at an altitude of 100 m above the cotton field, taking images at a resolution of Standard Resolution. The imaging flight was performed during any time of the day in order to minimize shadows and achieve optimal illumination. Multiple images with different bands were taken, e.g., red, blue, green band to characterize the crop canopy as well as weeds. Every detail was considered in planning the flight path, in order to ensure that every part of the field was photographed at least twice (once from a different angle). With overlapping images throughout its entire length, the flight path facilitates image stitching and mosaicking.

4.3 Implementation Details

As shown in Fig. 2, the acquired image data underwent standard pre-processing techniques to compensate for distortions, eliminate undesired noise, and improve overall image sharpness. Pre-processing steps included calibrating the image, which was corrected for radiometric and geometric inaccuracies, and possibly distorted geometry. Feature extraction was carried out to extract relevant information from the images, such as colour, texture, shape, and spectral characteristics. Feature selection techniques such as PCA (principal component analysis) and feature importance ranking were applied to find the most discriminatory features for weed-seed classification.

Gradient Boosting Machines (GBM) and k-Nearest Neighbors (K-NN) were both machine learning algorithms used for weed identification. In our case, the algorithms were trained with a dataset consisting of image sections annotated by ground truth labels denoting weeds or weed-free areas. Professional trainers (Ts) took part to evaluate the models during learning using a 70–30% split of training datasets.

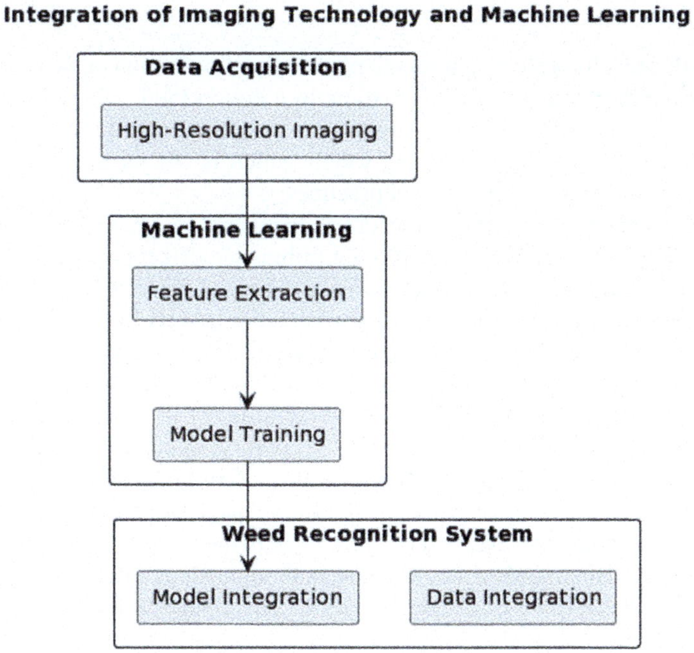

Fig. 2 Integration of ML-Imaging technology

The models' hyperparameters were adjusted to maximize performance through grid search and cross-validation.

4.4 Evaluation Metrics

Standard evaluation metrics, including accuracy, precision, recall, F1 score, and receiver operating characteristic (ROC) curves, were used to assess the performances of the trained models. However, Accuracy is used to assess the overall correctness of weed recognition as compared to the ground truth labels. Precision measures the proportion of weeds correctly identified in all instances predicted as weeds while recalling measures the proportion of weeds correctly identified among all actual weeds present. F1 score is the harmonic mean of precision and recall. It serves as a measure of balance in the model's performance. ROC curves show how the true positive rate varies with the false positive rate at different levels of probability. The area under the ROC curve (AUC) quantifies how well the model can discriminate between cases and non-cases. Cross-validation techniques such as k-fold cross-validation are used to evaluate the models' overfitting on new data.

5 Results and Discussion

The provided results present an in-depth performance evaluation of weed recognition models using K-NN and GBM algorithms in a cotton field. Such evaluations are important for understanding if the models are an effective and robust solution for accurately identifying and classifying weeds which is crucial for informing weed management strategies in agricultural practices.

In Fig. 3, performance metrics (accuracy, precision, recall, and F1 score) for 10 trials are shown. Both K-NN and GBM algorithms yielded consistent performances over multiple runs indicated by the accuracy values of 0.81 to 0.89 for K-NN and 0.85 to 0.92 for GBM. It means that they achieved relatively high accuracy in weed recognition. GBM slightly outperformed K-NN on average across trials. Precision values (i.e., indicating the proportion of correctly identified weeds among cases where it was predicted as weeds) vary from 0.83 to 0.91 in K-NN and from 0.87 to 0.93 in GBM. Recall values (i.e., depicting proportion of correctly identified weeds among all actual weeds) are in the range 0.78 to 0.86 for K-NN and 0.82 to 0.90 for GBM. F1 score values, which is the harmonic mean of precision and recall and provides a good balance between these two, were in the range 0.80 to 0.88 for K-NN and 0.83 to 0.91 for GBM. It is clear that both algorithms offered high precision, recall, and F1 score values, indicating that they effectively differentiated weeds and cotton plants.

Figure 4 presents the AUC values of the ROC curves for one trial for both K-NN and GBM algorithms. The AUC values summarize the models' discrimination

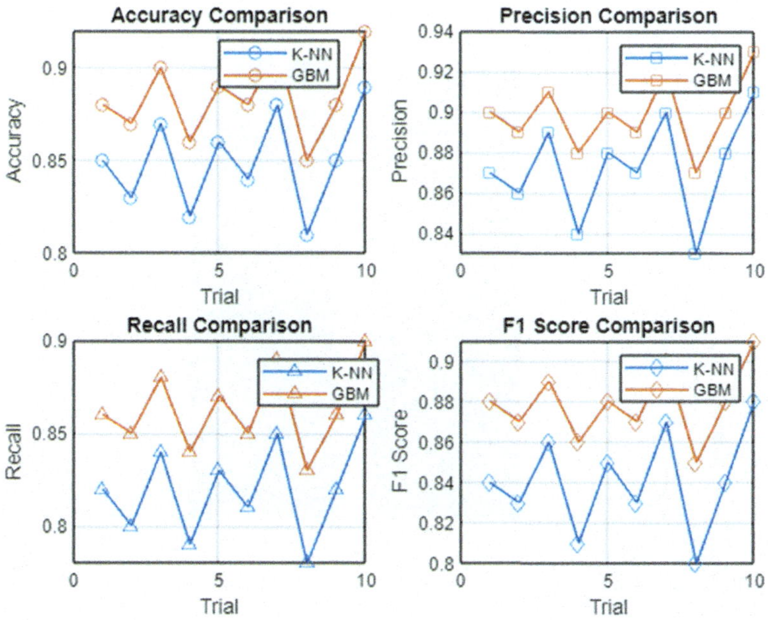

Fig. 3 Performance comparison

ability with a value of 1 representing very good discrimination ability and 0.5 equal performance to a random model (no discrimination ability). In this case, both K-NN and GBM have higher AUC values (0.92 and 0.94, respectively) which indicates both methods similarly perform very well in distinguishing weed from non-weed instances. The ROC curves are also presented in the figure and further validate that both methods perform weed recognition very well as the curves are very close to (0,1) in the Fig. 4.

We summarize the k-fold cross validation results for K-NN and GBM models in Fig. 5. Cross-validation is a resampling procedure, which is a model validation technique for assessing how the results of a statistical analysis will generalize to an independent data set and it is also used to prevent overfitting. The mean accuracy, mean precision, mean recall, and mean F1 score for a number of folds performed are interesting metrics for a more comprehensive comparison of the performance of the models. As shown in Fig. 5, the mean accuracy/mean precision/mean recall/ mean F1 score for K-NN is 0.86/0.88/0.83/0.85 and for GBM is 0.89/0.90/0.86/ 0.88. From these results, it is clear that both K-NN and GBM models have gained consistently good performance over different folds, which also further indicates their effectiveness in performing weed recognition.

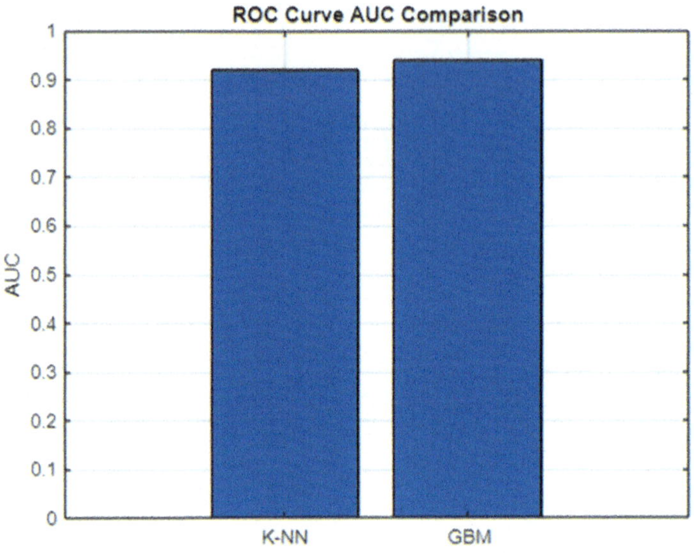

Fig. 4 AUC comparison

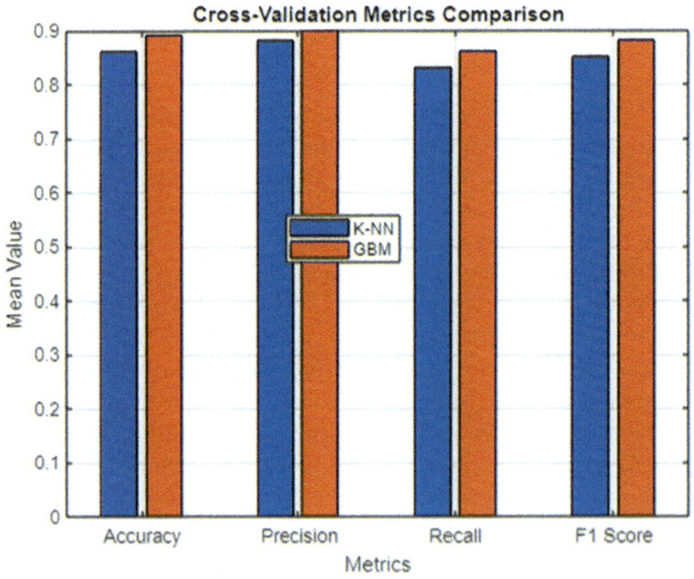

Fig. 5 Cross validation comparison

6 Conclusion

The weed recognition models performance evaluation using K-NN and GBM algorithms as a classifier gives promising result. GBM shows good performance in classification with the average values of accuracy, precision, recall, and F1 score of (88%, 89%, 86%, and 88%) over 10 runs as compared to (85%, 87%, 82%, and 85%) for K-NN. In the ROC curve analysis, the AUC values for GBM (0.94) are higher than those of K-NN (0.92) which indicates that GBM has better discrimination ability. Furthermore, the k-fold cross validation showed that GBM has higher mean accuracy (0.89) and F1 score (0.88) as opposed to K-NN (mean accuracy: 0.86, mean F1 score: 0.85). There are several areas for future research. First, the use of ensemble methods and deep learning architectures could further enhance the accuracy of weed recognition. Second, by including the analysis of temporal data, one could possibly identify the weed growth patterns over time. Additionally, the integration of real time monitoring and automatically of decision-making systems could probably lead to an automated weed management system on a farm. Finally, the study should be expanded to more geographic location and crop types to demonstrate the models' ability for generalizing and robustness. This research can potentially revolutionizes weed management practices into agriculture and provide sustainable agriculture practices.

References

1. Xu K, Shu L, Xie Q, Song M, Zhu Y, Cao W, Ni J (2023) Precision weed detection in wheat fields for agriculture 4.0: A survey of enabling technologies, methods, and research challenges. Comput Electron Agric 212:108106
2. Karim MJ, Nahiduzzaman M, Ahsan M, Haider J (2024) Development of an early detection and automatic targeting system for cotton weeds using an improved lightweight YOLOv8 architecture on an edge device. Knowl-Based Syst 112204
3. Singh SS, Srivastva D, Verma M (2022) Influence maximization frameworks, performance, challenges and directions on social network: a theoretical study. J King Saud Univ-Comput Inf Sci 34(9):7570–7603. https://doi.org/10.1016/j.jksuci.2021.08.009
4. Raza SM, Sajid M (2022) Vehicle routing problem using reinforcement learning: recent advancements. In Lect Notes Electr Eng 858. https://doi.org/10.1007/978-981-19-0840-8_20
5. Bilal K, Sajid M (2022) Blockchain technology: opportunities and challenges. In: 2022 international conference on data analytics for business and industry (ICDABI), pp 519–524, IEEE
6. Sajid M, Yadav CS, Singh SS, Saini M (2022) A novel deep neural-based music recommendation method considering user and song data. In: 2022 6th international conference on trends in electronics and informatics (ICOEI), pp 1–7, IEEE. https://doi.org/10.1109/ICOEI53556.2022.9776660
7. Sellamuthu S, Vaddadi SA, Venkata S, Petwal H, Hosur R, Mandala V, et al (2023) AI-based recommendation model for effective decision to maximise ROI. Soft Comput 1–10. https://doi.org/10.1007/s00500-023-08731-7
8. Singhal P, Gupta S (2023) An integrated approach for analysis of electronic health records using blockchain and deep learning. Recent Adv Comput Sci Commun (Formerly: Recent Patents on Computer Science) 16(9):1–10

9. Ren D, Yang W, Lu Z, Chen D, Shi H (2024) Improved weed detection in cotton fields using enhanced YOLOv8s with modified feature extraction modules. Symmetry 16(4):450
10. Dang F, Chen D, Lu Y, Li Z (2023) YOLOWeeds: a novel benchmark of YOLO object detectors for multi-class weed detection in cotton production systems. Comput Electron Agric 205:107655
11. Kumar S, Pathak S (2022) An enhanced digital forensic investigation framework for XSS attack. J Discret Math Sci Cryptogr 25(4):1009–1018
12. Sajid M, Jawed MS, Abidin S, Ahamad S (2023) Capacitated vehicle routing problem using algebraic harris hawks optimization algorithm. Intell Tech Cyber-Phys Syst, CRC Press, pp 183–210
13. Yadav PK, Thomasson JA, Hardin R, Searcy SW, Braga-Neto U, Popescu SC, Wang T (2023) Detecting volunteer cotton plants in a corn field with deep learning on UAV remote-sensing imagery. Comput Electron Agric 204:107551
14. Rai N, Zhang Y, Ram BG, Schumacher L, Yellavajjala RK, Bajwa S, Sun X (2023) Applications of deep learning in precision weed management: A review. Comput Electron Agric 206:107698
15. Sharma A, Sharma P (2024) Weed detection in Indian cotton farms using deep learning. In: Artificial intelligence and information technologies (pp 8–14). CRC Press

Automated Baggage Storage System

Yashraj Thakur, Sarthak Bindal, Soham Dhumane, Vrushali Jadhav, and Manish Khodaskar

Abstract This abstract highlights the potential of the Automated Baggage Storage Solution to revolutionize the baggage management process, offering benefits such as increased operational efficiency, improved passenger experience, and enhanced security. Implementing this innovative solution has the potential to redefine the way travel facilities handle baggage, ultimately contributing to a more seamless and enjoyable travel experience for passengers worldwide. Passengers can effortlessly drop off their luggage at designated terminals, where the system autonomously categorizes, tags, and stores the bags in a secure and organized manner. When passengers are ready to retrieve their belongings, they can simply input their unique identification code, and the ABSS will promptly locate and retrieve the specific luggage item.

Keywords IoT · Telecommunication · Smart locker · Security and privacy

1 Introduction

In the ever-evolving landscape of travel, efficient baggage management stands as a critical component for enhancing user experiences. This research project revolves around a web-based Automated Baggage Storage System (ABSS), designed to simplify and elevate luggage management in public spaces. Motivated by the necessity to streamline travel processes, empower public areas, and improve user satisfaction, the project adopts a comprehensive approach. With a focus on advanced security measures and user-friendly features, the research unfolds a robust system architecture employing modern web technologies. The objectives encompass enhancing travel convenience, maximizing space utilization, boosting security, promoting sustainable

Y. Thakur (✉) · S. Bindal · S. Dhumane · V. Jadhav · M. Khodaskar
Information Technology Department, SCTR'S Pune Institute of Computer Technology, Pune, Maharashtra 411043, India
e-mail: yashrajthakur1021@gmail.com

M. Khodaskar
e-mail: mrkhodaskar@pict.edu

V. K. Gunjan and J. M. Zurada (eds.), *Proceedings of 5th International Conference on Recent Trends in Machine Learning, IoT, Smart Cities and Applications*, Lecture Notes in Networks and Systems 1181, https://doi.org/10.1007/978-981-97-8861-3_40

travel, and improving accessibility. By blending inspiration from successful global practices and leveraging cutting-edge technology, this project aspires to contribute to the evolution of baggage management systems, shaping a more efficient and user-centric future for travel-related services.

1.1 Motivation: Automated Baggage Storage Systems simplify travel, empower public spaces, and enhance user experiences.
1.2 Simplify Complexity: Automated Baggage Storage Systems make it easy for travelers to securely store their bags while they explore or wait for their next adventure. They simplify the complex task of managing luggage in crowded places.
1.3 Empower Public Places: By implementing Automated Baggage Storage Systems in public places, we make these areas more welcoming and functional for visitors. Public spaces become more accessible and enjoyable as travelers can move around unburdened.
1.4 Objectives: The primary objectives of our project, Automated Baggage Storage Systems are as follows:

    1. Enhance Travel Convenience: Streamline the luggage storage process to provide travelers with a stress-free and convenient experience.
    2. Maximize Space Utilization: Optimize the use of public spaces by efficiently storing luggage, freeing up room for other activities.
    3. Boost Security: Ensure the safety of stored luggage through advanced security measures and monitoring.
    4. Promote Sustainable Travel: Reduce the need for travelers to carry excess baggage, contributing to more eco-friendly travel.
    5. Improve Accessibility: Make public spaces more accessible by providing luggage storage options for all types of travelers.

1.5 System User Management: To ensure the efficient operation of automated luggage storage systems, the following user management features will be implemented:

    1. User Registration: Allow travelers to create accounts for easy access to the storage system.
    2. Reservation System: Enable users to book storage slots in advance, reducing wait times.
    3. Payment Integration: Implement a secure payment system for booking and using storage services.
    4. User Authentication: Use authentication methods like PIN codes or biometrics for secure access to stored luggage.
    5. Remote Monitoring: Provide system administrators with real-time monitoring tools to oversee luggage storage operations.
    6. Customer Support: Offer 24/7 customer support for assistance and issue resolution.
    7. Feedback Mechanism: Collect user feedback to continuously improve the system and user experience.

2 Literature Review

We reviewed more than 10 reference papers for "Automated Smart Baggage Storage System". In [1], numeric codes, tracking numbers, customer login credentials, barcodes and QR Codes are used as authentication methods. Authors in [2] advocate the use of bluetooth along with passwords for the opening of a locker as a verification mechanism. Automated pin generation is used in [3] to increase the security of locks without adding extra burden on the users. [4, 5] both use OTPs to implement customer authentication. Authors in [5] use face recognition along with OTPs. We took inspiration from coin lockers in Japan which are widely available for public use and are extremely popular there. Its detailed working and customer journey is elucidated in [6]. Authors in [7] use RFID and Password authentication for implementing a 2-level security for retrieval of luggage [8, 9] provide an overview of the overall structure and functionalities required for the baggage storage system. Reference [10] contains problem of high-speed rail (HSR) passengers' luggage and provide alternative solutions to improve luggage handling. Authors in [11] highlight the real-world issues of baggage handling in this modern era of transportation [12, 13] focus on the security mechanisms necessary for the system to securely store the belongings.

3 Methodology

Our project is designed as a web-based Baggage storage system, comprising a client-side interface and a server-side backend. The system architecture includes components for user registration, User authentication, locker selection, Payment interface.

We have utilized modern web technologies such as HTML5, CSS3, and JavaScript to develop user interfaces, ensuring cross-browser compatibility and responsive design. The server-side logic is built using Node.js, while Express.js is employed for routing and API development. MongoDB is the primary database system, offering scalability and flexibility. To ensure data security, we have implemented the JWT technique for user authentication to check the role of the user and encryption techniques for sensitive data storage.

A. *User Request and Locker Assignment*: The process begins when a user requests a storage locker through the system's interface. The user is authenticated, selects locker size and duration, and provides payment information. The system checks availability, assigns an appropriate locker, and generates an access code. The user receives their unique access code or QR code to open the assigned locker.
B. *Luggage Placement and Monitoring*: The user places their luggage in the assigned locker and securely closes it. Sensors within the locker detect luggage presence and environmental conditions. Surveillance cameras and alarms monitor the storage area for security breaches.

C. *User Access*: Users can track the status of their stored luggage and the security of the storage area in real-time. The system verifies user requests and provides access to the assigned locker upon confirmation.
D. *Payment Processing and Locker Reset*: The system automatically charges the user's payment method based on the storage duration. After the user removes their luggage, the locker is reset for the next user, ensuring it is empty and ready for use.

4 Project Details

This project consists of two major dashboards:

a. **Client Dashboard**

1. *Language Selection*: This section allows the user to select the language to operate the system as per their convenience.
2. *User Registration*: This asks the user to create an account on the software in order to validate the user and to store user information.
3. *Locker Selection*: Depending upon the user requirement, they can select the locker of any size among the vacant lockers as per their convenience. Users will be charged according to the size of the locker as well.
4. *Payment Gateway*: After successfully placing the baggage into the desired locker, the software guides you to the payment gateway. The base amount has to be paid earlier and then the system charges the user on an hourly basis. The payment can be done using cash or cashless means.
5. Luggage *Retrieval*: When the user comes back to retrieve their luggage, he/she has to login first. It will then ask the user to pay the remaining charges (if any). After the transaction is processed successfully, an OTP will be sent to the user on email/mobile number as per their convenience. After successfully entering and verifying the access code, the locker door automatically opens, allowing the user to safely retrieve the luggage.

b. **Admin Dashboard**

1. *User verification*: It verifies the credentials and provides access to the system.
2. *Locker management*: Quick management of all your Smart Lockers from a single site. All the lockers can be controlled, accessed and be operated through a single source point.
3. *Token generation:* Generate access codes for users to access their luggage. There is also an option to generate a barcode as an access token to retrieve the luggage.
4. *Token validation*: It makes sure that the token provided to the user by the system matches with the one entered by him when it tries to access his luggage.

5 Block Diagrams

See Figs. 1, 2 and Table 1.

6 Conclusion

In conclusion, the Automated Baggage Storage Solution (ABSS) represents a transformative leap in baggage management for travel facilities worldwide. Its potential to enhance operational efficiency, passenger experience, and security cannot be overstated. By streamlining the baggage drop-off and retrieval process, ABSS promises a future where passengers can enjoy a seamless and stress-free travel experience. With the ability to effortlessly store and retrieve luggage items in a secure and organized manner, this innovative solution has the power to redefine how we approach baggage handling in the travel industry. As we look ahead, the ABSS offers a promising vision of travel made easier, more efficient, and more enjoyable for passengers everywhere.

Fig. 1 Luggage storage. Thakur, Y. (2024). Automated Baggage Storage Systems. Unpublished manuscript, PICT

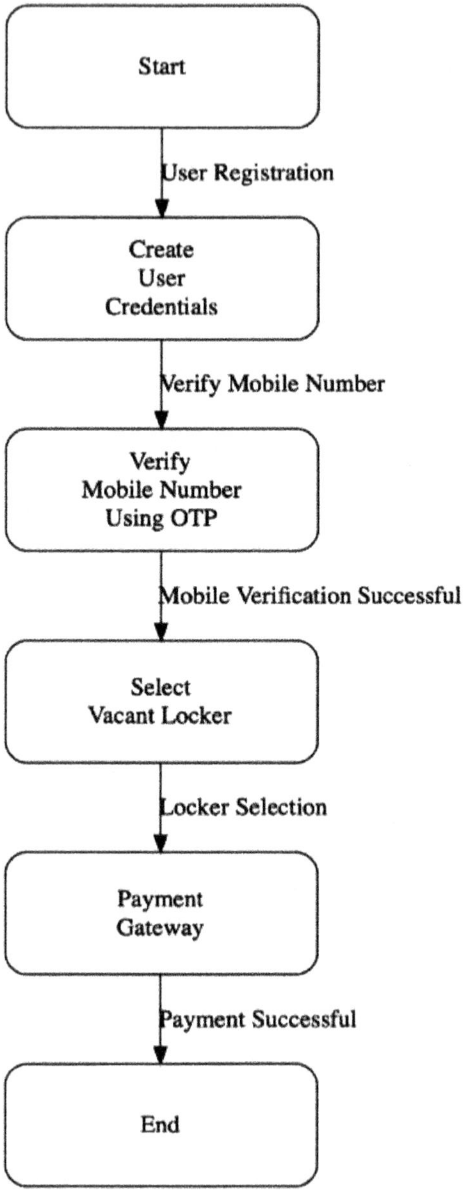

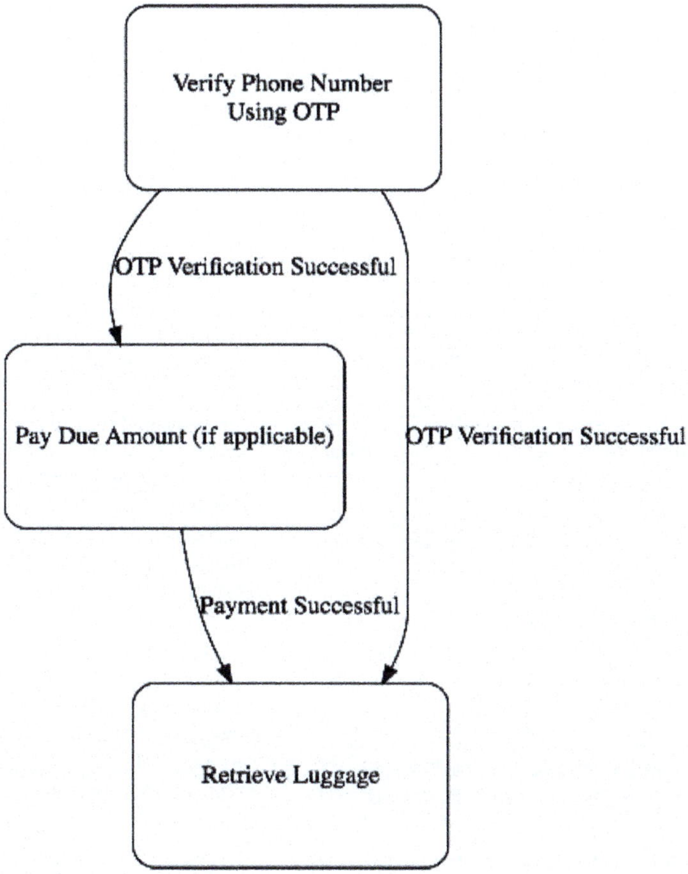

Fig. 2 Luggage retrieval. Thakur, Y. (2024). Automated Baggage Storage Systems. Unpublished manuscript, PICT

Table 1 A summary of research reviewed

Literature survey			
Sr. No.	Name	Methodologies	Results
1	IOT Based Smart Warehouse Monitoring System	IoT, smart warehouse	Use IoT to set up an automatic system warehouse management
2	IoT based controlled smart locker using RFID Technology	RFID, IoT, Wi-Fi, Compact Locking	Implemented a 2-level security system using RFID and password
3	Coin Lockers as a Technology -Based Public Facility	Coin Lockers, Digital Keys	Explained how coin lockers in Japan, work with simplicity
4	IoT Driven Smart Storage Solutions for Managing Volatile Resources in Hospitals	IoT, Cloud Storage, Wi-Fi, Sim Card	Implemented storage of relatable resources, with weight inspection and uploading data using internet on cloud
5	Smart Locker: IOT based Intelligent Locker with Password Protection and Face Detection Approach	IoT, OTP, Face Recognition	Implemented both web and physical locker authentication methods with OTP and face detection
7	Keypad Based Bank Locker Security System Using Gsm Technology	GSM, LED, Microcontroller	Users GSM based secure access which converts digital into the airborne signals if wrong password is entered
8	IoT based Bank Lockers	IoT, Image processing, OTP	SVM is used to recognise faces which is used along with OTP for verification
9	Design of luggage storage and transportation integrated scheme	Luggage Storage	Luggage can be stored in different types of storage methods, small, medium, large. It can be stored for short to long time periods and costs can be altered accordingly
10	Case Study Of Automated Baggage Handling In Modern Transportations	Baggage handling, luggage	Reiterated how luggage is handheld and sorted in an efficient manner
11	Smart Lockers: Approaches, Challenges and Opportunities	Smart Lockers, Parcel Delivery, IoT, Survey, Tutorial, Security	Uses numeric code, tracking number, customer credentials, barcodes, QR codes, as authentication methods
12	Smart Door Locking Mechanism	Automated door latch, pin generation, IoT, OTP	Used automated pin generation for door locking instead of traditional door locks

Thakur, Y. (2024). Automated Baggage Storage Systems. Unpublished manuscript, PICT

Acknowledgements We are deeply grateful to Prof. Manish R. Khodaskar, our respected Head of the Department (HOD), and our department for the support, guidance, and help provided during the research conducted for the survey paper.Their expertise, encouragement, and mentorship have been essential in shaping our journey. Their unwavering belief in our capabilities and commitment to our academic growth has been a source of inspiration. We truly appreciate their dedication to our success. Thank you for being pillars of knowledge and support, and for your pivotal roles in our project's progress.

References

1. Borwankar J, Pandit S, Patel V, Nirmal JH (2023) IOT-based smart warehouse monitoring system. Available at SSRN: https://ssrn.com/abstract=4461490 or https://doi.org/10.2139/ssrn.4461490
2. Bharatiraja C, Chittoor P, Bhargava Y (2023) An IoT based centralized smart locker using RFID technology. AIP Conf Proc 2427:020098. https://doi.org/10.1063/5.0101139
3. Raihanah N, Syafariani RF (2019) Coin lockers as a technology-based public facility. In: IOP conference series: materials science and engineering, vol 662, Issue 3, p 032047. IOP Publishing. https://doi.org/10.1088/1757-899x/662/3/032047
4. Naik C, Pandit S, Gokhale S, Kulkarni M (2021) IoT driven smart storage solutions for managing volatile resources in hospitals
5. Phani PVLN, Narendra Kumar Reddy O, Manisha Reddy R (2015) Keypad based bank locker security system using GSM technology
6. Mostakim N, Sarkar RR, Hossain MA (2019) Smart locker: IOT based intelligent locker with password protection and face detection approach. Int J Wirel Microw Technol 9:1–10. https://doi.org/10.5815/ijwmt.2019.03.0110.1088/1757-899x/662/3/032047
7. Kook J (2019) Design and implementation of a OTP-based IoT digital door-lock system and applications
8. Anand A (2021) IOT based bank locker. Int J Res Appl Sci Eng Technol 9:1248–1253. https://doi.org/10.22214/ijraset.2021.35257
9. Alqahtani HF et al (2020) Automated smart locker for college. In: 2020 3rd international conference on computer applications and information security (ICCAIS), Riyadh, Saudi Arabia, pp 1–6. https://doi.org/10.1109/ICCAIS48893.2020.9096868. https://doi.org/10.1109/ICCAIS48893.2020.9096868
10. Lo LP, Marinov M, Rüger B (2021) A qualitative study on providing alternative solutions for handling the HSR passenger's luggage. Urban Rail Transit 7:12–28. https://doi.org/10.1007/s40864-020-00139-y
11. Papoutsidakis M, Sfyroera A, Srivastrava A (2019) Case study of automated baggage handling in modern transportations. Int J Eng Appl Sci Technol 04. https://doi.org/10.33564/IJEAST.2019.v04i05.074. https://doi.org/10.33564/IJEAST.2019.v04i05.074
12. Luís AFS, Martins GMC, Caldeira JMLP, Soares VNGJ (2022) Smart lockers: approaches, challenges and opportunities. In: International journal of engineering and advanced technology, vol 11, Issue 3. Blue Eyes Intelligence Engineering and Sciences Engineering and Sciences Publication—BEIESP, pp 141–149. https://doi.org/10.35940/ijeat.c3374.0211322
13. Baikerikar J, Kavathekar V, Ghavate N, Sawant R, Madan K (2021) Smart door locking mechanism. In: 2021 4th biennial international conference on nascent technologies in engineering (ICNTE), Navi Mumbai, India, pp 1–4. https://doi.org/10.1109/ICNTE51185.2021.9487704. https://doi.org/10.1109/ICNTE51185.2021.9487704

Building Trust in Agriculture: Blockchain Powered Livestock Buying and Selling

Apurva Shinde, Sanjana Mamde, Gouri Umate, and M. A. Gangarde

Abstract This paper delves into the potential transformation brought about by blockchain technology within the livestock industry, specifically targeting issues such as transparency, traceability, and trust in the buying and selling processes. Conventional methods often grapple with inefficiencies and a deficit of trust, necessitating innovative solutions. The paper offers a comprehensive overview of blockchain technology, spotlighting its decentralized and immutable characteristics, along with key elements like smart contracts and consensus mechanisms. A thorough examination of existing literature and case studies underscores blockchain's applications in agriculture, emphasizing its role in bolstering trust and streamlining transactions in the livestock sector. The focus centers on decentralized marketplaces, livestock tokenization, and the integration of smart contracts. Illustrative case studies, including IBM Food Trust and Beef Chain, showcase practical implementations. The paper further explores the advantages, challenges, regulatory considerations, and future trajectories in the realm of blockchain-driven livestock transactions. It serves as a valuable resource for stakeholders interested in the convergence of blockchain and agriculture, offering insights into the current landscape and guiding future research and development endeavors.

Keywords Livestock industry · Agriculture · Blockchain technology · Livestock transactions · Innovative technologies · Livestock supply chain

A. Shinde · S. Mamde · G. Umate (✉) · M. A. Gangarde
SCTR's Pune Institute of Computer Technology Pune, Pune, India
e-mail: gouriumate@gmail.com

M. A. Gangarde
e-mail: magangarde@pict.edu

V. K. Gunjan and J. M. Zurada (eds.), *Proceedings of 5th International Conference on Recent Trends in Machine Learning, IoT, Smart Cities and Applications*, Lecture Notes in Networks and Systems 1181, https://doi.org/10.1007/978-981-97-8861-3_41

1 Introduction

The global livestock industry plays a crucial role in agriculture, providing essential protein sources and livelihoods for millions worldwide. However, the conventional methods of buying and selling livestock grapple with challenges like lack of transparency, inefficiencies, and a deficit of trust among stakeholders. These hurdles hinder the smooth operation of the livestock supply chain, demanding innovative solutions. Enter blockchain technology, a promising remedy for these challenges. Offering a decentralized and immutable framework, it has the potential to enhance transparency, traceability, and overall trust in livestock transactions. The decentralized nature of blockchain ensures secure data sharing among participants, and its immutability guarantees that once recorded, information remains unaltered. Smart contracts further streamline processes by automating predefined agreements.

This survey delves into blockchain's transformative potential in livestock transactions, covering fundamentals, existing applications, and specific implementations. It discusses benefits, challenges, and regulations, providing a comprehensive view for a transparent, efficient, and trustworthy livestock supply chain in the evolving agricultural industry.

2 Proposed Methods

Designing a blockchain-powered cattle trading system involves several components and considerations. Here's a proposed system architecture for such a system.

2.1 Smart Contracts

Implement smart contracts to automate and execute the terms of cattle transactions, including ownership transfer, payment settlements, and delivery confirmation [1].

2.2 Decentralized Ledger

Use a decentralized ledger, such as a blockchain, to record all cattle-related transactions. This ledger ensures transparency, immutability, and security.

2.3 *Consensus Mechanism*

Choose an appropriate consensus mechanism (e.g., Proof of Work, Proof of Stake) to secure the network and validate transactions. Consider a mechanism that aligns with the specific requirements and scale of the cattle trading system.

2.4 *Identity Management*

Implement a secure identity management system to uniquely identify cattle, farmers, buyers, and other stakeholders in the system. This can be achieved through cryptographic techniques to ensure data integrity and privacy.

2.5 *User Interface (UI) and User Experience (UX)*

Create a user-friendly interface for farmers, buyers, and participants, ensuring intuitive navigation for easy transactions, cattle monitoring, and access to relevant information.

2.6 *Payment Integration*

Enable secure and transparent payment transactions using cryptocurrency or a token system. Implement a payment gateway that integrates with the blockchain to facilitate seamless and trustless financial transactions.

2.7 *Regulatory Compliance*

Incorporate features for compliance with local regulations in cattle trading, including traceability, quality assurance, and reporting [2].

2.8 *Secure Communication*

Implement secure communication channels using encryption to protect sensitive data during transactions and interactions within the system.

2.9 *Analytics and Reporting*

Integrate analytics tools to provide insights into cattle trading trends, market conditions, and other relevant data. Generate reports for stakeholders to make informed decisions.

2.10 *Governance Mechanism*

Establish a governance model to manage system upgrades, resolve disputes, and adapt to changing market conditions. This could involve voting mechanisms among network participants.

2.11 *Security Audits*

Regularly conduct security audits to identify and address potential vulnerabilities. This helps ensure the integrity and security of the entire cattle trading system.

2.12 *Interoperability*

Consider interoperability with other blockchain networks or traditional systems to enhance connectivity and expand the reach of the cattle trading platform.

Implementing these components is vital for a strong blockchain-powered cattle trading system. Involving stakeholders, conducting thorough testing, and iterative improvements based on feedback are crucial steps.

3 Livestock Industry Challenges:

The global livestock industry faces challenges such as transparency issues, traceability concerns, and trust deficits, hindering optimal functioning. Inefficiencies in traditional business methods add complexity. Addressing these challenges through innovative solutions is crucial for reshaping and enhancing the industry, promoting resilience and efficiency [3].

3.1 Lack of Transparency

Transparency challenges in the livestock industry arise from manual record-keeping and paper-based documentation, limiting real-time access to accurate information, fostering misinformation, disputes, and reduced confidence among stakeholders, impacting industry efficiency [4].

3.2 Traceability Issues

Inadequate traceability in traditional livestock systems hinders tracking animal movement, posing challenges for disease control, food safety, and regulatory compliance.

3.3 Trust Deficits

Trust deficits arise in livestock transactions due to the complex and opaque nature of traditional processes, leading to uncertainties about information accuracy, livestock conditions, and ethical practices [5]. Establishing trust is crucial for long-term relationships and the integrity of the supply chain.

3.4 Inefficiencies in Traditional Transactions

Traditional livestock transactions: manual, time-consuming processes, paperwork, inspections, intermediaries cause delays, higher costs, reduced agility. Reliance on intermediaries adds complexity, potential errors [6].

From this we know that livestock industry faces transparency issues with manual record-keeping, traceability challenges, trust deficits, and inefficiencies in traditional transactions, hindering efficiency and reliability in the supply chain.

Flowchart:

From Fig. 1 the integrating blockchain into livestock trading enhances transparency, security, and efficiency. Farmers list livestock, buyers place offers, and smart contracts automate transactions, ensuring compliance and holding funds in escrow until delivery confirmation. This blockchain-based system creates a secure, transparent, and streamlined process for buying and selling livestock.

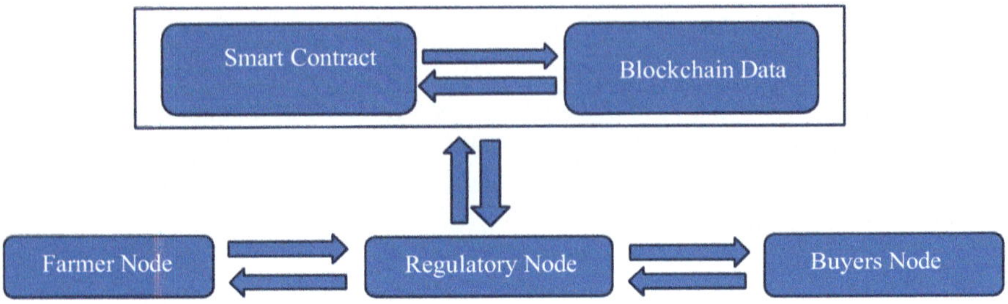

Fig. 1 System flowchart

4 The Role of Livestock in Economic Development

Livestock plays a crucial role in global agriculture, contributing 40% to the agricultural output and sustaining the livelihoods of 1.3 billion people [7]. While vital, improvements are needed for sustainability, equity, and health. In sustainable food systems, livestock serves as a multifaceted contributor, providing natural fertilizer, supporting vulnerable communities, and fostering cultural resilience [7, 8]. The environmental footprint, accounting for 14.5% of greenhouse gas emissions, necessitates comprehensive approaches to enhance efficiency in supply chains [9]. Livestock's financial role includes employment generation, farmer resilience, and global market integration, contributing to trade relations and foreign exchange earnings [10, 11]. Table 1 gives total income from farm in different state in different years.

5 Livestock Buying and Selling on the Blockchain

Blockchain is transforming livestock transactions, from decentralized marketplaces to tokenization, immutability, and real-time monitoring. Join us on this journey through innovative applications.

5.1 Decentralized Marketplaces

Decentralized marketplaces use blockchain for transparent, efficient peer-to-peer livestock transactions. Without intermediaries, blockchain ensures trust, providing real-time information and secure transactions, streamlining the process for inclusivity and efficiency [12].

Table 1 Income from animal husbandry as % of total farm income

States	2012–13[a]	2018–19[b]
Meghalaya	9.22	3.84
Chhattisgarh	0	10.78
Telangana	8.13	12.25
Mizoram	15.93	16.76
Karnataka	10.85	19.57
Odisha	48.29	20.96
Kerala	14	22.4
Madhya Pradesh	15.42	23.11
West Bengal	18.69	23.11
Maharashtra	12.26	24.49
Tripura	10.09	24.79
Sikkim	36.62	25.29
Assam	15.95	25.56
Punjab	13.24	26.13
Uttar Pradesh	15.98	29.32
Haryana	25.16	30.66
Uttarakhand	25.1	38.42
Rajasthan	23.56	38.71
Bihar	13.99	38.83
Himachal Pradesh	26.69	41.51
Andhra Pradesh	34.71	42.8
Jharkhand	45.12	42.87
Tamil Nadu	36.46	43.09
Gujarat	39.69	44.61
Arunachal Pradesh	16.46	44.63
Manipur	34.83	44.9
Jammu and Kashmir	20.73	53.48
Nagaland	30.11	65.41
ALL-INDIA	19.85	29.41

[a] *Source* Annual Report for 2012–2013, Department of Animal Husbandry and Dairying, Ministry of fisheries, Animal Husbandry and Dairying, Government of India

[b] *Source* Annual Report for 2018-2019, Department of Animal Husbandry and Dairying, Ministry of fisheries, Animal Husbandry and Dairying, Government of India

5.2 Tokenization of Livestock

Tokenization on blockchain represents livestock as digital tokens, each corresponding to a real animal, enabling fractional ownership for easier trading [13]. This enhances market liquidity, diversifies investment opportunities, and ensures transparent livestock provenance with recorded histories on the blockchain.

5.3 Immutable Records and Documentation

Blockchain's immutability ensures a secure, transparent history for livestock transactions, reducing fraud and disputes by guaranteeing information integrity in buying and selling [14].

5.4 Real-Time Monitoring and Data Sharing

Blockchain enables real-time livestock monitoring through integrated IoT devices, enhancing transparency for informed decisions and enabling early detection of health issues within the livestock ecosystem [15, 16].

6 Results and Discussion

6.1 Results

Figures 2, 3, 4, 5, 6, 7 and 8 illustrate various aspects of the livestock trading platform using blockchain technology. Specifically, Figure 2 is the home page, Figure 3 is the signup page, Figure 4 is the login page, Figure 5 is the cart page, Figure 6 is the buyer page, Figure 7 is the transaction page, and Figure 8 is the Ganache migration.

6.2 Discussion

The historical context of cattle prices and livestock trading underscores the economic significance of these practices throughout human history. Dating back to ancient agricultural societies, the trade of livestock, particularly cattle, has been an integral component of economic exchange. Cattle, serving as both a source of sustenance and a form of wealth, played a crucial role in early barter systems [17]. Over time, with the evolution of agricultural and trading practices, cattle trading has become

Fig. 2 Home page

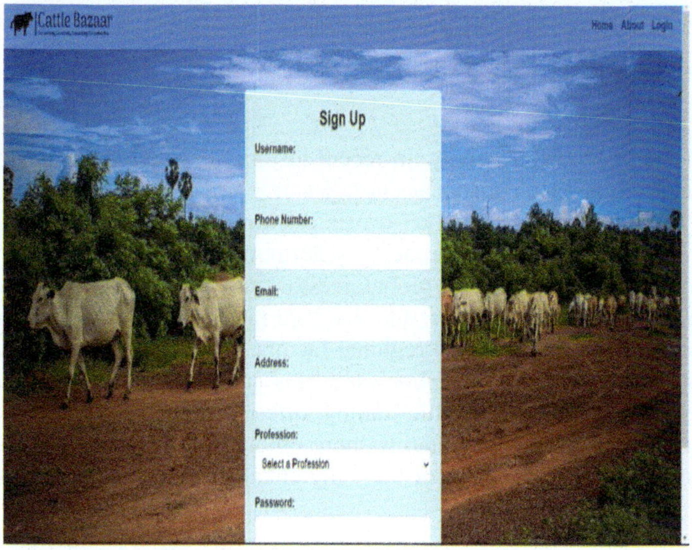

Fig. 3 Signup page

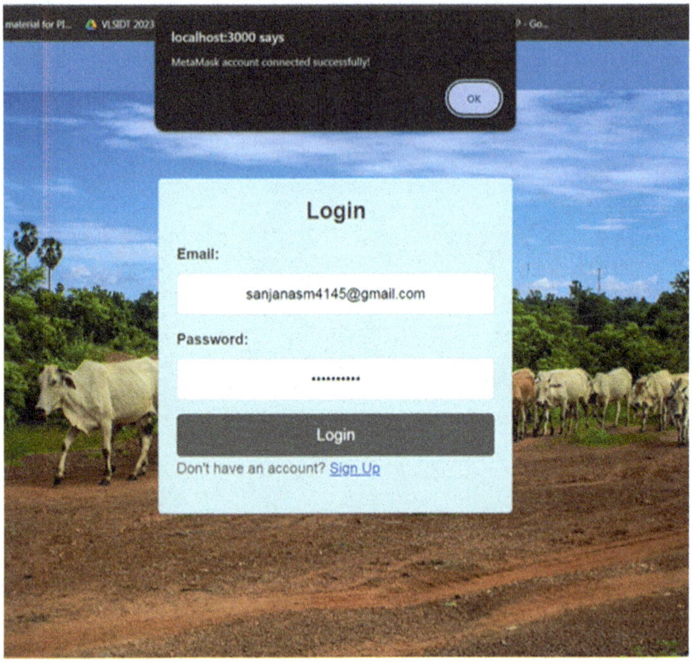

Fig. 4 Login page

a sophisticated economic activity. In recent history, the establishment of livestock markets and pricing mechanisms has brought structure to the cattle trading industry [18, 19]. The introduction of market-driven pricing systems has allowed for a more transparent and standardized approach to determining cattle values.

In contemporary contexts, the global livestock market has witnessed the integration of technology, data analytics, and financial instruments. Cattle prices are now influenced not only by local and regional factors but also by international market dynamics [20]. The advent of online platforms and electronic trading has further transformed the landscape of cattle trading, offering stakeholders real-time access to market information and facilitating more efficient transactions [21]. The given Table 2 and Fig. 9 provide data of the daily change in the cow prices in different markets.

6.3 Performance Parameter

Designing a Blockchain-powered cattle trading system involves various performance parameters and equations to ensure efficiency, security, and transparency. Below are some key performance parameters and potential equations relevant to such a system [22]:

Fig. 5 Cart page

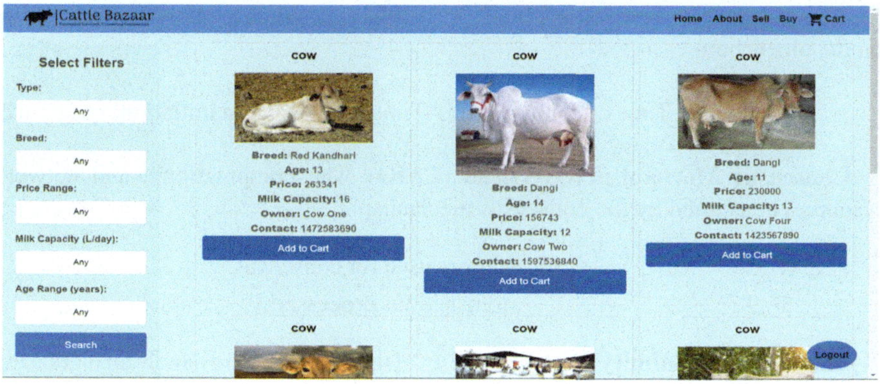

Fig. 6 Buyer page

Transaction Throughput (TP): The number of transactions processed per unit of time.

$$TP = \frac{\text{Total transactions processed}}{\text{Time}} \tag{1}$$

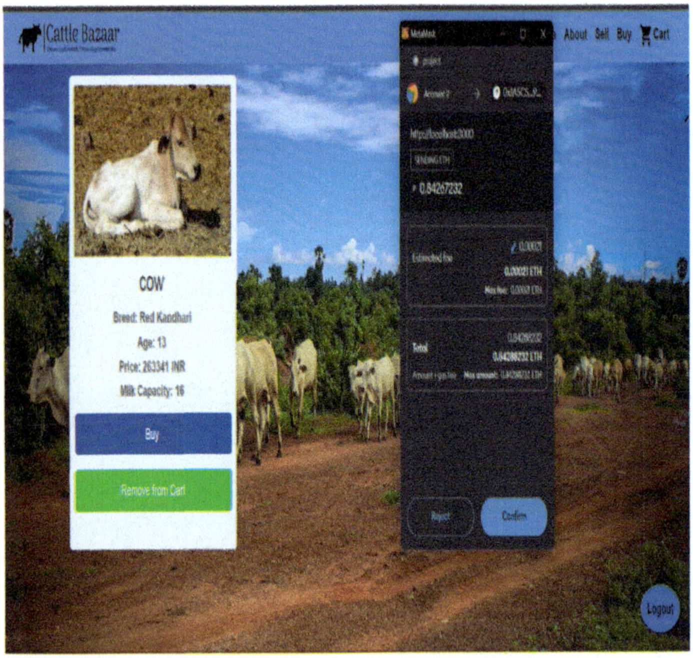

Fig. 7 Transaction page

Transaction Confirmation Time (TCT): The time taken to confirm a transaction on the blockchain.

$$\text{TCT} = \text{Time of confirmation} - \text{Time of transaction initiation} \tag{2}$$

Consensus Mechanism Overhead (CMO): The computational and network resources consumed by the consensus mechanism.

$$\text{CMO} = \frac{\text{Resources used for consensus}}{\text{transactions processed}} \tag{3}$$

Blockchain Scalability (BS): The ability of the blockchain to handle an increasing number of transactions.

$$\text{BS} = \frac{\text{TP1}}{\text{TP0}} \text{ where TP1} > \text{TP0} \tag{4}$$

Security Level (SL): The level of security provided by the blockchain network.

$$\text{SL} = \frac{\text{Nodes in agreement}}{\text{Total number of nodes}} \tag{5}$$

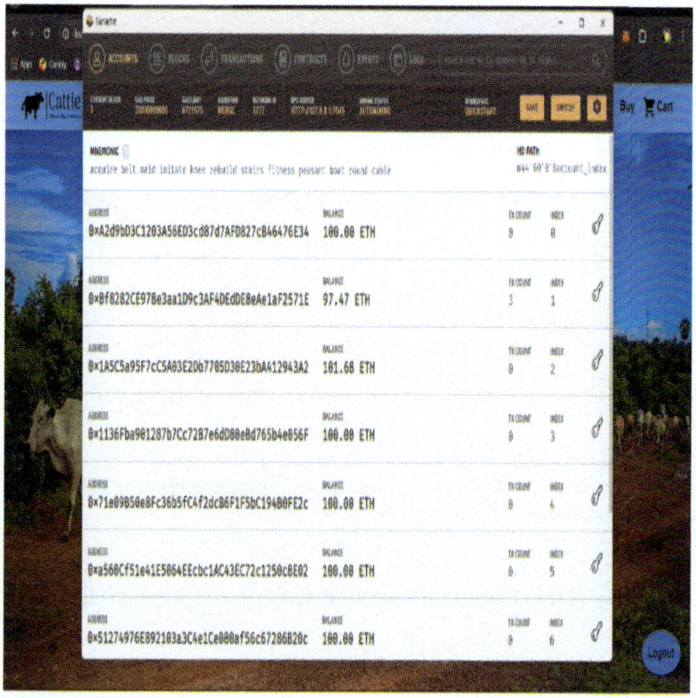

Fig. 8 Ganache migration

Table 2 Livestock current market prices

Commodity	Variety	Market details	Date	Min price	max price	Modal price
Cow	Other	Junnar (Alephata) State: Maharash Tra	10-Jan-24	20,000.00 INR/Quintal	60,000.00 INR/Quintal	40,000.00 INR/Quintal
Cow	Cow	Masli State: Tripura	09-Jan-24	15,000.00 INR/Quintal	25,000.00 INR/Quintal	20,000.00 INR/Quintal
Cow	Cow	Chintamani State: Karnataka	08-Jan-24	40,000.00 INR/Quintal	80,000.00 INR/Quintal	60,000.00 INR/Quintal
Cow	Cow	Ramdurga State: Karnataka	07-Jan-24	30,000.00 INR/Quintal	30,000.00 INR/Quintal	30,000.00 INR/Quintal

Source Commodityonline, 2024, *Commodityonline website,* Comex Agritech Pvt Ltd, 2024

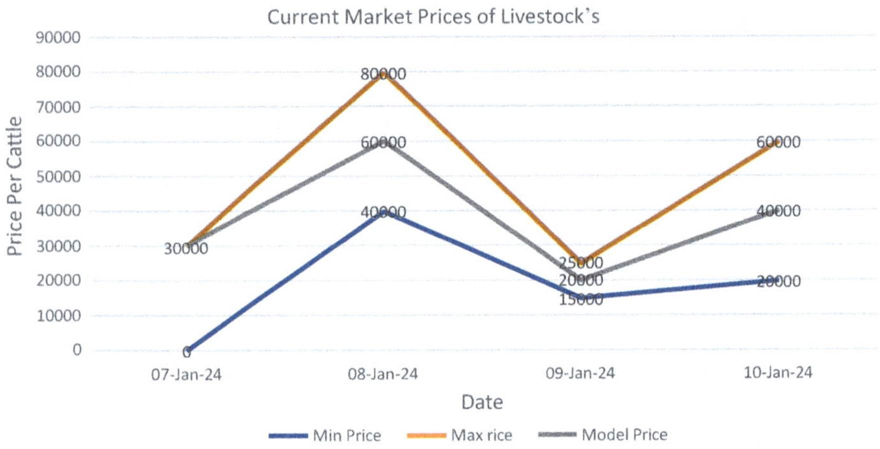

Fig. 9 Livestock current market prices. *Source* Commodityonline, 2024, *Commodityonline website,* Comex Agritech Pvt Ltd, 2024

Data Integrity (DI): The assurance that data on the blockchain is accurate and unchanged.

$$\text{DI} = \frac{\text{Confirmed transactions}}{\text{Total transactions}} \tag{6}$$

User Experience Index (UEI): An index reflecting the overall experience of users interacting with the system.

$$\text{UEI} = \frac{\text{Number of positive user experiences}}{\text{Total user interactions}} \tag{7}$$

Energy Consumption (EC): The total energy consumed by the blockchain network.

$$\text{EC} = \text{Energy consumed by nodes} + \text{Smart contract execution energy} \tag{8}$$

It's important to note that the specific equations and parameters may vary based on the implementation details, consensus algorithm, and other factors unique to your Blockchain-powered cattle trading system.

6.4 Cattle Population Variation

In modern agriculture, cattle population dynamics are shaped by climate, ecology, consumer preferences, and economic shifts [23]. Globally, some regions see growth

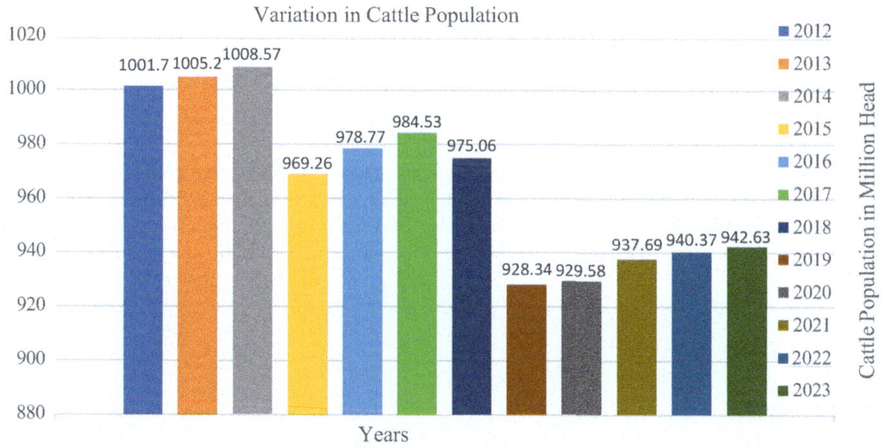

Fig. 10 Variation in cattle population. *Source* Blümmel et al. [18]

through selective breeding and enhanced veterinary interventions, while others contract due to challenges like droughts, disease outbreaks, and shifts in land use. From Fig. 10 the variation in cattle population are given from year 2012 to 2023.

Technological shifts like precision farming significantly impact cattle numbers. This research analyzes these dynamics, providing insights for sustainable practices, food security, and rural economies. Precision farming employs advanced technologies to optimize farming practices, leading to better livestock management. Blockchain technology enhances traceability, transparency, and trust in livestock buying and selling, ensuring secure data about cattle health, breeding, and movement. Integrating precision farming with blockchain creates a synergistic effect, improving cattle management and sustainability. Understanding the impact of these advancements on cattle numbers is essential for developing robust strategies for agricultural sustainability and economic resilience. By leveraging precision farming and blockchain technology, stakeholders can build a more trustworthy, efficient, and sustainable livestock industry

7 Conclusion

Livestock trading is a crucial economic force locally and globally, sustaining rural communities and contributing significantly to GDP. Embracing technology like digital platforms, IoT, blockchain, and data analytics, the industry undergoes a transformative era, enhancing efficiency, transparency, decision-making, and sustainability. Amid environmental concerns, the sector adapts with innovations in environmental monitoring, supply chain optimization, and resource management. Challenges like disease outbreaks and ethical considerations spur opportunities for research and innovation in disease prevention, market forecasting, and ethical trading. Globally connected, livestock trading fosters economic ties, foreign exchange earnings, and food security, emphasizing the industry's need for resilience, adaptability, continuous innovation, and responsible practices.

Acknowledgements I express profound gratitude to mentors, advisors, and colleagues for unwavering support in completing this work on blockchain technology in livestock transactions. Their guidance was pivotal. Thanks also to the extensive literature, research, and case studies providing a foundation. This work reflects the collaborative ethos of progress in the academic and research community.

References

1. Naess MW, Bårdsen BJ (2013) Why herd size matters—mitigating the effects of livestock crashes. PLoS ONE 8:e70161
2. Chantarat S, Mude AG, Barrett CB, Carter MR (2013) Designing index-based livestock insurance for managing asset risk in Northern Kenya. J Risk Insur 80:205–237
3. Renkow M, Hallstrom DG, Karanja DD (2004) Rural infrastructure, transactions costs and market participation in Kenya. J Dev Econ 73:349–367
4. Shikuku KM, Valdivia RO, Paul BK, Mwongera C, Winowiecki L, Läderach P, Herrero M, Silvestri S (2017) Prioritizing climate-smart livestock technologies in rural Tanzania: a minimum data approach. Agric Syst 151:204–216
5. Herrero M, Grace D, Njuki J, Johnson N, Enahoro D, Silvestri S, Rufino MC (2013) The roles of livestock in developing countries. Animal 7:3–18
6. Ahmadi BV, Moran D, Barnes AP, Baret PV (2015) Comparing decision-support systems in adopting sustainable intensification criteria. Front Genet 6:1–5
7. Berman A (2011) Invited review: are adaptations present to support dairy cattle productivity in warm climates? J Dairy Sci 94:2147–2158
8. Mondal RK, Sen S, Rayhan SJ (2010) A comparative economic analysis of local breed and cross breed milk cow in a selected area of Bangladesh. J Sci Found 8:23–29
9. Gill M, Smith P, Wilkinson JM (2010) Mitigating climate change: the role of domestic livestock. Animal 4:323–333
10. Havlik P, Valin H, Herrero M, Obersteiner M, Schmid E, Rufino MC, Mosnier A, Thornton PK, Bottcher H, Conant RT, Frank S, Fritz S, Fuss S, Kraxner F, Notenbaert A (2014) Climate change mitigation through livestock system transitions. PNAS 111:3709–3714
11. Hoffmann I (2010) Climate change and the characterization, breeding and conservation of animal genetic resources. Anim Genet 41:32–46
12. Bernard T, Spielman DJ (2009) Reaching the rural poor through rural producer organizations? a study of agricultural marketing cooperatives in Ethiopia. Food Policy 34:60–69
13. Kantor P (2013) Transforming gender relations: key to positive development outcomes in aquatic agricultural systems. Brief: AAS-2013-12; Antor P (2013) CGIAR research program on aquatic agricultural systems, Penang, Malaysia
14. Marshall K, Okeyo AM, Johnson N (2009) Translating animal breeding research into the real world. In: Van Der Werf JHJ, Nimbkar C, Gupta VS (eds) Proceedings of the Helen newton turner memorial international workshop. ACIAR, Pune, Maharashtra, India, pp 190–198
15. Andrieu N, Descheemaeker K, Sanou T, Chia E (2015) Effects of technical interventions on flexibility of farming systems in Burkina Faso: lessons for the design of innovations in West Africa. Agric Syst 136:125–137
16. Jensen N, Barrett C, Mude A (2015) The favourable impacts of index-based livestock insurance: evaluation results from Ethiopia and Kenya. ILRI Res. Brief 52

17. Lemke U, Kaufmann B, Thuy LT, Emrich K, Valle Zárate A (2006) Evaluation of smallholder pig production systems in North Vietnam: pig production management and pig performances. Livest Sci 105:229–243
18. Blümmel M, Zerbini E, Reddy BVS, Hash CT, Bidinger F, Khan AA (2003) Improving the production and utilization of sorghum and pearl millet as livestock feed: progress towards dual-purpose genotypes. Field Crops Res 84:143–158; Campbell BM, Thornton P, Zougmoré R, Van Asten P, Lipper L (2014)
19. Castellini C, Boggia A, Cortina C, Dal Bosco A, Paolotti L, Novelli E, Mugnai C (2012) A multicriteria approach for measuring the sustainability of different poultry production systems. J Clean Prod 37:192–201
20. McDermott JJ, Staal SJ, Freeman HA, Herrero M, Van De Steeg JA (2010) Sustaining 8intensification of smallholder livestock systems in the tropics. Livest Sci 130:95–109
21. Antle JM (2011) Parsimonious multidimensional impact assessment. Am J Agric Econ 93:1292–1311
22. Anderson JR (2003) Risk in rural development: challenges for managers and policy makers. Agric Syst 75:161–197
23. Ouma R, Njoroge L, Romney D, Ochungo P, Staal S, Baltenweck I (2007) Targeting dairy interventions in Kenya: a guide for development planners, researchers and extension workers. SDP/KDDP, Nairobi, Kenya

Author Index

V. K. Gunjan and J. M. Zurada (eds.), *Proceedings of 5th International Conference on Recent Trends in Machine Learning, IoT, Smart Cities and Applications*, Lecture Notes in Networks and Systems 1181, https://doi.org/10.1007/978-981-97-8861-3

The manufacturer's authorised representative in the EU is Springer Nature Customer Service Centre GmbH, Europaplatz 3, 69115 Heidelberg, Germany. If you have any concerns regarding our products, please contact ProductSafety@springernature.com

Printed and bound by CPI Group (UK) Ltd, Croydon, CR0 4YY
01/10/2026
02225901-0008